T0406280

A Differential Approach to Geometry

Francis Borceux

A Differential Approach to Geometry

Geometric Trilogy III

Francis Borceux
Université catholique de Louvain
Louvain-la-Neuve, Belgium

ISBN 978-3-319-01735-8 ISBN 978-3-319-01736-5 (eBook)
DOI 10.1007/978-3-319-01736-5
Springer Cham Heidelberg New York Dordrecht London

Library of Congress Control Number: 2013954709

Mathematics Subject Classification (2010): 53A04, 53A05, 53A45, 53A55, 53B20, 53C22, 53C45

Cover image: Carl Friedrich Gauss by Christian Albrecht Jensen

Printed on acid-free paper

Springer is part of Springer Science+Business Media (www.springer.com)

To Océane, Anaïs, Magali, Lucas, Cyprien,
Théophile, Constance, Léonard, Georges, . . .
and those still to come

Preface

The reader is invited to immerse himself in a "love story" which has been unfolding for 35 centuries: the love story between mathematicians and geometry. In addition to accompanying the reader up to the present state of the art, the purpose of this *Trilogy* is precisely to tell this story. The *Geometric Trilogy* will introduce the reader to the multiple complementary aspects of geometry, first paying tribute to the historical work on which it is based and then switching to a more contemporary treatment, making full use of modern logic, algebra and analysis. In this *Trilogy*, Geometry is definitely viewed as an autonomous discipline, never as a sub-product of algebra or analysis. The three volumes of the *Trilogy* have been written as three independent but complementary books, focusing respectively on the axiomatic, algebraic and differential approaches to geometry. They contain all the useful material for a wide range of possibly very different undergraduate geometry courses, depending on the choices made by the professor. They also provide the necessary geometrical background for researchers in other disciplines who need to master the geometric techniques.

In the 1630s Fermat and Descartes were already computing the tangents to some curves using arguments which today we would describe in terms of derivatives (see [4], *Trilogy II*). However, these arguments concerned algebraic curves, that is, curves whose equation is expressed by a polynomial, and the derivative of a polynomial is something that one can describe algebraically in terms of its coefficients and exponents, without having to handle limits. Some decades later, the development of differential calculus by Newton and Leibniz allowed these arguments to be formalized in terms of actual derivatives, for rather arbitrary curves. In the present book, we focus on this general setting of curves and surfaces described by functions which are no longer defined by polynomials, but are arbitrary functions having sufficiently well behaved properties with respect to differentiation.

We have deliberately chosen to restrict our attention to curves in the 2- and 3-dimensional real spaces and surfaces in the 3-dimensional real space. Although we occasionally give a hint on how to generalize several of our results to higher dimensions, our focus on lower dimensions provides the best possible intuition of the basic notions and techniques used today in advanced studies of differential geometry.

An important notion is the consideration of *parametric equations*, following an idea of *Euler* (see [4], *Trilogy II*). A closer look at such equations suggests that we should view a curve not as the set of points whose coordinates satisfy some equation(s), but as a continuous deformation of the real line in $\mathbb{R}^2$ or in $\mathbb{R}^3$, according to the case. When a parameter t varies on the real line, the parametric equations describe successively all the points of the curve. Analogously, a surface can be seen as a continuous deformation of the real plane in the space $\mathbb{R}^3$. This is the notion of a *parametric representation*, which is the basic tool that we shall use in our study.

Our first chapter is essentially historical: its purpose is to explain where the ideas of differential geometry came from and why we choose this or that precise definition and not another possible one.

The formalized study of curves then begins with Chap. 2, where we restrict our attention to the simplest case: the plane curves. We pay special attention to basic notions like tangency, length and curvature, but we also prove very deep theorems, such as the *Hopf theorem* for simple closed curves. Working in the plane makes certainly things easier to grasp in a first approach. However, it is also a matter of fact that the study of plane curves offers many interesting aspects, such as envelopes, evolutes, involutes, which have beautiful applications. Many of these aspects do not generalize elegantly to higher dimensions.

Our Chap. 3 is a kind of parenthesis in our theory of differential geometry: we present a *museum* of some specimens of curves which have played an important historical role in the development of the theory.

Chapter 4 is then devoted to the study of curves in three dimensional space: the so-called *skew curves*. We focus our attention on the main aspects of the theory, namely, the study of the *curvature* and the *torsion* of skew curves and the famous *Frenet trihedron*.

Next we switch to surfaces in $\mathbb{R}^3$. In Chap. 5 we concentrate our attention on the *local properties* of surfaces, that is, properties "in a neighborhood of a given point of the surface", such as the tangent plane at that point or the various notions of curvature at that point: normal curvature, Gaussian curvature, and the information that we can get from these on the shape of the surface in a neighborhood of the point.

Chapter 6 then begins by repeating many of the arguments of Chap. 5, but using a different notation: the notation of Riemannian geometry. Our objective is to provide in this way a good intuitive approach to notions such as the *metric tensor*, the *Christoffel symbols*, the *Riemann tensor*, and so on. We provide evidence that these apparently very technical notions reduce, in the case of surfaces in $\mathbb{R}^3$, to very familiar notions studied in Chap. 5. We also devote special attention to the case of *geodesics* and establish the main properties (including the existence) of the systems of geodesic coordinates.

The last chapter of this book is devoted to some *global properties* of surfaces: properties for which one has to consider the full surface, not just what happens in a neighborhood of one of its points. We start with a basic study of the surfaces of revolution, the ruled and the developable surfaces and the surfaces with constant curvature. Next we switch to results and notions such as the *Gauss–Bonnet* theorem and the *Euler characteristic*, which represent some first bridges between the elementary theory of surfaces and more advanced topics.

Each chapter ends with a section of "problems" and another section of "exercises". Problems are generally statements not treated in this book, but of theoretical interest, while exercises are more intended to allow the reader to practice the techniques and notions studied in the book.

Of course reading this book assumes some familiarity with the basic notions of linear algebra and differential calculus, but these can be found in all undergraduate courses on these topics. An appendix on general topology introduces the few ingredients of that theory which are needed to properly follow our approach to Riemannian geometry and the global theory of surfaces. A second appendix states with full precision (but without proofs this time) some theorems on the existence of solutions of differential equations and partial differential equations, which are required in some advanced geometrical results.

A selective bibliography for the topics discussed in this book is provided. Certain items, not otherwise mentioned in the book, have been included for further reading.

The author thanks the numerous collaborators who helped him, through the years, to improve the quality of his geometry courses and thus of this book. Among them a special thanks to *Pascal Dupont*, who also gave useful hints for drawing some of the illustrations, realized with *Mathematica* and *Tikz*.

The Geometric Trilogy

I. An Axiomatic Approach to Geometry

1. Pre-Hellenic Antiquity
2. Some Pioneers of Greek Geometry
3. Euclid's *Elements*
4. Some Masters of Greek Geometry
5. Post-Hellenic Euclidean Geometry
6. Projective Geometry
7. Non-Euclidean Geometry
8. Hilbert's Axiomatization of the Plane

   Appendices

   A. Constructibility

   B. The Three Classical Problems

   C. Regular Polygons

II. An Algebraic Approach to Geometry

1. The birth of Analytic Geometry
2. Affine Geometry
3. More on Real Affine Spaces
4. Euclidean Geometry
5. Hermitian Spaces
6. Projective Geometry
7. Algebraic Curves

   Appendices

   A. Polynomials over a Field

   B. Polynomials in Several Variables

   C. Homogeneous Polynomials

   D. Resultants

   E. Symmetric Polynomials

   F. Complex Numbers

Contents

Chapter 1
The Genesis of Differential Methods

This first chapter is intentionally provocative, and useless! By *useless* (besides being at once provocative) we mean: this first chapter is not formally needed to follow the systematic treatment of the theory of curves and surfaces developed in the subsequent chapters.

So what is this chapter about? Usually, when you open a book on—let us say—the theory of curves in the real plane, you expect to find first "the" precise definition of a plane curve, followed by a careful study of the properties of such a notion. We all have an intuitive idea of what a plane curve is. Everybody knows that the straight line, the circle or the parabola *are* curves, but a single point or the empty set *are not* curves! Nevertheless, all these "figures" can be described by an equation $F(x, y) = 0$, with F a polynomial: for example, $x^2 + y^2 = 0$ is an "equation of the origin" in $\mathbb{R}^2$ while $x^2 + y^2 = -1$ is "an equation" of the empty set. Thus a curve cannot simply be defined via an equation $F(x, y) = 0$, even when F is a "very good" function! For example, consider the picture comprised of 7 hyperbolas, thus 14 branches. Is this *one* curve, or *seven* curves, or *fourteen* curves? After all, it is not so clear what a curve should be!

Starting at once with a precise definition of a curve would give the false impression that this is *the* definition of a curve. Instead it should be stressed that such a definition is *a possible* definition. Discussing the advantages and disadvantages of the various possible definitions, in order to make a sensible choice, is an important aspect of every mathematical approach.

There is also a second aspect that we want to stress. For *Euclid*, a straight line was *What has a length and no width and is well-balanced at each of its points* (see Definition 3.1.1 in [3], *Trilogy I*). The intuition behind such a sentence is clear, but such a "definition" assumes that before beginning to develop geometry, we know what a *length* is. Of course what we want to do concerning a *length* is then to find a formula to compute it, such as $2\pi R$ for a circle of radius R.

With more than two thousand years of further mathematical developments and experience, we now feel quite uneasy about such an approach. How can we establish a formula to compute the *length of a curve* if we did not define first what the *length of a curve* is?

F. Borceux, *A Differential Approach to Geometry*, DOI 10.1007/978-3-319-01736-5_1,

For many centuries—essentially up to the 17th century—mathematicians could hardly handle problems of length for curves other than the straight line and the circle. Differential calculus, with the full power of the theories of derivatives and integrals, opened the door to the study of arbitrary curves. However, in some sense, one was still taking the notion of length (or surface or volume) as something "which exists and that one wants to calculate".

Like many authors today, we adopt in the following chapters a completely different approach: the theory of integration is a well-established part of analysis and we use it to define a length. Analogously the theory of derivatives is a well-established part of analysis and we use it to define a tangent. And so on.

This first chapter is intended to be a "bridge" between the "historical" and the "contemporary" approaches. We present typical arguments developed in the past (and sometimes, still today) to master some geometrical notions (like length, or tangent), but we do that in particular to develop an intuition for the contemporary definitions of these notions. In this introductory chapter, we refer freely to [3] and [4], *Trilogy I* and *II*, when the historical arguments that we have in mind have been developed there.

Various arguments in this chapter can appear quite disconcerting. We often rely on our intuition, without trying to formalize the argument. We freely apply many results borrowed from a first calculus course, taking as a blanket assumption that when we apply a theorem, the necessary assumptions for its validity should be satisfied, even if we have not tried to determine the precise context in which this is the case! This is not a very rigorous attitude, however our point in this chapter is not to *prove* results, but to *guess* what possible "good" definitions should be.

1.1 The Static Approach to Curves

Originally, Greek geometry (see [3], *Trilogy I*) was essentially concerned with the study of two curves: the line and the circle.

> *The line is what has length and no width and is well-balanced around each of its points.*
>
> *The circle is the locus of those points of the plane which are at a fixed distance R from a fixed point O of the plane.*

Passing analogously to three dimensional space, using a circle in a plane and a point not belonging to the plane of the circle, you can then—using lines—construct the cone on this circle with vertex the given point. "Cutting" this cone by another plane then yields new curves that, according to the position of the "cutting plane", you call *ellipse*, *hyperbola* or *parabola*. This is the origin of the theory of curves.

It is common practice to describe a curve by giving its equation with respect to some basis. In this book, we are interested in the study of curves in the real plane $\mathbb{R}^2$. For example a circle of radius R centered at the origin admits the equation (see Chap. 1 in [4], *Trilogy II*)

$$x^2 + y^2 = R^2$$

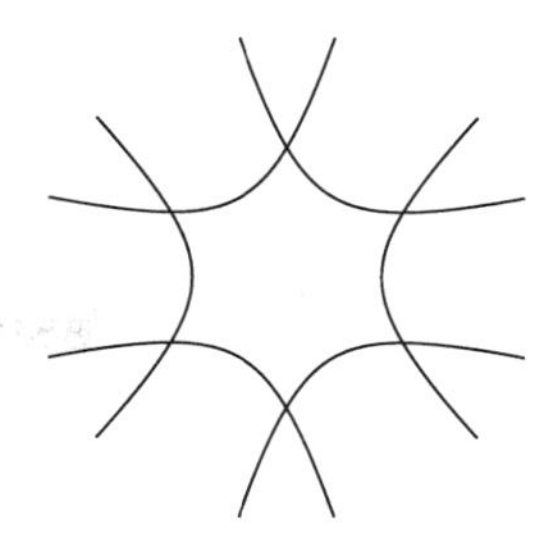

Fig. 1.1

which we can equivalently write as

$$x^2 + y^2 - R^2 = 0.$$

One might be tempted to introduce a general theory of curves by allowing equations of the form

$$F(x, y) = 0$$

where

$$F : \mathbb{R}^2 \longrightarrow \mathbb{R}$$

is an arbitrary function. But it does not take long to realize that:

- choosing $F(x, y) = x^2 + y^2$, we get the equation of a single point: the origin;
- choosing $F(x, y) = x^2 + y^2 + 1$, we even get the equation of the empty set!

In both cases the function $F(x, y)$ is certainly "a very good one": it is even a polynomial, but we do not want a point or the empty set to be considered as a curve.

For more food for thought, look at the picture in Fig. 1.1: should this be considered as *one* curve, or as *six* curves?

In fact, if you look carefully at Fig. 1.1, you will realize that it is comprised of three hyperbolas. The equation of this picture is "simply"

$$\big(x^2 - y^2 - 1\big)\big((x - \sqrt{3}y)^2 - (\sqrt{3}x + y)^2 - 4\big)\big((x + \sqrt{3}y)^2 - (\sqrt{3}x - y)^2 - 4\big) = 0$$

thus again an equation of the form $F(x, y) = 0$ with F a polynomial. But since this is the equation of three hyperbolas, should we consider that the picture represents three curves, not one or six?

If you decide that a hyperbola is *one* curve, then you accept that a curve can have several disjoint branches. Thus you should probably also consider that the picture of Fig. 1.1 represents *one* curve with six branches. Furthermore, you should also consider that a picture comprising 247 straight lines is *one* curve as well. Taking the opposite point of view, the hyperbola is no longer *one* curve, but the union of *two* curves.

If you have not yet given up, the following example may cause you to:

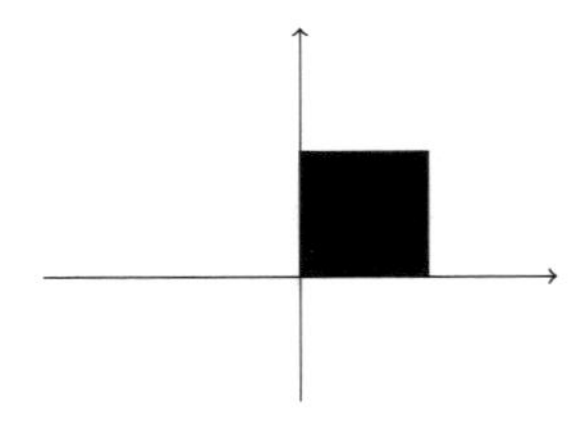

Fig. 1.2

Example 1.1.1 There exist continuous functions $F: \mathbb{R}^2 \longrightarrow \mathbb{R}$ such that

$$\{(x, y) \in \mathbb{R}^2 \mid F(x, y) = 0\}$$

is a full square.

Proof The following function is one among many examples:

$$F(x, y) = (x - |x|)^2 + (y - |y|)^2 + ((1 - x) - |1 - x|)^2 + ((1 - y) - |1 - y|)^2.$$

The condition $F(x, y) = 0$ is indeed equivalent to

$$x \geq 0, \qquad y \geq 0, \qquad 1 - x \geq 0, \qquad 1 - y \geq 0.$$

The corresponding "curve" is the full square of Fig. 1.2. Certainly, you do not want this to be called *a curve*! □

Should we thus give up our attempt to define a curve via a rather general equation of the form $F(x, y) = 0$? For the time being we shall abandon this idea, but we will come back to this problem later, with adequate differential tools.

Nevertheless, let us conclude this section with a comment. Every equation of the form $F(x, y) = 0$ determines a subset of $\mathbb{R}^2$

$$\{(x, y) \mid F(x, y) = 0\} \subseteq \mathbb{R}^2$$

and we would like to find conditions on F so that this subset is worthy of being called *a curve*. If we achieve this program, a curve will thus be *a subset of* $\mathbb{R}^2$. Being a subset is a *static* notion: no sense of *movement* is involved here. The full meaning of this comment will be expanded upon in the following Sect. 1.2.

1.2 The Dynamic Approach to Curves

The idea of "separating the variables" of an equation is due to the Swiss mathematician *Leonhard Euler* (1707–1783) (see Chap. 1 in [4], *Trilogy II*). In the case of the circle

$$x^2 + y^2 = R^2$$

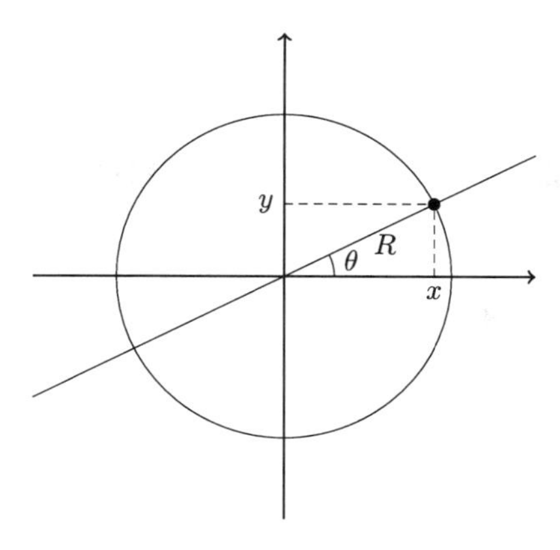

Fig. 1.3

this idea consists, for example, of describing the circle via the classical formulas

$$\begin{cases} x = R\cos\theta \\ y = R\sin\theta \end{cases}$$

where θ is the angle between the x-axis and the radius (see Fig. 1.3).

We thus obtain a *dynamical* description of the circle: when θ runs from $-\infty$ to $+\infty$, we repeatedly travel around the circle.

We are thus tempted to define a plane curve "dynamically" as a function

$$f\colon \mathbb{R} \longrightarrow \mathbb{R}^2, \qquad t \mapsto f(t).$$

In this spirit, a curve becomes a "deformation of the real line in the plane". Our intuition of a curve is that such a deformation should at least be continuous. Indeed we cannot imagine calling a "curve" a function such as

$$f(t) = \begin{cases} (t, 1) & \text{if } t \text{ is rational} \\ (t, 0) & \text{if } t \text{ is irrational.} \end{cases}$$

Let us observe that if we want to view a curve as a *continuous deformation of the real line*, then by continuity, every curve will have a single "branch". We discussed the case of the hyperbola in Sect. 1.1: the hyperbola is *not* a continuous deformation of the real line, but each of its two branches is. Thus we slowly begin to realize that choices have to be made and that probably, no optimal choice exists.

In a first "dynamic" approach, let us therefore view a curve as a continuous function

$$f\colon \mathbb{R} \longrightarrow \mathbb{R}^2, \qquad t \mapsto f(t)$$

as in Fig. 1.4.

The curve is thus thought of as the *trajectory of a point*, the trajectory expressed in terms of a parameter t which runs along the real line. This parameter t could be

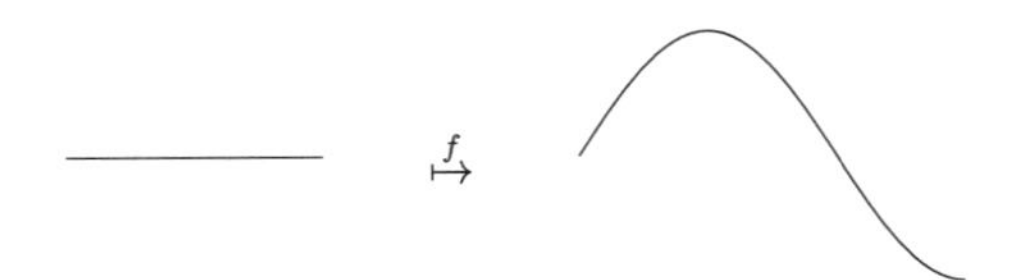

Fig. 1.4

regarded as the "time" calculated (positively or negatively) from a given origin of time: at the instant t, the point has reached the position $f(t)$ in the plane. Alternatively t could also be thought of as the "distance traveled on the curve" from a fixed origin on this curve: after having already traveled a distance t, the point has reached the position $f(t)$. And so on. When you prepare an itinerary for your holiday, you will probably say something like

After 247 km I shall be in Paris.

But when you comment on your travels afterwards, you will probably say

After 2 hours and 36 minutes I was in Paris.

In both cases you are commenting on the same itinerary, using different parameters.

Of course since various functions in terms of various parameters can describe the same curve, each of these functions should better be called a *parametric representation of a curve*.

However, we still have not avoided the "undesirable examples" encountered in the previous section. Simply choose for f the constant function on a point $(a, b) \in \mathbb{R}^2$ (not a particularly convincing "holiday itinerary": you spend your entire holiday at home)! Again we do not want to call this a "representation of a curve". We have a point, not a curve. More surprisingly:

Example 1.2.1 There exist continuous functions

$$f : \mathbb{R} \longrightarrow \mathbb{R}^2$$

whose image covers a full square.

Proof Let us sketch the construction of an example proposed by the Italian mathematician *Peano* in 1890. He defines a sequence

$$f_n : [0, 1] \longrightarrow \mathbb{R}^2, \quad n \in \mathbb{N}$$

of continuous functions, which converges uniformly to a continuous surjective function

$$f : [0, 1] \longrightarrow [0, 1] \times [0, 1].$$

Since moreover

$$f(0) = (0, 0), \qquad f(1) = (1, 1)$$

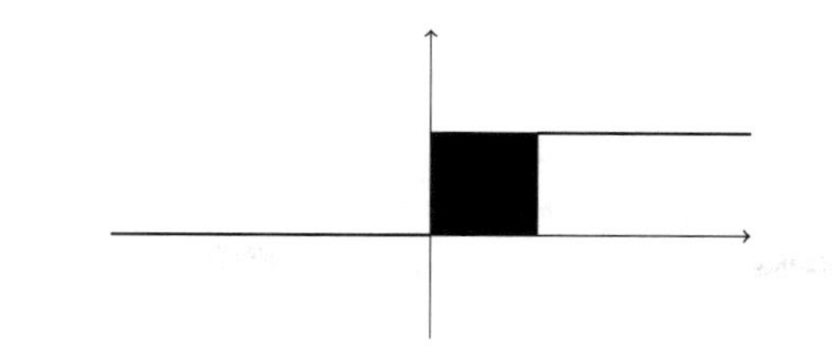

Fig. 1.5

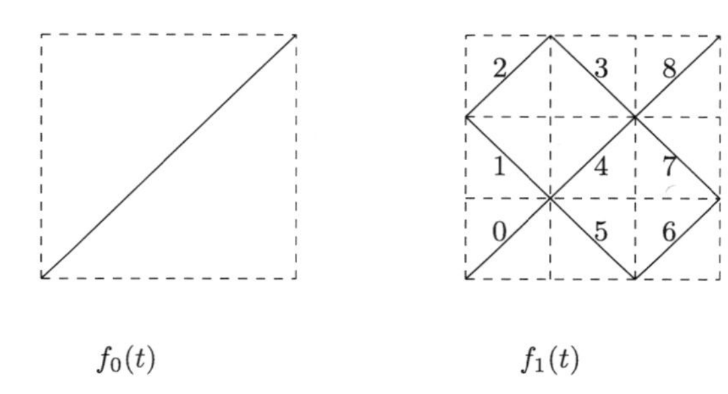

Fig. 1.6

it suffices to extend the definition by

$$\begin{cases} f(t) = (t, 0) & \text{if } t \leq 0, \\ f(t) = (t, 1) & \text{if } t \geq 1, \end{cases}$$

to get the expected counterexample as in Fig. 1.5.

The sequence begins with the identity function: $f_0(t) = t$. The graph of the function $f_1(t)$ is then given by the right hand picture in Fig. 1.6. Simply follow the path according to the numbering of the sub-squares 0 to 8.

To obtain $f_2(t)$, replace each diagonal of a small square in the graph of f_1 by an analogous zigzag of nine smaller segments, each starting and ending at the same points as the small diagonal. Repeat the process to pass from f_2 to f_3, and so on. Each function $f_n(t)$ is continuous and the sequence $(f_n)_{n \in \mathbb{N}}$ converges uniformly, since at each level, the further variations are at most the length of the diagonal of the smallest square already constructed. It is then a standard result in analysis that the limit function $f(t)$ is still continuous.

To prove that f is surjective, express t in base 9. The construction shows at once that, writing $a, b, c, d, \ldots$ for the successive digits of the expansion of t in base 9,

$$t = 0.abcde \ldots$$

then

- $f_1(t)$ is in the square numbered a;
- $f_2(t)$ is in the sub-square of the previous square numbered b;

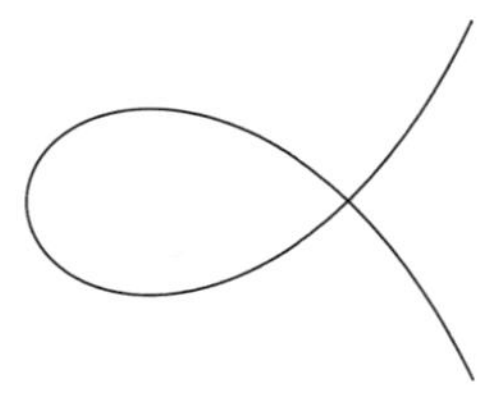

Fig. 1.7

- $f_3(t)$ is in the sub-sub-square numbered c;
- and so on.

It is obvious that every sequence of square–sub-square–sub-sub-square– $\cdots$ determines a unique point P of the square, and each point P of the square can be determined in this way. Such a sequence is by no means unique since (except for $(0,0)$ and $(1,1)$) each vertex of a small square, at whatever level, belongs to several squares. But nevertheless, choosing one of the possible sequences of square–sub-square–sub-sub-square– $\cdots$ which determines the point P, the list of the numbers 0 to 8 attached to each term of this sequence is then the base 9 expansion of a number $t \in [0,1]$ such that $f(t) = P$. Thus f is surjective. But as we have observed, such a number t is generally not unique, thus f is not injective. □

Again a dead end? Not really! We are now close to a solution. If we think of a parametric representation of a curve

$$f\colon \mathbb{R} \longrightarrow \mathbb{R}^2, \qquad t \mapsto f(t)$$

as being the trajectory of a point which "actually" moves in the plane, then when t varies, $f(t)$ should vary as well. Let us then simply impose that f is injective. This immediately eliminates the trivial case $f(t) = (a,b)$, but also Example 1.2.1, as we have seen.

The assumption "*f injective*" is perhaps a little too strong Fig. 1.7 depicts a "curve", even if the "trajectory" passes through the same point twice.

Considering the parametric representation of the circle

$$f(\theta) = (\cos\theta, \sin\theta)$$

as the parameter runs along the real line the corresponding point rotates around the circle infinitely many times. A single loop contains all the required information.

The following definition takes care of these "wishes".

Definition 1.2.2 A *parametric representation* of a plane curve is a continuous function

$$f\colon]a,b[\longrightarrow \mathbb{R}^2, \qquad t \mapsto f(t), \qquad a,b \in \mathbb{R} \cup \{-\infty, +\infty\}$$

which is *locally injective*, that is, injective in a neighborhood of each point.

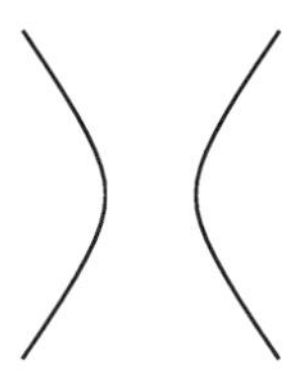

Fig. 1.8

More explicitly, the *local injectivity* means that for every $t_0 \in]a, b[$ one can find $\varepsilon > 0$ such that $]t_0 - \varepsilon, t_0 + \varepsilon[$ is still contained in $]a, b[$ and

$$f :]t_0 - \varepsilon, t_0 + \varepsilon[\longrightarrow \mathbb{R}^2$$

is injective. Allowing a and b to take "infinite values" is a quick way of saying that we allow f to be defined on $\mathbb{R}$ itself, on a half line or on an open interval.

Of course a constant function $f(t) = (a, b)$ is not locally injective. But what about the function f of Example 1.2.1?

Lemma 1.2.3 *The function f of Example* 1.2.1 *is not locally injective.*

Proof Consider the diagonal of a small square at the level n. This is the injective image under f_n of a small subsegment of $[0, 1]$. Let us say that this is the subsegment of origin u and length v. Observing the construction of the zigzag in Example 1.2.1, we conclude that all f_i with $i > n$ are such that

$$f_i\left(u + \frac{1}{9}v\right) = f_i\left(u + \frac{5}{9}v\right), \qquad f_i\left(u + \frac{4}{9}v\right) = f_i\left(u + \frac{8}{9}v\right).$$

Thus at the limit we still have

$$f\left(u + \frac{1}{9}v\right) = f\left(u + \frac{5}{9}v\right), \qquad f\left(u + \frac{4}{9}v\right) = f\left(u + \frac{8}{9}v\right).$$

This proves that one can always find points, everywhere in $[0, 1]$, "as close as one wants to each other", which are mapped by f onto the same point. Thus f is not *locally injective.* □

We conclude that the "non-examples" of curves that we gave earlier do not satisfy our Definition 1.2.2 of a curve. Does this mean that Definition 1.2.2 is *the* good one? The *only possible* good one? Certainly not. Nevertheless, the following chapters will give evidence that this is certainly *a possible* good definition.

For example, as already observed, our choice prevents us from considering the hyperbola (Fig. 1.8) as *one* curve, since it has two branches.

To overcome this problem, in the definition of a parametric representation of a curve, we could decide to allow as domain a *union of open intervals*, but probably not any kind of union. It would surely be wise to exclude such unions as

$$\bigcup_{n=1}^{\infty} \left] \frac{1}{2n+1}, \frac{1}{2n} \right[.$$

Reducing one's attention to a *finite* union of open intervals could be a reasonable compromise. However, as already mentioned in Sect. 1.1, do we really want the union of 247 straight lines to be considered as a single curve?

We could also decide to allow *closed* intervals as domains, not only *open* intervals. We would of course not allow these closed intervals to reduce to single points. But then every time we consider a construction using limits or derivatives, at the extremities of a closed interval, we would have to work with "one-sided" limits or derivatives. For example if we define a circle via

$$f\colon [0, 2\pi] \longrightarrow \mathbb{R}^2, \qquad f(\theta) = (\cos\theta, \sin\theta)$$

we have to treat separately the point $f(0) = f(2\pi)$, which by the way, in a circle, should have the same properties as any other point of the circle! As far as possible, we shall avoid entering into these considerations (nevertheless, see Definition 2.14.1).

In Definition 1.2.2 you may also want to impose that f is differentiable, or even of class $\mathcal{C}^\infty$, or some other class. We shall not do this: we will introduce these additional assumptions (or others) when they are needed for some results.

The conclusion of this discussion is thus

Defining a curve is a matter of choice!

But not all choices are sensible. Our choice is Definition 1.2.2.

1.3 Cartesian *Versus* Parametric

In Sect. 1.1 we have tried (without much success up to now) to determine a curve via a *Cartesian equation*

$$F(x, y) = 0$$

while in Sect. 1.2 we have focused our attention on parametric representations

$$f\colon]a, b[\longrightarrow \mathbb{R}^2, \qquad t \mapsto \big(f_1(t), f_2(t)\big)$$

that is, on a system of *parametric equations*

$$\begin{cases} x = f_1(t) \\ y = f_2(t). \end{cases}$$

Can we switch easily from one approach to the other, and perhaps guess what a good notion of *Cartesian equation of a curve* might be?

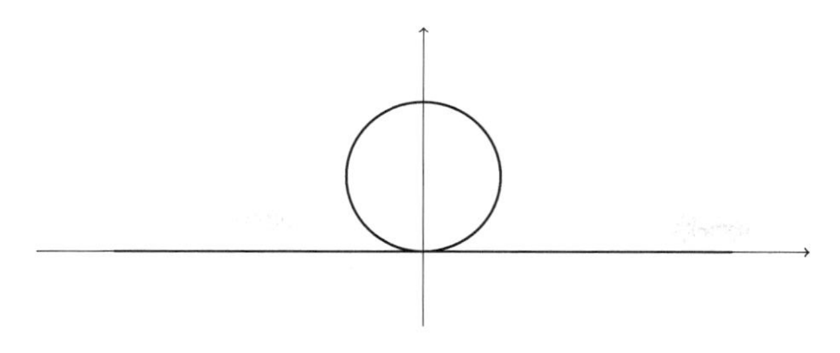

Fig. 1.9

An initial warning must be made. The "static" definition of a curve presents it as "a subset of the plane". The "dynamic" definition of a curve presents is as "a trajectory in the plane". However a curve, regarded as a subset of the plane, can easily be obtained via very different trajectories! Fig. 1.9 presents a curve comprising the x-axis and a circle of radius 1 with center $(0, 1)$. This "curve" admits the equation

$$y\big(x^2 + (y-1)^2 - 1\big) = 0.$$

You can view this as the "smooth" trajectory of a point coming from $(-\infty, 0)$, turning counter-clockwise around the circle, and proceeding next to $(+\infty, 0)$. Having arrived at the origin, you could also very well turn clockwise: as a trajectory, this is completely different! Of course you could also follow both trajectories in the reverse direction, but this is certainly not an essential difference.

Let us thus see how we can pass from a "static" description to a "dynamic" description, and vice-versa. In one direction, the idea is clear. Given the system of parametric equations

$$\begin{cases} x = f_1(t) \\ y = f_2(t) \end{cases}$$

we just need to eliminate the parameter t between the two equations ending up with a Cartesian equation! This is easy to say, but not always that easy to do when f_1 and f_2 are fairly involved functions.

However, analysis is there to help us, at least formally. Let us recall the following important result:

Theorem 1.3.1 (Local Inverse Theorem) *Consider a function $g\colon \mathbb{R}^n \longrightarrow \mathbb{R}^n$ of class C^k $(k \geq 1)$. If the matrix*

$$\left(\frac{\partial g_i}{\partial x_j}(a_1, \ldots, a_n)\right)_{1 \leq i,j \leq n}$$

is regular, then the function g is invertible on a neighborhood of the point $(a_1, \ldots, a_n)$ and its inverse is still of class C^k.

Of course when $n = 1$, the condition in Theorem 1.3.1 reduces to $g'(a) \neq 0$. This suggests the following definition:

Definition 1.3.2 A parametric representation of a curve

$$f\colon]a,b[\longrightarrow \mathbb{R}^2, \qquad t \mapsto \bigl(f_1(t), f_2(t)\bigr)$$

is *regular* when it is of class $\mathcal{C}^1$ and $f'(t) \neq (0,0)$ for each $t \in]a,b[$.

We obtain the following result:

Proposition 1.3.3 *Let $f\colon]a,b[\longrightarrow \mathbb{R}^2$ be a regular parametric representation of a curve. For every $t_0 \in]a,b[$:*

1. *there exists a neighborhood of t_0 on which the curve admits a Cartesian equation $F(x,y)=0$;*
2. *the function $F\colon \mathbb{R}^2 \longrightarrow \mathbb{R}$ is of class $\mathcal{C}^1$ on this neighborhood;*
3. *at each point of the curve in the given neighborhood, at least one of the partial derivatives of F is non-zero.*

Proof Assume—for example—that $f_1'(t_0) \neq 0$. By Proposition 1.3.1 we can write $t = f_1^{-1}(x)$ on a neighborhood of t_0. This yields

$$y = f_2\bigl(f_1^{-1}(x)\bigr)$$

and it suffices to define

$$F(x,y) = f_2\bigl(f_1^{-1}(x)\bigr) - y.$$

Notice in particular that

$$\frac{\partial F}{\partial y} = -1 \neq 0. \qquad \square$$

Proposition 1.3.3 suggests further to try the following definition:

Definition 1.3.4 (Temporary Definition; see 1.4.2) By a *Cartesian equation* of a plane curve is meant an equation

$$F(x,y)=0$$

satisfying the following requirements:

- this equation admits solutions;
- $F\colon \mathbb{R}^2 \longrightarrow \mathbb{R}$ is a function of class $\mathcal{C}^1$;
- at each point (x_0, y_0) such that $F(x_0,y_0)=0$, at least one of the partial derivatives of F is non-zero.

The corresponding *curve* is the set of those points (x,y) such that $F(x,y)=0$.

Let us now consider the opposite problem: how do we pass from a Cartesian equation to a parametric representation? Once more, analysis is there to help us solve our problem. Let us recall a celebrated result:

Theorem 1.3.5 (Implicit Function Theorem) *Consider a function $F\colon \mathbb{R}^n \longrightarrow \mathbb{R}$ of class $\mathcal{C}^k$ $(k \geq 1)$. If*

$$F(a_0, \ldots, a_n) = 0, \qquad \frac{\partial F}{\partial x_n}(a_0, \ldots, a_n) \neq 0$$

then there exists

- *a neighborhood V of $(a_0, \ldots, a_{n-1})$ and*
- *a function $\varphi\colon V \longrightarrow \mathbb{R}$ of class $\mathcal{C}^k$*

such that

- $\varphi(a_0, \ldots, a_{n-1}) = a_n$;
- $\forall (x_1, \ldots, x_{n-1}) \in V \;\; F(x_1, \ldots, x_{n-1}, \varphi(x_1, \ldots, x_{n-1})) = 0.$

Moreover, the neighborhood V can be chosen such that a function φ as in the statement is necessarily unique.

The *implicit function* inferred from F is thus

$$x_n = \varphi(x_1, \ldots, x_{n-1}).$$

Proposition 1.3.6 *Consider a Cartesian equation $F(x, y) = 0$ of a plane curve (as in Definition* 1.3.4*) and a point (x_0, y_0) satisfying this equation. In a neighborhood of (x_0, y_0), there exists a regular parametric representation of a curve*

$$f\colon]a, b[\longrightarrow \mathbb{R}^2$$

such that each point $(x, y) = (f_1(t), f_2(t))$ satisfies the equation $F(x, y) = 0$.

Proof Assume that $\frac{\partial F}{\partial y}(x_0, y_0) \neq 0$. With the notation of Theorem 1.3.5 it suffices to define

$$f(x) = \big(x, \varphi(x)\big).$$

The parameter is thus $t = x$ and $f'(x) = (1, \varphi'(x)) \neq (0, 0)$. Considering its first component, we notice that f is injective. □

The slogan suggested by Propositions 1.3.3 and 1.3.6 is thus:

> *In good cases, one can switch* locally *from a system of parametric equations to a Cartesian equation, and vice-versa.*

Locally is certainly *the* point to emphasize here, but it is not the only one. The two processes seem to be "the inverse of each other", but this is definitely a false impression. Let us demonstrate this with some examples.

Consider first the circle of Fig. 1.10 and its parametric equations

$$\begin{cases} x = R\cos\theta \\ y = R\sin\theta. \end{cases}$$

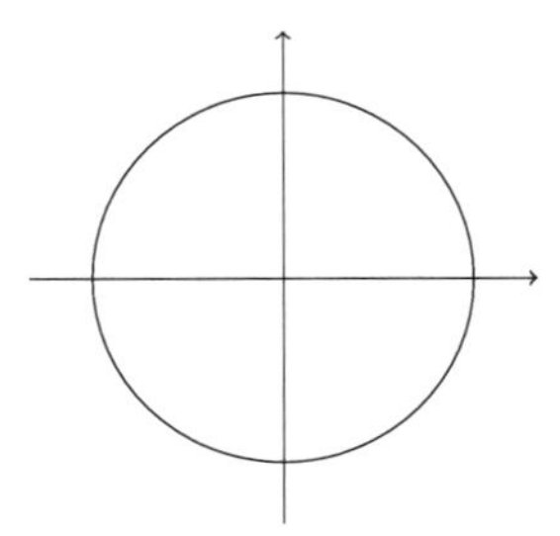

Fig. 1.10

To eliminate θ between the equations you will probably simply square each equation and add the results, to end up with

$$x^2 + y^2 = R^2.$$

Conversely, you will probably write

$$y = \pm\sqrt{R^2 - x^2}$$

and observe that each choice of the sign will give you half of the circle (the "upper half" or the "lower half"). You will then obtain

$$f :]-R, +R[\to \mathbb{R}^2, \qquad x \mapsto \left(x, \sqrt{R^2 - x^2}\right)$$

as a parametric representation of the upper half of the circle. Working with

$$x = \pm\sqrt{R^2 - y^2}$$

would give you the "left half" or the "right half". So from the parametric equations, you have obtained the "global" Cartesian equation of the circle, but from that Cartesian equation you have reconstructed—only locally—parametric equations of the circle. Moreover, these are completely different from the original parametric equations!

Let us now try the same with the parametric equations

$$\begin{cases} x = e^t \\ y = e^t \end{cases}$$

which represent the half-diagonal of Fig. 1.11. It suffices to subtract the two equations to eliminate t, and this yields the equation $x = y$ of the full diagonal! Of course one cannot possibly guess, given only the Cartesian equation $x = y$, that it comes from the original parametric equations.

Another slogan should thus be

Be careful ...

But we should perhaps also add *Be sorry!*, as the next section will explain.

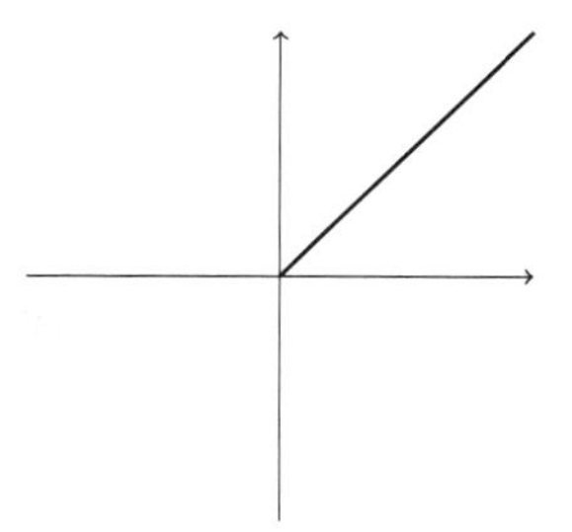

Fig. 1.11

Fig. 1.12

1.4 Singularities and Multiplicities

We have already observed in Sect. 1.1 that not all real polynomials $F(X,Y)$ yield the equation $F(x,y)=0$ of a curve in $\mathbb{R}^2$. But we know many examples where $F(x,y)=0$ does describe something which is worthy of being called a "curve". For example

$$x^3-3xy^2=\left(x^2+y^2\right)^2$$

which yields the "curve" of Fig. 1.12.

This equation can thus be written in the form $F(x,y)=0$ with

$$F(x,y)=x^3-3xy^2-\left(x^2+y^2\right)^2.$$

Since each term is of degree at least 2 (in fact, of degree 3 or 4),

$$\frac{\partial F}{\partial x}(0,0)=0, \qquad \frac{\partial F}{\partial y}(0,0)=0.$$

So unfortunately, this "curve" is not a curve in the sense of our Definition 1.3.4.

Not recapturing the "decent algebraic curves" in our theory is rather unsatisfactory. The present Section, deeply inspired by the considerations of Chap. 7 in [4], *Trilogy II*, will now discuss this "difficulty" further. We observe first that:

Proposition 1.4.1 *Let $f(X,Y) \in \mathbb{R}[X,Y]$ be a non-constant polynomial without any multiple factors. There are at most finitely many points (X,Y) such that $f(X,Y)=0$ and both derivatives of f vanish at (X,Y).*

Proof Let us consider the family $(a_i,b_i)_{i\in I}$ of those points such that

$$f(a_i,b_i)=0, \qquad \frac{\partial f}{\partial X}(a_i,b_i)=0, \qquad \frac{\partial f}{\partial Y}(a_i,b_i)=0.$$

We must prove that there are only finitely many of them.

Let us write $F(X,Y,Z)$ for the homogeneous polynomial associated with F (see [4], *Trilogy II*, Sect. C.2). Thus $f(X,Y)=F(X,Y,1)$ and the factors of f and F correspond to each other via the "homogenizing" process. Both polynomials f and F have the same degree: let us say, n. By Euler's formula (see C.1.5 in [4], *Trilogy II*)

$$nF = X\frac{\partial F}{\partial X} + Y\frac{\partial F}{\partial Y} + Z\frac{\partial F}{\partial Z}.$$

Applying this formula at the points $(a_i,b_i,1)$ yields

$$\frac{\partial F}{\partial Z}(a_i,b_i,1)=0.$$

Thus the points $(a_i,b_i,1)$ are *multiple points* (see Definition 7.4.5, [4], *Trilogy II*) of the complex projective curve $F(X,Y,Z)=0$.

By assumption, $f(X,Y)$ and thus $F(X,Y,Z)$ do not have any multiple factor as real polynomials. If we can prove that analogously $F(X,Y,Z)$ does not have any multiple factors in $\mathbb{C}[X,Y,Z]$, then the number of multiple points of $F(X,Y,Z)$ is bounded by $n(n-1)$ (see Sect. 7.9 in [4], *Trilogy II*). Thus there are at most $n(n-1)$ points (a_i,b_i) as above.

Let us recall that splitting all coefficients into their real and their imaginary parts, every complex polynomial $\alpha(X,Y,Z)$ can be written as

$$\alpha(X,Y,Z)=\beta(X,Y,Z)+i\gamma(X,Y,Z)$$

where α and β are polynomials with real coefficients. This shows at once that given a non-constant real polynomial $\delta(X,Y,Z)$, if $\alpha\delta$ is a real polynomial, then $\gamma(X,Y,Z)=0$. In other words, if a non-constant real polynomial $\delta(X,Y,Z)$ divides another real polynomial in $\mathbb{C}[X,Y,Z]$, it divides it in $\mathbb{R}[X,Y,Z]$.

Replacing the coefficients of α by their conjugates then yields

$$\overline{\alpha}(X,Y,Z)=\beta(X,Y,Z)-i\gamma(X,Y,Z).$$

It follows at once that, just as for complex numbers

$$\alpha(X,Y,Z)\overline{\alpha}(X,Y,Z)=\beta(X,Y,Z)^2+\gamma(X,Y,Z)^2$$

that is, a polynomial with real coefficients.

Write now

$$F(X, Y, Z) = G_1(X, Y, Z) \cdots G_m(X, Y, Z)$$

with the $G_k(X, Y, Z)$ irreducible. We must prove that each factor G_k is simple.

Since F has real coefficients, passing to the conjugates yields

$$F(X, Y, Z) = \overline{G_1}(X, Y, Z) \cdots \overline{G_m}(X, Y, Z).$$

Of course, the $\overline{G_k}$ are still irreducible, because conjugation is a homomorphism of fields.

If some G_k has real coefficients, it divides $F(X, Y, Z)$ in $\mathbb{R}[X, Y, Z]$ as we have just seen. Therefore by assumption, it is a simple factor.

Otherwise by uniqueness of the decomposition into irreducible factors, there exists an index $j \neq k$ such that $G_j = \overline{G_k}$. Then $G_k G_j = G_k \overline{G_k}$ is a polynomial with real coefficients which divides $F(X, Y, Z)$. By assumption, it is a simple factor. Dividing by this polynomial and repeating the argument allows as to conclude that all non-real G_k are simple factors as well. □

Of course replacing a multiple factor of $f(x, y)$ by the same factor with degree 1 does not modify the set of points (x, y) such that $f(x, y) = 0$. Thus the assumption in Proposition 1.4.1 is not really a restriction, as far as the study of curves is concerned.

All this suggests modifying Definition 1.3.4 in the following way:

Definition 1.4.2 By a *Cartesian equation* of a plane curve is meant an equation

$$F(x, y) = 0$$

where:

- $F: \mathbb{R}^2 \longrightarrow \mathbb{R}$ is a function of class $\mathcal{C}^1$;
- there exist solutions (x, y) where at least one partial derivative of F does not vanish;
- there are at most finitely many solutions (x, y) where both derivatives of F vanish.

The corresponding *curve* is the set of those points (x, y) such that $F(x, y) = 0$.

Of course now, Proposition 1.3.6 holds only for those points where at least one of the partial derivatives is not zero.

Still inspired by the considerations of Chap. 7 in [4], *Trilogy II*, it is also sensible to define:

Definition 1.4.3 Let $F(x, y) = 0$ be a Cartesian equation of a plane curve. The points (x, y) of the curve where both partial derivatives of F vanish are called the *multiple points* of the curve.

In the curve of Fig. 1.12 there is thus one single multiple point, namely, $(0,0)$.

However, if we define *multiple points* when the curve is given by a Cartesian equation, we are immediately faced with the challenge of defining a corresponding notion for a parametric representation. Since at a multiple point Proposition 1.3.6 does not hold, we would be tempted, in the case of a parametric representation, to consider those points where the "converse" Proposition 1.3.3 does not hold:

Definition 1.4.4 Given a parametric representation of class $\mathcal{C}^1$ of a curve, a point of parameter t is *singular* when $f'(t)=(0,0)$.

It is important to stress two facts concerning this notion:

- "being a singular point" is a property of the parametric representation which does not necessarily exhibit a "singularity" of the corresponding subset of $\mathbb{R}^2$;
- being a "singular point" for a parametric representation is by no means equivalent to being a "multiple point" for a corresponding Cartesian equation.

For example

$$f\colon \mathbb{R} \longrightarrow \mathbb{R}^2, \qquad t \mapsto (t^3,0)$$

is a parametric representation of class $\mathcal{C}^\infty$ of the x-axis, which on the other hand admits the Cartesian equation $y=0$. Observe that

$$f'(t)=(3t^2,0), \qquad f'(0)=(0,0)$$

thus the origin is a *singular point* of the parametric representation f. But the origin is by no means a *multiple point* of the x-axis, that is, the algebraic curve $y=0$.

If we consider the more usual parametric representation of the x-axis

$$g\colon \mathbb{R} \longrightarrow \mathbb{R}, \qquad t \mapsto (t,0)$$

then

$$g'(t)=(1,0)\neq(0,0)$$

and there is no singular point at all.

Next consider the curve of Fig. 1.12 and its multiple point at the origin. It is routine to verify that

$$\begin{cases} x=\cos\theta\cdot\cos 3\theta \\ y=\sin\theta\cdot\cos 3\theta \end{cases}$$

is a system of parametric equations of the same "curve". Since $\sin\theta$ and $\cos\theta$ do not vanish together, the origin $(0,0)$ is reached when $\cos 3\theta=0$, that is (up to 2π) for

$$\theta=\frac{\pi}{6}, \qquad \theta=\frac{\pi}{2}, \qquad \theta=\frac{5\pi}{6}.$$

A straightforward computation shows that, writing f for the parametric representation, $f'(\theta)\neq(0,0)$ at these three points (in fact, the parametric representation f

is regular!) Thus these three values of the parameter θ are *not singular*, while the corresponding point is *multiple*.

The conclusion is clear: parametric representations are not appropriate for the study of multiple points in the sense of the theory of algebraic curves! Intuitively, if you travel "regularly" along a curve, there is nothing special about passing through a point you have passed through earlier.

You might claim that passing through the same point several times *is* something special. When working with a parametric representation f, we might then try to say that a point $P \in \mathbb{R}^2$ is *multiple* if $P = f(t)$ for several values of the parameter t. But then all points of the circle

$$f : \mathbb{R} \longrightarrow \mathbb{R}^2, \qquad \theta \mapsto (\cos\theta, \sin\theta)$$

are multiple, and even of *multiplicity* ∞! I am sure this is not what you had in mind!

You should certainly now be convinced that defining *a curve* is definitely a matter of choice. If you strengthen the conditions in order to avoid some pathologies, then you eliminate some examples that you would like to keep, and conversely. Moreover, working with parametric equations or with a Cartesian equation lead rather naturally to non-equivalent choices of definitions.

In this book, we shall adopt Definitions 1.2.2 and 1.4.2, and we shall stop our endless search for possible improvements of these definitions.

1.5 Chasing the Tangents

The Greek geometers defined tangents in the following way:

Definition 1.5.1 A *tangent* to a circle at one of its points P is a line whose intersection with the circle is reduced to the point P.

They proved (see Proposition 3.3.2, [3], *Trilogy I*):

Proposition 1.5.2 *Given a point P of a circle, there exists a unique tangent at P to the circle, namely, the perpendicular to the radius at P (see Fig.* 1.13).

Very trivially, such a definition does not work at all for arbitrary curves. Just have a look at Fig. 1.14: a tangent can cut the curve at a second point, and a line which cuts the curve at exactly one point has no reason to be a tangent.

Consider the trivial case of a straight line: the tangent to a straight line should be the line itself, which certainly takes us very far from a "unique" point of intersection, globally or locally. Also keep in mind that a tangent can "cut" the curve at the point of tangency, as in Fig. 1.15: thus "touching without cutting" is an inadequate definition. Finally do not forget the case of multiple points, as in Fig. 1.12. At a multiple point, there could be several tangents, not just a single one.

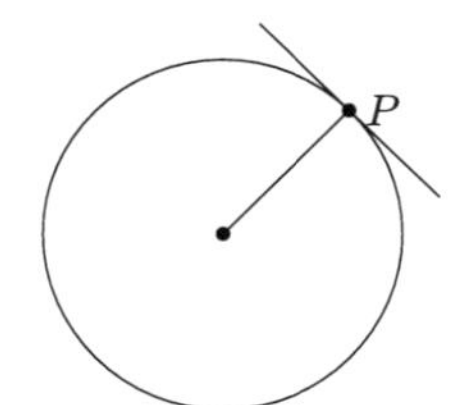

Fig. 1.13

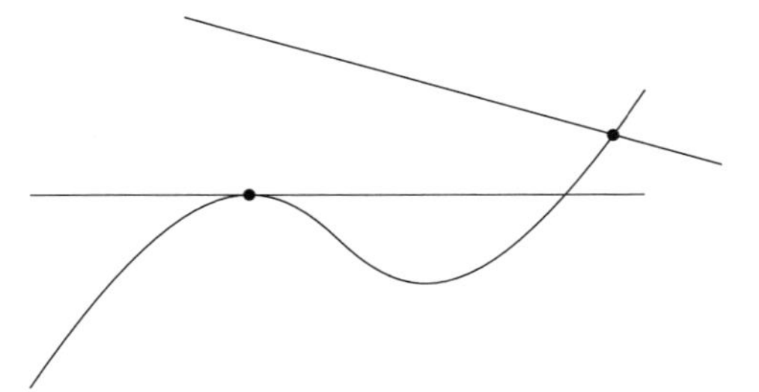

Fig. 1.14

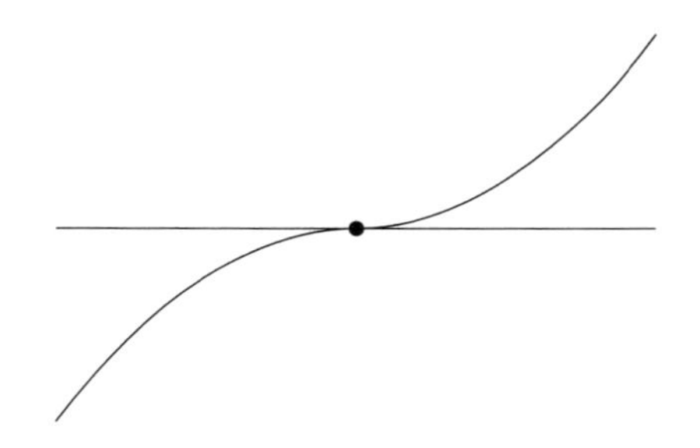

Fig. 1.15

Fortunately, the fact of not having a good definition of a tangent did not prevent mathematicians from calculating tangents!

In the 1630's, *Fermat* and *Descartes* proposed methods to calculate the tangent to a curve given by a polynomial equation $F(x, y) = 0$ (see Sect. 1.9 of [4], *Trilogy II*). The idea was that

A tangent is a line having a double point of intersection with the curve.

The notion of "double point of intersection" was in those days (1630–1640) more heuristic than precisely defined, but today it has been formalized in rigorous contemporary algebraic terms (see Definition 7.4.5 in [4], *Trilogy II*). In this book, we shall instead turn our attention to some attempts which prefigure contemporary differential methods.

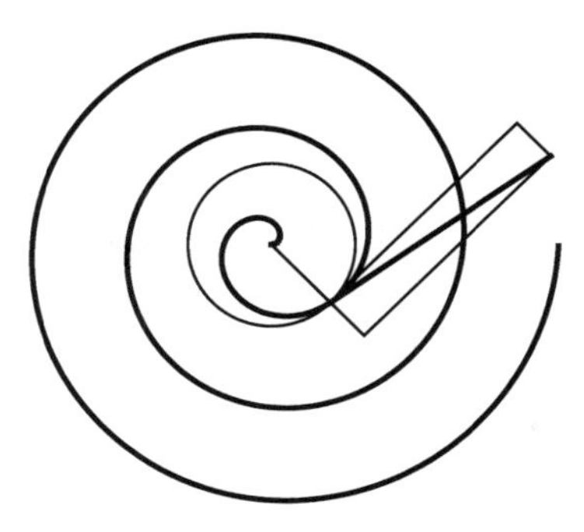

Fig. 1.16

Greek geometers could thus calculate the tangent to a circle and of course the tangent to a line *Archimedes* (287–212 AC) had already made the conceptual leap of regarding a curve as the trajectory of a moving point, and in this context he treated the tangent as follows:

Definition 1.5.3 The *tangent* to a curve is the line in the direction of the instantaneous movement of a point traveling on that curve.

Archimedes then computed the tangent to certain curves by "decomposing" the movement into a combination of linear and circular movements, and assuming that the direction of the tangent can be decomposed analogously. Let us follow his argument on a precise example (see Sect. 4.5 in [3], *Trilogy I*, for more comments on this curve).

Definition 1.5.4 The *spiral of Archimedes* (see Fig. 1.16) is the trajectory of a point in a plane, which moves at constant speed along a line, while the line turns at constant speed around one of its points, called the *center* of the spiral (see Fig. 1.16).

Example 1.5.5 Archimedes' construction of the tangent to his spiral.

Proof The global movement has two components: one resulting from the uniform linear movement of the point on the line, one resulting from the uniform circular movement of the line.

The component resulting from the uniform linear movement of the point on the line is expressed by a segment oriented along this line. Its length is the distance traveled on the line during a unit of time: let us say (to keep the picture on a page), during the time necessary for the line to make a half turn. This length is then half the distance between two turns of the spiral.

To obtain the component of the movement resulting from the uniform circular movement:

- consider the circle centered at the center of the spiral and which passes through the point at which you want to compute the tangent;

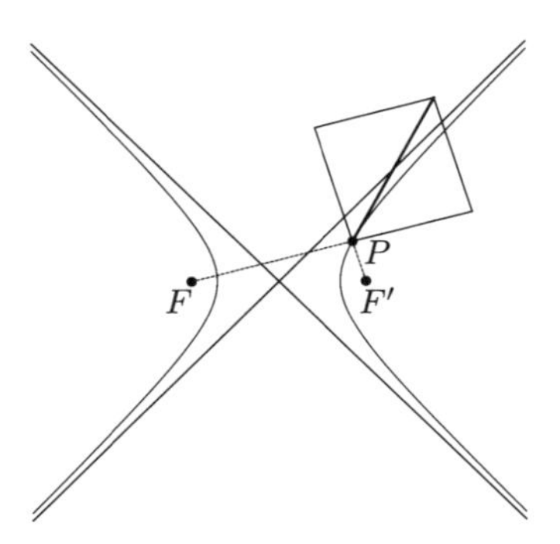

Fig. 1.17

- the circular component of the movement is a segment in the direction of the tangent to this circle;
- this component has a length equal to the distance traveled on this circle during a unit of time, that is, half the length of the circle.

Adding these two components via the parallelogram rule, *Archimedes* gets the direction of the tangent. □

This is a beautiful theoretical result, but of course, quite a disturbing one for a Greek geometer! Indeed Greek geometers could not, by ruler and compass constructions, draw a segment whose length is equal to the length of the circle, and we have known since the 19th century that this is in fact impossible: "circle squaring", that is, constructing π, is impossible by ruler and compass constructions (see Corollary B.3.3 in [3], *Trilogy I*).

In 1636, the French mathematician *Roberval* systematized *Archimedes*' idea to compute what he called *the touching line*. He applied this method to a wide variety of curves: various spirals, the conchoids, the cycloid, and so on (see Chap. 3 for a description of these curves). But let us focus here on his treatment of the tangent to a conic.

Proposition 1.5.6

1. *The tangent at a point P to a hyperbola with foci F, F' is a bisector of the two lines FP, $F'P$.*
2. *The tangent at a point P to an ellipse with foci F, F' is a bisector of the two lines FP, $F'P$.*
3. *The tangent at a point P to a parabola with focus F and directrix f is a bisector of the line FP and the perpendicular to f through P.*

Proof Consider first the case of the hyperbola (Fig. 1.17). As proved in Proposition 1.12.1 [4], *Trilogy II*, the hyperbola is the locus of those points P such that the difference of the distances

$$\left|d(P,F)-d(P,F')\right|$$

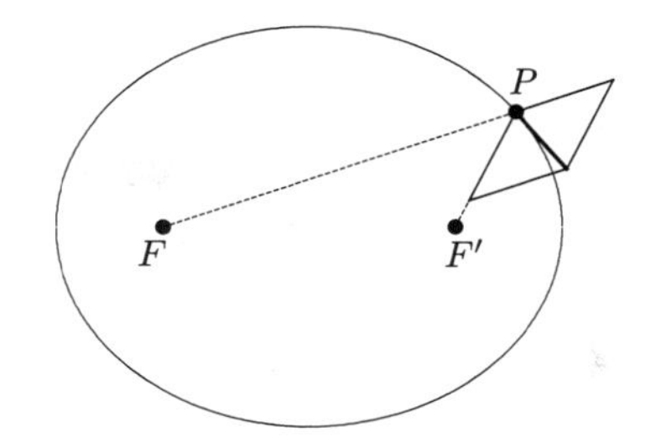

Fig. 1.18

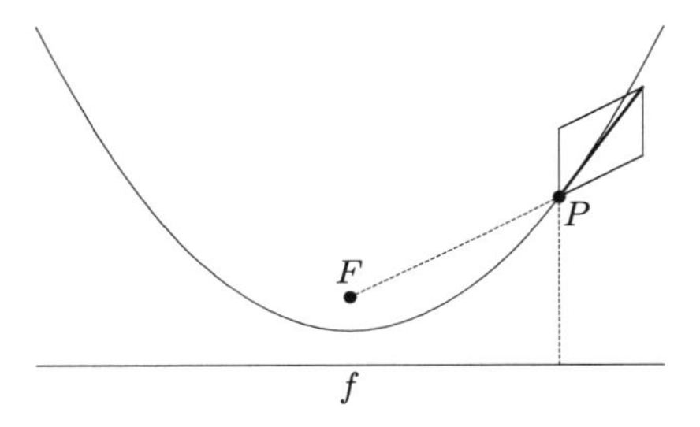

Fig. 1.19

to the two foci remains constant. When you move along a branch of the hyperbola—let us say—away from the origin, both distances increase. But since the difference between the two distances remains the same, both distances increase at the same rate. Roberval decomposes the movement into two instantaneous movements: one along the line FP, one along the line $F'P$. Since these two movements have equal amplitudes, the corresponding "parallelogram of movements" is a diamond, and the length of a side has no influence on the direction of the diagonal. Therefore, the tangent is simply the bisector of the two lines joining the point P and the two foci F and F'.

An analogous argument holds for the ellipse (see Fig. 1.18): this time, by Proposition 1.11.1 in [4], *Trilogy II*,

$$d(F, P) + d(F, P)$$

is constant. Thus one distance increases in the same way as the other one decreases. This again yields a diamond as "parallelogram of movements".

Finally for the parabola (see Fig. 1.19) with focus F and directrix f, when you move away from the origin, the two distances $d(F, P)$ and $d(f, P)$ increase at the same rate. Therefore the "parallelogram of movements" is a diamond with one side perpendicular to f and the other one in the direction FP. □

Think what you want of such arguments, they were nevertheless efficient in a period when differential calculus did not exist!

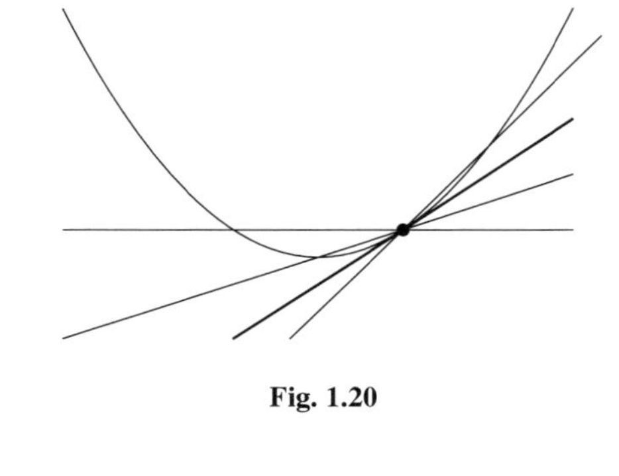

Fig. 1.20

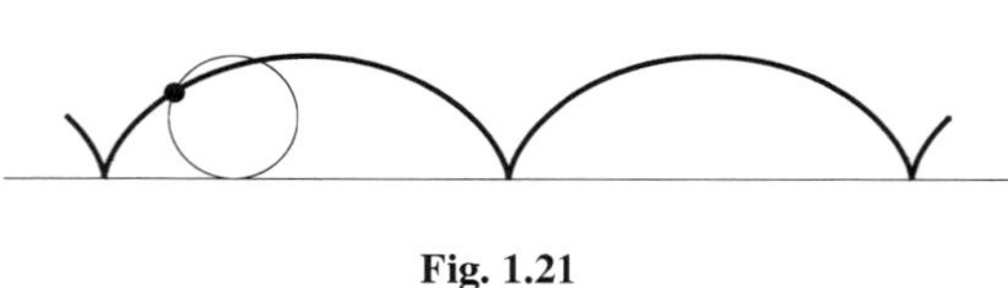

Fig. 1.21

1.6 Tangent: The Differential Approach

Although it is nice to see how tangents were computed historically, today everybody "with a basic mathematical culture" knows that:

> *The* tangent *to an arbitrary curve at one of its points P is the limit of the secant through P and another point Q of the curve, as Q converges to P (see Fig.* 1.20).

Indeed, this "dynamic" definition of the tangent, taking full advantage of the notion of limit, recaptures precisely our intuition of what a tangent should be (see Fig. 1.20).

Of course this is no longer the case at a multiple point (see Fig. 1.12): there we should consider separately the various "branches" of the curve, whatever that means! Perhaps we should decide if at a vertex of a square, there are two tangents, or no tangent at all. In the case—for example—of the cycloid: the trajectory of a point of a circle which rolls on a line (see Fig. 1.21), we should decide whether or not there is a tangent at each *cusp point*.

What might be a possible tangent at the origin for the curve with parametric representation

$$f: \mathbb{R} \longrightarrow \mathbb{R}^2, \qquad t \mapsto \begin{cases} \left(t, t^k \sin \frac{1}{t}\right) & \text{if } t \neq 0 \\ (0,0) & \text{if } t = 0, \end{cases} \qquad k \in \mathbb{N}?$$

It is no longer clear which curves have a tangent and which do not. We still need a precise definition. The trouble with the "definition" above is that we can define the limit of a family of points in $\mathbb{R}$, and the limit of a family of vectors in $\mathbb{R}^2$, but how are we to precisely define the limit of a family of lines?

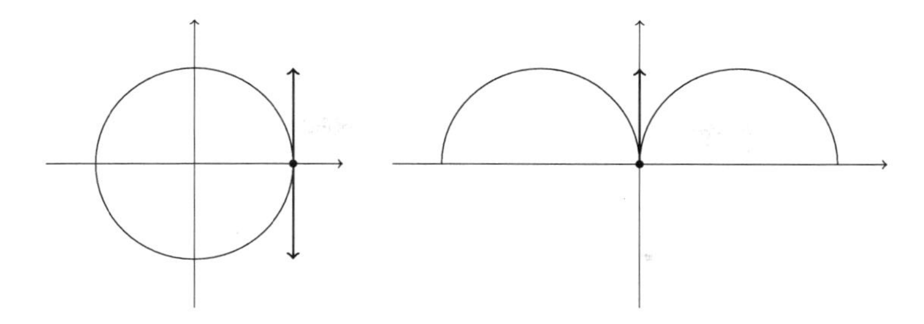

Fig. 1.22

Let us take a parametric representation of the curve

$$f\colon]a,b[\longrightarrow \mathbb{R}^2$$

and consider the point $P = f(t_0)$. When t converges to t_0, the point $Q = f(t)$ converges to $P = f(t_0)$. The secant through P and Q is the line

- passing through $P = f(t_0)$;
- of direction $\overrightarrow{PQ} = f(t) - f(t_0)$.

The tangent will thus be the line

- passing through $P = f(t_0)$;
- of direction $\lim_{t\to t_0}(f(t) - f(t_0))$.

Unfortunately this does not make any sense because by continuity of f, the limit is simply $f(t_0) - f(t_0)$, that is the zero vector.

But the difficulty is easy to overcome. Two vectors in the same direction define the same secant, thus let us simply work with vectors of length 1, so that the limit should remain of length 1. The tangent should thus be the line

- passing through $P = f(t_0)$;
- of direction

$$\lim_{t\to t_0} \frac{f(t) - f(t_0)}{\|f(t) - f(t_0)\|}.$$

When this limit exists, of course. Unfortunately, this limit most often does not exist.

Look at the following two curves, represented in Fig. 1.22.

- the circle with parametric representation

$$f\colon \mathbb{R} \longrightarrow \mathbb{R}^2, \qquad \theta \mapsto (\cos\theta, \sin\theta);$$

- the curve constituted of two half-circles, whose parametric representation is given by

$$g\colon]-1,+1[\longrightarrow \mathbb{R}^2, \qquad t \mapsto \bigl(t, \sqrt{1-x^2}\bigr).$$

In both cases, consider the point with parameter 0. What about the tangent, at these points, in the sense of the "definition" above?

- In the case of the circle, the expected limit does not exist! Indeed, we observe immediately that

$$\lim_{\substack{t\to 0\\ t<0}} \frac{f(t)-f(t_0)}{\|f(t)-f(t_0)\|} = (-1,0), \qquad \lim_{\substack{t\to 0\\ t>0}} \frac{f(t)-f(t_0)}{\|f(t)-f(t_0)\|} = (1,0).$$

  Both results are different, thus the limit does not exist.
- In the case of the two half-circles, the same kind of computation shows at once that

$$\lim_{t\to 0} \frac{g(t)-g(t_0)}{\|g(t)-g(t_0)\|} = (1,0)$$

  and the limit exists.

What does this mean? Although defining a tangent as "a limit of secants" is a good idea, when you try to make precise what "a limit of secants means", you easily run into severe problems. For example, with the attempt above, the circle does not have a tangent while the curve comprising of two half circles does! This first attempt to define a "limit of secants", because of the "counterexample" of the circle, is certainly unacceptable.

Note that in the case of the circle, the limits for $t < t_0$ and $t > t_0$ are opposite vectors, thus define the same direction, thus the same line. So one could modify our definition of the tangent by saying that both limits

$$\lim_{\substack{t\to t_0\\ t<t_0}} \frac{f(t)-f(t_0)}{\|f(t)-f(t_0)\|}, \qquad \lim_{\substack{t\to t_0\\ t>t_0}} \frac{f(t)-f(t_0)}{\|f(t)-f(t_0)\|}$$

on the right and on the left should exist, be non-zero, and be proportional vectors. However, one cannot expect to be able to prove elegant results and make computations with such a convoluted definition of the tangent.

The sensible thing to do is indeed to replace the vector $f(t)-f(t_0)$ by a vector proportional to it, but not a vector of length 1. Consider instead

$$\lim_{t\to t_0} \frac{f(t)-f(t_0)}{t-t_0}$$

which, when it exists, is simply $f'(t_0)$. Of course for such an approach to be efficient, not only must the derivative exist, but it must be non-zero! Therefore we make the following definition:

Definition 1.6.1 Consider a regular parametric representation of a curve

$$f\colon]a,b[\longrightarrow \mathbb{R}^2.$$

- The *tangent* to this curve at the point with parameter t_0 is the line containing $f(t_0)$ and of direction $f'(t_0)$.
- The *normal* to this curve at the point with parameter t_0 is the perpendicular to the tangent at this point.

Now of course, since the same curve can be described by various parametric representations, one should verify that Definition 1.6.1 of the tangent does not depend on the choice of the representation. We shall treat this question in Sect. 2.4. We recall that our point here is not to prove theorems, but to "guess" good definitions! Analogously one should see what happens to this definition when the curve is given by a Cartesian equation, but again this will be done in Sect. 2.4.

To conclude this section, let us insist once more on the fact that *defining the tangent is a matter of choice*.

Of course with Definition 1.6.1, our parametric representation f of the circle now yields a tangent at each point, because it is regular.

In our first attempt, the curve comprising two half circles also had a tangent at each point, but the parametric representation g of this curve is not differentiable at $t=0$. So using Definition 1.6.1, the curve represented by g does not have a tangent at the origin.

Therefore one might want to further modify the definition of a tangent, to get the best of the two attempts. For example, by requiring only the proportionality of the left derivative and the right derivative, not the existence of the derivative. Then the curve represented by g would also have a tangent at the origin.

If we consider the parametric representation

$$h\colon \mathbb{R} \longrightarrow \mathbb{R}^2, \qquad t \mapsto (t^3, 0)$$

of the x-axis, we observe that $h'(0) = (0,0)$, thus in view of Definition 1.6.1, the x-axis—when represented by h—does not have a tangent at the origin, which is less than satisfactory. Again one might want to revise the definition of a tangent to avoid such a situation.

However, we shall not enter into these considerations: we adopt once and for all Definition 1.6.1. We are now well aware that in doing so, we exclude examples where a more involved definition would have produced a "sensible tangent".

1.7 Rectification of a Curve

As far as the length of a curve is concerned, the greatest achievement of Greek geometry was (see Theorem 3.1.4 in [3], *Trilogy I*):

Theorem 1.7.1 *The ratio between the length of a circle and the length of its diameter is a constant, independent of the size of the circle. This constant is written* π.

Proof This result was proved by the so-called *exhaustion method* due to *Eudoxus* (around 380 AC): this method was the direct ancestor of the notion of *limit*. Greek geometers first proved a corresponding result for regular polygons inscribed in a circle and then "by a limit process", inferred the result for the circle. □

The importance of this result is often hidden by the systematic use of the well-know formulas $2\pi R$ and πR^2 for the length and the area of a circle of radius R. *These formulas hold because the number π involved is independent of the size of the circle!* Today many of us consider that these formulas answer the question fully. For Greek geometers, they were only a beginning: what is the precise value of this quantity π? The famous problem of *squaring the circle* consisted equivalently of finding a construction of a segment of length π. However, all attempts in this direction seemed to be hopeless.

Two thousand years later, in 1637, the French mathematician and philosopher *Descartes* wrote

> *The relations between straight lines and curves are not known and, I think, cannot be discovered by the human mind; for that reason, no conclusion at all based on such relations can be accepted as rigorous and exact.*

So, even if you say that the length of a circle is $2\pi R$, you still do not know the length of the circle since you do not know the precise value of π! You are unable to construct (with ruler and compass) a segment having the same length as the circle, and if you cannot do this for the circle, how could you possibly hope to do it for more complicated curves?

One year later Descartes studied the movement of a body falling on the Earth, while the Earth is itself was considered as a body in rotation. For that he introduced the so-called *logarithmic spiral* (see Fig. 1.23).

Definition 1.7.2 The *logarithmic spiral* is the trajectory of a point moving on a line, at a speed proportional to the distance already travelled on this line, while the line itself turns at constant speed around one of its points.

In an irony of history, the logarithmic spiral was the first curve to be *rectified*, that is, a precise construction was given to produce a segment whose length is equal to the length of a given arc of the curve. This result is due to the Italian mathematician *Torricelli* (1608–1647), a student of *Galilee Galileo*. At the same time, *Torricelli* rectified various other curves, such as the *cycloid* (see Definition 1.9.1 and Proposition 1.9.4).

Proposition 1.7.3 *The length of an arc of a logarithmic spiral, from its origin O to a given point P, is equal to the length of the segment joining P and the intersection Q of the tangent at P and the perpendicular at O to the radius OP.*

Proof Let R be the length of the radius OP. By definition of the spiral, the component of the movement at P along the radius is kR, for a fixed constant k. The

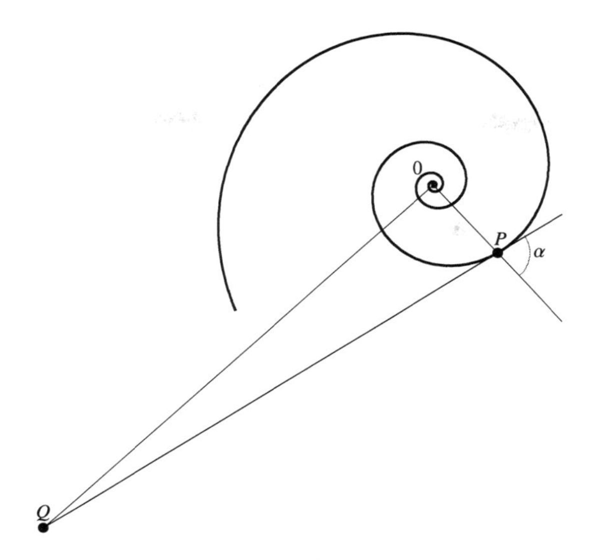

Fig. 1.23

circular component of the movement at P is oriented along the tangent to the circle with center O and radius R, that is, perpendicular to the radius OP; it has a length equal to the length of the circle, that is $2\pi R$. The parallelogram of movements is thus a rectangle with sides kR and $2\pi R$; its diagonal is thus oriented as that of a rectangle with sides k and 2π. Therefore the angle between the tangent and the radius OP is independent of R and is thus a constant α.

Now "unroll" the spiral along its tangent at P, starting from P. As we have just seen, the various successive radii OP' will keep forming an angle α with the tangent. Therefore the movement of the point O during this "unrolling" process is perpendicular to the direction of these radii. So the point O moves on the perpendicular to the radius OP at the point O. When O finally reaches the intersection point Q with the tangent at P, the spiral is entirely unrolled. □

Most probably, you are not fully convinced by these "dynamical" arguments. However, we must bear in mind that differential calculus did not exist in Toricelli's time. In modern terms, writing θ for the angle of rotation of the line, the logarithmic spiral is defined by the differential equation

$$\frac{dR(\theta)}{d\theta} = kR(\theta).$$

This yields as possible solution

$$R(\theta) = ae^{k\theta}, \quad a \in \mathbb{R}.$$

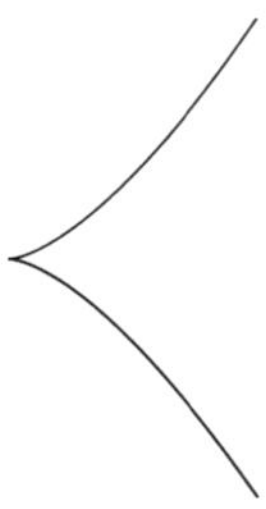

Fig. 1.24

A parametric representation of the logarithmic spiral is then

$$f(\theta) = \left(ae^{k\theta}\cos\theta, ae^{k\theta}\sin\theta\right).$$

One can now obtain the result by brute computation rather than imagination!

This result was doubly amazing for the mathematicians of the time:

- first, as already mentioned, the result finds a precisely defined segment whose length is equal to that of a piece of a curve;
- second, this segment has a finite length, while the piece of the curve winds infinitely many times before reaching the origin.

However, this is not really a counterexample to *Descartes*' statement. Indeed the curve itself was considered as "badly defined": the curve was described in a dynamic way, but its equation could not be written. Of course in those days, the exponential function could by no means be considered as a function and, even less, as a "well defined function".

Nevertheless this first attempt raised the hope of being able to rectify some curves and, perhaps, all curves. The British mathematician *Neil* (1659), the Dutch mathematician *van Heuraet* (1659) and the French mathematician *Fermat* (1660) were able to "rectify" the semi-cubic parabola, that is, the "well-defined" curve with equation

$$y^2 = x^3$$

(see Fig. 1.24).

The method of *Neil* and *van Heuraet* consisted of approaching the curve by a polygonal line inscribed to the curve and letting the distance between two consecutive points tend to zero. The method of *Fermat* consisted instead of approaching the curve by a polygonal line tangent to the curve (see Fig. 1.25). The most amazing point is the fact that all of them succeeded in computing the limit of the lengths of these polygonal lines before the birth of differential calculus in 1676. *Newton* (1642) and *Leibniz* (1646) were at the time already teenagers who were starting to get interested in these questions!

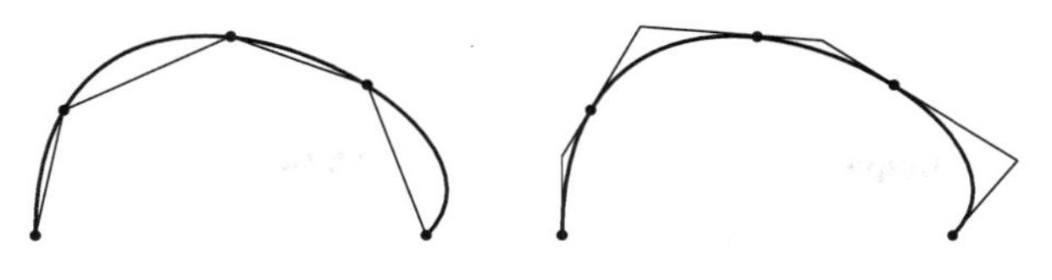

Fig. 1.25

1.8 Length *Versus* Curve Integral

The tricky computations of *Neil*, *van Heuraet* and *Fermat* (see Sect. 1.7) to compute the length of an arc of a cubic parabola are of course based on the following "definition":

> *Given a curve, we approximate it by a polygonal line as in Fig.* 1.25. *The length of the curve is the limit of the lengths of all possible polygonal lines as the length of all segments tends to zero.*

Once more, the intuition behind this "definition" is clear, but the terms contained in it should now be given a precise mathematical meaning. To achieve this, let us first work with this "definition" as such, without asking too many questions about its precise meaning and about the assumptions needed to develop the following proof.

Proposition 1.8.1 *Consider a parametric representation*

$$f\colon\,]a,b[\longrightarrow \mathbb{R}^2, \qquad t\mapsto f(t)$$

of class C^1 of a plane curve and two points $c<d$ in $]a,b[$. The length of the arc of the curve between the points of parameters c and d is equal to

$$\int_c^d \|f'(t)\|dt.$$

Proof We thus call the *length* of the arc of the curve the limit of the lengths of the polygonal lines inscribed to the curve, as the length of each side tends to zero. For a natural number $n\neq 0$, put

$$\Delta_n(t)=\frac{d-c}{n}$$

and consider the polygonal line determined by the values

$$t_i=c+i\,\Delta_n(t), \quad 0\leq i\leq n$$

of the parameter. The length of this polygonal line is equal to

$$\sum_{i=0}^{n}\|f(t_{i+1})-f(t_i)\|.$$

The function f is continuous, thus it is uniformly continuous on the compact interval $[c,d]$. Therefore when n tends to ∞, that is as $\Delta_n(t)$ tends to 0, each side of the polygonal line has a length which also tends to 0.

But for "good" functions f, the Taylor expansion of f tells us that

$$f(t_{i+1}) = f(t_i) + \Delta_n(t) f'(t_i) + \mathcal{O}_1\big(\Delta_n(t)\big)$$

where $\mathcal{O}_1$ has the property

$$\lim_{x \to 0} \frac{\mathcal{O}_1(x)}{x} = 0.$$

This suggests to re-write the length of the polygonal line as

$$\sum_{i=0}^{n} \left\| f'(t_i) + \frac{\mathcal{O}_1(\Delta_n(t))}{\Delta_n(t)} \right\| \cdot \Delta_n(t).$$

When the number n of sides of the polygonal line tends to ∞, $\Delta_n(t)$ tends to 0 and this sum should thus have as limit

$$\int_c^d \|f'(t)\| dt.$$

□

This "proof" is not very rigorous, and nor should we expect it to be, after all it concerns a definition whose terms have bot been given a precise meaning. Nevertheless, the formula in Proposition 1.8.1 should remind us of well-known result in analysis:

Theorem 1.8.2 *Consider an injective function of class $\mathcal{C}^1$*

$$f\colon [c,d] \longrightarrow \mathbb{R}^n$$

and another continuous function

$$g\colon [c,d] \longrightarrow \mathbb{R}.$$

Then the curve integral of g along f exists and is equal to

$$\int_c^d g(t) \cdot \|f'(t)\| \, dt.$$

This immediately suggests the following definition

Definition 1.8.3 Let $f\colon]a,b[\longrightarrow \mathbb{R}^2$ be an injective parametric representation of class $\mathcal{C}^1$ of a curve. Given $a < c < d < b$, the *length* of the arc of the curve between the points with parameters c and d is by definition the *curve integral* of the constant function 1 along $f\colon [c,d] \longrightarrow \mathbb{R}^2$.

Since a parametric representation is always locally injective, the restriction of injectivity can easily be overcome: it suffices to compute a length "by pieces".

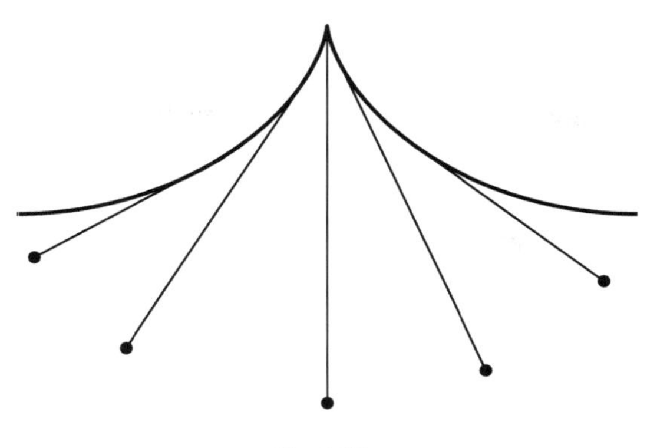

Fig. 1.26

1.9 Clocks, Cycloids and Envelopes

Still in the years 1650–1560, thus before the invention of differential calculus, the Dutch physicist *Huygens* was very much concerned with constructing the best clocks of his time. The most immediate way to construct a clock is based on the *pendulum principle*: attach a weight at the extremity of a chord and let it swing! Our physics courses tell us that the frequency of such a *pendulum* is "more or less" independent of the amplitude of the movement, at least in the case of oscillations of "small amplitude". We make these various qualifications, "more or less", "small amplitudes", and so on, but is it not possible to construct an *isochronal* pendulum: a pendulum which always swings at the same frequency, whatever the amplitude of the oscillations?

The frequency of a pendulum increases when the "chord" of the pendulum is shorter. When the amplitude of the oscillation increases, the frequency of the pendulum decreases. Thus to obtain a pendulum whose frequency is independent of the amplitude of the oscillation, it would "suffice" to have a chord of variable length: a length which diminishes as the pendulum moves away from its position of equilibrium, and gets longer again as the pendulum moves back towards its bottom position. How can one realise such a pendulum?

Huygens' idea was to attach the chord between two templates, so that the chord "rolls" on these templates (see Fig. 1.26) while the pendulum is swinging. The chord has its full length in vertical position and this length becomes shorter and shorter as the pendulum moves away from this equilibrium position. But what form should you give to the templates in order to get a pendulum whose frequency is independent of the amplitude of the oscillations?

Huygens discovers that the solution to the problem is obtained via the *cycloid*: a curve already mentioned in Sect. 1.6 (see Fig. 1.27).

Definition 1.9.1 The *cycloid* is the trajectory of a fixed point of a circle, as this circle rolls on a line.

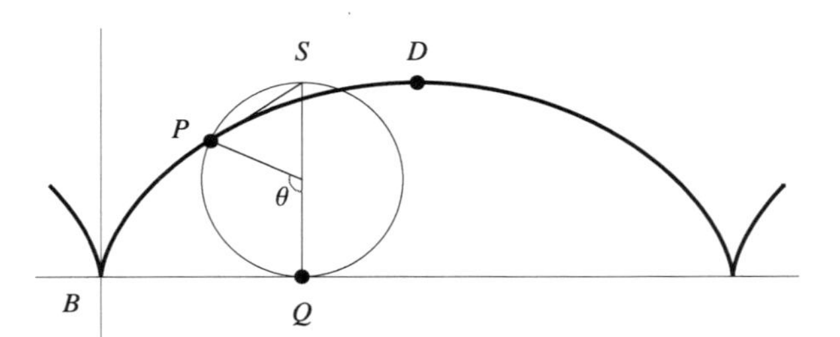

Fig. 1.27

The Italian mathematician *Torricelli* (1608–1647) and the French mathematician *Roberval* (1602–1675) had already studied problems of tangency, length and area for various curves, including the cycloid. *Huygens* knew these results and pursued the study of the cycloid further. Let us establish the necessary results with the contemporary techniques of Sects. 1.6 and 1.8.

Proposition 1.9.2 *Choosing the radius of the rolling circle as unit of length, a parametric representation of the cycloid is given by*

$$f\colon \mathbb{R} \longrightarrow \mathbb{R}^2, \qquad f(\theta) = (\theta - \sin\theta, 1 - \cos\theta).$$

Proof In Fig. 1.27, when the circle has turned by an angle θ, the length of the arc PQ is equal to θ. But by definition of the cycloid, this length is also that of the segment BQ. Observing further that the center of the circle moves on the line with equation $y = 1$, we obtain the announced formula. □

The following result was first discovered by *Roberval*, using his technique of *composition of movements*.

Proposition 1.9.3 *In Fig.* 1.27, *the tangent to the cycloid at a point P is the line joining P and the point S diametrically opposite to Q on the circle.*

Proof In view of Definition 1.6.1 and Proposition 1.9.2, the direction of the tangent at P is $(1 - \cos\theta, \sin\theta)$. The equation of this tangent is thus

$$\sin\theta\big(x - (\theta - \sin\theta)\big) - (1 - \cos\theta)\big(y - (1 - \cos\theta)\big) = 0.$$

The point Q has coordinates $(\theta, 0)$, thus S has coordinates $(\theta, 2)$. It is immediate that the coordinates of S satisfy the equation of the tangent. □

Here we present (with contemporary proof) *Toricelli's* result concerning the rectification of the cycloid.

Proposition 1.9.4 *In Fig.* 1.27, *write D for the middle point of the full arch of the cycloid. The length of the arc of the cycloid between P and D is equal to twice the* *length of the tangent segment PS (see Proposition 1.9.3). As a consequence, the length of a full arch of the cycloid is equal to eight times the radius of the rolling circle.*

Proof Going back to the proof of Proposition 1.9.3, we have

$$P = (\theta - \sin\theta, 1 - \cos\theta), \qquad S = (\theta, 2).$$

Therefore

$$\|\overrightarrow{PS}\| = \sqrt{2 + 2\cos\theta} = 2\cos\frac{\theta}{2}$$

since

$$1 + \cos\theta = 2\cos^2\frac{\theta}{2}.$$

On the other hand the length of the arc PD of the cycloid is given by (see Proposition 1.8.1)

$$\begin{aligned}\int_\theta^\pi \|f'\| &= \int_\theta^\pi \sqrt{2 - 2\cos\theta}\, d\theta \\ &= 2\int_\theta^\pi \sin\frac{\theta}{2}\, d\theta \\ &= -4\left(\cos\frac{\pi}{2} - \cos\frac{\theta}{2}\right) \\ &= 4\cos\frac{\theta}{2}\end{aligned}$$

where this time, we have used the formula

$$1 - \cos\theta = 2\sin^2\frac{\theta}{2}.$$

The length of the arc PD of the cycloid is thus indeed twice the length of the segment PS.

The length of a full arch is therefore four times the length of the tangent vector PS, when $P = B$. Except that at $P = B$, the argument above does not apply! Indeed for $\theta = 0$, the tangent vector $f'(\theta)$ becomes simply $(0, 0)$: the cusp points of the cycloid are singular points (see Definition 1.4.4). But taking the limit of the lengths of the arcs PD as P converges to B yields

$$\lim_{\theta \to 0} 4\cos\frac{\theta}{2} = 4$$

as expected. □

What *Huygens* proved about the cycloid is the following theorem.

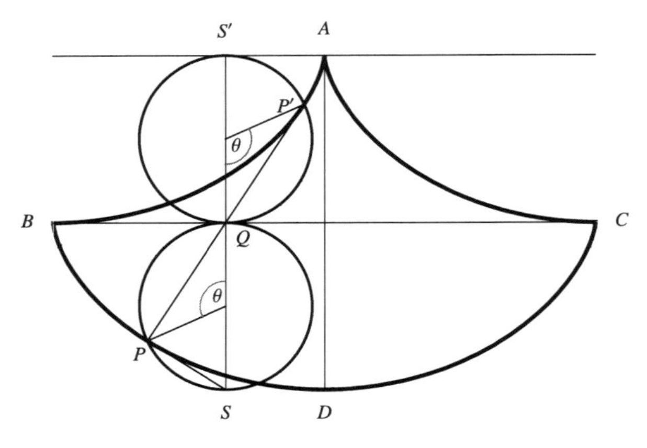

Fig. 1.28

Theorem 1.9.5 *Put a cycloid upside-down in a gravitational field. Attach at a cusp point of this cycloid a pendulum whose length is equal to half the length of an arch of the cycloid. The trajectory of this pendulum is another cycloid of the same size and the frequency of the pendulum is independent of the amplitude of the oscillation.*

Proof In Fig. 1.28, consider the lower cycloid, obtained when the lower circle of radius 1 rolls on the middle horizontal line. Analogously consider the upper cycloid, obtained when the upper circle with the same radius 1 rolls on the upper horizontal line. Write P, P' for the fixed points on these circles whose trajectories are the cycloids. Write further Q for the contact point of these two circles with the middle horizontal line and S, S' for the points of the circles diametrically opposite to Q.

By Proposition 1.9.3, PS is the tangent to the lower cycloid at P while $P'Q$ is the tangent to the upper cycloid at P'. Since both circles have already turned by the same angle θ after leaving B, the points P and P' are symmetric to each other with respect to Q. In other words, P, Q, P' are on the same line and both segments PQ, $P'Q$ have the same length. But by Proposition 1.9.4, the length of the segment $P'Q$ is half the length of the arc BP' of the cycloid. Thus the arc BP' of the cycloid has the same length as the segment PP'.

Thus the length of the arc AP' of the cycloid, augmented by the length of the segment $P'P$, yields the same result as the length of the full arc AB of the cycloid. In other words, if a pendulum is attached at A, with a length of cord equal to the arc AB of the upper cycloid, when this pendulum swings, its trajectory is exactly the lower cycloid.

Now call τ the angle between the tangent at P and the horizontal line. Writing g for the gravitational force, the acceleration of the pendulum along its trajectory is thus equal to $-g \sin \tau$. But, still by Proposition 1.9.4

$$\sin \tau = \frac{\|PS\|}{2} = \frac{\frac{1}{2}\text{arc } DP}{2}.$$

Write s for the length of the arc DP, viewed as the position of the pendulum on the cycloid. The acceleration of the pendulum along the cycloid is thus characterized by the differential equation

$$\ddot{s}(t) = -\frac{g}{4}s(t)$$

where t is the time. Integrating this differential equation yields

$$s(t) = \alpha \cos\left(\frac{1}{2}\sqrt{g}\,t\right) + \beta \sin\left(\frac{1}{2}\sqrt{g}\,t\right).$$

Let us assume that at the instant $t = 0$ the pendulum reaches its highest position $s = s_0$. The symmetry of the problem forces $s(t) = s(-t)$, which implies $\beta = 0$. Putting $t = 0$ yields $\alpha = s_0$, so that the equation of the movement is

$$s(t) = s_0 \cos\left(\frac{1}{2}\sqrt{g}\,t\right).$$

The point D, that is $s = 0$, is thus reached at the time t_0 such that

$$0 = s_0 \cos\left(\frac{1}{2}\sqrt{g}\,t_0\right)$$

that is

$$\frac{1}{2}\sqrt{g}\,t_0 = \frac{\pi}{2}.$$

The time necessary for the pendulum to reach its bottom position D is thus $\pi\sqrt{\frac{1}{g}}$: this time is indeed independent of s_0, the amplitude of the oscillation.

Not surprisingly, given that he was essentially trying to solve a differential equation before the invention of differential calculus, *Huygens*' argument for this last point was fairly convoluted. Amazingly, he nevertheless managed to solve the problem. □

This study of the cycloid underlines another important geometrical notion: the *envelope* of a family of curves.

Proposition 1.9.6 *In the situation depicted in Fig.* 1.28, *the upper cycloid is tangent to all the normals to the lower cycloid*: *therefore the upper cycloid is called the envelope of these normals* (*see Fig.* 1.29).

Proof The angle QPS is inscribed in a half circle, thus it is a right angle. Since PS is tangent to the lower cycloid, PQP' is normal to this cycloid, but it is also the tangent to the upper cycloid. □

The idea of considering the *envelope* of the normals to a given curve is much older that the work of *Huygens*. The first result in this direction is probably the case

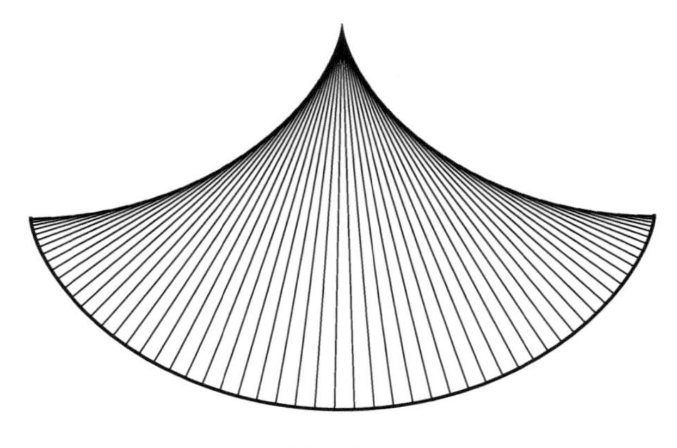

Fig. 1.29

of the normals to a parabola, studied once again by *Apollonius*, around 220 BC! The first case of an envelope of a family of arbitrary curves (not just straight lines) is probably due to *Torricelli* around 1642. We shall come back to these problems in Sect. 2.6.

1.10 Radius of Curvature and Evolute

The study of the cycloidal pendulum, as developed by *Huygens* (see Sect. 1.9) gives rise to a very interesting observation. An ordinary pendulum moves along a circle whose radius is the length of the chord of this pendulum. But the cycloidal pendulum has a chord of variable length, originating from a variable point: in Fig. 1.28, the chord is $P'P$ with origin P'. The instantaneous movement of the cycloidal pendulum at the point P thus coincides with the instantaneous movement of an ordinary pendulum whose chord would be $P'P$. Therefore the circle of radius $P'P$ is the best circular approximation of the cycloid at its point P. *Huygens* calls P' the *center of curvature* of the cycloid at P and $P'P$, the corresponding *radius of curvature*. The upper cycloid is thus the locus of the centers of curvature of the lower cycloid: what one calls the *evolute* of the lower cycloid.

Again it was *Huygens* who first succeeded in handling the problem of the *radius of curvature* in a quite general setting. Let us follow his argument, taking for granted that when we refer to a tangent, an intersection or a limit, it does exist!

> *Consider a fixed point P on a given curve and a variable point Q on the same curve.*

- The *center of curvature* at P is the limit of the intersection of the normal at P and the normal at Q, when Q converges to P.
- The distance between P and the corresponding center of curvature is called the *radius of curvature* at P.

- The locus of all the centers of curvature is called the *evolute* of the given curve.

As already mentioned, we take for granted that this definition makes sense, which is of course false, even for very good curves! For example if the curve is a straight line, the two normals at P and Q are parallel and you cannot even start the process! Our purpose is therefore once more to guess what a "good" contemporary definition should be. Let us translate *Huygens*' argument in contemporary terms.

Proposition 1.10.1 *Consider a plane curve with parametric representation f. In "good cases", the radius of curvature is given by the formula*

$$\rho = \frac{\|f'\|^3}{|f_2' f_1'' - f_1' f_2''|}.$$

Proof Let us consider the fixed point $P = f(t_0)$ and the variable point $Q = f(t)$. The normal vector at Q is thus $n(t) = (f_2'(t), -f_1'(t))$ and analogously at P. The intersection of the two normals is thus such that

$$f(t_0) + \alpha_t n(t_0) = f(t) + \beta_t n(t)$$

for two scalars α_t and β_t that we have now to determine.

This equality yields the system of equations

$$\begin{aligned} f_1(t_0) + \alpha_t f_2'(t_0) &= f_1(t) + \beta_t f_2'(t) \\ f_2(t_0) - \alpha_t f_1'(t_0) &= f_2(t) - \beta_t f_1'(t) \end{aligned}$$

so that, by Cramer's rule for solving such a system and the well-known properties of determinants

$$\alpha_t = \frac{\det\begin{pmatrix} f_1(t)-f_1(t_0) & -f_2'(t) \\ f_2(t)-f_2(t_0) & f_1'(t) \end{pmatrix}}{\det\begin{pmatrix} f_2'(t_0) & -f_2'(t) \\ -f_1'(t_0) & f_1'(t) \end{pmatrix}} = \frac{\det\begin{pmatrix} \frac{f_1(t)-f_1(t_0)}{t-t_0} & -f_2'(t) \\ \frac{f_2(t)-f_2(t_0)}{t-t_0} & f_1'(t) \end{pmatrix}}{\det\begin{pmatrix} f_2'(t_0) & \frac{f_2'(t_0)-f_2'(t)}{t-t_0} \\ -f_1'(t_0) & \frac{f_1'(t)-f_1'(t_0)}{t-t_0} \end{pmatrix}}.$$

When t converges to t_0, we obtain

$$\alpha = \lim_{t\to t_0} \alpha_t = \frac{\det\begin{pmatrix} f_1'(t_0) & -f_2'(t_0) \\ f_2'(t_0) & f_1'(t_0) \end{pmatrix}}{\det\begin{pmatrix} f_2'(t_0) & -f_2''(t_0) \\ -f_1'(t_0) & f_1''(t_0) \end{pmatrix}} = \frac{\|f'(t_0)\|^2}{f_2'(t_0) f_1''(t_0) - f_1'(t_0) f_2''(t_0)}.$$

The center of curvature at P is then the point

$$f(t_0) + \alpha\, n(t_0).$$

The radius of curvature is simply

$$\begin{aligned}\|\alpha\, n(t_0)\| &= \left| \frac{\|f'(t_0)\|^2}{f_2'(t_0) f_1''(t_0) - f_1'(t_0) f_2''(t_0)} \sqrt{\left(f_2'(t_0)\right)^2 + \left(f_1'(t_0)\right)^2} \right| \\ &= \frac{\|f'(t_0)\|^3}{|f_2'(t_0) f_1''(t_0) - f_1'(t_0) f_2''(t_0)|}\end{aligned}$$

which is indeed the announced formula. □

1.11 Curvature and Normality

The treatment of the *radius of curvature* as in Sect. 1.10 is certainly very intuitive, but raises many questions. It refers again to a notion of "limit computed on a family of lines" and as we have seen in Sect. 1.6, such a notion of limit is not always as simple as one might imagine. So we now want to translate the ideas of Sect. 1.10 in "decent differential terms".

The idea is the following. What measures the *curvature* of a curve is the speed at which the tangent to the curve changes direction as you travel along this curve. But the measure of the variation of a quantity is something well-known today: this is the derivative of the quantity. The direction of the tangent to a regular curve represented by $f(t)$ is given by the tangent vector $f'(t)$ (see Definition 1.6.1). One could thus be tempted to define the *curvature* as the derivative of this tangent vector, that is as the vector $f''(t)$.

However, $f'(t)$ is a vector in $\mathbb{R}^2$: it thus has a direction and a length. When this vector varies, it can vary both in direction and in length. What we are interested in, is only the *variation in direction*. Of course if it turns out that the vector $f'(t)$ has a constant length, for all values of the parameter t, then the derivative $f''(t)$ measures exactly the variation in direction of the tangent vector $f'(t)$, and we end up with an elegant way of defining the curvature. But is such a situation possible?

Proposition 1.11.1 *Choose as parameter for describing a curve, the length s of the arc of the curve from an arbitrary point on the curve chosen as origin. If the corresponding parametric representation $f(s)$ is regular, the tangent vector $f'(s)$ is of length 1, for every value of the parameter s. Such a parametric representation is called normal.*

Proof Applying Proposition 1.8.1 to such a special parametric representation, we get

$$s = \int_0^s \|f'\|.$$

Deriving both sides with respect to s yields $1 = \|f'\|$. □

We therefore make the following definition:

Definition 1.11.2 Let $f(s)$ be a normal representation of class $\mathcal{C}^2$ of a curve (see Proposition 1.11.1). The *curvature* at the point with parameter s is by definition the quantity $\|f''(s)\|$.

It is fairly immediate to observe that this definition does not depend on the normal representation chosen: we shall see this in more detail in Sect. 2.9.

To conclude this section, we should now exhibit the link between the notion of *curvature* in Definition 1.11.1 and the more intuitive idea of *radius of curvature* obtained via the intersection of normals. For that it suffices to remember that the derivative of a scalar product can be computed via a formula analogous to that of the derivative of an ordinary product.

Lemma 1.11.3 *Consider two functions of class* $\mathcal{C}^1$

$$f, g : \mathbb{R} \to \mathbb{R}^2$$

and the corresponding function

$$(f|g) : \mathbb{R} \longrightarrow \mathbb{R}, \qquad t \mapsto \big(f(t)\big|g(t)\big).$$

Under these conditions

$$(f|g)' = (f'|g) + (f|g').$$

Proof Indeed

$$\begin{aligned}
(f|g)' &= (f_1 g_1 + f_2 g_2)' \\
&= (f_1' g_1 + f_1 g_1') + (f_2' g_2 + f_2 g_2') \\
&= (f_1' g_1 + f_2' g_2) + (f_1 g_1' + f_2 g_2') \\
&= (f'|g) + (f|g')
\end{aligned}$$

by the ordinary formula for the derivative of a product. □

We then have:

Proposition 1.11.4 *Let $f(s)$ be a normal representation of class $\mathcal{C}^2$ of a curve. The radius of curvature at a given point is the inverse of the curvature at this point, provided of course that this curvature is not zero.*

Proof By Proposition 1.11.1, a normal representation f has a tangent vector f' of constant length 1. Differentiating the equality $(f'|f') = 1$, we obtain $2(f'|f'') = 0$. Thus f'' is orthogonal to f' and therefore, the vector $v = (-f_2'', f_1'')$ perpendicular to f'' is parallel to f'. It follows that

$$(f'|v) = \|f'\| \cdot \|v\| \cdot \cos k\pi = \pm\|v\| = \pm\sqrt{(f_2'')^2 + (f_1'')^2} = \pm\|f''\|.$$

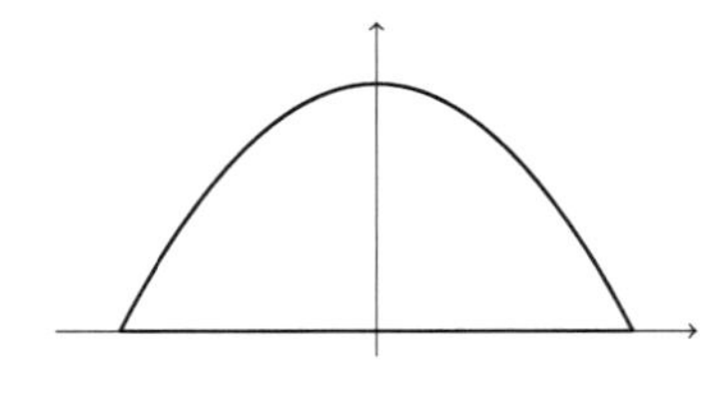

Fig. 1.30

But

$$(f' \mid v) = -f_1' f_2'' + f_2' f_1''.$$

Introducing these values into the formula of Proposition 1.10.1 gives

$$\rho = \frac{1}{\|f''\|}$$

as announced. □

1.12 Curve Squaring

"Circle squaring", that is, constructing by ruler and compass a square having the same area as a circle, is a problem which puzzled mathematicians for more than two millenniums. This problem has largely been discussed in Sect. 2.4 of [3], *Trilogy I*; its impossibility was finally proved in 1882 as a corollary of a famous result of *Lindemann*: the number π is transcendental, that is, it cannot be obtained as a solution of an equation with rational coefficients (see Sect. B.3 of [3], *Trilogy I*).

Of course, today, "squaring" a portion of the plane is no longer seen as a "ruler and compass" problem, but as a question of integral calculus. Therefore "curve squaring" is generally not considered as part of curve theory and is instead treated in an analysis course: we thus direct the reader towards an analysis book for a systematic treatment of these questions. Notice that making clear which curves can be "squared" is already a challenging problem.

Nevertheless, due to the historical importance of these questions, it is sensible to present here a short section on this curve squaring problem, focusing on some historically important examples. Our first example was treated by *Archimedes* (see Sect. 4.4 in [3], *Trilogy I*).

Example 1.12.1 The area of the portion of the plane delimited by the x-axis and the parabola of equation $y = 1 - x^2$ is equal to $\frac{4}{3}$ (see Fig. 1.30).

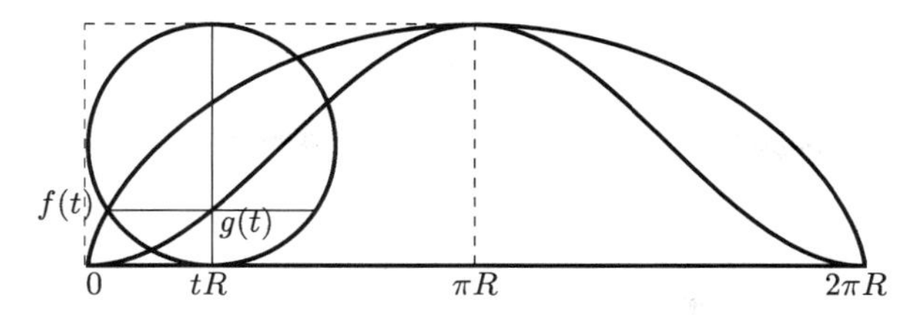

Fig. 1.31

Proof We know that this area is given by the integral

$$\int_{-1}^{+1} 1 - x^2\,dx = \left[x - \frac{x^3}{3}\right]_{-1}^{+1} = \left(1 - \frac{1}{3}\right) - \left(-1 + \frac{1}{3}\right) = \frac{4}{3}. \qquad \square$$

Don't forget that the parabola is a section of a circular cone by a plane. Greek geometers were able to "square" the parabola, but not the circle which for them was thus a priori a "more elementary" curve. A rather intriguing situation! Today we know the easy explanation for this phenomenon: the parabola admits an equation of the form $y = p(x)$ with $p(x)$ a polynomial: the integration yields another polynomial $\int p(x)\,dx$, that is a formula in terms of the four arithmetical operations, and the four arithmetical operations (as well as the square root) can be performed by ruler and compass constructions. On the other hand, viewing the upper half of a circle of radius R as the graph of the function $y = \sqrt{R^2 - x^2}$, the area of the circle is given by

$$2\int_{-R}^{+R} \sqrt{R^2 - x^2}\,dx$$

and such an integral involves inverse trigonometric functions!

Again it was during the 17th century that several mathematicians were able to compute—using precursors of differential methods—areas delimited by various curves. A celebrated achievement of this type, before the invention of differential calculus by *Newton* and *Leibniz*, is the squaring of the cycloid by *Roberval*, in 1634. *Roberval's* subtle computation of the integral, before the discovery of that notion, is described below.

Proposition 1.12.2 *The area of an arch of a cycloid, with generating circle of radius R, is equal to $3\pi R^2$.*

Proof We shall follow the argument in modern terms (see Fig. 1.31). We know already that a parametric representation of the cycloid is given by (see Proposition 1.9.2)

$$f(t) = R(t - \sin t, 1 - \cos t).$$

An arch of the cycloid is obtained when t varies from 0 to 2π.

Let us write $g(t)$ for the orthogonal projection of the point $f(t)$ on the instantaneous vertical radius of the rolling circle. Thus

$$g(t) = R(t, 1 - \cos t).$$

The function g represents the so-called *Roberval curve*. It is immediate that this curve g, for t varying from 0 to π, is symmetric with respect to its middle point $g(\frac{\pi}{2})$: that is

$$\frac{g(\frac{\pi}{2} + t) - g(\frac{\pi}{2} - t)}{2} = g\left(\frac{\pi}{2}\right).$$

Therefore the area under this curve g, between 0 and π, is equal to half the area of the corresponding rectangle; that is, $\frac{1}{2}(\pi R)(2R) = \pi R^2$. It follows that the area under the *Roberval curve*, between the points with parameters $t = 0$ and $t = 2\pi$, is equal to $2\pi R^2$.

It remains to compute the area between the *Roberval curve* and the cycloid. *Roberval* simply observes that the segment joining $f(t)$ and $g(t)$ is equal to half the corresponding chord of the circle. "Pushing" all the segments $f(t)g(t)$ to the left, in order to align their right extremities $g(t)$ vertically, for t varying from 0 to π, *Roberval* concludes that these segments fill the left half of the generating circle. This half circle has area $\frac{1}{2}R^2$. Thus as t varies from 0 to 2π, the area between the *Roberval curve* and the cycloid is equal to R^2.

Putting all of this together, an arch of the cycloid has an area equal to $3\pi R^2$. □

This proof is very interesting for two reasons. First—choosing the radius R as unit length—the *Roberval curve* is simply the curve

$$y = 1 - \cos x.$$

Up to a translation, the *Roberval curve* is thus the graph of a cosine (or sine) function. It seems that this is the first time that the graph of the sine function appears in the mathematical treatment of a problem. The way the corresponding integral is computed is particularly imaginative.

Second, the "pushing" argument of *Roberval* may appear rather strange to us. Let us nevertheless observe that writing the equations of the first halves of the cycloid and the *Roberval* curve in the form

$$x = c(y), \quad x = r(y), \quad 0 \le x \le R\pi,\ 0 \le y \le 2R$$

this "pushing" argument corresponds precisely to the modern formula

$$\int (c - r) = \int c - \int r.$$

To conclude this short section, let us recall a celebrated result of integral calculus: the so-called *Green–Riemann* formula.

Theorem 1.12.3 (Green–Riemann) *Let $K \subseteq \mathbb{R}^2$ be a compact subset whose boundary is constituted of finitely many curves of class $\mathcal{C}^1$. Moreover assume that the boundary of K is oriented in such a way that K is always on the left of its boundary. Given a differential form $P(x,y)dx + Q(x,y)dy$ of class $\mathcal{C}^1$ defined on an open subset containing K, one has*

$$\int_{\partial K} \big(P(x,y)dx + Q(x,y)dy\big) = \int_K \left(\frac{\partial Q}{\partial x}(x,y) - \frac{\partial P}{\partial y}(x,y)\right) dx\,dy.$$

Corollary 1.12.4 *The area delimited by a closed plane curve $\mathcal{C}$, the boundary of a compact subset $K \subseteq \mathbb{R}^2$ as in the* Green–Riemann *theorem, is equal to either of the following equal quantities*

$$\left|\int_{\mathcal{C}} x\,dy\right|, \qquad \left|\int_{\mathcal{C}} y\,dx\right|.$$

Proof The area is simply the integral

$$\int\int_K dx\,dy$$

of the constant function 1 on K. Putting $P = 0$ and $Q = x$ in the *Green–Riemann* formula yields the first formula of the statement; putting $P = y$ and $Q = 0$ yields the second formula. □

Example 1.12.5 The area delimited by an ellipse of half axis a and b is equal to πab.

Proof A parametric representation of the ellipse $\mathcal{E}$ is given by

$$f(\theta) = (a\cos\theta, b\sin\theta).$$

By Corollary 1.12.4, the corresponding area is thus

$$\begin{aligned}
\left|\int_{\mathcal{E}} a\cos\theta\, d(b\sin\theta)\right| &= \left|\int_0^{2\pi} a\cos\theta\, b\cos\theta\, d\theta\right| \\
&= \left|ab\int_0^{2\pi} \cos^2\theta\, d\theta\right| \\
&= \left|ab\left[\frac{\theta}{2} + \frac{\sin 2\theta}{4}\right]_0^{2\pi}\right| \\
&= ab\pi.
\end{aligned}$$

When $a = R = b$, we recapture the usual formula πR^2 for the area of a circle of radius R. □

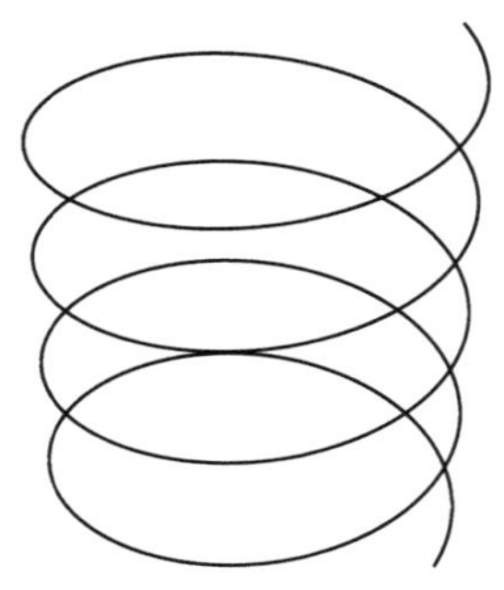

Fig. 1.32

1.13 Skew Curves

Let us now switch to the case of *skew curves*, or *space curves*, that is: curves in the three dimensional space $\mathbb{R}^3$.

The systematic study of skew curves was initiated in 1731 by the French mathematician *Clairaut*. His idea is to present a skew curve as the intersection of two surfaces, just as a line can be presented as the intersection of two planes. A skew curve is thus described by a system of two equations

$$\begin{cases} F(x,y,z)=0 \\ G(x,y,z)=0. \end{cases}$$

The tangent line to the skew curve at a given point is then obtained as the intersection of the tangent planes to the surfaces

$$F(x,y,z)=0, \qquad G(x,y,z)=0$$

at this same point. As you might suspect, the technicalities inherent to such an approach are quite heavy.

Let us for example focus on the question of the curvature. In the plane, a curve with constant curvature is a circle. However, in three dimensional space, there are many more curves with constant curvature. The easiest example is that of the *circular helix* (see Fig. 1.32): the curve having the shape of a screw. Of course this curve must be "identical" at all points: this is why you can screw a bolt through a nut! In particular, the curvature must be the same at all points. There are many more examples of curves with constant curvature, as we shall see later.

The study of the curvature of a skew curve was initiated by the French mathematician *Monge* in 1771, thus still before the introduction of parametric equations. In three dimensional space, the *normal* to the curve now becomes a *normal plane* to the curve: the plane perpendicular to the tangent at a given point. Therefore *Monge* makes the following definitions:

- *The* axis of curvature *at a point P of the skew curve is the limit of the intersection of the normal plane at P and the normal plane at a point Q of the curve, as Q converges to P.*
- *The* radius of curvature *at P is the distance between P and the axis of curvature at P.*

The fact of having an *axis* of curvature instead of a *center* of curvature explains in particular why curves with the same curvature can have very different shapes. The orientation of the axis of curvature is generally not constant and its variations in direction affect in an essential way the shape of the curve.

It remains an excellent exercise of technical virtuosity to compute the axis of curvature, starting from a system of Cartesian equations as in the time of *Clairaut* and *Monge*.

The ideas of *Monge* were clarified and developed in 1805 by his student *Lancret*, who introduced what we call today the *osculating plane* and the *torsion*, in order to study the variations in direction of the *axis of curvature*. The *osculating* plane is in a sense the *tangent plane* to the curve: the plane in which the curve tends to fit locally. Provided the following can be made precise, the idea is this:

> *The* osculating plane *at a fixed point P of a skew curve is the limit of the planes through P, Q, R, when Q and R are two other points of the curve converging to P.*

Lancret observed that the axis of curvature is perpendicular to the osculating plane.

Following the comments of the previous sections, we need not re-emphasize the fact that the definitions of *Monge* and *Lancret*, however intuitive, raise endless difficulties! Again, this is not the point here: in the "good cases" these definitions should recapture what we have in mind. What remains to be done is to work out these unpolished ideas to end up with a rigorous alternative presentation in decent differential terms!

Let us recall, as already mentioned in Sect. 1.2, that *Euler* introduced in 1775 his idea of *separating the variables*. This allows us to define a skew curve via three parametric equations

$$\begin{cases} x = f_1(t) \\ y = f_2(t) \\ z = f_3(t) \end{cases}$$

that is, finally, via a parametric representation

$$f\colon \mathbb{R} \longrightarrow \mathbb{R}^3, \qquad t \mapsto f(t) = \big(f_1(t), f_2(t), f_3(t)\big).$$

This approach, together with the full strength differential calculus introduced a century earlier by *Newton* and *Leibniz*, allows us to transpose to skew curves most of the considerations developed in the previous sections. We do this immediately.

- A *parametric representation* of a skew curve is a locally injective continuous function

$$f\colon\,]a,b[\, \longrightarrow \mathbb{R}^3.$$

- The parametric representation is *regular* when it is of class $\mathcal{C}^1$ and $f'(t) \neq 0$ at each point.
- In the regular case, the *tangent* to the curve at the point with parameter t is the line through $f(t)$ and of direction $f'(t)$.
- The *normal plane* to the curve at a point is the plane perpendicular to the tangent at this point.
- When f is injective of class $\mathcal{C}^1$, the *length* of the arc of the curve between the points with parameters $c < d$ is the integral of the constant function 1 along this arc; it is also equal to $\int_c^d \|f'\|$.
- The parametric representation f is *normal* when the parameter is the length traveled on the curve from an arbitrary origin.
- Given a normal representation of class $\mathcal{C}^2$, $\|f'\| = 1$ and f' is orthogonal to f''.
- Given a normal representation of class $\mathcal{C}^2$, the *curvature* is the quantity $\|f''\|$.

Let us follow *Lancret's idea* and investigate first the case of the *osculating plane*.

Proposition 1.13.1 *Let $f(t)$ be a parametric representation of a skew curve. "Under suitable assumptions", the* osculating plane *at a point $f(t)$ is the plane through $f(t)$ whose direction is determined by the vectors $f'(t)$ and $f''(t)$.*

Proof Of course for this statement to make sense, f should be at least of class $\mathcal{C}^2$, with $f'(t)$ and $f''(t)$ linearly independent, in order to determine a plane. But our point here is not to exhibit all the "suitable" assumptions.

We thus fix a point $P = f(t_0)$ and consider two variable points $Q = f(t_1)$, $R = f(t_2)$ on the curve. We are interested in the plane

- containing the point $f(t_0)$;
- whose direction contains the vectors $f(t_1) - f(t_0)$ and $f(t_2) - f(t_0)$.

We now have to let t_1 and t_2 converge to t_0. With the considerations of Sect. 1.6 on the tangent in mind, we might be tempted to divide the two vectors $f(t_i) - f(t_0)$ by $t_i - t_0$ and let t_i converge to t. But of course this cannot possibly work since in both cases, the limit would be the same vector $f'(t_0)$. *One* vector no longer determines a plane! So let us handle separately the points Q and R. We consider first that the direction of the plane is equivalently given by

$$\frac{f(t_1) - f(t_0)}{t_1 - t_0}, \qquad f(t_2) - f(t_0)$$

and we let t_1 tend to t_0. This yields a plane whose direction contains the vectors

$$f'(t_0), \qquad f(t_2) - f(t_0).$$

Using a Taylor expansion we write

$$f(t_2) = f(t_0) + (t_2 - t_0)f'(t_0) + \frac{1}{2}(t_2 - t_0)^2 f''(t_0) + \mathcal{O}(t_2 - t_0)$$

where

$$\lim_{x\to 0}\frac{\mathcal{O}(x)}{x^2}=0.$$

This allows us to further characterize the direction of the plane by the vectors

$$f'(t_0),\qquad (t_2-t_0)f'(t_0)+\frac{1}{2}(t_2t_0)^2f''(t_0)+\mathcal{O}(t_2-t_0).$$

Working on linear combinations of these two vectors, the direction is also determined by the vectors

$$f'(t_0),\qquad f''(t_0)+2\frac{\mathcal{O}(t_2-t_0)}{t_2-t_0}.$$

Letting t_2 tend to t_1, we get

$$f'(t_0),\qquad f''(t_0).$$

□

Next, let us make the link with the ideas of *Monge*.

Proposition 1.13.2 *"Under suitable assumptions", the axis of curvature is perpendicular to the osculating plane and the radius of curvature is the inverse of the curvature.*

Proof Let us work with a parametric representation

$$f\colon]a,b]\longrightarrow \mathbb{R}^3,\qquad t\mapsto f(t)=\big(f_1(t),f_2(t),f_3(t)\big)$$

and let us assume that Proposition 1.13.1 applies: in particular $f'(t)$ and $f''(t)$ are linearly independent at each point. We study the curvature at $f(t_0)$ and there is no loss of generality in choosing a rectangular system of coordinates with origin $f(t_0)$ and such that $f'(t_0)$ is oriented along the third axis. Thus

$$f(t_0)=(0,0,0),\qquad f'(t_0)=\big(0,0,f_3'(t_0)\big).$$

The normal plane at $f(t_0)$ is thus the plane with equation $z=0$.

The normal plane at $f(t)$ admits the equation

$$f_1'(t)\big(x-f_1(t)\big)+f_2'(t)\big(y-f_2(t)\big)+f_3'(t)\big(z-f_3(t)\big)=0.$$

Its intersection with the plane $z=0$ is thus the line with equation

$$f_1'(t)\big(x-f_1(t)\big)+f_2'(t)\big(y-f_2(t)\big)=f_3'(t)f_3(t)$$

in the (x,y)-plane. Keeping in mind the very special form of the coordinates of $f(t_0)$ and $f'(t_0)$, and dividing by $t-t_0$, this equation can equivalently be re-written as

$$\frac{f_1'(t)-f_1'(t_0)}{t-t_0}\big(x-f_1(t)\big)+\frac{f_2'(t)-f_2'(t_0)}{t-t_0}\big(y-f_2(t)\big)=f_3'(t)\frac{f_3(t)-f_3(t_0)}{t-t_0}.$$

Passing to the limit when t converges to t_0 yields

$$f_1''(t_0)x+f_2''(t_0)y=\big(f_3'(t_0)\big)^2.$$

Assume now that f is a normal representation. Then $\|f'\|=1$, thus $f_3'(t_0)=1$. Moreover f'' is perpendicular to f':

$$f_1'f_1''+f_2'f_2''+f_3'f_3''=0$$

thus, $f_3''(t_0)=0$. The system of equations of the axis of curvature can thus be written as

$$\begin{cases} f_1''(t_0)x+f_2''(t_0)y+f_3''(t_0)z=1\\ z=0. \end{cases}$$

The first plane is orthogonal to $f''(t_0)$ and the second one is orthogonal to $f'(t_0)$, thus their intersection—the axis of curvature—is perpendicular to these two vectors, which by Proposition 1.13.1 span the osculating plane.

The radius of curvature is the distance, in the (x,y)-plane, between the origin and the line with equation

$$f_1''(t_0)x+f_2''(t_0)y=1.$$

This distance is simply

$$\sqrt{\big(f_1''(t_0)\big)^2+\big(f_2''(t_0)\big)^2}=\sqrt{\big(f_1''(t_0)\big)^2\big(f_2''(t_0)\big)^2\big(f_3''(t_0)\big)^2}=\big\|f''(t_0)\big\|. \qquad \square$$

Let us conclude with the definition of the *torsion* of the curve which measures *the variation of the osculating plane*, that is, the *variation of the axis of curvature*. The symbol $\times$ indicates the cross product (see Sect. 1.7 in of [4], *Trilogy II*).

Definition 1.13.3 The *torsion* of a skew curve is the variation of its osculating plane. More precisely, given a normal representation $f(s)$ of the curve, it is the quantity $\|n'(s)\|$ where

$$n(s)=\frac{f'(s)\times f''(s)}{\|f'(s)\times f''(s)\|}$$

is the vector of length 1 perpendicular to the osculating plane at $f(s)$.

Subsequent work of the French mathematicians *Cauchy*, *Frenet*, *Serret* and *Darboux* (among others) then established the modern bases of the theory of skew curves, studied in our Chap. 4.

1.14 Problems

1.14.1 Show that the "non-curve" of Example 1.1.1 can also be obtained via a function $F(x,y)$ of class $\mathcal{C}^1$.

1.14.2 Show that the curve with equation

$$y^2(1-x)=x^2(1+x)$$

has the shape of the curve in Fig. 1.7. Find a corresponding parametric representation.

1.14.3 It is possible to modify Example 1.2.1 so that the continuous functions f_n are all injective, and still converge uniformly to a function which is surjective from the "unit interval" to the "unit square", but the limit function is nevertheless not locally injective. Can you imagine such an example?

1.14.4 When $n=1$, the *Local inverse theorem* (see 1.3.1) reduces to the simple fact that a function $g\colon \mathbb{R} \longrightarrow \mathbb{R}$ of class $\mathcal{C}^1$ whose derivative is non-zero at a point is *monotone*, and thus bijective, in a neighborhood of this point.

1.14.5 An irreducible polynomial $F(X,Y)$ yields a curve $F(x,y)=0$ in the sense of Definition 1.4.2 as soon as this "curve" is not empty.

1.14.6 Determine the tangent to a cycloid using the method of "composition of movements".

1.14.7 Prove Proposition 1.7.3 using the differential definitions of a tangent and a length.

1.14.8 Calculate the length of an arc of the cubic parabola, the curve with equation $y^2=x^3$.

1.14.9 In Propositions 1.8.1 and 1.13.1, what about the case where f is of class $\mathcal{C}^\infty$ but the Taylor expansion of f does not converge to f on a neighborhood of t_0?

1.14.10 Prove *Apollonius*' result attesting that every normal to the parabola $y^2=2px$ is tangent to the semi-cubic parabola $27py^2=8(x-p)^3$.

1.14.11 Using the differential techniques, prove that the evolute of a cycloid is another cycloid of the same size.

1.14.12 Is there a normal parametric representation of the circle of radius R?

1.14.13 Prove that when a curve admits a regular parametric representation, all its normal representations are regular.

1.14.14 Calculate the area of an arch of a cycloid using the *Green–Riemann* formula.

1.14.15 Find a system of two Cartesian equations describing the circular helix.

1.14.16 Calculate the axis of curvature of the circular helix.

1.14.17 Prove that the circular helix has constant curvature and constant torsion.

1.14.18 Consider a *logarithmic helix* with parametric representation

$$f(\theta) = \left(e^{\theta}\cos\theta, e^{\theta}\sin\theta, \alpha(\theta)\right).$$

Determine the differential equation that α must satisfy in order for this spiral to have a constant curvature.

1.14.19 Given two functions of class $\mathcal{C}^1$

$$f, g\colon]a,b[\longrightarrow \mathbb{R}^3$$

and their cross product

$$f \times g\colon]a,b[\longrightarrow \mathbb{R}^3$$

(see Sect. 1.7, Vol. 2), prove that $f \times g$ is still of class $\mathcal{C}^1$ and

$$(f \times g)' = (f' \times g) + (f \times g').$$

1.15 Exercises

1.15.1 Find a polynomial equation $F(x, y) = 0$ whose set of solutions comprises n points (a_i, b_i), $i = 1, \ldots, n$.

1.15.2 Find an equation $F(x, y) = 0$, with F a continuous function, whose set of solutions is the full circle of radius 1 centered at the origin.

1.15.3 Find an open subset U of the real line and an injective function of class $\mathcal{C}^\infty$,

$$f\colon U \longrightarrow \mathbb{R}^2$$

whose image is the hyperbola with equation

$$x^2 - y^2 = 1.$$

1.15.4 Give a parametric representation of the curve in Fig. 1.9.

1.15.5 Consider the curve represented by

$$f: \mathbb{R} \longrightarrow \mathbb{R}^2, \qquad \theta \mapsto (\cos k\theta \cos\theta, \cos k\theta \sin\theta), \quad k \in \mathbb{R}.$$

Draw a picture of this curve for $k = 0$, $k = 1$, $k = 2$ and $k = 3$. Find the corresponding Cartesian equations. Are there multiple points? Can you guess what the shape of the curve becomes in the case of an irrational parameter k?

1.15.6 Find the two tangents to the ellipse of equation $x^2 + 2y^2 = 1$, passing through the point $(3, 3)$. Determine the position of the two foci and observe the bisector property mentioned in Proposition 1.5.6.2.

1.15.7 Find the two tangents to the hyperbola with equation $x^2 - 2y^2 = 3$, passing through the point $(1, 2)$. Determine the position of the two foci and observe the bisector property mentioned in Proposition 1.5.6.1.

1.15.8 Find the two tangents to the parabola with equation $y = 3x^2 + 2x - 1$, passing through the point $(-2, -3)$. Determine the position of the focus and that of the directrix and observe the bisector property mentioned in Proposition 1.5.6.3.

1.15.9 In $\mathbb{R}^2$, consider an ellipse, a hyperbola or a parabola with equation $p(x, y) = 0$, with p a polynomial of degree 2. Consider further the two subsets

$$\mathcal{Q}_- = \{(x, y) | p(x, y) < 0\}, \qquad \mathcal{Q}_+ = \{(x, y) | p(x, y) > 0\}.$$

Show that through a point in one of these subsets, there are always two distinct tangents (or asymptotes) to the conic, while through a point in the other subset, there is no tangent at all.

1.15.10 Consider the cycloidal pendulum as in Fig. 1.26. Find a parametric equation of the trajectory of the pendulum when the two arches of the cycloid are substituted by quarters of circles.

1.15.11 In a coordinate system of your choice, find a parametric representation of the skew curve obtained as the intersection of a sphere and a cone "in arbitrary positions".

Chapter 2
Plane Curves

Inspired by the intuitive and historical considerations of Chap. 1, we now begin our systematic study of differential geometry with the case of curves in the real plane $\mathbb{R}^2$.

After setting precise definitions, both in terms of parametric and Cartesian equations, we begin with the study of the tangent. We pay special attention to the case of *envelopes*: those curves which are tangent to a continuous family of curves. This notion has numerous applications: we treat some of them as examples.

Next we switch to the main characteristic of a curve: its *curvature*. For that we first need to be able to compute the length of an arc of a curve: the curvature is then the variation in direction of the tangent, the variation with respect to the distance traveled on the curve. This allows us to recapture notions such as the *radius of curvature* or the *center of curvature*. The locus of the centers of curvature, called the *evolute* of the curve, has then rather striking properties. *Huygens* used these properties to construct his cycloidal pendulum (see Sect. 1.9).

We complete the chapter with some other interesting properties of the curvature. First, knowledge of the curvature allows us to reconstruct the curve, up to its position in the plane. Second, in the case of a closed curve, integrating the curvature along one loop of the curve allows us to compute the number of times that the curve has been "rolled up". But above all, we provide a proof of the famous *Hopf theorem* on simple closed curves and pay special attention to the case of convex curves, in particular to the *four vertices theorem*.

2.1 Parametric Representations

With the considerations of Sect. 1.2 in mind, we first define:

Definition 2.1.1 A function

$$f \colon U \longrightarrow \mathbb{R}^m, \quad U \subseteq \mathbb{R}^n$$

F. Borceux, *A Differential Approach to Geometry*, DOI 10.1007/978-3-319-01736-5_2,

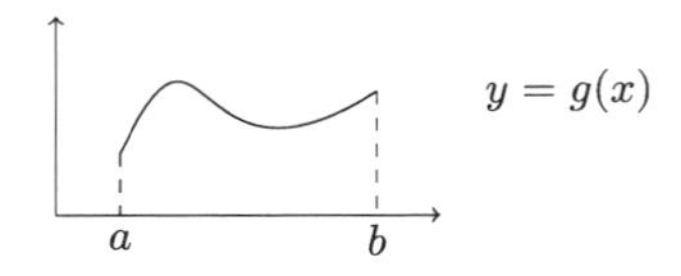

Fig. 2.1

is *locally injective* when every point $x \in U$ possesses a neighborhood on which f is injective.

The basic definition of the theory of plane curves is then:

Definition 2.1.2 By a *parametric representation* of class $\mathcal{C}^k$, $k \in \mathbb{N} \cup \{\infty\}$, of a plane curve is meant a mapping

$$f\colon]a, b[\longrightarrow \mathbb{R}^2, \quad a, b \in \mathbb{R} \cup \{-\infty, +\infty\}$$

which is locally injective and of class $\mathcal{C}^k$. The set

$$\{f(t) \big| t \in]a, b[\}$$

is called the *support* of the curve represented by f.

In this definition, we have taken the notion of an *open interval* in its more general sense. This allows us to view the full real line, as well as open half lines, as open intervals. This useful convention allows maximal flexibility in examples.

The reader should be aware that in Definition 2.1.2, we have defined expressions such as *parametric representation of a plane curve* or *support of the curve represented by f* but we have not (yet) defined what a *curve* is.

Obviously, a parametric representation of class $\mathcal{C}^k$ is also a parametric representation of class $\mathcal{C}^{k'}$, for every $k' \leq k$.

Example 2.1.3 The graph of every continuous function $g\colon]a, b[\longrightarrow \mathbb{R}$ is the support of a parametric representation

$$f\colon]a, b[\longrightarrow \mathbb{R}^2, \qquad x \mapsto (x, g(x))$$

of a plane curve. When g is of class $\mathcal{C}^k$, so is f (see Fig. 2.1).

Proof The function f is locally injective because its first component is injective. When g is of class $\mathcal{C}^k$ then so is f, because the first component of f is of class $\mathcal{C}^\infty$. □

Of course there are many more parametric representations than those given by the graph of a function!

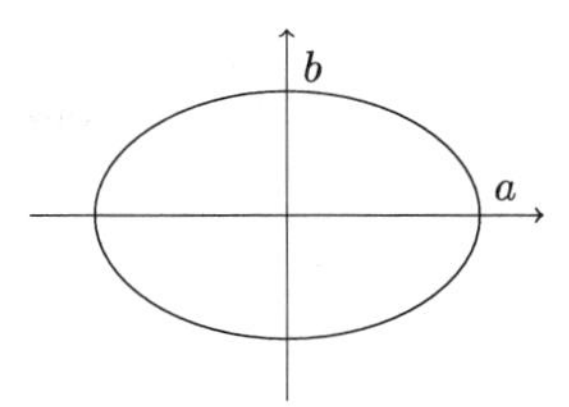

Fig. 2.2

Example 2.1.4 The function

$$f : \mathbb{R} \longrightarrow \mathbb{R}^2, \qquad \theta \mapsto (a\cos\theta, b\sin\theta), \quad a, b > 0$$

is a parametric representation of class $\mathcal{C}^\infty$ of a curve whose support is an ellipse of radii a and b. (See Fig. 2.2.)

Proof Since the functions $\cos\theta$ and $\sin\theta$ are of class $\mathcal{C}^\infty$, then so is f. On the other hand

$$\begin{aligned}(a\cos\theta, b\sin\theta) = \big(a\cos\theta', b\sin\theta'\big) \quad &\Longrightarrow \quad (\cos\theta, \sin\theta) = \big(\cos\theta', \sin\theta'\big) \\ &\Longrightarrow \quad \theta' = \theta + 2k\pi, \quad k \in \mathbb{N}.\end{aligned}$$

Thus f is injective on each interval of length strictly less than 2π. We therefore have a parametric representation of class $\mathcal{C}^\infty$.

It is immediate that each point

$$(x, y) = (a\cos\theta, b\sin\theta)$$

satisfies the equation

$$\frac{x^2}{a^2} + \frac{y^2}{b^2} = 1$$

which is that of an ellipse of radii a and b (see Sect. 1.14 in of [4], *Trilogy II*). Conversely every point (x, y) of this ellipse is such that

$$\left(\frac{x}{a}\right)^2 + \left(\frac{y}{b}\right)^2 = 1$$

thus

$$\frac{x}{a} = \cos\theta, \qquad \frac{y}{b} = \sin\theta$$

for a unique $\theta \in [0, 2\pi[$. Therefore the support of f is comprised precisely of the points of the ellipse. □

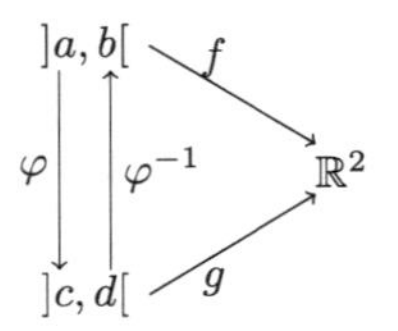

Fig. 2.3

Counterexample 2.1.5 The mapping of class $\mathcal{C}^\infty$

$$f\colon \mathbb{R} \mapsto \mathbb{R}^2, \qquad t \mapsto \left(t^2, t^2\right)$$

is not a parametric representation of a plane curve, while the other mapping of class $\mathcal{C}^\infty$

$$g\colon \mathbb{R} \mapsto \mathbb{R}^2, \qquad t \mapsto \left(t^3, t^3\right)$$

is a parametric representation of the first diagonal.

Proof One always has $f(t) = f(-t)$, thus f is not locally injective in a neighborhood of 0. Notice that the "support" of f (i.e. its image) is one half of the first diagonal (compare with the example in Fig. 1.11).

On the other hand $h(t) = t^3$ is an injective function, thus g is a parametric representation of the first diagonal. □

Let us now proceed to the definition of a *curve*.

Definition 2.1.6 Two parametric representations of class $\mathcal{C}^k$

$$f\colon]a,b[\longrightarrow \mathbb{R}^2, \qquad t \mapsto f(t); \qquad g\colon]c,d[\longrightarrow \mathbb{R}^2, \qquad s \mapsto g(s)$$

are *equivalent in class* $\mathcal{C}^k$ when there exist inverse bijections of class $\mathcal{C}^k$

$$\varphi\colon]a,b[\longrightarrow]c,d[, \qquad \varphi^{-1}\colon]c,d[\longrightarrow]a,b[$$

such that (see Fig. 2.3)

$$g \circ \varphi = f, \qquad f \circ \varphi^{-1} = g.$$

Of course in this definition, the equality $f \circ \varphi^{-1} = g$ is redundant since it follows at once from $g \circ \varphi = f$ and the bijectivity of φ. But φ^{-1} being of class $\mathcal{C}^k$ is certainly not a redundant condition, as Counterexample 2.1.11 shows.

Definition 2.1.6 thus tells us that the two parametric representations f and g are equivalent when you can switch in both ways from one to the other by a simple change of parameter

$$s = \varphi(t), \qquad t = \varphi^{-1}(s)$$

respecting the class of differentiability. Trivially:

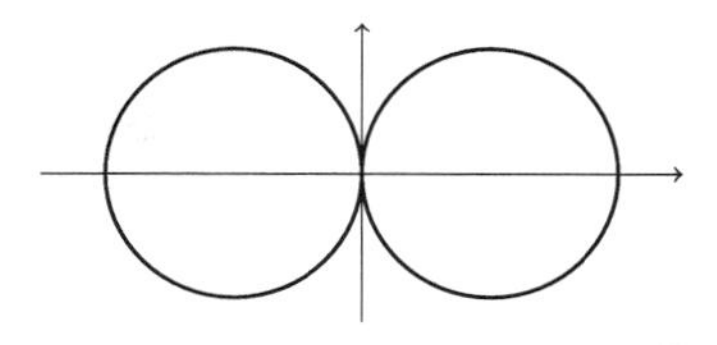

Fig. 2.4

Proposition 2.1.7 *Being equivalent in class $\mathcal{C}^k$ is an equivalence relation on the parametric representations of class $\mathcal{C}^k$ of plane curves.*

We can now define a *curve*:

Definition 2.1.8 By a *plane curve of class $\mathcal{C}^k$* is meant an equivalence class of parametric representations of class $\mathcal{C}^k$ of plane curves, for the equivalence relation of Proposition 2.1.7.

We observe at once that:

Proposition 2.1.9 *Two equivalent parametric representations, whatever the class of differentiability, always have the same support, which therefore is called the* support of the corresponding curve.

Proof With the notation of Definition 2.1.6, the support of f is contained in that of g because $f(t) = g(\varphi(t))$. Analogously the support of g is contained in that of f because $g(s) = f(\varphi^{-1}(s))$. □

It is very important to note that different curves, in the sense of Definition 2.1.8, can very well have the same support.

Counterexample 2.1.10 Consider the two functions

$$f\colon]0, 2\pi[\longrightarrow \mathbb{R}^2, \qquad f(\theta) = \begin{cases} (\cos 2\theta - 1, \sin 2\theta) & \text{if } \theta \le \pi \\ (1 - \cos 2\theta, \sin 2\theta) & \text{if } \theta > \pi \end{cases}$$

$$g\colon]0, 2\pi[\longrightarrow \mathbb{R}^2, \qquad g(\tau) = \begin{cases} (\cos 2\tau - 1, \sin 2\tau) & \text{if } \tau \le \pi \\ (1 - \cos 2\tau, -\sin 2\tau) & \text{if } \tau > \pi. \end{cases}$$

The functions f and g are parametric representations, respectively of class $\mathcal{C}^1$ and $\mathcal{C}^0$, of two plane curves having the same support. These two curves are different, even as curves of class $\mathcal{C}^0$. (See Fig. 2.4.)

Proof When θ runs from 0 to π, $f(\theta)$ travels counter-clockwise around the left hand circle, starting from the origin (or "just after it" since we are working on an open interval). When θ varies from π to 2π, $f(\theta)$ travels clockwise around the right hand circle. In particular, f is injective; it is also of class $\mathcal{C}^1$, with $f'(\pi) = (0, 1)$.

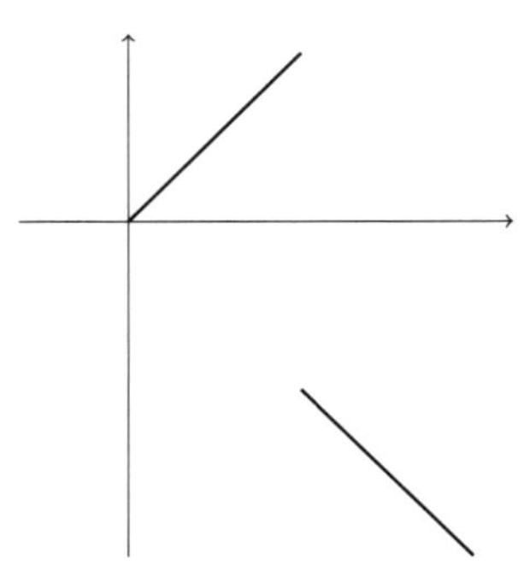

Fig. 2.5

When τ runs from 0 to π, $g(\tau)$ travels counter-clockwise around the left hand circle, starting again from the origin. But when τ varies from π to 2π, $g(\tau)$ now travels counter-clockwise around the right hand circle. In particular, g is injective and continuous.

The change of parameter interchanging f and g is thus

$$\tau = \varphi(\theta) = \begin{cases} \theta & \text{if } \theta \leq \pi \\ -\theta & \text{if } \theta > \pi. \end{cases}$$

This is definitely not a continuous function (see Fig. 2.5). □

Let us think of a parametric representation f as describing the trajectory of the point $f(t)$ when the parameter t varies. Then Counterexample 2.1.10 becomes intuitively clear: f and g describe two different trajectories, even if these trajectories eventually pass through the same points, but in a different order.

The following counterexample is more subtle.

Counterexample 2.1.11 The following two functions

$$f : \mathbb{R} \longrightarrow \mathbb{R}^2, \qquad t \mapsto (t, 0)$$

$$g : \mathbb{R} \longrightarrow \mathbb{R}^2, \qquad s \mapsto (s^3, 0)$$

are two parametric representations of class $\mathcal{C}^\infty$ of the x-axis. They are equivalent in class $\mathcal{C}^0$, but not in class $\mathcal{C}^k$, for every $k \neq 0$.

Proof The change of parameter formulas are simply

$$\varphi(t) = \sqrt[3]{t}, \qquad \varphi^{-1}(s) = s^3.$$

These two bijections are continuous, with φ^{-1} of class $\mathcal{C}^\infty$, but φ is not differentiable for $t = 0$. □

2.2 Regular Representations

Of course in order to obtain relevant results, the parametric representations involved should have sufficiently good properties. Being of class $\mathcal{C}^1$ appears as a first sensible assumption, but we shall often need more.

Definition 2.2.1 Consider a parametric representation $f\colon]a,b[\longrightarrow \mathbb{R}^2$ of class $\mathcal{C}^1$ of a plane curve.

- The point with parameter $t \in]a,b[$ is *regular* when $f'(t) \neq 0$.
- The representation itself is *regular* when $f'(t) \neq 0$ for all $t \in]a,b[$.
- The point with parameter $t \in]a,b[$ is *singular* when $f'(t) = 0$.

Being *regular* is in fact a property of the corresponding curve of class $\mathcal{C}^1$, as our next result proves.

Proposition 2.2.2 *With the notation of Definition* 2.1.6, *consider two parametric representations f, g of class $\mathcal{C}^1$, equivalent in class $\mathcal{C}^1$.*

- *The point with parameter $t \in]a,b[$ is regular for f if and only if the point with parameter $\varphi(t)$ is regular for g.*
- *The parametric representation f is regular if and only if the parametric representation g is regular.*

Proof From $f = g \circ \varphi$ we deduce

$$f'(t) = g'\big(\varphi(t)\big) \cdot \varphi'(t).$$

If $f'(t) \neq 0$, necessarily $g'(\varphi(t)) \neq 0$. □

It should be clear that for a parametric representation, being regular is definitely stronger than being of class $\mathcal{C}^1$.

Counterexample 2.2.3 The following function

$$f\colon \mathbb{R} \longrightarrow \mathbb{R}^2, \qquad x \mapsto \big(x^3, |x^3|\big)$$

is a non-regular parametric representation of class $\mathcal{C}^2$ of a plane curve, whose support is the graph of the function $y = |x|$, which is only of class $\mathcal{C}^0$ (see Fig. 2.6).

Proof Consider the functions

$$g_-(x) = -x^3, \qquad g_+(x) = x^3$$

whose successive derivatives are

$$g'_-(x) = -3x^2, \qquad g'_+(x) = 3x^2, \qquad g''_-(x) = -6x, \qquad g''_+(x) = 6x.$$

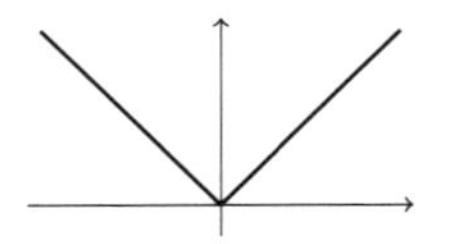

Fig. 2.6

The functions g_- and g_+ have the same first and second derivatives for $x = 0$, proving that the function $h(x) = |x^3|$ is of class $\mathcal{C}^2$. Thus f is of class $\mathcal{C}^2$; it is also injective since the function $k(x) = x^3$ is injective. The corresponding support is trivially comprised of the points $(u, |u|)$, that is, of the graph of the function $y = |x|$.

Observe further that the function $y = |x|$ is not differentiable at $x = 0$, thus the parametric representation

$$x' \mapsto (x', |x'|)$$

of its graph (see Example 2.1.3) is only of class $\mathcal{C}^0$. This parametric representation is equivalent to f in class $\mathcal{C}^0$ via the change of parameter

$$x' = x^3, \qquad x = \sqrt[3]{x'}$$

which, as already observed in Counterexample 2.1.11, is only of class $\mathcal{C}^0$. □

Let us conclude this section by observing that the "regularity condition" immediately implies "local injectivity".

Lemma 2.2.4 *Consider a function of class* $\mathcal{C}^1$

$$f :]a, b[\longrightarrow \mathbb{R}^n$$

such that $f'(t) \neq 0$ *for each* $t \in]a, b[$. *Then* f *is locally injective.*

Proof Fix a value $t_0 \in]a, b[$. At least one of the components of f' is non-zero at this point: let us say, $f_1'(t_0) > 0$. By continuity, $f_1'(t) > 0$ on some neighborhood $]t_0 - \delta, t_0 + \delta[$ of t_0. On this neighborhood of t_0, f must be injective. Otherwise one would have $t_1 < t_2$ in this interval such that $f(t_1) = f(t_2)$. The *mean value theorem* (see an analysis course) applied to f_1 would then imply the existence of $t_1 < t_3 < t_2$ such that

$$0 = \frac{f_1(t_2) - f_1(t_1)}{t_2 - t_1} = f_1'(t_3).$$

Since $t_3 \in]t_0 - \delta, t_0 + \delta[$, this contradicts $f_1'(t_3) \neq 0$. □

This lemma immediately implies:

Proposition 2.2.5 *For a function*

$$f\colon]a,b[\longrightarrow \mathbb{R}^2, \qquad t \mapsto f(t)$$

the following conditions are equivalent:

1. f *is a regular parametric representation of a plane curve*;
2. f *is of class* $\mathcal{C}^1$ *and* $f'(t) \neq 0$ *for all* $t \in]a,b[$.

2.3 The Cartesian Equation of a Curve

Following the considerations of Sects. 1.3 and 1.4, we now investigate those equations of the form

$$F(x,y)=0$$

whose set of solutions is—at least locally—the support of some curve, in the sense of Sect. 2.1. We first rephrase our Definition 1.4.2.

Definition 2.3.1 By a *Cartesian equation* of a *Cartesian plane curve* we mean an equation

$$F(x,y)=0$$

where:

- $F\colon \mathbb{R}^2 \longrightarrow \mathbb{R}$ is a function of class $\mathcal{C}^1$;
- the equation admits infinitely many solutions;
- there are at most finitely many solutions (x,y) of the equation where both partial derivatives of F vanish.

The corresponding *Cartesian curve* is the set of those points (x,y) which are solutions of the equation $F(x,y)=0$.

We define further (see Definition 1.4.3).

Definition 2.3.2 Consider a Cartesian equation $F(x,y)=0$ of a Cartesian plane curve.

- A point of the Cartesian curve is *multiple* when both partial derivatives of F vanish at this point.
- A point of the Cartesian curve is *simple* when at least one partial derivative of F does not vanish at this point.

Example 2.3.3 The Cartesian curve with Cartesian equation

$$y^2(1-x)=x^2(1+x)$$

—the so-called *right strophoid* (see Fig. 2.7)—admits the origin as a multiple point.

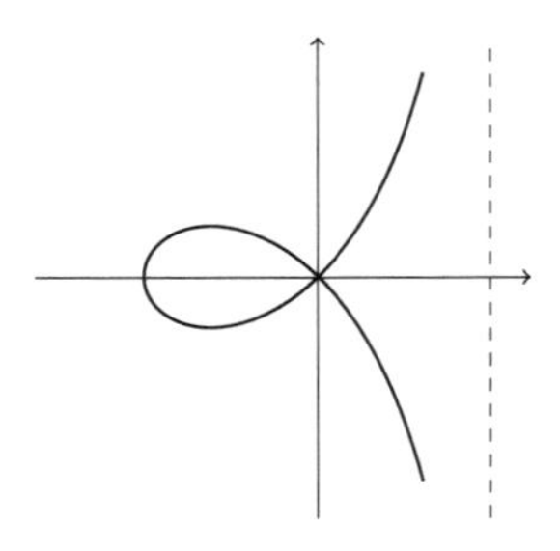

Fig. 2.7

Proof The two partial derivatives of the function

$$F(x, y) = y^2(1-x) - x^2(1+x)$$

are

$$\frac{\partial F}{\partial x} = -y^2 - x(2+3x), \qquad \frac{\partial F}{\partial y} = 2y(1-x).$$

Both vanish at $(0, 0)$, which is thus a multiple point.

This is in fact the only multiple point. Indeed $\frac{\partial F}{\partial y}$ vanishes when $y = 0$ or $x = 1$.

- If $y = 0$, $\frac{\partial F}{\partial x}$ vanishes for $x = 0$ or $x = -\frac{2}{3}$, but this last point is not a point of the curve.
- If $x = 1$, $\frac{\partial F}{\partial x}$ is strictly negative, whatever the value of y.

Thus $F(x, y) = 0$ is indeed the Cartesian equation of a Cartesian curve. □

It should be clear that "being a multiple point" is a property of the Cartesian equation $F(x, y) = 0$, not of the corresponding Cartesian curve. In particular, for an arbitrary Cartesian equation $F(x, y) = 0$, being a multiple point is certainly not equivalent to the intuitive idea of "a curve passing through this point several times". For example, $y^2 = 0$ is a Cartesian equation of the x-axis admitting the origin as a multiple point. A more precise study of these questions, in the projective complex case, with F an irreducible polynomial, can be found in Sect. 7.4 of [4], *Trilogy II*.

We now "repeat" our Propositions 1.3.3 and 1.3.6.

Proposition 2.3.4 *Consider a plane curve of class $\mathcal{C}^1$. In a neighborhood of each regular point, the support of the curve can be described by a Cartesian equation.*

Proof Consider a parametric representation $f : I \longrightarrow \mathbb{R}^2$ of class $\mathcal{C}^1$ of the curve and a regular point of parameter $t_0 \in I$. We then have (see Definition 2.2.1)

$$f'(t_0) = \left(f_1'(t_0), f_2'(t_0)\right) \neq (0, 0);$$

we consider the case $f_2'(t_0) \neq 0$.

By the *Local Inverse Theorem* (see Theorem 1.3.1), f_2' is invertible on some neighborhood of t_0, with an inverse still of class $\mathcal{C}^1$. For t in this neighborhood, we can thus write $t = f_2^{-1}(y)$ and consequently, $x = f_1(f_2^{-1}(y))$. Putting

$$F(x, y) = x - f_1\bigl(f_2^{-1}(y)\bigr)$$

we conclude that on the given neighborhood of t_0, the points of the curve are solutions of the equation $F(x, y) = 0$.

Notice that since f_1 and f_2^{-1} are of class $\mathcal{C}^1$ then so is F. On the other hand $\frac{\partial f}{\partial x} = 1 \neq 0$ at all points where the argument above applies. □

Proposition 2.3.5 *Let $F(x, y) = 0$ be a Cartesian equation of a Cartesian plane curve. On a neighborhood of a simple point (x_0, y_0) of this curve, the Cartesian curve is the support of a regular curve.*

Proof At (x_0, y_0), one of the partial derivatives of F does not vanish (see Definition 2.3.2); we consider the case $\frac{\partial F}{\partial y} \neq 0$. By the *Implicit Function Theorem* (see Theorem 1.3.5), there exist a neighborhood $]a, b[$ of x_0 and a mapping

$$\varphi \colon]a, b[\longrightarrow \mathbb{R}$$

of class $\mathcal{C}^1$, such that

$$\varphi(x_0) = y_0, \qquad \forall x \in]a, b[\quad F\bigl(x, \varphi(x)\bigr) = 0.$$

The graph of φ (see Example 2.1.3)

$$f \colon]a, b[\longrightarrow \mathbb{R}^2, \qquad x \mapsto \bigl(x, \varphi(x)\bigr)$$

is then a parametric representation of class $\mathcal{C}^1$ of a plane curve whose support coincides with the Cartesian curve with equation $F(x, y) = 0$, in a neighborhood of x_0. □

It is certainly useful to underline the fact that the functions F in Proposition 2.3.4 and f in Proposition 2.3.5 are by no means unique. Let us demonstrate this with two examples.

- In Example 2.1.4, we can of course choose

$$F(x, y) = \frac{x^2}{a^2} + \frac{y^2}{b^2} - 1.$$

But in a neighborhood of the point $(0, b)$ of the ellipse—in fact, on the entire upper half of the ellipse—we have

$$y = b\sqrt{1 - \frac{x^2}{a^2}}$$

and thus we could as well choose

$$F(x, y) = y - b\sqrt{1 - \frac{x^2}{a^2}}.$$

- In Counterexample 2.1.10, the common support of the two curves is a Cartesian curve whose Cartesian equation is the product of the Cartesian equations of the two individual circles, that is,

$$\big((x+1)^2 + y^2 - 1\big)\big((x-1)^2 + y^2 - 1\big) = 0.$$

  As we have seen, f and g are parametric representations of different curves having precisely this support.

It is also useful to draw attention to the fact that Propositions 2.3.4 and 2.3.5 give the false impression that *regular point* and *simple point* are closely related notions. The following counterexamples throw more light on this question.

Counterexample 2.3.6 The *right strophoid* (see Example 2.3.3 and Fig. 2.7) can be seen as a regular curve whose support admits a Cartesian equation with a multiple point.

Proof The following function f is of class $\mathcal{C}^\infty$.

$$f: \mathbb{R} \longrightarrow \mathbb{R}^2, \qquad t \mapsto \left(\frac{t^2-1}{t^2+1}, t\frac{t^2-1}{t^2+1}\right).$$

It is also locally injective. Indeed the equality

$$\left(\frac{t_0^2-1}{t_0^2+1}, t_0\frac{t_0^2-1}{t_0^2+1}\right) = \left(\frac{t_1^2-1}{t_1^2+1}, t_1\frac{t_1^2-1}{t_1^2+1}\right)$$

forces $t_0 = t_1$ as soon as $t_0^2 - 1 \neq 0$, which forces further $t_1^2 - 1 \neq 0$. Thus f is "almost injective", that is, the only points identified by f are $f(-1) = f(+1)$. Thus f is locally injective and is a parametric representation of class $\mathcal{C}^\infty$ of a plane curve.

The derivative of f is given by

$$f'(t) = \left(\frac{4t}{(t^2+1)^2}, \frac{t^4+4t^2-1}{(t^2+1)^2}\right).$$

The first component vanishes only for $t = 0$, in which case the second component equals -1. Thus f is also regular.

It is then routine to check that $F(x, y) = 0$ with

$$F(x, y) = y^2(1-x) - x^2(1+x)$$

is a possible Cartesian equation. This is precisely the Cartesian equation of the right strophoid of Example 2.3.3. □

Counterexample 2.3.6 can be interpreted intuitively as follows: the origin is a regular point "on each branch of the curve passing through the origin", but it is a multiple point precisely because the curve passes through the origin twice.

Counterexample 2.3.7 There exist curves with singular points whose support admits a Cartesian equation with respect to which all points are simple.

Proof The parametric representation g of the x-axis as described in Counterexample 2.1.11 admits the origin as singular point, and of course $y = 0$ is a Cartesian equation of the x-axis for which all points are simple. □

2.4 Tangents

With the considerations of Sect. 1.6, in mind, we first observe:

Lemma 2.4.1 *Consider two parametric representations f, g of class $\mathcal{C}^1$ of a plane curve, equivalent in class $\mathcal{C}^1$. With the notation of Proposition 2.2.2, at a regular point of the corresponding curve, the two vectors $f'(t)$ and $g'(\varphi(t))$ are proportional.*

Proof This is precisely what the proof of Proposition 2.2.2 shows. □

Definition 2.4.2 Consider a parametric representation

$$f\colon]a,b[\longrightarrow \mathbb{R}^2$$

of class $\mathcal{C}^1$ of a plane curve.

- The *tangent* to the curve at a regular point with parameter t_0 is the line passing through $f(t_0)$ and of direction $f'(t_0)$.
- The *normal* to the curve at a regular point is the perpendicular to the tangent through this point.

Lemma 2.4.1 shows that Definition 2.4.2 is independent of the choice of the parametric representation f.

Proposition 2.4.3 *Consider a parametric representation*

$$f\colon]a,b[\longrightarrow \mathbb{R}^2$$

of class $\mathcal{C}^1$ of a plane curve.

- *The tangent to the curve at a regular point of parameter t_0 admits the Cartesian equation*

$$f_2'(t_0)\big(x - f_1(t_0)\big) - f_1'(t_0)\big(y - f_2(t_0)\big) = 0.$$

- *The normal to the curve at a regular point of parameter t_0 admits the Cartesian equation*

$$f_1'(t_0)\big(x - f_1(t_0)\big) + f_2'(t_0)\big(y - f_2(t_0)\big) = 0.$$

Proof The line through (a, b) in the direction orthogonal to the vector (c, d) has the equation

$$c(x - a) + d(y - b) = 0$$

(see Corollary 4.8.4 in [4], *Trilogy II*). □

Proposition 2.4.4 *Consider a Cartesian equation $F(x, y) = 0$ of a Cartesian plane curve. The tangent to that curve at a simple point (x_0, y_0) is the line with equation*

$$\frac{\partial F}{\partial x}(x_0, y_0)(x - x_0) + \frac{\partial F}{\partial y}(x_0, y_0)(y - y_0) = 0.$$

Proof Let us consider the case $\frac{\partial F}{\partial y}(x_0, y_0) \neq 0$ and use the notation of Proposition 2.3.5. The curve admits the regular parametric representation $f(x) = (x, \varphi(x))$ on a neighborhood of x_0. Thus the tangent is the line through (x_0, y_0) with direction $(1, \varphi'(x_0))$ (see Definition 2.4.2). We must determine the value of $\varphi'(x_0)$.

Differentiating the equality $F(x, \varphi(x)) = 0$ with respect to x yields

$$\frac{\partial F}{\partial x}(x_0, y_0) \cdot \frac{dx}{dx}(x_0) + \frac{\partial F}{\partial y}(x_0, y_0) \cdot \frac{d\varphi}{dx}(x_0) = 0$$

that is,

$$\varphi'(x_0) = -\frac{\frac{\partial F}{\partial x}(x_0, y_0)}{\frac{\partial F}{\partial y}(x_0, y_0)}.$$

The parametric equations of the tangent are thus

$$x = x_0 + \alpha \cdot 1$$

$$y = y_0 - \alpha \frac{\frac{\partial F}{\partial x}(x_0, y_0)}{\frac{\partial F}{\partial y}(x_0, y_0)}.$$

Introducing the value $\alpha = x - x_0$ into the second equation yields the formula of the statement. □

Going back to the example of the right strophoid (Counterexample 2.3.6), working with the parametric representation f allows us to consider two tangents at the

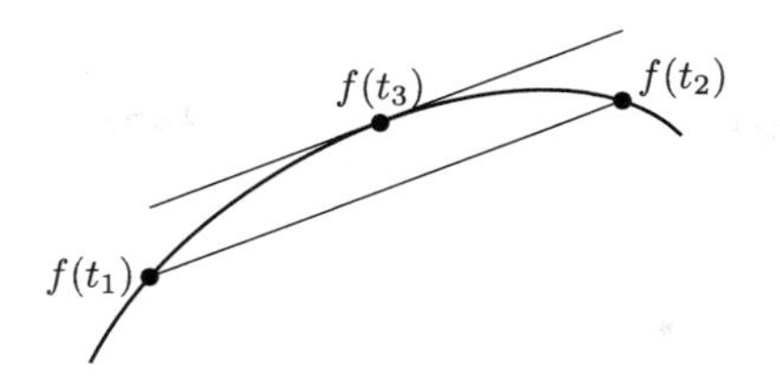

Fig. 2.8

origin: one for the value $t=+1$ and one for the value $t=-1$. But working with the Cartesian equation, Proposition 2.4.4 does not allow the consideration of any tangent at the origin. The question of multiple tangents at multiple points for algebraic curves is treated algebraically in Sect. 7.4 of [4], *Trilogy II*.

Let us conclude by rephrasing a well-known theorem in analysis:

Proposition 2.4.5 *Consider a regular parametric representation of a plane curve*

$$f\colon]a,b[\longrightarrow \mathbb{R}^2.$$

In a neighborhood of each point $t_0 \in [a,b[$, the line joining two points $f(t_1)$ and $f(t_2)$ is parallel to the tangent at some point with parameter t_3, with $t_1 < t_3 < t_2$ (see Fig. 2.8).

Proof As observed in the proof of Proposition 2.3.4, assuming $f_2'(t_0)\neq 0$, we can locally describe the curve by an equation of the form $y=g(x)$. The conclusion follows at once from the *mean value theorem* for functions of a real variable. □

2.5 Asymptotes

The notion of an *asymptote* is better treated in the context of projective geometry, where it reduces simply to a *tangent at a point at infinity* (see Chap. 6 in [4], *Trilogy II*). Let us nevertheless give a short account of this notion in $\mathbb{R}^2$, in the most simple (and restricted) cases.

Definition 2.5.1 Consider a regular parametric representation

$$f\colon]a,\infty[\longrightarrow \mathbb{R}^2,\qquad t\mapsto f(t)$$

of a plane curve. A line is an *asymptote* at $+\infty$ of the corresponding curve when this line admits a parametric representation of the form

$$g\colon]-\infty,\infty[\longrightarrow \mathbb{R}^2,\qquad t\mapsto (\alpha t+\beta,\gamma t+\delta)$$

such that when t tends to $+\infty$

$$\lim_{t\to\infty}\big(f(t)-g(t)\big)=(0,0).$$

An analogous definition holds in the case $]-\infty,b[$.

Example 2.5.2 The asymptotes to the hyperbola.

Proof Let us consider a branch of the hyperbola with Cartesian equation $xy=1$. A parametric representation is simply

$$f\colon]0,+\infty[\longrightarrow \mathbb{R}^2, \qquad x\mapsto \left(x,\frac{1}{x}\right).$$

The x-axis admits the parametric representation

$$g\colon]-\infty,+\infty[\longrightarrow \mathbb{R}^2, \qquad x\mapsto (x,0).$$

It is immediate that

$$\lim_{x\to\infty}\big(f(x)-g(x)\big)=\lim_{x\to\infty}\left(0,\frac{1}{x}\right)=(0,0).$$

Working instead with the parametric representation

$$h\colon]0,+\infty[\longrightarrow \mathbb{R}^2, \qquad y\mapsto \left(\frac{1}{y},y\right)$$

allows us to present the y-axis as the second asymptote. □

Example 2.5.3 The *right strophoid* of Counterexample 2.3.6 admits the vertical line of equation $x=1$ as asymptote, both at $+\infty$ and at $-\infty$ (see Fig. 2.7).

Proof With the notation of Counterexample 2.3.6, simply consider the parametric representation $g(t)=(1,t)$ of the line $x=1$. The result follows at once from the fact that

$$\lim_{t\to\pm\infty}\frac{t^2-1}{t^2+1}=1.$$

□

Our next example is more involved, but will play a significant role in surface theory (see Example 5.16.7). Imagine a horse, starting from the origin and moving up along the y-axis. The horse pulls a weight attached at the extremity of a rope of length R and this weight was originally situated at the point $(R,0)$ on the x-axis. While the horse moves, the weight follows a trajectory to which the rope is tangent at each instant. Such a trajectory is called a *tractrix* (see Fig. 2.9).

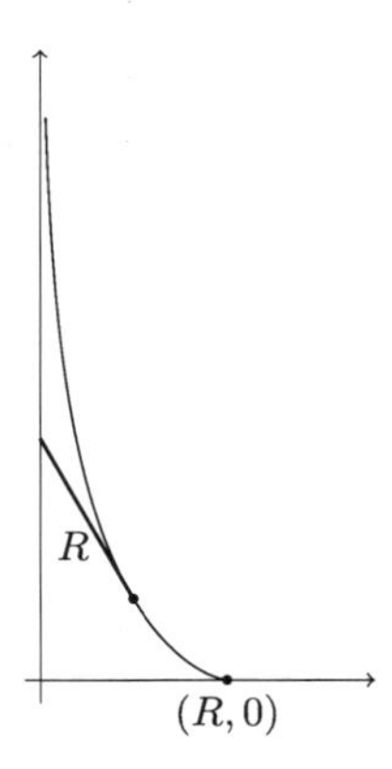

Fig. 2.9

Example 2.5.4 A *tractrix* is a plane curve with the property that the length of the segment of the tangent between the tangency point and the y-axis has constant length R. The graph of the function

$$g\colon]0, R[\longrightarrow \mathbb{R}, \qquad g(x) = \int_x^R \sqrt{\frac{R^2}{t^2} - 1}\, dt$$

is a tractrix converging at $x = R$ to the point $(0, R)$ and admitting the y-axis as asymptote.

Proof A function $g(x)$ yielding a tractrix as in the statement corresponds to a parametric representation $f(x) = (x, g(x))$ (see Example 2.1.3). The tangent at the point $f(x)$ is given by (see Definition 2.4.2)

$$\bigl(x, g(x)\bigr) + \alpha\bigl(1, g'(x)\bigr).$$

It intersects the y-axis when $\alpha = -x$, that is at the point

$$\bigl(0, g(x) - x\, g'(x)\bigr).$$

The length of the corresponding tangent segment up to the y-axis is thus

$$R = \sqrt{x^2 + x^2 g'(x)^2}.$$

This can be re-written as

$$g'(x) = \pm\sqrt{\frac{R^2}{x^2} - 1}.$$

We shall choose the "minus" sign in order to recapture the situation described just before the example (the "plus" sign would correspond to the horse moving down along the y-axis). The general solution of this differential equation can be written

$$g(x) = \int_x^R \sqrt{\frac{R^2}{t^2} - 1}\, dt + k$$

where k is a constant. Further imposing the initial condition $g(R) = 0$ forces $k = 0$.

Observe that for each $x \in]0, R[$

$$g'(x) = \sqrt{\frac{R^2}{x^2} - 1} < 0.$$

Thus the function g, which is of class $\mathcal{C}^\infty$ on $]0, R[$, is strictly decreasing on this interval. Therefore it is bijective and its inverse g^{-1} is still of class $\mathcal{C}^\infty$. We obtain an alternative parametric representation of the tractrix

$$h\colon]0, \infty[\longrightarrow \mathbb{R}^2, \qquad y \mapsto \big(g^{-1}(y), y\big).$$

On the other hand the y-axis admits the parametric representation

$$k\colon]0, \infty[\longrightarrow \mathbb{R}^2, \qquad y \mapsto (0, y).$$

We thus have

$$\lim_{y\to\infty} \big(k(y) - h(y)\big) = \lim_{y\to\infty} \big(g^{-1}(y), 0\big) = \lim_{x\to 0} (x, 0) = (0, 0).$$

Therefore the y-axis is an asymptote (see Definition 2.5.1). □

2.6 Envelopes

The problem of *envelopes* is another rather tricky question, if one wants to treat it in the most general and precise setting. We restrict ourselves to the proof of a rather classical (but restricted) result, which—with a little bit of effort and imagination—proves to be sufficient in most cases of interest.

Definition 2.6.1 Consider a function of class $\mathcal{C}^1$:

$$F\colon \mathbb{R} \times \mathbb{R} \times]a, b[\longrightarrow \mathbb{R}, \qquad (x, y, \alpha) \mapsto F(x, y, \alpha).$$

Assume that for each value $\alpha_0 \in]a, b[$, $F(x, y, \alpha_0) = 0$ is the Cartesian equation of a Cartesian plane curve $\mathcal{C}_\alpha$. A regular plane curve $\mathcal{C}$ represented by

$$f\colon]a, b[\longrightarrow \mathbb{R}^2, \qquad \alpha \mapsto f(\alpha)$$

is an *envelope* of the family of curves $(\mathcal{C}_\alpha)_\alpha$ when for each value $\alpha \in]a, b[$:

1. $f(\alpha)$ is a simple point of the curve $\mathcal{C}_\alpha$;
2. the curves $\mathcal{C}$ and $\mathcal{C}_\alpha$ have the same tangent at this common point.

Let us only prove the following necessary condition, due to *Leibniz*, which considerably facilitates the study of envelopes.

Proposition 2.6.2 *Under the conditions of Definition* 2.6.1, *all the points of the envelope satisfy the system of equations*

$$\begin{cases} F(x,y,\alpha)=0, \\ \frac{\partial F}{\partial \alpha}(x,y,\alpha)=0. \end{cases}$$

Proof The first equation expresses the fact that the point

$$\big(f_1(\alpha), f_2(\alpha)\big)$$

lies on the curve $\mathcal{C}_\alpha$ with Cartesian equation $F(x,y,\alpha)=0$.

By Definition 2.4.2 and Proposition 2.4.4,

$$\big(f_1'(\alpha), f_2'(\alpha)\big) \quad \text{and} \quad \left(\frac{\partial F}{\partial x}\big(f_1(\alpha), f_2(\alpha), \alpha\big), \frac{\partial F}{\partial y}\big(f_1(\alpha), f_2(\alpha), \alpha\big)\right)$$

are respectively the directions of the tangent to the curve $\mathcal{C}$ and the normal to the curve $\mathcal{C}_\alpha$. By definition of an envelope, these two directions are perpendicular.

Differentiating with respect to α the equation

$$F\big(f_1(\alpha), f_2(\alpha), \alpha\big)=0$$

yields

$$\frac{\partial F}{\partial x}\big(f_1(\alpha), f_2(\alpha), \alpha\big)\cdot f_1'(\alpha)+\frac{\partial F}{\partial y}\big(f_1(\alpha), f_2(\alpha), \alpha\big)\cdot f_2'(\alpha)$$
$$+\frac{\partial F}{\partial \alpha}\big(f_1(\alpha), f_2(\alpha), \alpha\big)=0.$$

Together with the perpendicularity condition above, this gives the second equation of the statement. □

Proposition 2.6.2 is often sufficient to investigate problems of envelopes: solve the system given by that proposition, and use your imagination to investigate which solutions of this system are indeed points of the envelope. Let us illustrate this with some examples.

Example 2.6.3 The envelope of the family of circles of radius R whose center is at a distance R of the origin is the circle of radius $2R$ centered at the origin (see Fig. 2.10).

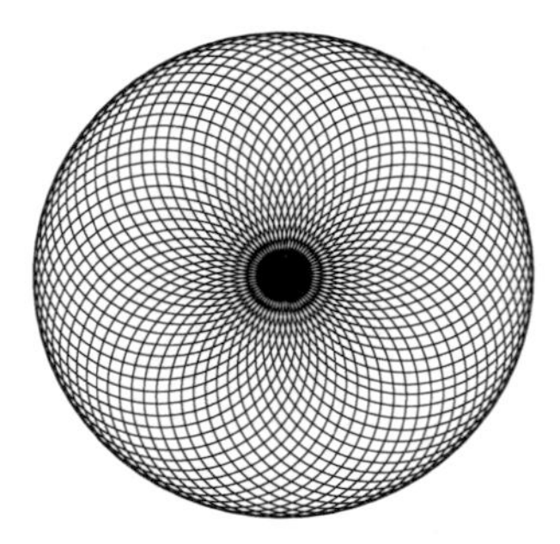

Fig. 2.10

Proof The circle of radius R centered at the point $(R\cos\theta, R\sin\theta)$ admits the equation

$$F(x,y,\theta) = (x - R\cos\theta)^2 + (y - R\sin\theta)^2 - R^2 = 0$$

which reduces to

$$x^2 - 2R(x\cos\theta + y\sin\theta) + y^2 = 0.$$

Differentiating with respect to θ yields

$$x\sin\theta - y\cos\theta = 0.$$

The system of equations

$$\begin{cases} x\cos\theta + y\sin\theta = \frac{x^2+y2}{2R} \\ -y\cos\theta + x\sin\theta = 0 \end{cases}$$

admits the trivial solution $(x,y) = (0,0)$. When $(x,y) \neq (0,0)$, this system with unknowns $\sin\theta$, $\cos\theta$ has determinant $x^2 + y^2 \neq 0$. The solutions are then

$$\cos\theta = \frac{x}{2R}, \qquad \sin\theta = \frac{y}{2R}$$

that is

$$(x,y) = (2R\cos\theta, 2R\sin\theta)$$

which is the parametric representation of a circle of radius $2R$ centered at the origin.

The solutions of the system given by Proposition 2.6.2 are thus constituted of the origin and the circle of radius $2R$ centered at the origin. The origin is certainly not an envelope: it is a point, not a curve. On the other hand the circle of radius $2R$ at this point with parameter θ trivially has the same tangent as the circle of radius R centered at $(R\cos\theta, R\sin\theta)$. Thus the envelope is the circle of radius $2R$ centered at the origin. □

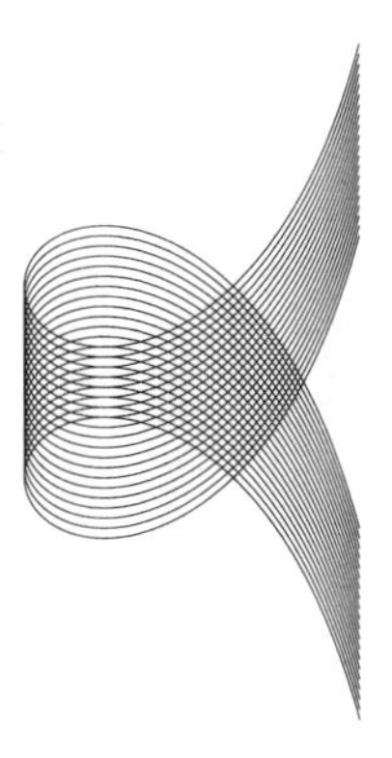

Fig. 2.11

In Example 2.6.3, the appearance of the undesired point $(0,0)$ in the solutions of the system of equations is of course due to the fact that we have precisely chosen circles of radius r centered on a circle of radius R, with $R=r$! Choosing $R \neq r$ would result in an envelope comprising two circles (see Problem 2.18.3).

Here is another phenomena which can cause the appearance of undesired solutions of the system of equations in Proposition 2.6.2.

Example 2.6.4 Consider the family of all the "vertically translated" right strophoids of Example 2.3.3 (see Fig. 2.11)

$$(y-\alpha)^2(1-x)=x^2(1+x).$$

The envelope is the vertical line with equation $x=-1$.

Proof The system of equations given by Proposition 2.6.2 is

$$\begin{cases} (y-\alpha)^2(1-x)-x^2(1+x)=0 \\ -2(y-\alpha)(1-x)=0. \end{cases}$$

The second equation is satisfied for $y-\alpha=0$ or $x=1$. But $x=1$ is not a solution of the first equation. On the other hand when $y-\alpha=0$, the first equation reduces to $x=0$ or $x=-1$. It is obvious that the vertical line $x=-1$ with parametric representation

$$\alpha \mapsto (-1,\alpha)$$

is tangent to all the strophoids of the family, thus it is certainly an envelope. But the line $x=0$ is definitely not an envelope: it is not tangent to any of the curves of the family!

The point here is the fact that the line with equation $x=0$ is the locus of the multiple points of the family of strophoids (see Example 2.3.3). With the notation

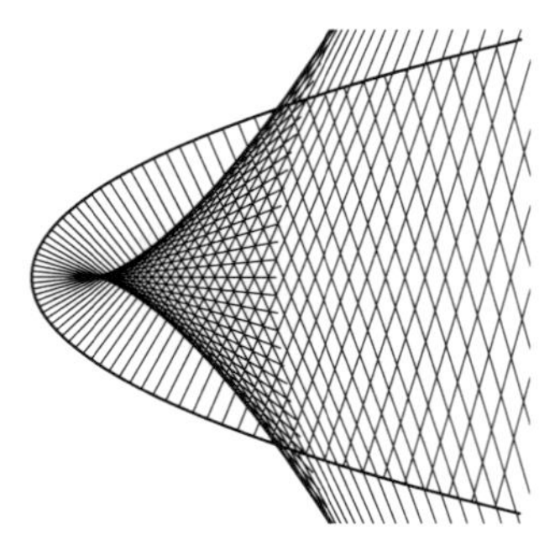

Fig. 2.12

of Proposition 2.6.2, at these points we thus have

$$\frac{\partial F}{\partial x}\big(f_1(\alpha), f_2(\alpha), \alpha\big) = 0, \qquad \frac{\partial F}{\partial y}\big(f_1(\alpha), f_2(\alpha), \alpha\big) = 0.$$

Introducing these values into the last formula of the proof of Proposition 2.6.2 then yields

$$\frac{\partial F}{\partial \alpha}\big(f_1(\alpha), f_2(\alpha), \alpha\big) = 0.$$

This is thus a general phenomenon: if we have a family of curves admitting a multiple point, the locus of these multiple points will appear as a solution of the system of equations of Proposition 2.6.2. □

The first study of the *envelope* of a family of curves was probably due to *Apollonius*, around 220 BC! He studied the envelope of the family of the normals to a parabola. It is hard to believe, but despite being confined to the tools of antiquity, *Apollonius* was able to compute (in the language of the time) the equation of this envelope!

Example 2.6.5 The envelope of the family of the normals to the parabola of equation $y^2 = 2x$ is the so-called *semi-cubic parabola* with equation

$$27y^2 = 8(x-1)^3$$

(see Fig. 2.12).

Proof Write

$$G(x, y) = y^2 - 2x = 0$$

for the Cartesian equation of the parabola.

For a value $y = \alpha$, the corresponding point of the parabola has coordinates

$$\left(\frac{\alpha^2}{2}, \alpha\right).$$

By Proposition 2.4.4, the normal to the parabola at this point admits the equation

$$-\frac{\partial G}{\partial y}\left(\frac{\alpha^2}{2}, \alpha\right)\left(x - \frac{\alpha^2}{2}\right) + \frac{\partial G}{\partial x}\left(\frac{\alpha^2}{2}, \alpha\right)(y - \alpha) = 0$$

that is,

$$F(x, y, \alpha) = 0 \quad \text{with } F(x, y, \alpha) = \alpha\left(x - \frac{\alpha^2}{2}\right) + (y - \alpha).$$

Proposition 2.6.2 tells us that the points of the envelope satisfy the equations

$$\begin{cases} \alpha(x - \frac{\alpha^2}{2}) + (y - \alpha) = 0 \\ (x - \frac{\alpha^2}{2}) - \alpha^2 - 1 = 0. \end{cases}$$

The second equation yields

$$x - \frac{\alpha^2}{2} = \alpha^2 + 1.$$

Introducing this value into the first equation gives first

$$\alpha^3 = -y.$$

The first equation can thus be re-written

$$\frac{3}{2}y = \alpha(1 - x).$$

Taking the third power of each side and dividing by y yields the expected equation

$$27y^2 = 8(x - 1)^3.$$

A quick look at Fig. 2.12 suggests that this should indeed be the envelope, perhaps with a problem of regularity at the point $(0, 1)$. But verifying the conditions in Definition 2.6.2 "from scratch" would be a serious challenge! Fortunately, the considerations of Sect. 2.11 will take care of that. □

The first study of the envelope of an arbitrary family of curves (not just straight lines) was probably due to *Torricelli*, around 1642: this student of *Galileo* has already been mentioned several times in Chap. 1. *Galileo* suggested to *Torricelli* that he should study the region of the space which can be reached by a projectile thrown in all possible directions with a fixed initial speed. All the possible trajectories are parabolas, and—treating the problem in a plane—*Torricelli* concluded that the corresponding envelope is again a parabola (see Fig. 2.13).

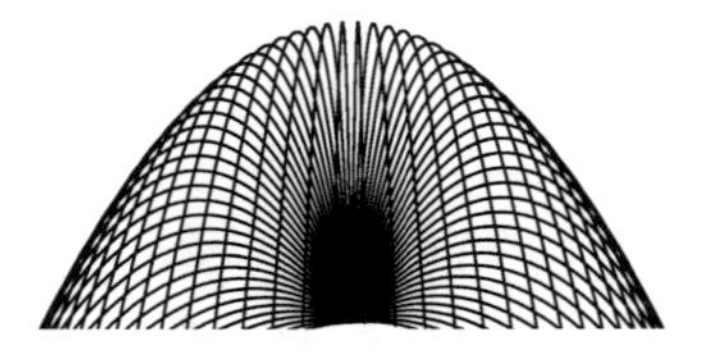

Fig. 2.13

Example 2.6.6 The envelope of the trajectories of a bullet fired by a gun, with a constant initial speed, in a field of constant gravity, in all possible directions in a vertical plane, is a parabola.

Proof Write g for the gravitational force and v_0 for the initial speed of the bullet. If the bullet is fired in a direction making an angle θ with the horizontal, then:

- the horizontal component of the speed is $v_0 \cos\theta$;
- the vertical component of the initial speed is $v_0 \sin\theta$, while the constant vertical acceleration is $-g$; at the instant t the vertical component of the speed will thus be

$$v_0 \sin\theta + \int_0^t -g\,dt = v_0 \sin\theta - gt.$$

The equations of the movement are obtained by integrating these equations for the speed: choosing the gun as origin, i.e. choosing $(0,0)$ as initial values, this yields

$$\begin{cases} x = v_0 t \cos\theta \\ y = v_0 t \sin\theta - g\frac{t^2}{2}. \end{cases}$$

Eliminating t between these two equations immediately gives

$$y = x\tan\theta - \frac{g}{2v_0^2}x^2\bigl(1+\tan^2\theta\bigr)$$

as equation of the trajectory. This is the equation of a parabola.

The *Leibniz* system of equations (see Proposition 2.6.2) is then

$$\begin{cases} y = x\tan\theta - \frac{g}{2v_0^2}x^2(1+\tan^2\theta) \\ 0 = x(1+\tan^2\theta) - \frac{g}{2v_0}2\tan\theta(1+\tan^2\theta). \end{cases}$$

- A first solution of this system is given by $x=0$, that is the y-axis. This curve is not tangent to the various trajectories, thus is not part of the envelope. In fact, this pathology corresponds to the absurd idea of firing the bullet vertically above the gun!
- When $x \neq 0$, the second equation yields

$$x = \frac{v_0^2}{g\tan\theta}$$

and the solution of the system becomes

$$y = \frac{v_0^2}{2g} - \frac{g}{2v_0^2}x^2$$

which is the equation of the parabola discovered by *Torricelli.*

Figure 2.13 provides evidence that this last parabola is indeed the expected envelope! More rigorously, let us check the conditions of Definition 2.6.1. For a given value of x, the angular coefficient of the tangent to the envelope is $-\frac{g}{v_0^2}x$. Comparing this value with the angular coefficient at x of the curve with parameter θ, we conclude that in order to have the same tangent, we need

$$\tan\theta - \frac{g}{v_0^2}x\left(1 + \tan^2\theta\right) = -\frac{g}{v_0^2}x$$

that is $x = \frac{v_0^2}{g\tan\theta}$. It suffices to introduce this change of parameter in the solution of the Leibniz system, to conclude by routine computation that

$$f(\theta) = \left(\frac{v_0^2}{g\tan\theta}, \frac{v_0^2}{2g} - \frac{v_0^2}{2g\tan^2\theta}\right)$$

is an envelope of the given family of parabolas, in the sense of Definition 2.6.1. In fact, since f is not defined for $\theta = \frac{\pi}{2}$, we get *two* envelopes in the sense of Definition 2.6.1: they are represented by f defined, respectively, on the intervals $]0, \frac{\pi}{2}[$ and $]\frac{\pi}{2}, \pi[$. Thus we get the two halves of the parabola discovered by *Torricelli*: once more the point $x = 0$ appears as a singularity of the problem. □

Example 2.6.7 Fronts of waves: the *Tcherenkov* effect in nuclear physics.

Proof With just an ordinary protractor, you can measure speeds close to that of light! Admittedly, you probably need more sophisticated instruments to accelerate a particle to such speeds.

Write c_0 for the speed of light in a vacuum (which is more or less 300,000 km per second). The refraction law allows you, just by measuring an angle, to infer the speed c_1 of light in another transparent medium (see Fig. 2.14)—let us say *glass*—via the formula

$$\frac{c_1}{c_0} = \frac{\sin\theta_1}{\sin\theta_0}.$$

In glass, you will find that the speed of light is more or less 200,000 km per second.

In the vacuum, accelerate a given particle to a very high speed v close to the light speed c_0 and in any case, a speed higher than the speed c_1 of light in the transparent medium.

$$c_1 < v < c_0.$$

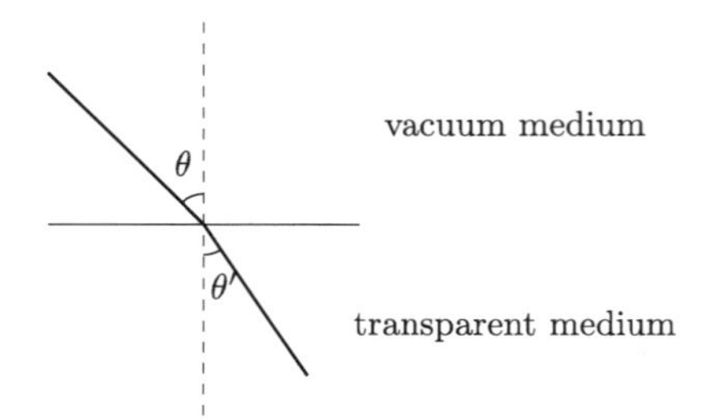

Fig. 2.14

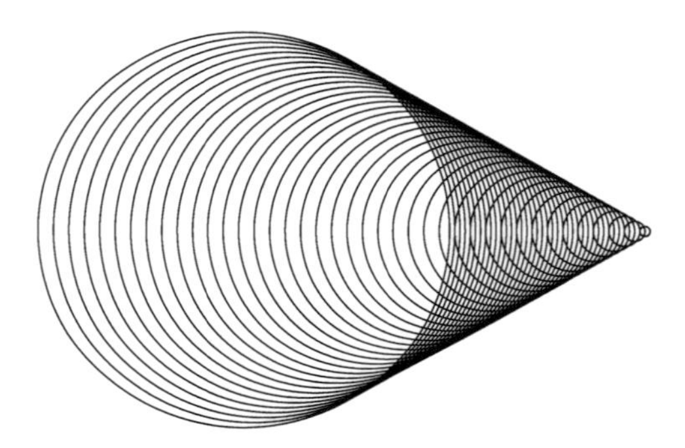

Fig. 2.15

For example, suppose you accelerate the particle to some speed exceeding 250,000 km per second, a speed that you want to measure. To realize this measure, you send the particle through a piece of glass. Because of its mass, its high speed, and thus its considerable inertia, the particle will cross a thin piece of glass almost without being slowed down. Thus it will cross the piece of glass at a speed higher than the speed of light in glass. The *Tcherenkov effect* tells us that this will result in an emission of light by this particle while moving in the piece of glass.

Let $t = 0$ be the instant when the particle enters the piece of glass; let us take its trajectory as the x-axis. At the fixed instant τ, the light wave emitted at a former instant t will be (considering a planar section) a circle of radius $c_1(\tau - t)$ centered at the position $(vt, 0)$ of the particle at the instant t (see Fig. 2.15):

$$(x - vt)^2 + y^2 = c_1^2(\tau - t)^2.$$

The *Leibniz* system of Proposition 2.6.2 for this family of circles indexed by the parameter t is then

$$\begin{cases} (x - vt)^2 + y^2 = v^2(\tau - t)^2 \\ -2v(x - vt) = -2c_1^2(\tau - t). \end{cases}$$

The second equation gives

$$t = \frac{vx - c_1^2\tau}{v^2 - c_1^2}.$$

Introducing this value into the first equation we obtain:

$$y^2 - \frac{c_1^2}{v^2 - c_1^2}(v\tau - x)^2 = 0.$$

This is the equation of two intersecting lines, namely

$$y = \pm\frac{c_1}{\sqrt{v^2 - c_1^2}}(v\tau - x);$$

this formula makes sense since by assumption $v > c_1$. Writing θ for the angle between these lines and the x-axis, we have

$$\tan\theta = \frac{c_1}{\sqrt{v^2 - c_1^2}}.$$

So indeed, the value of v can be obtained from that of c_1 by simply measuring the angle 2θ of the "light cone".

The reader will immediately observe that the expression of t in terms of x also yields

$$x = \frac{t(v^2 - c_1^2) + c_1^2\tau}{v}.$$

Introducing this change of parameter into the equations of the two enveloping lines, we obtain parametric representations of the two pieces of the envelope

$$f\colon]0,\tau[\longrightarrow \mathbb{R}^2, \qquad t \mapsto \left(\frac{t(v^2 - c_1^2) + c_1^2\tau}{v}, \pm\frac{c_1}{v}\sqrt{v^2 - c_1^2}(\tau - t)\right)$$

as required by Definition 2.6.1. □

Of course the considerations of Example 2.6.7 apply to calculate the behaviour of arbitrary waves emitted in a medium by a source moving more rapidly than the waves: for example, the wake of a boat moving rapidly on the water, or the slipstream of a plane flying at a speed higher than *Mach 1*.

The examples already considered in this section certainly justify the name *envelope*: the so-called *envelope curve* actually *envelopes* the given family of curves. However, this is not generally true.

Example 2.6.8 Consider the graph of the function $y = x^3$ and translate it horizontally. The x-axis is an envelope of the corresponding family of curves (see Fig. 2.16).

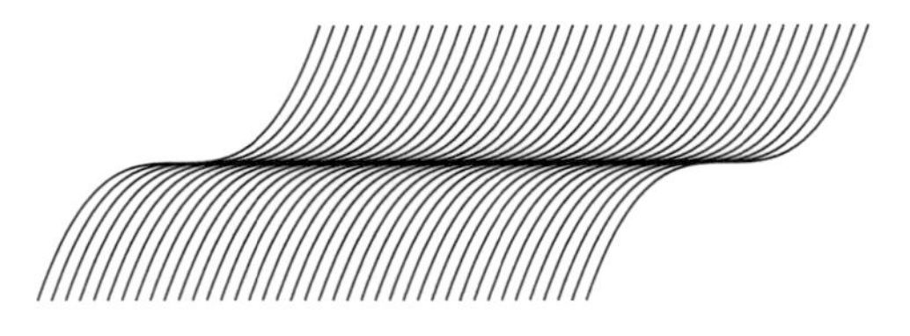

Fig. 2.16

Proof The various curves are thus

$$y = (x - \alpha)^3$$

and admit $x = 0$ as tangent at $(\alpha, 0)$. The x-axis

$$f(\alpha) = (\alpha, 0)$$

is thus an envelope in the precise sense of Definition 2.6.1. □

The study of envelopes is in particular a key ingredient in the theory of stability of ships with respect to the movements of waves.

2.7 The Length of an Arc of a Curve

Most often, *curve integrals* are only defined along an "injective" curve. Let us begin with an important observation concerning the "locally injective case".

Lemma 2.7.1 *Consider a locally injective function of class C^1 defined on an open interval $]a, b[\subseteq \mathbb{R}$:*

$$f :]a, b[\longrightarrow \mathbb{R}^n.$$

For all points $c, d \in]a, b[$, there exist finitely many values

$$c = r_0 < r_1 < \cdots < r_{n-1} < r_n = d$$

such that on each interval $[r_i, r_{i+1}]$ the function f is injective.

Proof By definition, f is continuous on an open neighborhood in $]a, b[$ of each point $t \in [c, d]$; in particular these open neighborhoods cover $[c, d]$. By compactness of $[c, d]$, the covering can already be realized by a finite number of these open neighborhoods. The rest is easy.

We first put $r_0 = c$. The number c belongs to one of the finitely many open neighborhoods: let us say $]u_1, v_1[$. If $d < v_1$, we are done: just put $r_1 = d$. Otherwise consider a second of these open neighborhoods, let us say $]u_2, v_2[$, which contains v_1. Then choose r_1 such that $c < u_2 < r_1 < v_1$. If $d < v_2$ it remains to put $r_2 = d$. Otherwise repeat the process, which will stop after finitely many steps. □

We are thus tempted to make the following definition:

Definition 2.7.2 Consider a locally injective continuous function

$$f :]a, b[\longrightarrow \mathbb{R}^n.$$

Consider further $c, d \in]a, b[$ and a continuous function

$$g : [c, d] \longrightarrow \mathbb{R}.$$

By Lemma 2.7.1, consider values

$$c = r_0 < r_1 < \cdots < r_{n-1} < r_n = d$$

such that on each interval $[r_i, r_{i+1}]$, f is injective. The *curve integral* of g along

$$f : [c, d] \longrightarrow \mathbb{R}^n$$

is the sum of the curve integrals of g along the various

$$f : [r_i, r_{i+1}] \longrightarrow \mathbb{R}^n.$$

This definition makes perfect sense because:

Lemma 2.7.3 *Definition* 2.7.2 *is independent of the choice of the points* r_i.

Proof By Theorem 1.8.2, the curve integral of Definition 2.7.2 is equal to

$$\sum_{i=0}^{n-1} \int_{r_i}^{r_{i+1}} g \cdot \|f'\| = \int_{r_0}^{r_n} g \cdot \|f'\| = \int_c^d g \cdot \|f'\|.$$ □

With the considerations of Sect. 1.8 in mind, we define:

Definition 2.7.4 Consider a parametric representation of class C^1 of a plane curve

$$f :]a, b[\longrightarrow \mathbb{R}^2.$$

The *length* of the arc of curve between the points of parameters c, d, with $a < c < d < b$, is the curve integral of the constant function 1 along

$$f : [c, d] \longrightarrow \mathbb{R}^2.$$

Again by Theorem 1.8.2 we have:

Proposition 2.7.5 *Consider a parametric representation of class* C^1 *of a plane curve*

$$f :]a, b[\longrightarrow \mathbb{R}^2.$$

The length of the arc of the curve between the points with parameters c, d, with $a < c < d < b$, is equal to the absolute value of

$$\int_c^d \|f'\|.$$

It is important to notice that the *length* is a notion depending only on the curve of class $\mathcal{C}^1$ which is considered, not on a particular parametric representation.

Proposition 2.7.6 *Let f, g be two regular parametric representations of class $\mathcal{C}^1$ of a plane curve, equivalent in class $\mathcal{C}^1$ via a change of parameter φ as in Fig.* 2.3. *Given two points*

$$f(t_0) = g\big(\varphi(t_0)\big) = g(s_0), \qquad f(t_1) = g\big(\varphi(t_1)\big) = g(s_1)$$

of this curve, the length of the corresponding arc is the same when computed using f or using g.

Proof From $f = g \circ \varphi$ we deduce $f' = (g' \circ \varphi)\varphi'$. Since $f'(t) \neq 0$ at each point, $\varphi'(t) \neq 0$ at each point. Since φ' is continuous, it is always strictly positive or always strictly negative. We consider the positive case: the negative one is analogous.

$$\int_{t_0}^{t_1} \|f'\| = \int_{t_0}^{t_1} \big\|(g' \circ \varphi)(t)\big\| \cdot \varphi'(t)\,dt = \int_{s_0}^{s_1} \big\|g'(s)\big\|\,ds$$

via the change of variable $s = \varphi(t)$. □

Let us apply Proposition 2.7.5 to some of the curves which have played an important historical role in the problem of "rectification" (see Sect. 1.7).

Example 2.7.7 The length of a circle of radius R is equal to $2\pi R$.

Proof Using the parametric representation

$$f(\theta) = (R\cos\theta, R\sin\theta)$$

we obtain

$$\big\|f'(\theta)\big\| = \big\|(-R\sin\theta, R\cos\theta)\big\| = R.$$

The length of the circle is thus

$$\int_0^{2\pi} R\,d\theta = [R\theta]_0^{2\pi} = R2\pi - R0 = 2\pi R.$$

□

Example 2.7.8 The length of an arc of the logarithmic spiral

$$f(\theta) = \big(a\,e^{k\theta}\cos\theta, a\,e^{k\theta}\sin\theta\big)$$

(see Proposition 1.7.3 and Fig. 1.23).

Proof In this case we have

$$\begin{aligned}\|f'(\theta)\| &= \|(ka\,e^{k\theta}\cos\theta - a\,e^{k\theta}\sin\theta, ka\,e^{k\theta}\sin\theta + a\,e^{k\theta}\cos\theta)\| \\ &= a\sqrt{1+k^2}\,e^{k\theta}.\end{aligned}$$

The length of an arc starting from the origin is thus equal to

$$\int_0^\theta \sqrt{1+k^2}\,e^{k\theta} = \left[a\sqrt{\frac{1+k^2}{k^2}}\,e^{k\theta}\right]_0^\theta = a\sqrt{\frac{1+k^2}{k^2}}\left(e^{k\theta}-1\right). \qquad \square$$

Example 2.7.9 The length of an arc of the semi-cubic parabola $y^2 = x^3$ (see Fig. 1.24).

Proof It is immediate that

$$f(t) = \left(t^2, t^3\right)$$

is a parametric representation of the semi-cubic parabola, with the cusp point $(0,0)$ as a singular point. We then have

$$\|f'(t)\| = \|\left(2t, 3t^2\right)\| = \sqrt{4t^2+9t^4}.$$

The length of an arc starting from the cusp point is thus equal to

$$\int_0^{\sqrt{x}} \sqrt{4t^2+9t^4}\,dt = \left[\frac{1}{27}\left(4+9t^2\right)^{\frac{3}{2}}\right]_0^{\sqrt{x}} = \frac{(4+9x)^{\frac{3}{2}}-8}{27}. \qquad \square$$

Example 2.7.10 The length of an arch of a cycloid generated by a circle of radius R is equal to $8R$ (see Definition 1.9.1 and Fig. 1.27).

Proof By Proposition 1.9.2, a parametric representation of the cycloid is given by

$$f(\theta) = R(\theta - \sin\theta, 1 - \cos\theta).$$

This yields

$$\|f'(\theta)\| = \|R(1-\cos\theta, \sin\theta)\| = R\sqrt{2-2\cos\theta} = 2R\sin\frac{\theta}{2}.$$

The length of an arch is thus equal to

$$\int_0^{2\pi} 2R \sin\frac{\theta}{2}\, d\theta = \left[-4R\cos\frac{\theta}{2}\right]_0^{2\pi} = 4R + 4R = 8R. \qquad \square$$

So the length of an arch of a cycloid can be expressed as a rational quantity in terms of the radius, while the length of a circle—an *a priori* simpler curve—cannot. *Torricelli*, who found this very simple formula $8R$ for the cycloid around 1640, had certainly been quite amazed by his own result.

2.8 Normal Representation

We shall now investigate the particular properties of a parametric representation where the parameter is the "length traveled on the curve from a fixed arbitrary origin".

Definition 2.8.1 By a *normal* representation of a plane curve is meant a parametric representation of class $\mathcal{C}^1$

$$f\colon]a,b[\longrightarrow \mathbb{R}^2, \qquad s \mapsto f(s)$$

such that $\|f'(s)\| = 1$ for each $s \in]a,b[$.

It follows immediately from Definition 2.2.1 that a normal representation is in particular regular. Choosing the "length traveled on the curve" (see Proposition 2.7.5) as a parameter, we obtain a normal representation:

Proposition 2.8.2 *Consider a regular representation of class $\mathcal{C}^k$ of a plane curve*

$$f\colon]a,b[\longrightarrow \mathbb{R}^2, \qquad t \mapsto f(t).$$

Given $t_0 \in]a,b[$, the function

$$\sigma\colon]a,b[\longrightarrow \big]\sigma(a),\sigma(b)\big[, \qquad \sigma(t) = \int_{t_0}^{t} \|f'\|$$

is a change of parameter of class $\mathcal{C}^k$. The corresponding parametric representation

$$\overline{f} = f \circ \sigma^{-1}\colon \big]\sigma(a),\sigma(b)\big[\longrightarrow \mathbb{R}^2$$

is normal of class $\mathcal{C}^k$.

Proof Of course, in this statement,

$$\sigma(a) \in \mathbb{R} \cup \{-\infty\}, \qquad \sigma(b) \in \mathbb{R} \cup \{+\infty\}.$$

Since f is regular (see Definition 2.2.1), $\sigma' = \|f'\| > 0$ and thus σ is a strictly increasing function: as a consequence, it is a bijection.

This bijection is of class $\mathcal{C}^k$ since it has a derivative $\|f'\|$ which is a function of class $\mathcal{C}^{k-1}$

$$\sqrt{(f_1')^2 + (f_2')^2}$$

again because each f_i' is of class $\mathcal{C}^{k-1}$ and the argument of the square root is never zero.

The inverse bijection σ^{-1} admits the first derivative

$$(\sigma^{-1})' = \frac{1}{\sigma' \circ \sigma^{-1}},$$

which makes sense, since $\sigma' = \|f'\| \neq 0$. Next

$$(\sigma^{-1})'' = -\frac{(\sigma'' \circ \sigma^{-1})(\sigma^{-1})'}{(\sigma' \circ \sigma^{-1})^2} = -\frac{\sigma'' \circ \sigma^{-1}}{(\sigma' \circ \sigma^{-1})^3}$$

and so on. So σ^{-1} itself is of class $\mathcal{C}^k$ and therefore, $\overline{f} = f \circ \sigma^{-1}$ is of class $\mathcal{C}^k$.

The representation $\overline{f}$ is normal because

$$\overline{f}' = (f \circ \sigma^{-1})' = (f' \circ \sigma^{-1}) \cdot (\sigma^{-1})' = \frac{f' \circ \sigma^{-1}}{\|f' \circ \sigma^{-1}\|}.$$

Trivially, this quantity has norm 1. □

Proposition 2.8.3 *Consider a normal representation of class* $\mathcal{C}^2$

$$f\colon]a,b[\longrightarrow \mathbb{R}^2, \qquad s \to f(s)$$

of a plane curve. Then $(f'|f'') = 0$, *that is* f'' *is orthogonal to* f' *(or is equal to* 0*) at each point of the curve.*

Proof We have $(f'|f') = 1$ by Definition 2.8.1. Then $2(f'|f'') = 0$ by Lemma 1.11.3. □

Proposition 2.8.2 admits—in a sense—a converse statement: *in a normal representation the parameter—up to a possible translation of the origin—is necessarily the length traveled on the curve.*

Proposition 2.8.4 *Let*

$$f\colon]a,b[\longrightarrow \mathbb{R}^2, \qquad s \mapsto f(s)$$

be a normal representation of a plane curve. For every two values $a < s_0 < s_1 < b$ *of the parameter,* $s_1 - s_0$ *is equal to the length of the arc of the curve between the points with parameters* s_0 *and* s_1.

Proof By Proposition 2.7.5 and Definition 2.8.1, the length between the points with parameters s_0, s_1 is equal to

$$\int_{s_0}^{s_1} \|f'\| = \int_{s_0}^{s_1} 1 = [s]_{s_0}^{s_1} = s_1 - s_0. \qquad \square$$

Corollary 2.8.5 *Two normal representations $f(t)$ and $g(s)$ of the same regular curve differ only by a change of parameter of the form $s = \pm t + k$, where k is a constant.*

Proof We use the notation of Definition 2.1.6. Differentiating the equality $f = g \circ \varphi$ we obtain $f' = (g' \circ \varphi) \cdot \varphi'$. Taking the norms of both sides we obtain, by Definition 2.8.1

$$1 = \|f'\| = \|(g' \circ \varphi) \cdot \varphi'\| = \|g' \circ \varphi\| \cdot |\varphi'| = 1 \cdot |\varphi'| = |\varphi'|.$$

Thus φ is a real function of a real variable such that $\varphi' = \pm 1$. Integrating this equality yields the expected result. $\square$

Example 2.8.6 The usual parametric representation of a circle with center (a, b) and of radius 1

$$f(\theta) = (a + \cos\theta, b + \sin\theta)$$

is a normal one. But a normal representation of a circle of radius R and center (a, b) is

$$g(\theta) = \left(a + R\cos\frac{\theta}{R}, b + R\sin\frac{\theta}{R}\right).$$

Proof Simply observe that $\|f'\| = 1$ and $\|g'\| = 1$. $\square$

2.9 Curvature

At last we can now study of the main characteristic of a curve: the fact of being *curved*!

The derivative of a function somehow "measures" the variation of this function. Consider for example a regular parametric representation of class $\mathcal{C}^2$ of a plane curve

$$f\colon]a, b[\longrightarrow \mathbb{R}^2.$$

The function

$$f'\colon]a, b[\longrightarrow \mathbb{R}^2$$

yields at each point a vector $f'(t)$ tangent to the curve (see Definition 2.4.2). When t varies, this tangent vector $f'(t)$ varies both in length and direction. But when f

is a normal representation, $f'(t)$ has constant length 1 (see Definition 2.8.1) and its variation is thus only in direction. In that case, $f''(t)$ "measures" the variation in direction of the tangent vector, that is, the variation in direction of the tangent line. The "extent" of this variation is what we shall call the *curvature* of the curve.

Definition 2.9.1 Let

$$f:]a,b[\longrightarrow \mathbb{R}^2, \qquad s \mapsto f(s)$$

be a normal representation of class $\mathcal{C}^2$ of a plane curve. The *curvature* at the point with parameter s is the quantity $\|f''(s)\|$.

For Definition 2.9.1 to make sense, we must of course observe that:

Lemma 2.9.2 *Two normal representations of a given plane curve of class $\mathcal{C}^2$ admit the same second derivative at each point. In particular the definition of the curvature of a plane curve is independent of the chosen normal parametric representation.*

Proof Given two normal representations as in Corollary 2.8.5, we have a corresponding change of parameter $\varphi(t) = \pm t + k$, thus

$$\begin{aligned}
f &= g \circ \varphi \\
f' &= (g' \circ \varphi)\varphi' = (g' \circ \varphi)(\pm 1) \\
f'' &= (g'' \circ \varphi)(\pm 1)\varphi' = (g'' \circ \varphi)(\pm 1)^2 = g'' \circ \varphi.
\end{aligned}$$

This is the expected result. □

The following result allows us to calculate the curvature more easily:

Proposition 2.9.3 *Let $f(t)$ be an arbitrary parametric representation of class $\mathcal{C}^2$ of a plane curve. The curvature is equal to*

$$\kappa = \frac{|f_1' f_2'' - f_2' f_1''|}{\|f'\|^3}.$$

Proof We use the notation of Proposition 2.8.2 and the various observations made in its proof. We have $\kappa = \|\overline{f}''\|$ by Definition 2.9.1, and furthermore

$$\begin{aligned}
\overline{f} &= f \circ \sigma^{-1} \\
\overline{f}' &= (f' \circ \sigma^{-1})(\sigma^{-1})' \\
\overline{f}'' &= (f'' \circ \sigma^{-1})((\sigma^{-1})')^2 + (f' \circ \sigma^{-1})(\sigma^{-1})''.
\end{aligned}$$

The vector $\overline{f}'$ is thus equal to

$$\overline{f}' = (\sigma^{-1})'(f_1' \circ \sigma^{-1}, f_2' \circ \sigma^{-1}).$$

Therefore the vector

$$n = (\sigma^{-1})'(-f_2' \circ \sigma^{-1}, f_1' \circ \sigma^{-1})$$

is a vector of length 1 in the direction perpendicular to $\overline{f}'$, that is, a vector in the direction of the normal to the curve. But by Proposition 2.8.3, this direction is also that of the vector $\overline{f}''$. Therefore

$$(n|\overline{f}'') = \|n\| \cdot \|\overline{f}''\| \cdot \cos\measuredangle(n, \overline{f}'') = \pm\|\overline{f}''\| = \pm\kappa.$$

Thus we also have

$$\begin{aligned}\kappa &= \left|(n|\overline{f}'')\right| \\ &= \left|(n|(f'' \circ \sigma^{-1})((\sigma^{-1})')^2 + (f' \circ \sigma^{-1})(\sigma^{-1})'')\right| \\ &= \left|(n|(f'' \circ \sigma^{-1})((\sigma^{-1})')^2)\right|\end{aligned}$$

since n is orthogonal to f'.

We conclude that

$$\begin{aligned}\kappa &= \left|(n|(f'' \circ \sigma^{-1})((\sigma^{-1})')^2)\right| \\ &= (\sigma^{-1})'^3 \cdot \left|((-f_2' \circ \sigma^{-1}, f_1' \circ \sigma^{-1})|(f_1'' \circ \sigma^{-1}, f_2'' \circ \sigma^{-1}))\right| \\ &= \frac{|((-f_2' \circ \sigma^{-1}, f_1' \circ \sigma^{-1})|(f_1'' \circ \sigma^{-1}, f_2'' \circ \sigma^{-1}))|}{\|f' \circ \sigma^{-1}\|^3} \\ &= \frac{-(f_2' \circ \sigma^{-1})(f_1'' \circ \sigma^{-1}) + (f_1' \circ \sigma^{-1})(f_2'' \circ \sigma^{-1})}{\|f'\|^3}.\end{aligned}$$

In terms of the parameter $t = \sigma^{-1}(s)$, this is precisely the formula of the statement. □

Let us begin with the most obvious examples:

Example 2.9.4 The curvature of a straight line is everywhere zero.

Proof A parametric representation of a line has the form

$$f(t) = (a + bt, c + dt), \quad a, b, c, d \in \mathbb{R}.$$

Thus $f'(t) = (b, d)$ and $f''(t) = (0, 0)$. The result follows by Proposition 2.9.3. □

Example 2.9.5 The curvature of a circle of radius R is constant and equal to the inverse $\frac{1}{R}$ of the radius.

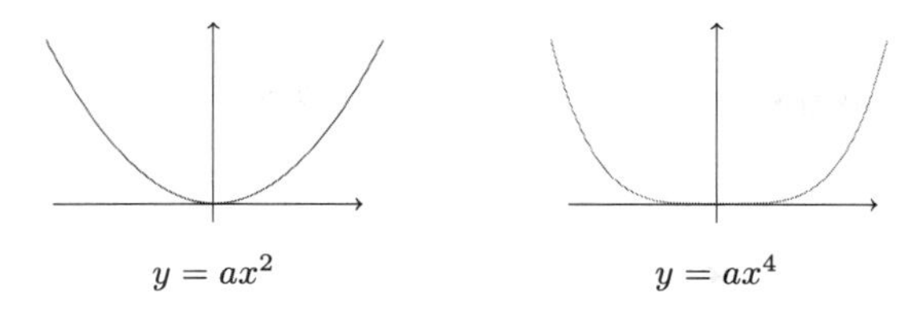

Fig. 2.17

Proof With the notation of Example 2.8.6 we get

$$g'(\theta) = \left(-\sin\frac{\theta}{R}, \cos\frac{\theta}{R}\right), \qquad g''(\theta) = \left(-\frac{1}{R}\cos\frac{\theta}{R}, -\frac{1}{R}\sin\frac{\theta}{R}\right)$$

from which $\|g''(\theta)\| = \frac{1}{R}$. □

One should be aware that "not being straight" does not mean "being curved". The following example (see Fig. 2.17) illustrates the limits of our intuition.

Example 2.9.6 At the origin, the parabola of equation $y = ax^2$ has curvature $2a$, while the graph of the function $y = ax^4$ has curvature 0.

Proof Just apply Proposition 2.9.3. □

Example 2.9.7 At its "vertex" $(0, b)$, the curvature of the ellipse with equation

$$\frac{x^2}{a^2} + \frac{y^2}{b^2} = 1$$

is equal to $\frac{b}{a^2}$.

Proof A parametric representation of the ellipse is given by (see Example 2.1.4)

$$f(\theta) = (a\cos\theta, b\sin\theta).$$

It follows at once that

$$f'(\theta) = (-a\sin\theta, b\cos\theta), \qquad f''(\theta) = (-a\cos\theta, -b\sin\theta).$$

The point $(0, b)$ corresponds to the value $\theta = \frac{\pi}{2}$, thus

$$f'\left(\frac{\pi}{2}\right) = (-a, 0), \qquad f''\left(\frac{\pi}{2}\right) = (0, -b).$$

The result then follows immediately from Proposition 2.9.3. □

Let us conclude this section with the observation that in the plane, a curve can bend away from the tangent on one side or the other in Fig. 2.18.

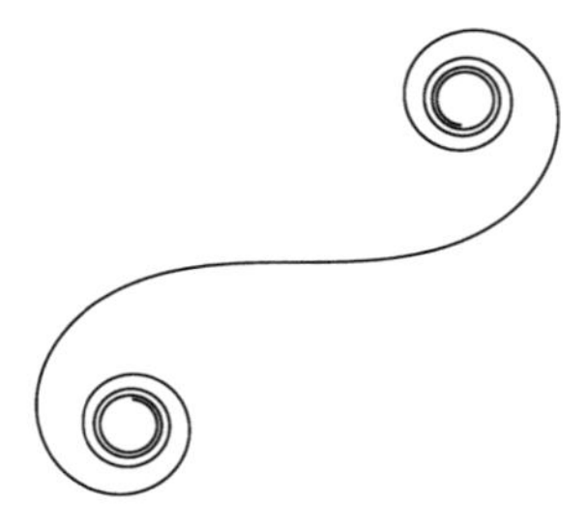

Fig. 2.18

For that reason, for various purposes, it is useful to give a sign to the curvature.

Definition 2.9.8 Let $f(s)$ be a normal representation of class $\mathcal{C}^2$ of a plane curve; write $\kappa(s)$ for its curvature at the point with parameter s. The *relative curvature* with respect to this normal representation f is the quantity

- 0 if $f''(s) = 0$;
- $+\kappa(s)$ if $(f'(s), f''(s))$ is a basis having the same orientation as the canonical basis of $\mathbb{R}^2$;
- $-\kappa(s)$ if $(f'(s), f''(s))$ is a basis having the orientation opposite to that of the canonical basis of $\mathbb{R}^2$.

Let us recall that two bases have the same orientation when the change of base matrix has a positive determinant (see Sect. 3.2 in [4], *Trilogy II*).

Warning 2.9.9 *The sign of the curvature, as in Definition* 2.9.8, *is by no means an intrinsic property of the curve*: *it depends on the chosen normal representation.*

Proof In the proof of Lemma 2.9.2 (and with its notation) we have indeed observed that, given two normal representations of the same curve

$$f' = \pm(g' \circ \varphi), \qquad f'' = g'' \circ \varphi.$$

So when the sign -1 applies in the first equality, the two bases (f', f'') and (g', g'') have opposite orientations at a point where the curvature does not vanish.

As an example, consider the "counter-clockwise" representation of a circle

$$f(\theta) = (R\cos\theta, R\sin\theta).$$

The first and second derivatives are

$$f'(\theta) = (-R\sin\theta, R\cos\theta), \qquad f''(\theta) = (-R\cos\theta, -R\sin\theta).$$

The change of basis matrix with respect to the canonical basis is then

$$\begin{pmatrix} -R\sin\theta & -R\cos\theta \\ R\cos\theta & -R\sin\theta \end{pmatrix}$$

whose determinant is equal to R^2. The relative curvature of this circle is thus equal to $\frac{1}{R}$ (see Example 2.9.5).

Now consider the same circle described in "clockwise" orientation

$$g(\theta) = (R\cos\theta, -R\sin\theta).$$

The determinant of the change of base matrix is now equal to $-R^2$ and therefore the relative curvature is equal to $-\frac{1}{R}$. □

Nevertheless, let us make clear that:

Proposition 2.9.10 *Consider two normal representations $f(t)$, $g(s)$ of class C^2 of a plane curve. One of the following two possibilities holds:*

- *both representations yield the same relative curvature at all points;*
- *the two representations yield opposite relative curvatures at all points.*

Proof By Proposition 2.8.5, the corresponding change of parameter has the form $s = \pm t + k$, which yields $g'(s) = \pm f'(t)$ at a given point of the curve. On the other hand $g''(s) = f''(t)$ by Lemma 2.9.2. □

Proposition 2.9.11 *Consider a parametric representation $f(t)$ of class C^2 of a plane curve and write $\overline{f}(s)$ for the corresponding normal representation as in Proposition 2.8.2. When the curvature is non-zero, both bases (f', f'') and $(\overline{f}', \overline{f}'')$ have the same orientation and the relative curvature, expressed in terms of the parametric representation f, is given by*

$$\kappa = \frac{f_1' f_2'' - f_2' f_1''}{\|f'\|^3}.$$

Proof By Proposition 2.9.3, the curvature is given by the absolute value of the quantity in the statement. This quantity is positive precisely when the determinant

$$\det\begin{pmatrix} f_1' & f_1' \\ f_1'' & f_2'' \end{pmatrix}$$

is positive. Considering the normal representation $\overline{f} = f \circ \sigma^{-1}$ as in Proposition 2.8.2, the corresponding determinant is that of the matrix

$$\begin{pmatrix} (f_1' \circ \sigma^{-1})(\sigma^{-1})' & (f_2' \circ \sigma^{-1})(\sigma^{-1})' \\ (f_1'' \circ \sigma^{-1})(\sigma^{-1})'^2 + (f_1'\sigma^{-1})(\sigma^{-1})'' & (f_2'' \circ \sigma^{-1})(\sigma^{-1})'^2 + (f_2'\sigma^{-1})(\sigma^{-1})'' \end{pmatrix}$$

which is thus equal to

$$(\sigma^{-1})'^3 \big((f_1' \circ \sigma^{-1})(f_2'' \circ \sigma^{-1}) - (f_2' \circ \sigma^{-1})(f_1'' \circ \sigma^{-1})\big).$$

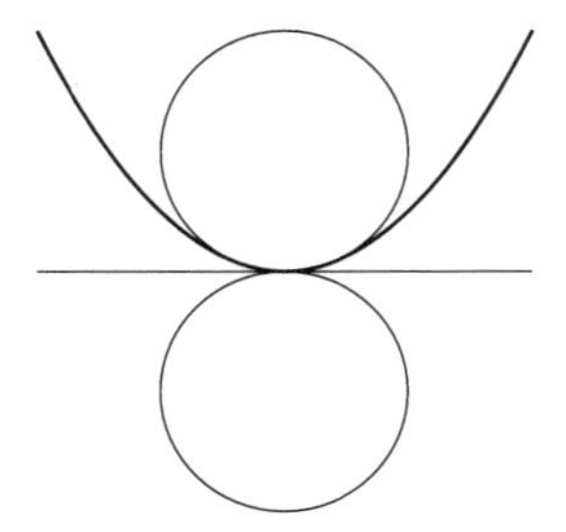

Fig. 2.19

Since

$$(\sigma^{-1})' = \frac{1}{\sigma' \circ \sigma^{-1}} = \frac{1}{\|f' \circ \sigma^{-1}\|} \geq 0$$

we conclude that both determinants are simultaneously positive or negative. Provided that the curvature is not zero, both bases (f', f'') and $(\overline{f}', \overline{f}'')$ thus have the same orientation. The determinant in the case of the normal representation is precisely the relative curvature, since $\|\overline{f}'\| = 0$ (see Definition 2.8.1). □

So indeed, if the sign of the relative curvature is not relevant (see Counterexample 2.9.9), the changes of sign of the relative curvature are highly significant.

2.10 Osculating Circle

With the example of the circle in mind (see Example 2.9.5), we are tempted to "approximate" a curve at a given point by the circle passing through that point and having the same tangent and the same curvature. However, one should be aware that there are two such circles: one on each side of the tangent! (see Fig. 2.19). Lemma 2.10.1 helps us to make the "right" choice.

Lemma 2.10.1 *Consider a normal representation $f(s)$ of class C^2 of a plane curve. Given on this curve a point $f(s_0)$ with non-zero curvature, there exist two circles with normal representation $g(\theta)$ such that:*

1. *each circle passes through the point $f(s_0)$, for a value θ_0 of the parameter;*
2. *the curve and each circle have the same tangent at $f(s_0) = g(\theta_0)$;*
3. *the curve and each circle have the same curvature $\kappa(s)$ at $f(s_0) = g(t_0)$.*

For one of these two circles, one has further

$$f''(s_0) = g''(\theta_0)$$

while for the other one, $f''(s_0) = -g''(s_0)$. The center of the first circle is the point

$$f(s_0) + \frac{1}{\kappa(s_0)^2} f''(s_0).$$

Proof Let us first recall that for a normal representation of class $\mathcal{C}^2$, the value of the second derivative is independent of the chosen representation (see Lemma 2.9.2).

The radius of a circle is perpendicular to its tangent. Thus by Proposition 2.8.3, the center of a circle as in the statement must be of the form $f(s_0) + k\, f''(s_0)$, with k some constant.

The radius of such a circle is then

$$R = \left\| k\, f''(s_0) \right\| = |k| \cdot \left\| f''(s_0) \right\| = |k| \cdot \kappa(s_0).$$

Since the circle and the curve must have the same curvature. Example 2.9.5 yields

$$\frac{1}{R} = \kappa(s_0).$$

Therefore

$$R = \frac{1}{\kappa(s_0)}, \qquad k = \frac{\pm 1}{\kappa(s_0)^2}.$$

The normal representations of the two circles corresponding to the two possible values of k, and the derivatives of these, are thus (see Example 2.8.6)

$$\begin{aligned} g(\theta) = \Bigg(& f_1(s_0) \pm \frac{f_1''(s_0)}{\kappa(s_0)^2} + \frac{1}{\kappa(s_0)} \cos\big(\kappa(s_0)\theta\big), \\ & f_2(s_0) \pm \frac{f_2''(s_0)}{\kappa(s_0)^2} + \frac{1}{\kappa(s_0)} \sin\big(\kappa(s_0)\theta\big)\Bigg) \\ g'(\theta) = \big(& -\sin\big(\kappa(s_0)\theta\big), \cos\big(\kappa(s_0)\theta\big)\big) \\ g''(\theta) = \big(& -\kappa(s_0)\cos\big(\kappa(s_0)\theta\big), -\kappa(s_0)\sin\big(\kappa(s_0)\theta\big)\big). \end{aligned}$$

The first equality, together with the requirement $f(s_0) = g(\theta_0)$, and the last equality, together with the requirement $f''(s_0) = g''(\theta_0)$, then yield

$$\mp \frac{f''(s_0)}{\kappa(s_0)} = \big(\cos(\kappa(s_0)\theta), \sin(\kappa(s_0)\theta)\big) = -\frac{f''(s_0)}{\kappa(s_0)}.$$

This forces the choice of the sign $-$, which comes from the sign $+$ in the description of g. □

With Example 2.9.5 in mind, we then complete our terminology in the following way.

Definition 2.10.2 Consider a normal representation $f(s)$ of class $\mathcal{C}^2$ of a plane curve. Given on this curve a point $f(s_0)$ with non-zero curvature:

1. the *radius of curvature* at the point $f(s_0)$ is the quantity $\frac{1}{\kappa(s_0)}$, where $\kappa(s_0)$ is the curvature;
2. the *center of curvature* at the point $f(s_0)$ is the point

$$f(s_0) + \frac{1}{\kappa(s_0)^2} f''(s_0);$$

3. the circle centered at the center of curvature, with radius the radius of curvature, is called the *osculating circle* to the curve at the point $f(s_0)$.

2.11 Evolutes and Involutes

In view of Definition 2.10.2, we make the following definition:

Definition 2.11.1 Consider a regular curve of class $\mathcal{C}^2$ whose curvature is non-zero at each point. The locus of the centers of curvature is called the *evolute* of the curve.

Of course there is no reason that the *evolute* of a curve will be a curve: for example, the evolute of a circle is reduced to a point, namely the center of the circle!

Proposition 2.11.2 *Consider a normal representation*

$$f\colon]a,b[\longrightarrow \mathbb{R}^2, \qquad s \mapsto f(s)$$

of class $\mathcal{C}^2$ of a plane curve, whose curvature $\kappa(s)$ is non-zero at each point. When the function

$$f + \frac{f''}{\kappa^2}\colon]a,b[\longrightarrow \mathbb{R}^2, \qquad s \mapsto f(s) + \frac{f''(s)}{\kappa(s)^2}$$

is locally injective, it is a parametric representation of the evolute of the curve represented by f.

Proof Since f'' is continuous, so is the function of the statement and, by assumption, $f''(s) \neq 0$ for each s. The result follows by Definition 2.10.2 and Lemma 2.10.1. □

Here are two important properties of the evolute, when the evolute turns out to be a curve.

Proposition 2.11.3 *Consider a normal representation $f(s)$ of class $\mathcal{C}^3$ of a plane curve. Assume that the curvature is non-zero at each point and that the function*

$$g(s) = f(s) + \frac{f''(s)}{\kappa(s)^2} f''(s)$$

whose support is the evolute, is a regular parametric representation (see Proposition 2.11.2). Under these conditions, the evolute is the envelope of the family of all normals to the curve.

Proof In view of Definition 2.6.1, we must prove that for a given parameter s, the tangent to the evolute at $g(s)$ is normal to the original curve at $f(s)$.

The normal to the curve admits the equation (see Proposition 2.4.3)

$$f_1'(s)\big(x - f_1(s)\big) + f_2'(s)\big(y - f_2(s)\big) = 0.$$

Proving that the point $g(s)$ satisfies this equation reduces to

$$f_1'(s)\frac{f_1''(s)}{\kappa(s)^2} + f_2'(s)\frac{f_2''(s)}{\kappa(s)^2} = 0.$$

This reduces further to $(f'|f'') = 0$, which is the case by Proposition 2.8.3.

Differentiating the equality $(f'|f'') = 0$ yields further

$$(f''|f'') + (f'|f''') = 0$$

(see Lemma 1.11.3). The direction of the tangent to the evolute is

$$g' = f' - 2\frac{\kappa'}{\kappa^3}f'' + \frac{f'''}{\kappa^2}.$$

Again since f' and f'' are perpendicular (Proposition 2.8.3), the scalar product with f' is then simply

$$(f'|f') + \frac{(f'|f''')}{\kappa^2} = (f'|f') - \frac{(f''|f'')}{\kappa^2} = 0.$$

Thus the tangent to the evolute is indeed the normal to the curve. □

Proposition 2.11.4 *Consider a normal representation $f(s)$ of class $\mathcal{C}^3$ of a plane curve. Assume that the curvature is non-zero at each point and is a monotone function on the interval considered. Assume further that the function*

$$g(s) = f(s) + \frac{1}{\kappa(s)^2}f''(s)$$

whose support is the evolute is a regular parametric representation (see Proposition 2.11.2). Under these conditions, the difference between the radii of curvature at two points with parameters s_1 and s_2 equals the length of the arc of the evolute between the points $g(s_1)$ and $g(s_2)$.

Proof Definition 2.8.1, Proposition 2.8.3 and the proof of Proposition 2.11.3 already imply that

$$(f'|f') = 1, \qquad (f'|f'') = 0, \qquad (f''|f'') = \kappa^2, \qquad (f'|f''') = -\kappa^2.$$

Differentiating the third equality yields further

$$2(f''|f''') = 2\kappa \cdot \kappa'.$$

By Proposition 2.8.3, the basis constituted of f' and $\frac{f''}{\kappa}$ is orthonormal. We thus obtain, by Proposition 4.6.2 in [4], *Trilogy II*:

$$f''' = (f'''|f')f' + \left(f'''\Big|\frac{f''}{\kappa}\right)\frac{f''}{\kappa} = -\kappa^2 f' + \frac{\kappa'}{\kappa} f''.$$

This yields further

$$\begin{aligned} g' &= f' - 2\frac{\kappa'}{\kappa^3} f'' + \frac{1}{\kappa^2} f''' \\ &= f' - 2\frac{\kappa'}{\kappa^3} f'' - f' + \frac{\kappa'}{\kappa^3} f'' \\ &= -\frac{\kappa'}{\kappa^3} f''. \end{aligned}$$

Thus

$$\|g'\| = \left|\frac{\kappa'}{\kappa^2}\right|.$$

Since by assumption the curvature κ is a monotone function, its derivative has constant sign; let us say, is positive (the negative case is analogous). Then

$$\int \|g'\| = \int \frac{\kappa'}{\kappa^2} = -\frac{1}{\kappa}.$$

The rest of the proof is now easy. The length of the arc of the evolute is equal to (see Proposition 2.7.5)

$$\int_{s_1}^{s_2} \|g'\| = \left[\frac{-1}{\kappa}\right]_{s_1}^{s_2} = \frac{1}{\kappa(s_1)} - \frac{1}{\kappa(s_2)}$$

that is, by Definition 2.10.2, the difference between the two radii of curvature. □

This last result admits an interesting practical interpretation, illustrated in Fig. 2.20.

> View the tangent at the point $g(s)$ as a chord rolling along the evolute with extremity at $f(s)$. As this chord rolls or unrolls further along the evolute, the extremity of the chord describes the curve $f(s)$.

It is this property that *Huygens* discovered in the case of the cycloid, in view of constructing a pendulum whose frequency is independent of the amplitude of the oscillations (see Sect. 1.9).

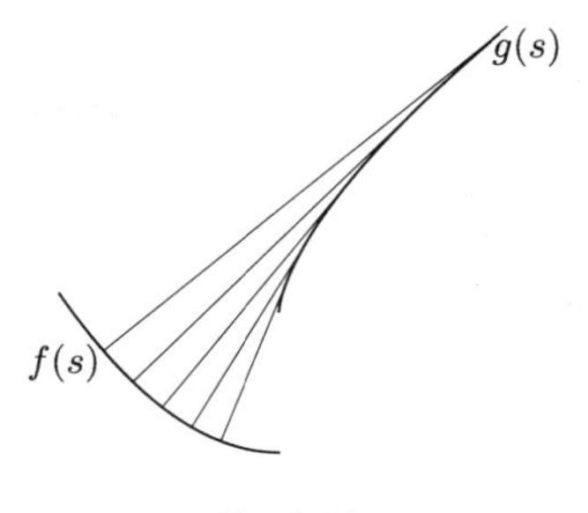

Fig. 2.20

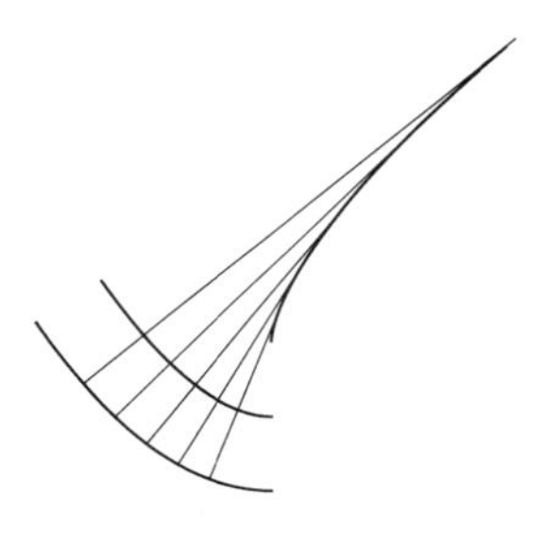

Fig. 2.21

Example 2.11.5 The evolute of a parabola is a semi-cubic parabola (see Fig. 2.12).

Proof This follows by Example 2.6.5 and Proposition 2.11.3. □

Example 2.11.6 The evolute of a cycloid is another cycloid of the same size (see Fig. 1.29).

Proof This follows from Proposition 2.11.3, via Proposition 1.9.6, which has been proved via "historical" methods. □

Let us conclude with another point of terminology.

Definition 2.11.7 By an *involute* of a curve $g(s)$ is meant a curve $f(s)$ admitting $g(s)$ as evolute.

It should be observed that a given curve can admit many involutes. In fact, in the "practical interpretation" above, it suffices to modify the "length of the chord" to get all possible involutes (see Fig. 2.21). More precisely:

Example 2.11.8 Under the conditions of Proposition 2.11.4, consider the vector $n(s) = (f_2'(s), -f_1'(s))$ of length 1 normal to the curve. For every constant $k \in \mathbb{R}$

such that

$$h(s) = f(s) + k\,n(s)$$

is still a regular parametric representation of a curve, that curve still admits $g(s)$ as evolute.

Proof It suffices to check that the tangent to $g(s)$ is still normal to $h(s)$, that is $(h'|g') = 0$. But differentiating $(n|n) = 1$ yields $2(n'|n) = 0$. Since n is parallel to g', this also proves $(g'|n') = 0$. Then

$$\big(h'\big|g'\big) = \big(f' + kn'\big|g'\big) = \big(f'\big|g'\big) + k\big(n'\big|g'\big) = 0 + k\,0 = 0. \qquad \square$$

2.12 Intrinsic Equation of a Plane Curve

When we speak of an ellipse with axes of lengths 2 and 3, we know what we mean, even if we are unable to give a precise equation for this ellipse from the information given. Indeed we did not say where the center of this ellipse lies nor in which directions the axis are oriented.

The idea of an *intrinsic equation* of a curve is precisely this: being able to describe a curve "intrinsically", without having to refer to a precise position of the curve in the plane. To this end we shall follow another simple idea: if we want to give to an iron wire the shape of a precise curve, it suffices to know how to "curve" this wire at each point. Thus the knowledge of the curvature at each point should be sufficient to describe the curve.

Of course going back to Fig. 2.18, we immediately realize that it is important to know when we have to curve the wire "one way or the other with respect to the tangent". We shall thus work with the *relative curvature* as in Definition 2.9.8.

We first need a lemma.

Lemma 2.12.1 *Consider a function of class* $\mathcal{C}^1$

$$f: I \longrightarrow \mathbb{R}^2 \quad (I \subseteq \mathbb{R} \textit{ generalized interval; see Example A.10.9})$$

with the property that $\|f(t)\| = 1$ *for all* $t \in I$. *Then there exists a function*

$$\theta: I \longrightarrow \mathbb{R}$$

of class $\mathcal{C}^1$ *such that for all* $t \in [a,b]$

$$f(t) = \big(\cos\theta(t), \sin\theta(t)\big).$$

Moreover, given two such continuous functions θ_1 *and* θ_2, *there exists an integer* k *such that for all* $t \in [a,b]$

$$\theta_2(t) = \theta_1(t) + 2k\pi.$$

Proof Of course since $\|f(t)\| = 1$, we have

$$f(t) = \big(\cos\theta(t), \sin\theta(t)\big)$$

for some angle $\theta(t)$. But the angle θ is only defined up to a multiple of 2π and the whole point is to make sure that a choice can be made which forces θ to be continuous and even, of class $\mathcal{C}^1$.

Let us write $\theta_0 \in [0, 2\pi[$ for the angle corresponding to a fixed value $t = t_0$ of the parameter. We define

$$\theta(t) = \int_{t_0}^{t} f_1(t)f_2'(t) - f_2(t)f_1'(t)\,dt + \theta_0.$$

By construction, this function θ is of class $\mathcal{C}^1$ since f and f' are of class $\mathcal{C}^0$. Moreover $\theta(t_0) = \theta_0$.

Consider now the function

$$\psi = (f_1 - \cos\theta)^2 + (f_2 - \sin\theta)^2.$$

Proving

$$f(t) = \big(\cos\theta(t), \sin\theta(t)\big).$$

is equivalent to proving that $\psi(t) = 0$ for all t. Since $\theta(t_0) = \theta_0$, we have already

$$\psi(0) = \big(f_1(t_0) - \cos\theta_0\big)^2 + \big(f_2(t_0) - \sin\theta_0\big)^2 = 0.$$

Thus it remains to prove that ψ is constant or equivalently, that $\psi' = 0$.

Differentiating the equality $f_1^2 + f_2^2 = 1$ we get $f_1 f_1' + f_2 f_2' = 0$. On the other hand $\theta' = f_2 f_2' - f_2 f_1'$. Therefore

$$\begin{aligned}
\frac{1}{2}\psi' &= (f_1 - \cos\theta)\big(f_1' + \theta'\sin\theta\big) + (f_2 - \sin\theta)\big(f_2' - \theta'\cos\theta\big) \\
&= \big(f_1 f_1' + f_2 f_2'\big) - \big(f_1'\cos\theta + f_2'\sin\theta\big) + \theta'(f_1\sin\theta - f_2\cos\theta) \\
&= -\big(f_1^2 + f_2^2\big)\big(f_1'\cos\theta + f_2'\sin\theta\big) + \big(f_1 f_2' - f_2 f_1'\big)\big(f_1'\cos\theta + f_2'\sin\theta\big) \\
&= -f_1\cos\theta\big(f_1 f_1' + f_2 f_2'\big) - f_2\sin\theta\big(f_1 f_1' + f_2 f_2'\big) \\
&= 0.
\end{aligned}$$

This concludes the proof that θ has the expected property.

Next let $\overline{\theta}$ be another function of class $\mathcal{C}^1$ satisfying

$$f(t) = \big(\cos\overline{\theta}(t), \sin\overline{\theta}(t)\big).$$

Differentiating we obtain

$$\big(f_1', f_2'\big) = \overline{\theta}'(-\sin\overline{\theta}, \cos\overline{\theta}) = \overline{\theta}'(-f_2, f_1).$$

This yields further

$$f_1' f_2 = -\overline{\theta}' f_2^2, \qquad f_2' f_1 = \overline{\theta}' f_1^2.$$

Subtracting these equalities yields

$$f_1 f_2' + f_2 f_1' = \overline{\theta}'\left(f_1^2 + f_2^2\right) = \overline{\theta}'.$$

Thus θ and $\overline{\theta}$ have the same derivative and therefore, differ only by a constant. Since at t_0 both express the angle between the x-axis and the vector $f(t_0)$, this constant is a multiple of 2π. □

Corollary 2.12.2 *Consider a normal representation of class $\mathcal{C}^k$*

$$f\colon]a,b[\longrightarrow \mathbb{R}^2, \qquad s \mapsto f(s)$$

of a plane curve.

1. *There exists a function of class $\mathcal{C}^{k-1}$*

$$\theta\colon]a,b[\longrightarrow \mathbb{R}$$

such that $\theta(s)$ is at each point, up to a multiple of 2π, the angle between the x-axis and the tangent vector $f'(s)$.

2. *Writing $\kappa(s)$ for the relative curvature, one has further*

$$\kappa(s) = \theta'(s).$$

Two possible such functions θ differ by a constant multiple of 2π. Such a function θ is called an angular function of the curve.

Proof Since $f'(s)$ has length 1, one can apply Lemma 2.12.1 and write

$$f'(s) = \bigl(\cos\theta(s), \sin\theta(s)\bigr)$$

with θ a function of class $\mathcal{C}^1$. Differentiating this equality yields

$$f''(s) = \theta'(s)\bigl(-\sin\theta(s), \cos\theta(s)\bigr).$$

Taking the norms of both sides yields

$$\bigl\|\kappa(s)\bigr\| = \bigl\|f''(s)\bigr\| = \bigl|\theta'(s)\bigr|.$$

The same equality can be re-written in the form

$$\bigl(f_1'', f_2''\bigr) = \theta'\bigl(-f_2', f_1'\bigr).$$

Observe that $(-f_2', f_1')$ is a vector of length 1 orthogonal to f', and such that the basis $(f', \overrightarrow{n})$ has direct orientation. Comparing with Definition 2.9.8, we conclude that $\theta' \geq 0$ precisely when the basis (f', f'') has direct orientation, that is, when the relative curvature $\kappa(s)$ is given the positive sign. □

Theorem 2.12.3 *Consider a continuous function*

$$\kappa\colon\,]a,b[\longrightarrow \mathbb{R}, \qquad s \mapsto \kappa(s).$$

All the normal representations of all the curves admitting κ as relative curvature have the form

$$f(s) = \left(x_0 + \int_{s_0}^{s} \cos\theta(s)\,ds,\ y_0 + \int_{s_0}^{s} \sin\theta(s)\,ds\right)$$

where

$$\theta(s) = \theta_0 + \int_{s_0}^{s} \kappa(s)\,ds$$

and $s_0 \in]a,b[$, $\theta_0 \in [0, 2\pi[$, $x_0, y_0 \in \mathbb{R}$ are arbitrary constants.

Proof With the notation of Corollary 2.12.2 we must have $\theta'(s) = \kappa(s)$. Integrating this equality yields

$$\theta(s) = \theta_0 + \int_{s_0}^{s} \kappa(s)\,ds$$

for some arbitrary choices $s_0 \in]a,b[$ and $\theta_0 \in [0, 2\pi[$.

However, by the considerations in the proof of Corollary 2.12.2, we must have further

$$f'(s) = \big(\cos\theta(s), \sin\theta(s)\big).$$

Integrating this equality yields the formula in the statement. □

Theorem 2.12.4 *Two regular plane curves of class $\mathcal{C}^2$ admitting the same relative curvature function with respect to the arc length are necessarily isometric.*

Proof By Proposition 2.12.3, the two curves admit the parametric representations

$$f(s) = \left(x_0 + \int_{s_0}^{s} \cos\theta(s)\,ds,\ y_0 + \int_{s_0}^{s} \sin\theta(s)\,ds\right)$$

$$g(t) = \left(u_0 + \int_{s_1}^{s} \cos\tau(s)\,ds,\ v_0 + \int_{s_1}^{s} \sin\tau(s)\,ds\right)$$

with

$$\theta(s) = \theta_0 + \int_{s_0}^{s} \kappa(s)\,ds, \qquad \tau(s) = \tau_0 + \int_{s_1}^{s} \kappa(s)\,ds.$$

Let us first write the first curve as

$$f(s) = \left(x_0 + \int_{s_0}^{s_1} \cos\theta(s)\,ds + \int_{s_1}^{s} \cos\theta(s)\,ds,\right.$$
$$\left. y_0 + \int_{s_0}^{s_1} \sin\theta(s)\,ds + \int_{s_1}^{s} \sin\theta(s)\,ds\right).$$

The translation mapping

$$\left(x_0 + \int_{s_0}^{s_1} \cos\theta(s)\,ds,\; y_0 + \int_{s_0}^{s_1} \sin\theta(s)\,ds\right)$$

on (u_0, v_0) transforms this first curve into

$$h(t) = \left(u_0 + \int_{s_1}^{s} \cos\theta(s)\,ds,\; v_0 + \int_{s_1}^{s} \sin\theta(s)\,ds\right).$$

Write further

$$\theta(s) = \theta_0 + \int_{s_0}^{s_1} \kappa(s)\,ds + \int_{s_1}^{s} \kappa(s)\,ds.$$

Apply now the rotation with center (u_0, v_0) transforming the angle

$$\theta_0 + \int_{s_0}^{s_1} \kappa(s)\,ds$$

into τ_0. The curve becomes further

$$k(s) = \left(u_0 + \int_{s_1}^{s} \cos\widetilde{\theta}(s)\,ds, v_0 + \int_{s_1}^{s} \sin\widetilde{\theta}(s)\,ds\right)$$

where

$$\widetilde{\theta}(s) = \tau_0 + \int_{s_1}^{s} \kappa(s)\,ds.$$

This proves the result since translations, rotations and orthogonal symmetries are examples of isometries (see Sects. 4.11 and 4.12 in [4], *Trilogy II*). □

Corollary 2.12.5 *Two regular plane curves of class $\mathcal{C}^2$ admitting opposite relative curvature functions with respect to the arc length are necessarily isometric.*

Proof Apply first to one the curves an orthogonal symmetry with respect to some arbitrary line in order to get an isometric curve (see Example 4.11.3 in [4], *Trilogy II*). This isometry changes the orientation of each base (see Proposition 3.3.2 in [4], *Trilogy II*), thus changes the sign of the curvature function. We are then back in the situation of Theorem 2.12.5. □

Definition 2.12.6 By an *intrinsic equation* of a regular plane curve of class $\mathcal{C}^2$ is meant a function

$$\kappa\colon\,]a,b[\longrightarrow \mathbb{R}, \qquad s \mapsto \kappa(s)$$

giving the relative curvature $\kappa(s)$ of the curve in terms of the arc length s.

Example 2.12.7 The intrinsic equation $\kappa(s) = 0$ is that of a straight line.

Proof This follows by Example 2.9.4 and Theorem 2.12.4. □

Example 2.12.8 The intrinsic equation $\kappa(s) = k$, with $k > 0$ a constant, is that of a circle of radius $\frac{1}{k}$.

Proof This follows by Example 2.9.5 and Theorem 2.12.4. □

Example 2.12.9 The intrinsic equation $\kappa(s) = ks$, with $0 \neq k \in \mathbb{R}$, is that of a *clothoid*, also called the *spiral of Cornu*, the curve depicted in Fig. 2.18.

Proof This curve—where the curvature is proportional to the distance traveled on the curve—is often used to construct roads: when the road makes a turn, it suffices to progressively turn the steering wheel at a constant speed.

Choosing the various constants of integration to be 0, a possible equation is thus obtained via

$$\theta = \frac{ks^2}{2}$$

and thus

$$f(s) = \left(\int_0^s \cos\frac{ks^2}{2}\,ds, \int_0^s \sin\frac{ks^2}{2}\,ds\right).$$

The integrals of the functions $\cos x^2$ and $\sin x^2$ do not admit expressions in terms of elementary functions. □

2.13 Closed Curves

Up to now, we have essentially considered results valid at a given point of the curve: the tangent, the curvature, and so on. These are *local* results, that is, the result is not affected when restricting the parametric representation to a neighborhood of the point considered. In this section, we begin the study of a *global result*, the so-called *Umlaufsatz*: a result which makes sense only when one considers the full curve.

Definition 2.13.1 By a *closed curve* is meant a curve admitting a periodic function

$$f : \mathbb{R} \longrightarrow \mathbb{R}^2$$

as parametric representation.

Thus if ω is a period of the function, we have $f(a) = f(a + \omega)$ for each value $a \in \mathbb{R}$. Of course given a period of length ω, any quantity $k\omega$, with k a strictly positive integer, is a period as well. As usual, we are mainly interested in the smallest period.

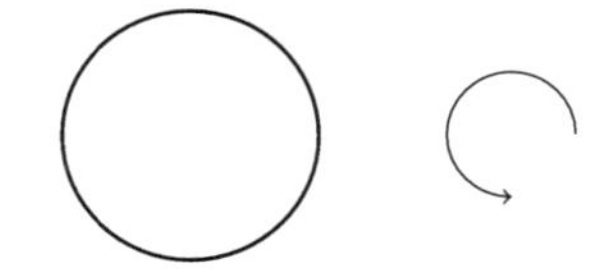

Fig. 2.22

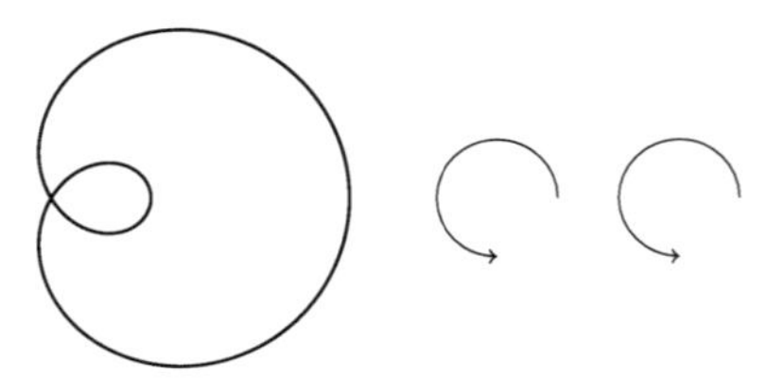

Fig. 2.23

Example 2.13.2 The circle with parametric equation

$$f: \mathbb{R} \longrightarrow \mathbb{R}^2, \qquad \theta \mapsto (R\cos\theta, R\sin\theta)$$

is a closed curve with minimal period 2π (see Fig. 2.22).

Proof The result is obvious. Consider the following experiment (we repeat this for various curves below, anticipating the *Umlaufsatz*): place your hand on Fig. 2.22, representing the tangent vector to the circle with your forefinger. Move around the circle, keeping your forefinger tangent to the curve. If you go around the whole circle your arm will be badly twisted a full turn! □

Example 2.13.3 The *Limaçon of Pascal* is the curve admitting the parametric representation

$$f: \mathbb{R} \longrightarrow \mathbb{R}^2, \qquad \theta \mapsto (\cos\theta + \cos 2\theta, \sin\theta + \sin 2\theta).$$

This is a closed curve with minimal period 2π (see Fig. 2.23).

Proof Again, this is immediate to check. Try the same experiment with your forefinger as in Example 2.13.2. If you succeed, you could confidently begin a promising career as a contortionist! Indeed, to follow the full curve with your forefinger remaining tangent to the curve, you would have to fully twist your arm twice! (And do not cheat: do not walk around the table: stay seated on your chair!) □

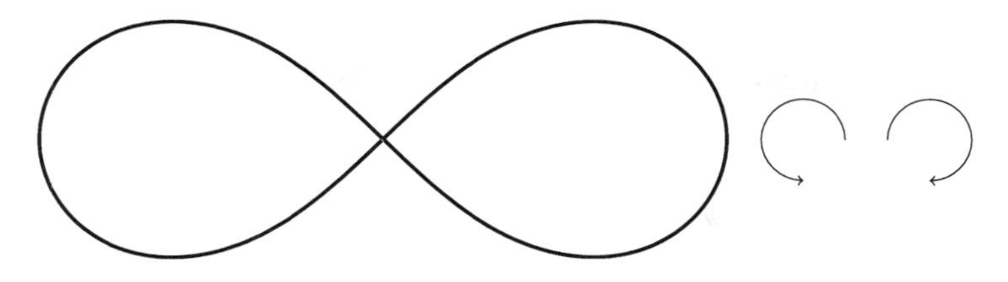

Fig. 2.24

Example 2.13.4 The *lemniscate* of *Bernoulli* is the curve with parametric representation

$$f\colon \mathbb{R} \longrightarrow \mathbb{R}^2, \qquad t \mapsto \left(\frac{\sin t}{1+\cos^2 t}, \frac{\sin t \cdot \cos t}{1+\cos^2 t}\right).$$

This is a closed curve with minimal period 2π (see Fig. 2.24).

Proof Once more the result is immediate. But if you try the experiment of Example 2.13.2 with your forefinger, you will realize that you can easily follow the full curve as many times as you wish. Only boredom can stop you. The point is now that every time that you twist your arm when following one "loop" of the curve, you untwist it when following the second loop. After traveling along a full period of the curve, you arm and your forefinger are back in their original positions. □

The *Umlaufsatz* (the "rotation theorem") is a result which gives a formula for counting the number of times that you will twist your arm, along a period of the curve, when performing the experiment of Example 2.13.2.

Theorem 2.13.5 (Umlaufsatz) *Consider a regular closed curve of class $\mathcal{C}^2$ represented by the function*

$$f\colon \mathbb{R} \longrightarrow \mathbb{R}^2, \qquad t \mapsto f(t)$$

which is periodic with minimal period ω. Write

$$f_\omega\colon [a, a+\omega] \longrightarrow \mathbb{R}^2, \quad a \in \mathbb{R}$$

for the restriction of f to an arbitrary period and $\kappa(t)$ for the curvature. Then

$$\frac{1}{2\pi}\int_{f_\omega} \kappa$$

is an integer, independent of the choice of a. This integer is called the rotation number *of the closed curve. For every $a \in \mathbb{R}$, it is also equal to the quantity*

$$\frac{1}{2\pi}\big(\theta(a+\omega) - \theta(a)\big)$$

where $\theta(t)$ is an angular function of the curve (see Corollary 2.12.2).

Proof We consider the change of parameter

$$s=\sigma(t)=\int_{t_0}^{t}\|f'\|$$

for some arbitrary origin $t_0\in\mathbb{R}$. It follows that $\overline{f}=f\circ\sigma$ is a normal representation (see Proposition 2.8.2). We write $\overline{\kappa}(s)$ for the curvature expressed as a function of the arc length s.

We shall also use our Lemma 2.12.2: writing $\theta(s)$ for the angle between the x-axis and the vector $\overline{f}'(s)$, one has $\overline{\kappa}(s)=\pm\theta'(s)$.

The usual formula for curve integrals then yields

$$\begin{aligned}\int_{f_\omega}\kappa &= \int_a^{a+\omega}\kappa(t)\|f'(t)\|\,dt\\ &= \int_a^{a+\omega}\overline{\kappa}\big(\sigma(t)\big)\sigma'(t)\,dt\\ &= \int_{\sigma(a)}^{\sigma(a+\omega)}\overline{\kappa}(s)\,ds\\ &= \pm\int_{\sigma(a)}^{\sigma(a+\omega)}\theta'(s)\,ds\\ &= \pm\big[\theta(s)\big]_{\sigma(a)}^{\sigma(a+\omega)}\\ &= \pm\Big(\theta\big(\sigma(a+\omega)\big)-\theta\big(\sigma(a)\big)\Big)\\ &= 2k\pi.\end{aligned}$$

The last equality holds by periodicity of f: the tangent vectors at $f(a)$ and $f(a+\omega)$ are the same, thus the corresponding angles with the x-axis are equal "up to a multiple of 2π". □

Corollary 2.13.6 *Consider a closed regular curve of class $\mathcal{C}^2$ represented by the function*

$$f:\mathbb{R}\longrightarrow\mathbb{R}^2,\qquad t\mapsto f(t)$$

which is periodic with minimal period ω. The rotation number of the curve is equal to

$$\frac{1}{2\pi}\int_0^{\omega}\frac{f_1'f_2''-f_2'f_1''}{f_1'^2+f_2'^2}.$$

Proof Going back to the proof of Theorem 2.13.5, we have

$$\frac{1}{2\pi}\int_{f_\omega}\kappa = \frac{1}{2\pi}\int_0^\omega \kappa(t)\|f'(t)\|\,dt = \frac{1}{2\pi}\int_0^\omega \frac{f_1'f_2''-f_2'f_1''}{f_1'^2+f_2'^2}$$

(see Proposition 2.9.11). □

Example 2.13.7 Applying Proposition 2.13.5 to the examples above, one finds as rotation numbers:

- ± 1 in the case of the circle;
- ± 2 in the case of the limaçon;
- 0 in the case of the lemniscate.

Proof In the case of the circle, we have

$$f'(t) = (-R\sin t, R\cos t)$$

thus $\|f'\| = R$ while $\kappa = \frac{1}{R}$ by Example 2.9.5. It follows that

$$\frac{1}{2\pi}\int_{f\omega}\kappa = \frac{1}{2\pi}\int_0^{2\pi}\kappa(t)\,\|f'(t)\|\,dt = \frac{1}{2\pi}\int_0^{2\pi}\frac{R}{R} = 1.$$

In the case of the *Limaçon*, the formula in Corollary 2.13.6 reduces to

$$\frac{1}{2\pi}\int_0^{2\pi}\frac{9+6\cos\theta}{5+4\cos\theta}\,d\theta = \frac{1}{2\pi}\cdot 4\pi = 2.$$

Indeed, a routine calculation shows that the indefinite integral involved in this formula is given by

$$\frac{3\theta}{2} + \arctan\left(\frac{1}{3}\tan\frac{\theta}{2}\right).$$

Applying Corollary 2.13.6 to the case of the *lemniscate* leads to lengthy calculations. However, observing that the parametric representation f of Example 2.13.4 is such that $f(-t) = -f(t)$, we get at once that

$$\int_{-\pi}^0 \frac{f_1'f_2''-f_2'f_1''}{(f_1')^2+(f_2')^2} = -\int_0^\pi \frac{f_1'f_2''-f_2'f_1''}{(f_1')^2+(f_2')^2}$$

from which

$$\int_{-\pi}^\pi \frac{f_1'f_2''-f_2'f_1''}{(f_1')^2+(f_2')^2} = 0.$$

Of course changing the orientation (i.e. introducing the change of variable $t' = -t$) changes the sign of the rotation number. □

2.14 Piecewise Regular Curves

Up to now in this book, we have essentially worked with regular curves (see Definition 2.2.1) defined on generalized open intervals. However, there are non-regular curves of great interest in geometry, such as the perimeter of a triangle, of a square, of an arbitrary polygon. Of course we can also imagine "curve polygons", that is, polygons whose sides are no longer straight lines, but arbitrary regular curves.

Doing this will in particular force us to consider the various "sides" of the curve polygon, each of which we shall rather naturally assume to be a "regular curve". But each such side—joining two vertices A and B of the curve polygon—should now be a curve defined on a *closed* interval:

$$f\colon [a,b] \longrightarrow \mathbb{R}^2, \qquad f(a) = A, \qquad f(b) = B.$$

The *continuity* of such a mapping is just a special case of the general notion of continuity (see Definition A.8.1). By a trivial adaptation of the proof of Proposition A.4.1, this reduces to the usual continuity of f at each point $a < t < b$ together with the usual right and left continuity of f at a and b, that is

$$f(a) = \lim_{t \to a, t > a} f(t), \qquad f(b) = \lim_{t \to b, t < b} f(t).$$

Analogously, the *differentiability* of f means its usual differentiability for each $a < t < b$ together with the usual left and right differentiability of f at a and b, that is, the existence of

$$f'(a) = \lim_{t \to a, t > a} \frac{f(t) - f(a)}{t - a}, \qquad f'(b) = \lim_{t \to b, t < b} \frac{f(t) - f(b)}{t - b}.$$

When f' itself is continuous on $[a,b]$, one says that f is of class $\mathcal{C}^1$ on $[a,b]$. Iterating these definitions yields the notion of a function of class $\mathcal{C}^k$ on $[a,b]$. Of course f is called *regular* when it is of class $\mathcal{C}^1$ and $f'(t) \neq 0$ for all $a \leq t \leq b$.

Definition 2.14.1 A parametric representation of a *piecewise regular plane curve of class $\mathcal{C}^k$* is a continuous locally injective function $\mathcal{C}^0$

$$f\colon [a,b] \longrightarrow \mathbb{R}^2, \qquad t \mapsto f(t), \quad a, b \in \mathbb{R};$$

together with finitely many values $a = a_0 < a_1 < \cdots < a_n < a_{n+1} = b$ of the parameter such that f is regular of class $\mathcal{C}^k$ on each interval $[a_i, a_{i+1}]$.

The points $f(a_i)$, $i = 1, \ldots, n$, are called the *corners* of the piecewise regular curve; the pieces of curve

$$f\colon [a_i, a_{i+1}] \longrightarrow \mathbb{R}^2$$

are called the *sides* of the piecewise regular curve.

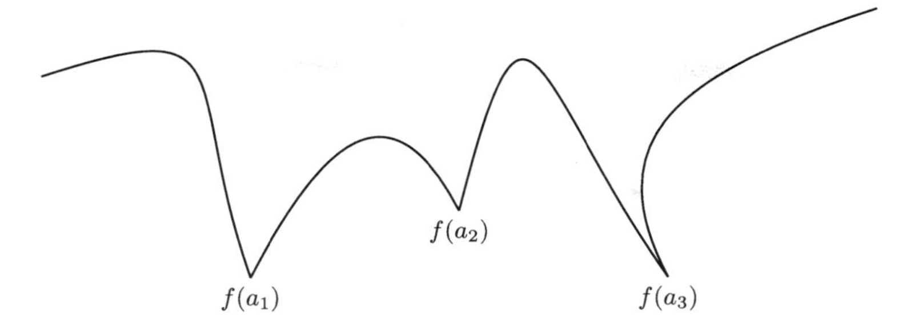

Fig. 2.25

The term *vertex* is sometimes preferred to the term *corner* in Definition 2.14.1, in particular in the case of *curve polygons*. Our choice intends to avoid any ambiguity with another totally different but still very classical notion of *vertex* of a curve, studied in Sect. 2.17.

Figure 2.25 shows an example of a *piecewise regular curve*. Under the conditions of Definition 2.14.1 it is common practice to write

$$f_0\colon\ ]a,a_1] \longrightarrow \mathbb{R}^2, \qquad f_i\colon [a_i,a_{i+1}] \longrightarrow \mathbb{R}^2, \qquad f_n\colon [a_n,b[\ \longrightarrow \mathbb{R}^2$$

for the various restrictions of f to the sub-intervals. It is also common practice to simply write

$$f'_{i-1}(a_i) = \lim_{t\to a_i,\, t<a_i} \frac{f(t)-f(a_i)}{t-a_i}, \qquad f'_i(a_i) = \lim_{t\to a_i,\, t>a_i} \frac{f(t)-f(a_i)}{t-a_i}$$

for the left and right derivatives of f at a_i.

Let us first clarify a few points:

Definition 2.14.2 By a *normal representation* of a piecewise regular curve is meant a parametric representation whose restriction to each side is *normal*, that is, with the notation above, $\|f'_i(t)\| = 1$ for all i and t.

Lemma 2.14.3 *Every piecewise regular curve admits normal representations.*

Proof This follows at once from Proposition 2.8.2. The continuity of the representation forces the behaviour at each corner. □

Definition 2.14.4 By a *closed piecewise regular curve* is meant a closed curve which is piecewise regular on each period.

Thus the requirement here is that there are only finitely many corner *in a period*. Of course in Definition 2.14.4, it is equivalent to assume that the curve is piecewise regular on one specified minimal period $[a, a+\omega]$, provided a corresponds to a corner.

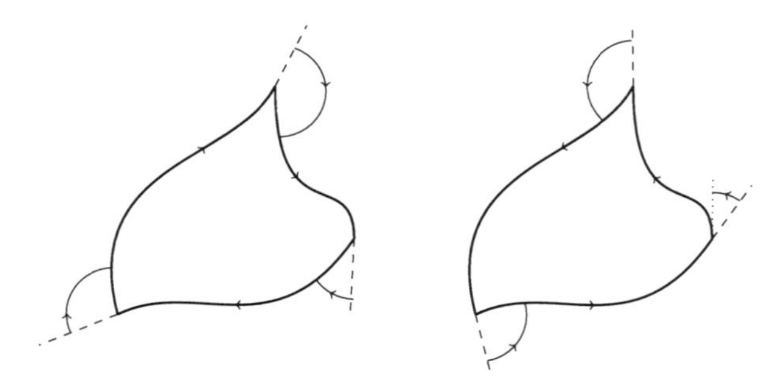

Fig. 2.26

Definition 2.14.5 With the notation indicated above, the *exterior angle* at a_i of a piecewise regular plane curve as in Definition 2.14.1 is the angle between the vector $f'_{i-1}(a_i)$ and the vector $f'_i(a_i)$.

It should be clear that the sign of an exterior angle at some corner of a piecewise regular plane curve is not an intrinsic property of the curve: a change of parameter $t' = -t$ in Definition 2.14.5 trivially results in changing the sign of all exterior angles (see Fig. 2.26).

We shall now generalize the *Umlaufsatz* (see Proposition 2.13.5) to the case of closed piecewise regular curves.

Definition 2.14.6 Consider a piecewise regular plane curve $c(s)$ of class $\mathcal{C}^2$, given in normal representation. An *angular function* for this curve consists of an angular function θ_i on each side of the curve, defined in such a way that at each corner $c(s_i)$, the corresponding external angle α_i remains given by

$$\alpha_i = \theta_i(s_i) - \theta_{i-1}(s_i).$$

Lemma 2.14.7 *Consider a piecewise regular plane curve $c(s)$ of class $\mathcal{C}^2$ given in normal representation. This curve admits an angular function. Two angular functions differ by a constant multiple of 2π.*

Proof By Proposition 7.10.1, on each side we have infinitely many possible angular functions, one for each multiple of 2π. Fixing the value of θ_i at one point thus fixes its value on the whole side. Choose arbitrarily θ_{i_0} on an arbitrarily chosen side of the curve. Then proceed side by side, using the requirement

$$\alpha_i = \theta_i(s_i) - \theta_{i-1}(s_i)$$

to fix the initial value of the function θ_i when passing from one side to the next. □

Theorem 2.14.8 (Umlaufsatz) *Consider a closed piecewise regular curve of class $\mathcal{C}^2$ given in normal representation by a function*

$$f : \mathbb{R} \longrightarrow \mathbb{R}^2, \qquad s \mapsto f(s).$$

Consider a minimal period of this function, defined on an interval

$$a_0 < a_1 < \cdots < a_{n-1} < a_n$$

with $f(a_0) = f(a_n)$ *and the various* $f(a_i)$ *the corners of the piecewise curve. Write further*

- $\kappa(s)$ *for the curvature at each point of class* $\mathcal{C}^2$;
- α_i *for the exterior angle of the curve (see Definition* 2.14.5) *at the corner* $f(a_i)$.

Under these conditions the quantity

$$\frac{1}{2\pi}\left(\sum_{i=0}^{n-1}\int_{a_i}^{a_{i+1}} \kappa(s)\,ds + \sum_{i=0}^{n-1}\alpha_i\right)$$

is an integer called the rotation number *of the piecewise regular closed curve. Writing* θ *for an angular function of the curve (see Corollary* 2.12.2), *the rotation number is also equal to*

$$\frac{1}{2\pi}\sum_{i=0}^{n-1}\big(\theta_(a_{i+1}) - \theta_i(a_i)\big) + \sum_{i=0}^{n-1}\alpha_i.$$

Proof Adopting the convention following Definition 2.14.1, we write $\theta_i(s)$ for the angle between the x-axis and $f_i'(s)$. Of course, each such function θ_i is *a priori* defined up to a multiple of 2π. As a convention, we fix $\theta_i(a_i) \in [0, 2\pi[$; this forces, by the required continuity of θ_i, the value of $\theta_i(s)$ on the whole interval $[a_i, a_{i+1}]$. In particular, $\theta_i(a_{i+1})$ is—up to a multiple of 2π—the angle between $f_i'(a_{i+1})$ and the x-axis. The external angle α_i at the point $f(a_i)$ thus has the form

$$\theta_i(a_i) - \theta_{i-1}(a_i) + 2k_i\pi.$$

We know by Lemma 2.12.2 that $\theta'(s) = \kappa(s)$ at all points of class $\mathcal{C}^2$. Therefore, since the integral on a closed interval is the same as on the corresponding open interval,

$$\int_{a_i}^{a_{i+1}} \kappa(s)\,ds = \int_{a_i}^{a_{i+1}} \theta_i'(s)\,ds = \theta_i(a_{i+1}) - \theta_i(a_i).$$

Summing all these equalities we obtain

$$\sum_{i=0}^{n-1}\int_{a_i}^{a_{i+1}} \kappa(s)\,ds = -\sum_{i=0}^{n-1}\big(\theta_i(a_i) - \theta_{i-1}(a_i)\big) = -\sum_{i=0}^{n-1}\alpha_i + 2k\pi$$

where in this formula we have used the notation $\theta_0 = \theta_n$. This proves the announced result. □

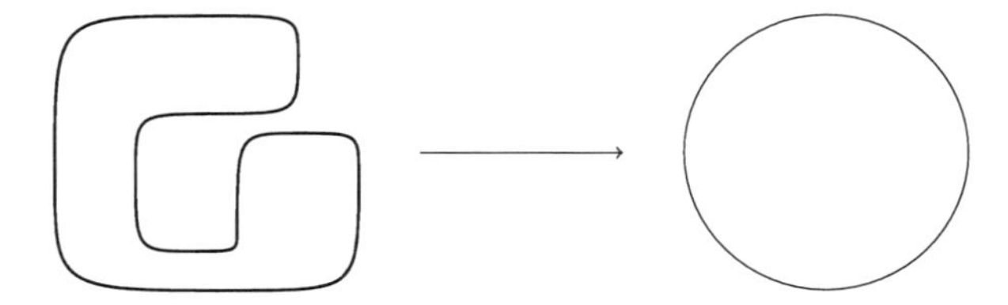

Fig. 2.27

2.15 Simple Closed Curves

The *simple closed curves* are those closed curves which "do not intersect themselves":

Definition 2.15.1 A *simple closed* curve is a closed curve admitting a periodic parametric representation which, restricted to a minimal period, is injective.

Clearly, this definition means that if

$$f: \mathbb{R} \longrightarrow \mathbb{R}^2$$

is a periodic parametric representation with minimal period ω, then f is injective on each interval $[a, a+\omega[$. Of course the circle is a simple closed curve, but the closed curves of Figs. 2.23 and 2.24 are not simple.

Our considerations concerning the *Umlaufsatz* and the *rotation number* in Sect. 2.13 probably convince you that the rotation number of a simple closed curve is necessarily ± 1 (the sign depends on proceeding clockwise or counter-clockwise along the curve). This is indeed true, but rather hard to establish. Here is the idea of the proof.

Imagine (see Fig. 2.27) that you have modeled the simple closed curve using some rope, set down on your table. Your intuition of what a simple closed curve is suggest that smoothly, continuously, you can modify the shape of the rope—paying attention to maintain all the time a simple closed curve—eventually obtaining a circle of radius R. A continuity argument should then imply that the curvature of the varying curve varies continuously during the whole process. Therefore by the *Umlaufsatz* (see Proposition 2.13.5), the rotation number of this varying curve should also vary continuously. But since the rotation number is an integer, the only way for it to vary continuously is to remain constant. Thus the rotation number of the original curve should be equal to the rotation number of the circle that you have eventually obtained: it is thus equal to ± 1, by Example 2.13.7. This argument is certainly intuitive and convincing, but the difficulty is to express mathematically how to continuously transform an arbitrary simple closed curve into a circle! Our proof of Theorem 2.15.2 is very close to the original proof of *Hopf* (see [14]).

Theorem 2.15.2 (Hopf) *The rotation number of a plane, closed, simple, and piecewise regular curve of class $\mathcal{C}^2$ is equal to ± 1.*

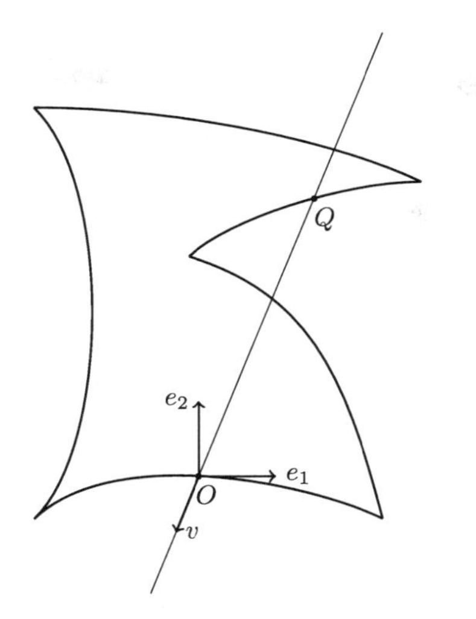

Fig. 2.28

Proof For the sake of clarity, let us split the proof into several steps.

Step 1: *Choice of a basis*

The support $\mathcal{C}$ of the curve contains infinitely many points and only finitely many corners. We can thus choose a straight line ℓ intersecting the curve $\mathcal{C}$ at a point Q but not containing any corner of it. Let us first observe that the intersection $\mathcal{C} \cap \ell$ is compact.

$\mathcal{C}$ is the continuous image, by a periodic parametric representation, of a closed interval corresponding to a period (see Definition 2.13.1). It is thus compact by Proposition A.8.3. In particular, $\mathcal{C}$ is contained in some closed ball $\overline{B}(P, r)$ (see Definition A.3.1). In $\mathcal{C} \cap d$, we can thus equivalently replace ℓ by the closed segment $\ell' = \ell \cap \overline{B}(P, r)$, which is compact. Thus

$$\mathcal{C} \cap \ell = \mathcal{C} \cap \ell'$$

is compact as an intersection of two compact (i.e. bounded, closed) subsets.

Consider now the continuous function

$$d(Q, -) \colon \mathcal{C} \cap d \longrightarrow \mathbb{R}, \qquad A \mapsto d(Q, A)$$

where d indicates the distance. By Corollary A.8.4, this positive function attains a maximum for some point $O \in \mathcal{C} \cap \ell$. We choose that point O as the origin of the axes (see Fig. 2.28). By the maximality property of O, either

- $O = Q$ is the only point in $\mathcal{C} \cap \ell$;
- or all the other points of $\mathcal{C} \cap \ell$ are on the half line of ℓ starting at O and passing through Q.

Let us fix a vector $\overrightarrow{v}$ with origin O, in the direction of a half line on ℓ which does not contain any other point of $\mathcal{C}$.

We choose an orthonormal basis $(0, e_1, e_2)$ of the plane whose origin is the point O and whose first vector is oriented along the tangent to the curve $\mathcal{C}$ at this regular point O. Moreover there is no loss of generality in assuming further that the curve is given in "normal representation" $f(s)$, that is, the parameter s is the arc length along the curve, again with O as origin for the arc lengths. Since $\|f'(0)\| = 1$, we have $f'(0) = \pm e_1$ and we choose further the orientation of e_1 to get the sign $+$: $f'(0) = e_1$. The considerations of Sect. 2.8 indicate that $\|f'(s)\| = 1$ at each regular point; these arguments extend immediately to the left and right derivatives at each corner.

Step 2: A "star-shaped" subset of $\mathbb{R}^2$

Again because the support $\mathcal{C}$ of the curve is compact and the "length function" from the origin O is continuous, this function reaches a maximum ω after a period (see Corollary A.8.4). Thus the closed curve has bounded length $\omega \in \mathbb{R}$. The various corners of the curve are obtained for some values s_i

$$0 < s_1 < \cdots < s_n < \omega$$

of the parameter.

We consider first the following subset of $\mathbb{R}^2$

$$\mathcal{S} = \{(u, v) \in \mathbb{R}^2 \,|\, 0 \le u \le v \le \omega,\ \forall i\ (u, v) \ne (s_i, s_i)\}.$$

Observe that $(0, \omega) \in \mathcal{S}$ because O is a regular point.

Let us observe that $\mathcal{S}$ is "star-shaped" with respect to $(0, \omega)$, that is, if $(u, v) \in \mathcal{S}$, then all points of the segment joining $(0, \omega)$ to (u, v) are still in $\mathcal{S}$. Such a point has the form

$$(0, \omega) + t\big((u, v) - (0, \omega)\big), \quad 0 \le t \le 1.$$

The assumptions on u, v, t imply at once

$$0 \le tu \le tv \le (1 - t)\omega + tv = \omega - t(\omega - v) \le \omega.$$

On the other hand

$$(0, \omega) + t\big((u, v) - (0, \omega)\big) = (s_i, s_i)$$

would imply

$$tu = s_i = \omega + t(v - \omega)$$

that is

$$t(u - v) = \omega(1 - t).$$

Since $t \ge 0$, $1 - t \ge 0$, $u - v \le 0$ while $\omega > 0$, this forces $1 - t = 0$, thus $t = 1 \ne 0$ and $u = v$. But then $u = s_i = v$, which is not the case.

Step 3: A normed continuous function on $\mathcal{S}$

Let us now define a function

$$\varphi\colon \mathcal{S} \longrightarrow \mathbb{R}^2, \qquad \forall (u,v) \in \mathcal{S} \quad \|\varphi(u,v)\| = 1,$$

which expresses, for all values (u,v) of the parameter (except possibly at the corners), the direction of the vector joining $f(u)$ to $f(v)$. More precisely

$$\varphi(u,v) = \begin{cases} f'(u) & \text{if } u = v; \\ -f'(0) & \text{if } (u,v) = (0,\omega); \\ \frac{f(v)-f(u)}{\|f(v)-f(u)\|} & \text{otherwise.} \end{cases}$$

Notice that this definition makes perfect sense. Indeed when $u = v$, by definition of $\mathcal{S}$ this value of the parameter does not correspond to a corner, thus f is regular at that point and $\|f'(u)\| = 1$. Since $O = f(0)$ is a regular point as well, $\|f'(0)\| = 1$. The form of the first case forces $u \neq v$ in the third case; since the curve is simple, this implies $f(u) \neq f(v)$, except if $(u,v) = (0,\omega)$: but this is excluded by the second case. Thus in the third case $f(u) \neq f(v)$ and the quotient makes sense; of course, it is of norm 1.

Let us first prove that this function φ is continuous, that is, the three pieces of the definition—which, separately, are trivially continuous—"glue together" continuously. Of course a pair (u,u) cannot converge to $(0,\omega)$ or to a pair (u,v) with $u < v$. Thus it suffices to prove the continuity when a pair (u,v) of condition 3 converges to a pair in conditions 1 or 2 of the definition of φ.

It is a straightforward variation on the definition of a derivative to observe that

$$\lim_{\substack{(u,v)\to(w,w)\\ u<v}} \frac{f(v)-f(u)}{\|f(v)-f(u)\|} = \frac{f'(w)}{\|f'(w)\|}$$

and

$$\lim_{\substack{(u,v)\to(w,w)\\ u>v}} \frac{f(v)-f(u)}{\|f(v)-f(u)\|} = -\frac{f'(w)}{\|f'(w)\|}.$$

Since we are in normal representation, we get at once, for $(w,w) \in \mathcal{S}$

$$\lim_{\substack{(u,v)\to(w,w)\\ u<v}} \frac{f(v)-f(u)}{\|f(v)-f(u)\|} = f'(w).$$

An analogous argument applies when (u,v) converges to $(0,\omega)$, but this time the sign is reversed. Indeed when v converges to ω, considering f as a periodic function, we have

$$f\colon \mathbb{R} \longrightarrow \mathbb{R}^2, \qquad f(v) = f(v-\omega)$$

with this time $(u, v - \omega)$ converging to $(0, 0)$. Since

$$\lim_{\substack{(u,v)\to(0,\omega)\\ u<v}} \frac{f(v) - f(u)}{\|f(v) - f(u)\|} = \lim_{\substack{(u,v-\omega)\to(0,0)\\ u>v-\omega}} \frac{f(v-\omega) - f(u)}{\|f(v-\omega) - f(u)\|}$$

we get

$$\lim_{\substack{(u,v)\to(0,\omega)\\ u<v}} \frac{f(v) - f(u)}{\|f(v) - f(u)\|} = -f'(0).$$

But in fact, φ is of class $\mathcal{C}^1$. Of course again, by differentiability of f, the function φ is of class $\mathcal{C}^1$ at all points in case 3 of its definition.

When (u, v) as in the third case converges to a point (w, w) as in the first case, we know that when u, v are sufficiently close to w, $f(v) - f(u)$ is parallel to $f'(w')$ for some $u < w' < v$ (see Proposition 2.4.5). Since we are in normal representation, this implies

$$\varphi(u, v) = \frac{f(v) - f(u)}{\|f(v) - f(u)\|} = f'(w').$$

As a consequence the limit of the derivative when (u, v) converges to (w, w), and thus when w' converges to w, will simply be $f''(w)$. This is also the derivative of φ along the diagonal (i.e. the first case). As above, an analogous argument (up to a change of sign) holds for the limit to $(0, \omega)$.

Step 4: *Trigonometric form of the function φ*

At each point $(u, v) \in \mathcal{S}$ we have $\|\varphi(u, v)\| = 1$. It follows that

$$\varphi(u, v) = \big(\cos\theta(u, v), \sin\theta(u, v)\big)$$

for some angle $\theta(u, v)$. Of course such an angle is only defined up to a multiple of 2π. The present crucial step of the proof consists of showing that a *continuous* choice of θ can be made. This is essentially the formalization of the intuitive continuity argument described at the beginning of this section. This is also an extension of Lemma 2.12.1 to the case of a function $\varphi(u, v)$ with two variables.

Let us keep the notation θ for the angle above, defined up to a multiple of 2π. We shall write Θ for the corresponding continuous function that we are looking for. Let us first notice that

$$\varphi(0, \omega) = -f'(0) = -e_1 = (-1, 0).$$

Let us then choose $\Theta(0, \omega) = \pi$ as "initial condition".

Now fix a point $(u, v) \neq (0, \omega)$ in $\mathcal{S}$. The function

$$\tau\colon [0, 1] \longrightarrow \mathbb{R}^2, \qquad t \mapsto \varphi\big((0, \omega) + t(u, v)\big)$$

is continuous. Using instead the notation

$$(u_t, v_t) = (0, \omega) + t(u, v),$$

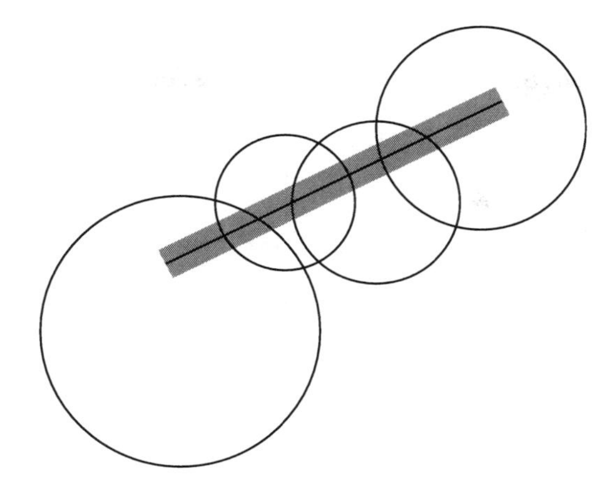

Fig. 2.29

by Lemma 2.12.1 we obtain a continuous function $\Theta(u_t, v_t)$ defined on the interval joining $(0, \omega)$ to (u, v) and such that

$$\varphi(u_t, v_t) = \tau(t) = \big(\cos\Theta(u_t, v_t), \sin\Theta(u_t, v_t)\big).$$

But this defines the function Θ on the whole of $\mathcal{S}$ and it remains to prove its continuity.

Let us fix a point $(u, v) \in \mathcal{S}$ and a "small" real number $\varepsilon > 0$ (it will appear eventually that "small" should mean $\varepsilon < \frac{\pi}{7}$). Since a closed segment is compact, the function Θ is in fact *uniformly continuous* on the ray joining $(0, \omega)$ to (u, v) (see Theorem A.4.2). By uniform continuity of Θ, there exists a $\delta > 0$ such that for two points of the ray

$$\big\|(u_1, v_1) - (u_2, v_2)\big\| \leq \delta \quad \Longrightarrow \quad \big|\Theta(u_1, v_1) - \Theta(u_2, v_2)\big| \leq \varepsilon.$$

The open balls $B((u_i, v_i), \delta)$ cover the ray, which is compact. Thus a finite number of them already suffices to cover this ray (see Sect. A.3).

But in $\mathbb{R}^2$, if a finite number of open balls covers a segment, a whole open rectangle admitting the segment as median line is still contained in these balls (see Fig. 2.29). Let us write w for the width of this rectangle; we thus have $\frac{w}{2} \leq \delta_i \leq \delta$ for each index i. Let us put

$$\overline{\delta} = \min\left\{\delta, \frac{w}{2}\right\}.$$

We shall prove that given a point $(\overline{u}, \overline{v}) \in \mathcal{S}$ such that

$$\big\|(\overline{u}, \overline{v}) - (u, v)\big\| \leq \overline{\delta}$$

then

$$\big\|\Theta(\overline{u}, \overline{v}) - \Theta(u, v)\big\| \leq \varepsilon.$$

This will prove the continuity of Θ.

Notice first that by choice of $\overline{\delta}$,

$$\left\|\theta(\overline{u},\overline{v})-\theta(u,v)\right\|\leq\varepsilon$$

that is

$$\left\|\Theta(\overline{u},\overline{v})-\Theta(u,v)\right\|=2\overline{k}\pi\overline{\varepsilon},\quad|\overline{\varepsilon}|<\varepsilon$$

for some integer $\overline{k}$. We therefore need to prove that $\overline{k}=0$.

To that end, consider first a point (u',v') on the segment joining $(0,\omega)$ and (u,v) and a point $(u'',v'')\in\mathcal{S}$ such that

$$\left\|(u'',v'')-(u',v')\right\|<\overline{\delta}.$$

We know that

$$(u',v')\in B\big((u_i,v_i),\delta_i\big),\qquad(u'',v'')\in B\big((u_j,v_j),\delta_j\big)$$

for some indices i and j. Comparing these points with the centers of the corresponding balls, this forces

$$\left|\theta(u',v')-\theta(u_i,v_i)\right|\leq 2\varepsilon,\qquad\left|\theta(u'',v'')-\theta(u_j,v_j)\right|\leq 2\varepsilon.$$

On the other hand

$$\begin{aligned}\left\|(u_i,v_i)-(u_j,v_j)\right\|&\leq\left\|(u_i,v_i)-(u',v')\right\|+\left\|(u',v')-(u'',v'')\right\|\\&\quad+\left\|(u'',v'')-(u_j,v_j)\right\|\\&\leq 3\overline{\delta}.\end{aligned}$$

By uniform continuity of Θ on the ray, this forces

$$\left\|\theta(u_i,v_i)-\theta(u_j,v_j)\right\|\leq 3\varepsilon.$$

Together these inequalities yield

$$\left\|\theta(u'',v'')-\theta(u',v')\right\|\leq 7\varepsilon.$$

We now consider the function

$$\psi(t)=\Theta\big((0,\omega)+t\big((u,v)-(0,\omega)\big)\big)-\Theta\big((0,\omega)+t\big((\overline{u},\overline{v})-(0,\omega)\big)\big).$$

Since Θ is continuous on both "rays" through (u,v) and $(\overline{u},\overline{v})$, the two terms are continuous in t and thus ψ is a continuous function of t.

Observe next that

$$\left\|\big((0,\omega)+t\big((u,v)-(0,\omega)\big)\big)-\big((0,\omega)+t\big((\overline{u},\overline{v})-(0,\omega)\big)\big)\right\|$$

$$
\begin{aligned}
&= \left\| t\big((\overline{u}, \overline{v}) - (u, v)\big) \right\| \\
&\leq \left\| (\overline{u}, \overline{v}) - (u, v) \right\| \\
&\leq \overline{\delta}.
\end{aligned}
$$

Since the two points

$$(0, \omega) + t\big((u, v) - (0, \omega)\big), \qquad (0, \omega) + t\big((\overline{u}, \overline{v}) - (0, \omega)\big)$$

are at a distance less than $\overline{\delta}$ apart, with the first one on the ray from $(0, \omega)$ to (u, v), we know already that

$$\left\| \theta\big((0, \omega) + t\big((u, v) - (0, \omega)\big)\big) - \theta\big((0, \omega) + t\big((\overline{u}, \overline{v}) - (0, \omega)\big)\big) \right\| \leq 7\varepsilon < \pi$$

since we started with a sufficiently small ε.

This last inequality can be rephrased by saying that for each t

$$\left| \psi(t) \right| = 2k_t\pi + \varepsilon_t, \qquad \left| \varepsilon'_t \right| < \pi$$

for some integer k_t. But by continuity, since $|\varepsilon_t| < \pi$, the integer k_t is constant. When $t = 1$ we find

$$\psi(1) = \Theta(\overline{u}, \overline{v}) - \Theta(u, v) = 2\overline{k}\pi + \overline{\varepsilon}.$$

Thus $k_t = \overline{k}$ for all values of t. But

$$\psi(0) = \Theta(0, \omega) - \Theta(0, \omega) = 0.$$

This proves that $k_0 = 0$ thus $k = 0$.

Step 5: $\Theta(\omega, \omega) - \Theta(0, 0) = \pm 2\pi$

Consider $0 < s < \omega$. By choice of the vector $\overrightarrow{v}$ in *Step 1*, $\overrightarrow{O f(s)} = \overrightarrow{f(\omega) f(s)}$ cannot be in the direction of the vector $\overrightarrow{v}$. Thus $\varphi(s, \omega)$, which is in the direction of $\overrightarrow{f(s) f(\omega)}$, cannot possibly be the vector $-\overrightarrow{v}$. Thus when s varies in $]0, \omega[$, $\Theta(s, \omega)$ never crosses the direction of $-\overrightarrow{v}$ and therefore, one always has

$$\left| \Theta(s, \omega) - \Theta(0, \omega) \right| < 2\pi$$

and thus by continuity of Θ

$$\left| \Theta(\omega, \omega) - \Theta(0, \omega) \right| \leq 2\pi.$$

But by definition of φ and Θ,

$$\Theta(0, \omega) = \pi, \qquad \varphi(\omega, \omega) = f'(\omega) = e_1, \qquad \Theta(\omega, \omega) = 2k\pi.$$

Putting together these two requirements, we obtain

$$\Theta(\omega, \omega) = 0 \quad \text{or} \quad \Theta(\omega, \omega) = 2\pi.$$

In other words

$$\Theta(\omega, \omega) - \Theta(0, \omega) = \pm\pi.$$

In a completely analogous way, one proves that

$$\Theta(0, \omega) - \Theta(0, 0) = \pm\pi.$$

An important point is to observe that in both cases, the same sign must be chosen. Indeed

$$\Theta(t, \omega) - \Theta(0, \omega)$$

represents the angle from the vector $-e_1$ to the vector $-\overrightarrow{Of(s)}$ while

$$\Theta(0, t) - \Theta(0, 0)$$

represents the angle from the vector e_1 to the vector $\overrightarrow{Of(s)}$: these two angles are equal.

It then remains to compute

$$\Theta(\omega, \omega) - \Theta(0, 0) = \big(\Theta(\omega, \omega) - \Theta(0, \omega)\big) + \big(\Theta(0, \omega) - \Theta(0, 0)\big) = \pm 2\pi.$$

Step 6: *The external angles*

Our next concern is to express the external angles α_i in terms of the continuous function Θ. By Definition 2.14.5 and using its notation, the external angle α_i is that between

$$f'_{i_1}(s_i) \quad \text{and} \quad f'_i(s_i).$$

By *Step 3* and the continuity of f'_i and f_{i-1}, this is the same as the angle between

$$f'_{i-1}(s_i) = \lim_{s \to s_i, s < s_i} f'_{i-1}(s_i) = \lim_{s \to s_i, s < s_i} \varphi(s, s)$$

and

$$f'_i(s_i) = \lim_{s \to s_i, s > s_i} f'_i(s_i) = \lim_{s \to s_i, s > s_i} \varphi(s, s).$$

Since these limits exist and the inverse trigonometric functions are continuous, the limits

$$\Theta_{i-1}(s_i, s_i) = \lim_{s \to s_i, s < s_i} \Theta(s, s), \qquad \Theta_i(s_i, s_i) = \lim_{s \to s_i, s > s_i} \Theta(s, s)$$

exist as well and we have

$$\alpha_i = \Theta_i(s_i, s_i) - \Theta_{i-1}(s_i, s_i) + 2k_i\pi$$

because $\Theta_i(s_i, s_i)$ and $\Theta_{i-1}(s_i, s_i)$ are equal—up to a multiple of 2π—to the angles between the x-axis and $f'_i(s_i)$, $f'_{i-1}(s_i)$. Our aim in this step is to prove that

$$\alpha_i = \Theta_i(s_i, s_i) - \Theta_{i-1}(s_i, s_i)$$

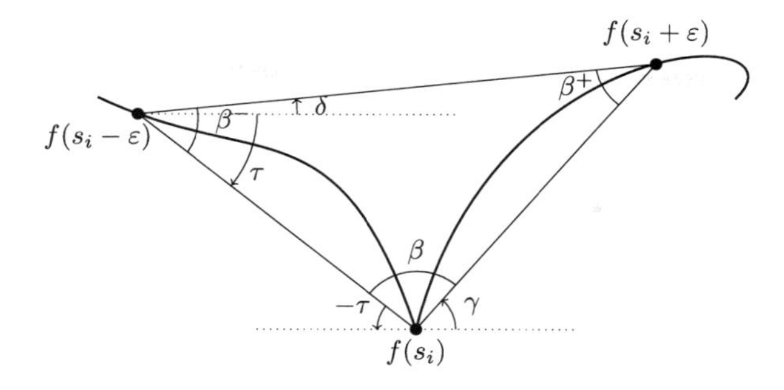

Fig. 2.30

that is, $k_i = 0$.

To achieve this, for sufficiently small $\varepsilon > 0$, consider the triangle whose vertices are

$$f(s_i - \varepsilon), \qquad f(s_i), \qquad f(s_i + \varepsilon)$$

(see Fig. 2.30). We assume that when running along the triangle, following the order in which these vertices have been introduced, the external angles are positive; an analogous proof holds in the other case. Notice that there is no loss of generality in assuming that we can always choose ε to be arbitrarily small so that the three points constitute an actual triangle: otherwise the external angle at s_i would be 0 and the point $f(s_i)$ would be regular: we could thus avoid treating it as a corner.

We write $\beta_i^-(\varepsilon)$, $\beta_i(\varepsilon)$ and $\beta_i^+(\varepsilon)$ for the corresponding (positive) interior angles of the triangle; for short, let us for now simply write β^-, β, β^+ for these angles. Let us write further

$$\Theta(s_i - \varepsilon, s_i) = \tau + 2r\pi, \qquad \Theta(s_i, s_i + \varepsilon) = \gamma + 2p\pi,$$
$$\Theta(s_i - \varepsilon, s_i + \varepsilon) = \delta + 2q\pi$$

with

$$\tau, \gamma, \delta \in]-\pi, +\pi[, \qquad p, q, r \in \mathbb{Z}.$$

Observe that

$$\beta^- + \beta + \beta^+ = \pi, \qquad \beta^- = \delta - \tau, \qquad \gamma + \beta - \tau = \pi$$

from which

$$\beta^+ = \pi - \beta^- - \beta = \pi - \delta + \tau - \beta = \gamma - \delta.$$

This proves that

$$\Theta(s_i, s_i + \varepsilon) - \Theta(s_i - \varepsilon, s_i + \varepsilon) = \beta^+ + 2m\pi.$$

By continuity of Θ, when ε is chosen sufficiently small, the difference

$$\left|\Theta(s_i, s_i + \varepsilon) - \Theta(t, s_i + \varepsilon)\right|$$

can be made arbitrarily small, thus in any case strictly smaller than π. This forces $m = 0$. This proves that

$$\Theta(s_i, s_i + \varepsilon) - \Theta(t, s_i + \varepsilon) = \beta^+.$$

Analogously one proves that

$$\Theta(s_i - \varepsilon, s_i + \varepsilon) - \Theta(s_i - \varepsilon, s_i) = \beta^-.$$

Putting these equalities together yields

$$\Theta(s_i, s_i + \varepsilon) - \Theta(s_i - \varepsilon, s_i) = \beta^+ + \beta^- = \pi - \beta.$$

Switching back to the more precise notation, we have trivially

$$\lim_{\varepsilon \to 0} \beta_i(\varepsilon) = \pi - \alpha_i.$$

This indicates that

$$\lim_{\varepsilon \to 0} \Theta(s_i, s_i + \varepsilon) - \lim_{\varepsilon \to 0} \Theta(s_i - \varepsilon, s_i) = \pi - \lim_{\varepsilon \to 0} \beta_i(\varepsilon),$$

that is

$$\Theta_i(s_i, s_i) - \Theta_{i-1}(s_i, s_i) = \alpha_i.$$

Step 7: *Conclusion of the proof*

The curve that we consider is periodic, with first corner at s_1 and last corner at s_n; as usual, to make the formulas easier to write down, we take $f(s_n)$ as "the corner $f(s_0)$" starting a new period. In other words, we introduce the convention $s_0 = s_n$ and analogously, $\Theta_0 = \Theta_n$.

The conclusion follows immediately from the proof of Theorem 2.14.8 and the various steps of the present proof. Indeed

$$\begin{aligned}
&\sum_{i=1}^{n} \int_{s_i}^{s_{i+1}} \kappa + \sum_{i=1}^{n} \alpha_i \\
&\quad = \int_0^{s_1} \kappa + \sum_{i=1}^{n-1} \int_{s_i}^{s_{i+1}} \kappa + \int_{s_n}^{\omega} \kappa + \sum_{i=1}^{n} \alpha_i \\
&\quad = \int_0^{s_1} \Theta_0'(s, s) + \sum_{i=1}^{n-1} \int_{s_i}^{s_{i+1}} \Theta_i'(s, s) + \int_{s_n}^{\omega} \Theta_n'(s, s) + \sum_{i=1}^{n} \alpha_i
\end{aligned}$$

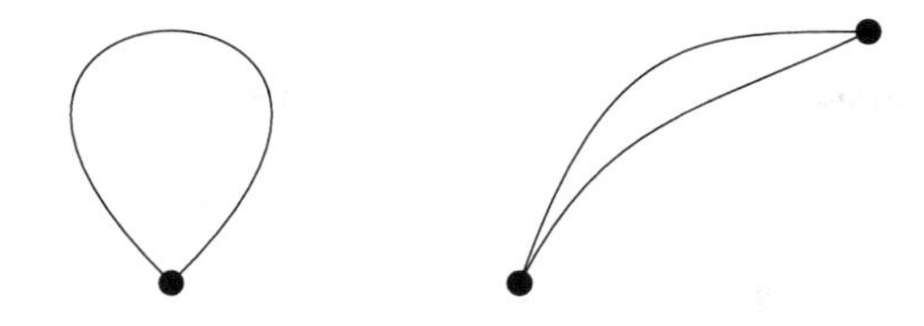

Fig. 2.31

$$
\begin{aligned}
&= \big(\Theta_0(s_1, s_1) - \Theta(0,0)\big) + \sum_{i=1}^{n-1} \big(\Theta_i(s_{i+1}, s_{i+1}) - \Theta_i(s_i, s_i)\big) \\
&\quad + \big(\Theta(\omega,\omega) - \Theta_n(s_n, s_n)\big) + \sum_{i=1}^{n} \big(\Theta_i(s_{i+1}, s_{i+1}) - \Theta_i(s_i, s_i)\big) \\
&= \Theta(\omega,\omega) - \theta(0,0) \\
&= \pm 2\pi
\end{aligned}
$$

and this forces the conclusion. □

Let us emphasize a special case of interest:

Definition 2.15.3 A *polygon* is a simple, closed, piecewise regular plane curve of class $\mathcal{C}^\infty$ all of whose sides are segments of straight lines (see Definition 2.14.1).

Of course we get at once (even if in this particular case, easier proofs can be given):

Proposition 2.15.4 *The sum of the exterior angles of a polygon equals* 2π.

Proof This follows by Theorems 2.14.8, 2.15.2 and Example 2.9.4. □

Certainly everybody knows that:

Corollary 2.15.5 *A polygon has at least three sides.*

Proof Since the sides are segments of straight lines, an exterior angle equal to π would mean that the two sides at the corresponding corner are superposed; this would contradict the simplicity of the curve. But if each exterior angle is strictly less than π, at least three of them are necessary to reach a sum of 2π. □

Of course when one switches to "curve polygons", that is, piecewise regular simple closed curves, Corollary 2.15.5 no longer holds true (see Fig. 2.31 for curve polygons with one or two sides). Notice also that curve polygons can also very well have zero interior angles, that is, the two consecutive sides are tangent at their common corner.

Later (see Sect. 7.11) we shall remark upon the properties of simple closed curves in the spirit of the *Jordan curve theorem* (see Theorem 7.11.3).

2.16 Convex Curves

We adopt the following definition:

Definition 2.16.1 A regular plane curve is *convex* when, at each point, it lies entirely on one side of the tangent to that point.

Trivially:

Example 2.16.2 An *ellipse*, a *parabola* and a branch of a *hyperbola* are all convex curves. But the *Limaçon of Pascal* (see Example 2.13.3), the *lemniscate* (see Example 2.13.4) and the curve $y = x^3$ are not convex.

Lemma 2.16.3 *Consider a simple, closed, convex regular plane curve which lies entirely on one side of some line ℓ. When the curve contains two distinct points of ℓ, the whole segment joining these two points is entirely contained in the curve.*

Proof Write $f(s)$ for a normal parametric representation of the curve and assume that two distinct points $f(s')$ and $f(s'')$ are on the line ℓ. If the segment joining $f(s')$ and $f(s'')$ is not entirely contained in the support of the curve, choose a point P on this segment which is not on the curve. The line $\ell^\perp$ through P, perpendicular to the segment, determines two open half planes which we write as $\mathcal{P}'$ and $\mathcal{P}''$. The curve has points in these two half planes, namely, the extremities $f(s')$ and $f(s'')$ of the segment. The two pieces of the closed curve with extremities these points $f(s')$ and $f(s'')$ are thus continuous lines connecting a point of one half plane to a point of the other half plane. They must therefore cut the line ℓ. Indeed the continuous function

$$f : [s', s''] \longrightarrow \mathbb{R}^2$$

cannot take values only in $\mathcal{P}' \cup \mathcal{P}''$, otherwise we would have a partition

$$[s', s''] = f^{-1}(\mathcal{P}') \cup f^{-1}(\mathcal{P}'')$$

of the closed interval $[s', s'']$ into two disjoint open subsets, which is impossible (see Propositions A.4.1 and Lemma A.10.3). An analogous argument holds for the other arc. Notice that since the curve is simple, the two intersections of the two arcs with the line $\ell^\perp$ are necessarily distinct.

So we have at least two intersections of the curve and the line $\ell^\perp$. As in *Step 1* of the proof of Theorem 2.15.2, we know that there are only finitely many intersection

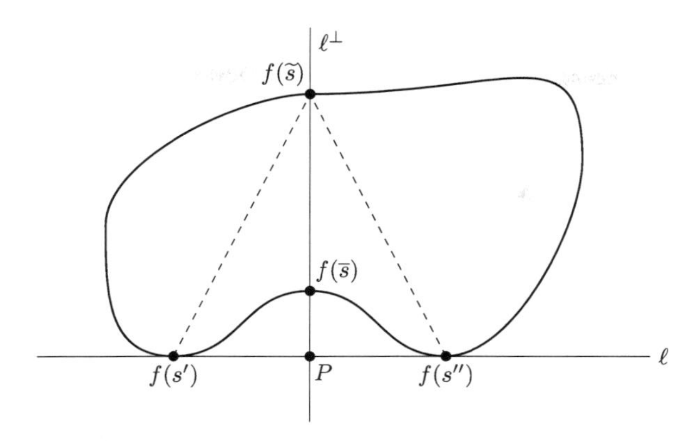

Fig. 2.32

points of the curve with $\ell^\perp$. By assumption, all these points lie on the same side of the line ℓ. We call $f(\overline{s})$ the point which is closest to the line ℓ and $f(\widetilde{s})$ the point which is furthest from d. But then $f(\overline{s})$ lies inside the triangle with vertices $f(s')$, $f(s'')$ and $f(\widetilde{s})$ (see Fig. 2.32). Thus whatever the line that we draw through $f(\overline{s})$, there are vertices of the triangle on both sides of it. Choosing as line the tangent at $f(\overline{s})$ contradicts the assumption of convexity. □

Theorem 2.16.4 *A simple closed regular curve is convex if and only if its relative curvature has constant sign.*

Proof We choose a normal representation

$$f(s)\colon [0,\omega] \longrightarrow \mathbb{R}^2$$

where ω is the minimal period of the curve. There is no loss of generality in assuming that we are working in an orthonormal basis (O, e_1, e_2) with $O = f(0)$ and $e_1 = f'(0)$. We consider the angular function θ of Corollary 2.12.2.

Assume first that the curve is convex. We know that the derivative of a monotone function has constant sign. Since $\kappa = \theta'$, to prove that κ has constant sign it suffices to prove that θ is monotone. (To be precise, let us make clear that we call a function h *monotone* when $a \le b$ implies $h(a) \le h(b)$.) Proving monotonicity is clearly equivalent to proving, for $s_1 < s_2$

$$\theta(s_1) = \theta(s_2) \quad \Longrightarrow \quad \theta(s) \text{ is constant on } [s_1, s_2].$$

Let us fix two such points s_1, s_2.

Since the curve is simple and closed, by Theorem 2.15.2 its rotation number is ± 1; thus by the *Umlaufsatz* (see Theorem 2.13.5)

$$\pm 2\pi = \int_0^\omega \kappa(s)\,ds = \int_0^\omega \theta'(s)\,ds = \theta(\omega) - \theta(0).$$

Since the variation of θ is 2π on a period, there exists a point s_3 where the variation is only π with respect to the value $\theta(s_1) = \theta(s_2)$. Thus

$$\theta(s_3) = \theta(s_2) \pm \pi$$

and the tangents at these three points are parallel. Of course these three tangents cannot be three different lines, otherwise the curve would be on both sides of the one which is between the other two; this would contradict the convexity of the curve. Thus at least two of these three tangents coincide. Let $s', s'' \in \{s_1, s_2, s_3\}$ be the two values yielding the same tangent. Since $f(s')$ and $f'(s'')$ are on this tangent, the common tangent is the line d through these two points. We choose $s' < s''$.

By convexity, the curve is entirely on one side of the tangent d and since $f(s') \neq f(s'')$, the whole segment joining these two points lies on the curve (see Lemma 2.16.3). As a consequence, $\theta(s)$ is constant on $[s', s'']$. Since

$$\theta(s_3) \neq \theta(s_1), \qquad \theta(s_3) \neq \theta(s_2)$$

we conclude that $s_1 = s'$ and $s_2 = s''$. But then $\theta(s)$ is constant on $[s_1, s_2]$ as expected.

Conversely, assume that the curvature does not change sign; let us say, $\kappa(s) \geq 0$. If the curve is not convex, there is a point of the curve such that the curve has points on both sides of the corresponding tangent. There is no loss of generality in assuming that this is the point that we have chosen as origin of the axis, with the x-axis as tangent at this point. The curve thus has points on both sides of the x-axis. This means that the function $f_2(s)$, the second component of f, changes sign. As a continuous function on the interval $[0, \omega]$, this function attains a maximum for a value $s = s_1$ and a minimum for a value $s = s_2$ (see Corollary A.8.4). Of course since f_2 takes both signs

$$f_2(s_1) < 0 = f_2(0) < f_2(s_2).$$

At an extremum, the derivative of f_2 is zero. Thus

$$f_2'(s_1) = 0 = f_2'(s_2).$$

This proves that $f'(s_1)$ and $f'(s_2)$ are both parallel to e_1. So at least two of the following three vectors are equal

$$f'(s_1), \qquad e_1 = f'(0), \qquad f'(s_2).$$

Let us write $s', s'' \in \{s_1, 0, s_2\}$ for two values of the parameter such that the corresponding two tangent vectors are equal. Choose further $s' < s''$.

We have, by Corollary 2.12.2

$$\int_{s'}^{s''} \kappa = \int_{s'}^{s''} \theta' = \theta(s'') - \theta(s') = 2k\pi, \quad k \in \mathbb{Z}.$$

Let us prove that $k > 0$.

First, $k \geq 0$ because $\kappa \geq 0$. On the other hand $k = 0$ would imply that the integral above is zero: since the function $\kappa(s)$ is always positive, this would force $\kappa(s)$ to be zero on the whole interval $[s', s'']$. The arc of the curve between $f(s')$ and $f(s'')$ would then be a segment (see Example 2.12.7) and since the tangent at $f(s')$ and $f(s'')$ is horizontal, this would be a horizontal segment. But this would contradict the inequality $f_2(s') \neq f_2(s'')$. Thus indeed $k \neq 0$ and therefore $k > 0$.

Analogously one proves that

$$\theta(s' + \omega) - \theta(s'') = 2k'\pi, \quad k' > 0.$$

Adding these equalities we obtain

$$\theta(s' + \omega) - \theta(s') = 2(k + k')\pi.$$

By Corollary 2.12.2 again, we have

$$\int_{s'}^{s'+\omega} \kappa = \int_{s'}^{s'+\omega} \theta' = \theta(s' + \omega) - \theta(s') = 2(k + k')\pi$$

and since the curve is simple and closed, Theorem 2.15.2 forces $k + k' = \pm 1$. This is a contradiction, since $k > 1$ and $k' > 1$. □

It should be noted that the assumption of being a *simple* curve is definitely necessary in Theorem 2.16.4. Indeed:

Counterexample 2.16.5 The *limaçon of Pascal* has a relative curvature with constant sign but is not convex.

2.17 Vertices of a Plane Curve

When considering the ellipse with equation

$$\left(\frac{x}{a}\right)^2 + \left(\frac{y}{b}\right) = 1$$

it is common practice to refer to the four points

$$(\pm a, 0), \qquad (0, \pm b)$$

as the *vertices* of the ellipse; analogously the origin of the axes is often called the *vertex* of the parabola with equation $y = x^2$ (see Fig. 2.33). Let us give a formal and general definition of what a *vertex* of a curve is.

Definition 2.17.1 A *vertex* of a regular plane curve of class C^3 is a point where the derivative of the relative curvature function vanishes (see Definition 2.9.8).

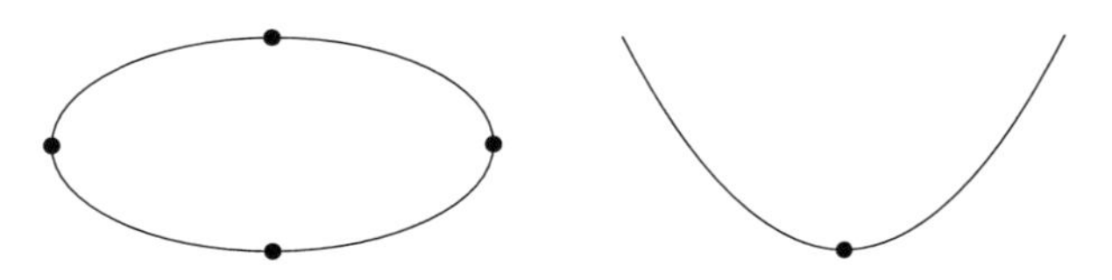

Fig. 2.33

Observe first that:

Lemma 2.17.2 *In Definition* 2.17.1, *the notion of vertex is independent of the choice of the parametric representation of class* C^3.

Proof Consider two equivalent representations $f(t)$, $g(s)$ as in the statement and write $\kappa(t)$, $\widetilde{\kappa}(s)$ for the corresponding curvature functions. Let $s=\varphi(t)$ express the equivalence between f and g. Differentiating the equality

$$\kappa(t)=\widetilde{\kappa}\big(\varphi(t)\big)$$

we obtain

$$\kappa'(t)=\widetilde{\kappa}'\big(\varphi(t)\big)\cdot\varphi'(t).$$

Since moreover

$$f'(t)=g'\big(\varphi(t)\big)\cdot\varphi'(t),$$

by regularity $\varphi'(t)\neq 0$ and thus κ' vanishes precisely when $\widetilde{\kappa}$ does. □

Example 2.17.3 The ellipse with equation

$$\left(\frac{x}{a}\right)^2+\left(\frac{y}{b}\right)=1,\quad a>b>0$$

admits the four vertices

$$(\pm a,0),\qquad (0,\pm b).$$

Proof Considering the parametric representation of the ellipse as in Example 2.1.4, the relative curvature is given by (see Proposition 2.9.11)

$$\kappa=\frac{ab}{(a^2\sin^2\theta+b^2\cos^2\theta)^{\frac{3}{2}}}.$$

The derivative of this function vanishes precisely when the derivative of the expression between parentheses vanishes, that is

$$2\big(a^2-b^2\big)\sin\theta\cos\theta=0.$$

Since $a\neq b$, this is the case precisely when $\sin\theta=0$ or $\cos\theta=0$. □

Example 2.17.4 The origin is the only vertex of the parabola with equation $y = x^2$.

Proof The parabola admits the parametric representation $f(x) = (x, x^2)$. By Proposition 2.9.11, the relative curvature is given by

$$\kappa = \frac{2}{(1+2x^2)^{\frac{3}{2}}}.$$

The derivative of this function vanishes precisely when

$$(1+2x^2)' = 4x = 0$$

that is when $x = 0$. □

Of course, we also have:

Example 2.17.5 All the points of the circle are vertices.

Proof The curvature is constant (see Example 2.9.5), thus its derivative is everywhere zero. □

Let us conclude this section with a celebrated result concerning vertices:

Theorem 2.17.6 (Four Vertices Theorem) *A convex simple regular closed curve of class $\mathcal{C}^3$ admits at least four vertices.*

Proof Let us work in normal representation $f(s)$ and assume that the minimal period is ω. Since $[0, \omega]$ is compact, the relative curvature function admits at least a maximum and a minimum (see Corollary A.8.4), thus the derivative of the relative curvature vanishes at at least two points. If the minimum equals the maximum, the relative curvature is constant, we have a circle (see Example 2.12.8) and all points are vertices. In that case the result is proved. Let us now assume that the curvature is not constant.

Up to a possible change of the origin for computing arc lengths, there is no loss of generality in assuming that the relative curvature reaches its minimal value at $s = 0$. Write s_0 for a value of the parameter where $\kappa(s)$ attains its global maximum; since the relative curvature is not constant, we have $0 < s_0 < \omega$. Up to a possible translation of the origin, there is also no loss of generality in assuming that $f(0) = (0, 0)$. Up to a possible rotation of the axis, we can further assume that the x-axis is the line joining $f(0)$ to $f(s_0)$. Since $\kappa(0) \neq \kappa(s_0)$, the curve is not contained in the x-axis as s runs through $[0, s_0]$ (see Example 2.9.4); thus there exists a point $0 < s_1 < s_0$ such that $f_2(s_1) \neq 0$. If $f_2(s_1) < 0$, rotate further by an angle π the system of axes so that in all cases we end up with $f_2(s_1) > 0$. Notice that translations and rotations do not change the orientation, thus do not affect the sign of the relative curvature.

The points $f(0)$ and $f(s_0)$ are thus on the x-axis: let us prove that no other point of the curve lies on the x-axis. Indeed if $f(s_2)$ is another point of the curve on the x-axis, we have three points $f(0)$, $f(s_0)$, $f(s_2)$ on the x-axis: thus one of these points is between the other two. Let us say, $f(s'')$ is between $f(s')$ and $f(s''')$. Since the curve is convex, it is entirely on one side of the tangent at $f(s'')$. But since $f(s')$ and $f(s''')$ are on the curve and on both sides of $f(s'')$ on the x-axis, this forces the tangent at $f(s'')$ to be the x-axis itself. This tangent then contains all three points $f(s')$, $f(s'')$, $f(s''')$ of the curve, thus the whole segment between $f(s')$ and $f(s''')$ lies entirely on the curve (see Lemma 2.16.3). But then, again by Example 2.9.4, the curvature is zero at all three points with parameters s', s'', s''', thus in particular at 0 and s_0, which is not the case. Thus indeed, the only points of the curve on the x-axis are $f(0)$ and $f(s_0)$.

Since $f_2(s_1) > 0$ and the curve does not cross the x-axis between $f(0)$ and $f(s_0)$, we conclude that this portion of the curve is entirely contained in the upper half plane: thus $f_2(s) > 0$ for all $0 < s < s_0$. Analogously, the whole portion of the curve between $f(s_0)$ and $f(\omega)$ is entirely on one side of the x-axis. But it cannot be in the upper half plane, otherwise the whole curve would be in the upper half plane and again by Lemma 2.16.3, the whole segment from $f(0)$ to $f(s_0)$ would lie on the curve. Thus $f_2(s) < 0$ for all $s_0 < s < \omega$.

Let us now prove that $\kappa'(s)$ must change sign (thus by continuity, must also vanish) at some other point than $s = 0$ and $s = s_0$. If not, then $\kappa'(s)$ has constant sign on each interval $]0, s_0[$ and $]s_0, \omega[$. Thus κ is monotone on each of these intervals. Since

$$\kappa(0) = \kappa(\omega) < \kappa(s_0),$$

the function $\kappa(s)$ is increasing on the interval $[0, s_0]$ and decreasing on the interval $[s_0, \omega]$. Therefore

$$\kappa'(s) \geq 0 \quad \text{for } 0 \leq s \leq s_0, \qquad \kappa'(s) \leq 0 \quad \text{for } s_0 \leq s \leq \omega.$$

Together with the sign of f_2 on these intervals as determined above, we obtain that $\kappa'(s) f_2(s) \geq 0$ at all points $s \in [0, \omega]$. Therefore, integrating by parts,

$$0 \leq \int_0^\omega \kappa'(s) f_2(s)\, ds = -\int_0^\omega \kappa(s) f_2'(s)\, ds.$$

But since f is a normal representation, Propositions 2.9.11 and 2.8.3 yield

$$\kappa = f_1' f_2'' - f_2' f_1'', \qquad 0 = f_1' f_1'' + f_2' f_2''.$$

It follows that

$$\kappa f_2' = f_1' f_2' f_2'' - f_2'^2 f_1'' = -f_1'^2 f_1'' - f_2'^2 f_1'' = -f_1''(f_1'^2 + f_2'^2) = -f_1''.$$

Therefore

$$\int_0^\omega \kappa(s) f_2'(s)\, ds = -\int_0^\omega f_1''(s)\, ds = f_1'(\omega) - f_1'(0) = 0.$$

Putting these results together, we obtain

$$\int_0^\omega \kappa'(s) f_2(s)\, ds = 0.$$

This forces $\kappa'(s) f_2(s) = 0$ for all $s \in [0, \omega]$. Since $f_2(s) \neq 0$ except for $s = 0$ and $s = s_0$, this forces $\kappa'(s) = 0$ on the whole interval $[0, \omega]$, that is, $\kappa(s)$ is a constant. We have already treated and excluded this case. This contradiction implies that there is indeed at least a third point s_3 where $\kappa'(s)$ changes sign.

But $s = s_0$ corresponds to a local minimum of $\kappa(s)$: thus $\kappa'(s)$ changes sign at $s = 0$, passing from negative to positive. Analogously $\kappa(s_0)$ is a local maximum, thus $\kappa'(s)$ changes sign at $s = s_0$, passing from positive to negative. If $0 < s_3 < s_0$, then since κ' changes sign at $s = s_3$, it becomes negative on one side of s_3 and since it must be positive when approaching 0 or s_0, it must change sign a second time in the interval $]0, s_0[$ to become positive again. In that case we have found a fourth point $0 < s_4 < s_0$ where $\kappa'(s) = 0$. An analogous argument holds if $s_0 < s_3 < \omega$. □

2.18 Problems

2.18.1 Two equivalent parametric representations of class $\mathcal{C}^0$ of a plane curve have *the same orientation* when the corresponding change of parameter is a strictly increasing function. Prove that this is an equivalence relation, yielding two equivalence classes. Give an example showing that this result no longer holds when one allows arbitrary open subsets in Definition 2.1.2.

2.18.2 Up to an appropriate choice of affine basis, in a neighborhood of a regular point, a curve of class $\mathcal{C}^1$ can always be presented as the graph of a continuous function with the tangent as x-axis.

2.18.3 Prove that the envelope of the family of circles of fixed radius r, centered on a circle of radius $R \neq r$, is constituted of two circles.

2.18.4 Show that the locus of intersection points of the pairs of orthogonal tangents to an ellipse is a circle.

2.18.5 Using the formula in Example 2.7.8, prove Proposition 1.7.3 concerning the logarithmic spiral.

2.18.6 Compute the evolute of a cycloid using the general formula of Proposition 2.11.2.

2.18.7 Consider two plane curves defined in terms of the same parameter

$$f\colon]a, b[\longrightarrow \mathbb{R}^2, \qquad \widetilde{f}\colon]a, b[\longrightarrow \mathbb{R}^2.$$

Assume that these curves are regular of class $\mathcal{C}^2$. These curves are called *Bertrand curves* when their normal lines coincide at corresponding points. Prove that the distance between corresponding points is constant. Show that two involutes of the same curve are *Bertrand* curves.

2.18.8 A regular closed curve of class $\mathcal{C}^2$ is *strictly convex* when its curvature is non-zero at each point. When the curve is given in normal representation $f(s)$, prove that for each vector $\overrightarrow{v}$ of norm 1 in $\mathbb{R}^2$, there is a unique value $s \in [a, a+\omega[$ in a minimal period such that $f'(s) = \overrightarrow{v}$.

2.18.9 If a closed strictly convex curve has exactly four vertices, then any circle has at most four points of intersection with that curve.

2.18.10 If a closed, strictly convex curve intersects a circle in $2n$ points, then it has at least $2n$ vertices.

2.18.11 Given a simple closed regular curve of length L, with an interior of area A, then $4\pi A \leq L^2$. (This is the so-called *isoperimetric inequality*, proved by *Steiner* and *Chern*; see [31] and [6].)

2.19 Exercises

2.19.1 Determine if the following mappings are parametric representations of plane curves. Are they regular?

1. $f: \mathbb{R} \longrightarrow \mathbb{R}^2$, $t \mapsto (t^2, \cos t^2)$;
2. $f: \mathbb{R} \longrightarrow \mathbb{R}^2$, $t \mapsto (|t|, \sin t)$;
3. $f:]-1, 1[\longrightarrow \mathbb{R}^2$, $t \mapsto (0, t^n)$ (discuss in terms of $n \in \mathbb{N}$);
4. $f:]-1, 1[\longrightarrow \mathbb{R}^2$, $t \mapsto (t, t^n)$ (discuss in terms of $n \in \mathbb{N}$);
5. $f: [0, 2\pi] \longrightarrow \mathbb{R}^2$, $t \mapsto (\cos \frac{t}{2}, \sin \frac{t}{2})$.

2.19.2 For each odd $n \in \mathbb{N}$, consider

$$f_n: \mathbb{R} \longrightarrow \mathbb{R}^2, \qquad t \mapsto (t^n, t^n).$$

Determine if, when $n \neq m$, the parametric representations f_n and f_m are equivalent and if so, in which class of differentiability.

2.19.3 Give a parametric representation of a curve whose support is:

1. an *epicycloid*: the path of a chosen point of a circle of radius r, which rolls without slipping around a fixed circle of radius R (Fig. 2.34).
2. a *hypocycloid*: the path of a chosen point of a circle of radius r, which rolls without slipping within a fixed circle of radius $R > r$ (Fig. 2.35).

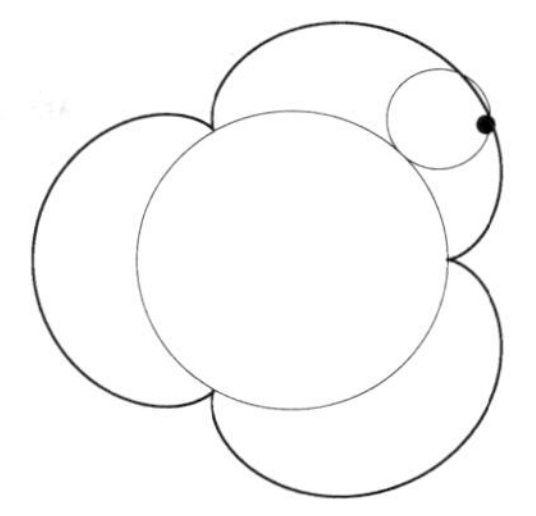

Fig. 2.34

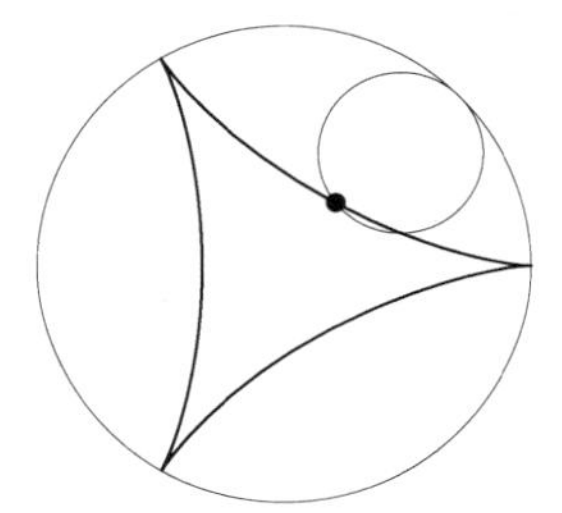

Fig. 2.35

Discuss the situation when the ratio $\frac{R}{r}$ is:

- an integer;
- a rational number;
- an irrational number.

2.19.4 Determine the points of the ellipse

$$\begin{cases} x = 2\cos t + \sin t \\ y = 2\cos t - \sin t \end{cases}$$

for which the normal line contains the origin.

2.19.5 Let P be an arbitrary point of the parabola

$$\begin{cases} x = 2t \\ y = t^2 - 1. \end{cases}$$

Through P, draw the line d_1 parallel to the y-axis and the line d_2 containing the origin. Show that the tangent at P to this parabola makes equal angles with d_1 and d_2.

2.19.6 Let $0 < a, b < \infty$; we work in the canonical basis of $E^2(\mathbb{R})$. Consider the circle Γ with center $(a, 0)$ and radius a and the line p with equation $y = b$.

Fig. 2.36

Through the origin $O = (0, 0)$, consider a line d, distinct from the axis, cutting the circle Γ at a point M and the line d at a point N. Through M, draw the parallel d_1 to the x-axis; through N, draw the parallel d_2 to the y-axis. Write P for the intersection of d_1 and d_2.

The locus of the point P, as d rotates around the origin, is called the *serpentine curve*; it was first studied by *L'Hospital* and *Huygens* in 1692, and more extensively by *Newton* in 1701.

1. Sketch the shape of the *serpentine*.
2. Give a parametric representation of the *serpentine*.

2.19.7 From a parabola $\mathcal{P}$, construct the *cissoid of Diocles* as the locus of the orthogonal projections of the vertex of this parabola on the tangents to the parabola (see Fig. 2.36). Determine a parametric representation of the cissoid corresponding to the parabola with equation $y^2 = 2px$, for $p > 0$.

2.19.8 Calculate the length of the *cardioid* (see Fig. 2.37)

$$\begin{cases} x = a(2\cos t - \cos 2t) \\ y = a(2\sin t - \sin 2t) \end{cases} \qquad (a \in \mathbb{R}_+^*).$$

Compare with Exercise 2.19.3.

2.19.9 Give a normal parametric representation of the logarithmic spiral (see Fig. 1.23)

$$\begin{cases} x = e^t \cos t \\ y = e^t \sin t. \end{cases}$$

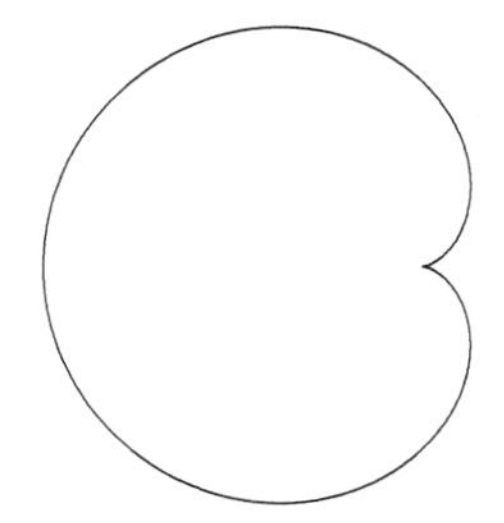

Fig. 2.37

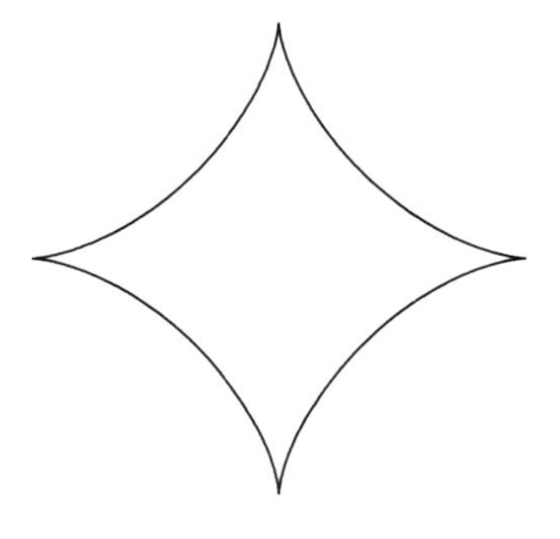

Fig. 2.38

2.19.10 Compute the curvature of the logarithmic spiral

$$\begin{cases} x = e^t \cos t \\ y = e^t \sin t \end{cases}$$

at an arbitrary point.

2.19.11 A plane curve $\mathcal{C}$ admits the parametric representation

$$f\colon \mathbb{R} \longrightarrow \mathbb{R}^2, \qquad t \mapsto \left(\sqrt{t^2+1}, t\right).$$

1. Give a Cartesian equation of $\mathcal{C}$.
2. Calculate the coordinates of the center of curvature of $\mathcal{C}$ at the point $(\sqrt{2}, -1)$.

2.19.12 Consider the *astroid* (see Fig. 2.38) with parametric representation

$$f\colon \mathbb{R} \longrightarrow \mathbb{R}^2, \qquad t \mapsto \left(2\cos^3 t, 2\sin^3 t\right).$$

1. Compute a Cartesian equation of the osculating circle to the astroid at each regular point.

2. Draw the evolute of the astroid; justify geometrically.
3. Give a parametric representation of this evolute.
4. Calculate the length of the evolute.
5. Explain geometrically why the sign of the curvature, computed from the given parametric representation, is necessarily negative at each regular point.
6. Give a Cartesian equation of the astroid.
7. Do you see a link with Exercise 2.19.3?

2.19.13 Give an intrinsic equation of the logarithmic spiral.

2.19.14 Determine the curves whose intrinsic equations are:

1. $\kappa(s) = 1/\sqrt{2as}$ with $a > 0$.
2. $\kappa(s) = 1/\sqrt{16a^2 - s^2}$ with $(a > 0)$.

2.19.15 Calculate the envelope of:

1. the family of lines $\alpha x - y = \frac{\alpha^2}{2}$;
2. the family of circles $(x - \alpha)^2 + y^2 - 2\alpha = 0$;
3. the normal lines to the parabola $y = (\frac{x}{2})^2 - 1$;
4. the circles centered on the unit circle $x^2 + y^2 = 1$ and tangent to the y-axis.

2.19.16 Find the vertices of the *Limaçon* (see Example 2.13.3) and the *Lemniscate* (see Example 2.13.4).

Chapter 3
A Museum of Curves

This chapter presents a list of some interesting properties of some famous plane curves. The lists—both of curves and properties of each curve—are far from being exhaustive. Some properties indicated have already been established elsewhere in this book; the other properties are cited without proof, but the omitted proofs reduce each time to routine calculations. Of course in these conditions, we do not introduce further sections of problems or exercises.

3.1 Some Terminology

Through the centuries, an incredible number of generic methods have been used to construct new curves from given ones, often with the with often the objective of inferring properties of some curves from well-known properties of other curves. We list here a few of these methods.

Definition 3.1.1 The *caustic* of a regular plane curve with respect to a fixed point P is the envelope of the family of lines which are, for each point Q of the curve, symmetric to the line PQ with respect to the normal to the curve at Q (see Fig. 3.1).

In other words, consider the point P as a light source and the curve as a mirror. The "light rays" PQ are then reflected by the mirror into those lines of which one considers the envelope. The envelope thus delimits the portion of the plane which is lit by the reflected rays. The interested reader will compare this situation with Proposition 1.13.2 in [4], *Trilogy II*: the case of the parabola and its focus P, a situation where the corresponding envelope does not exist! In that case, the reflected rays are parallel, a property which is widely used in optics.

Of course Definition 3.1.1 does not make sense at the point $Q = P$, when P turns out to be a point on the curve.

F. Borceux, *A Differential Approach to Geometry*, DOI 10.1007/978-3-319-01736-5_3,

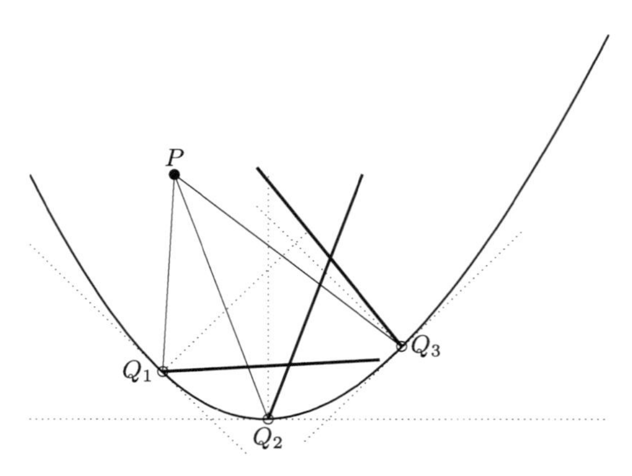

Fig. 3.1

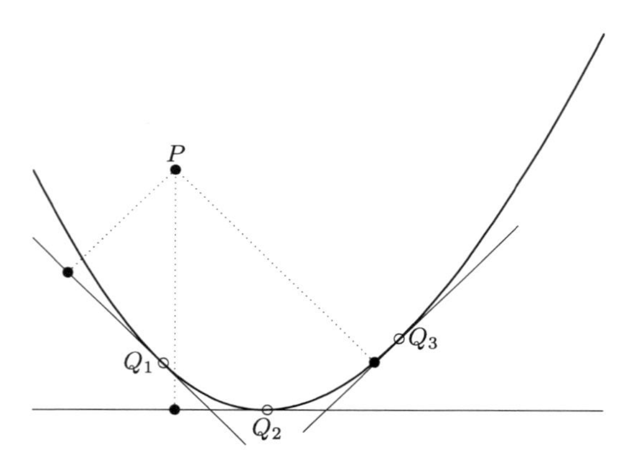

Fig. 3.2

Definition 3.1.2 The *pedal curve* of a regular plane curve with respect to a point P is the locus of the orthogonal projections of P on the tangents to the curve. (See Fig. 3.2.)

Definition 3.1.3 The *conchoid* of a regular plane curve with respect to a point P and a real number $\ell > 0$ is the locus of those points Q', Q'' obtained, for each point Q of the curve, by subtracting or adding to $\overrightarrow{PQ}$ a vector in the same direction and of fixed length ℓ (see Fig. 3.3).

$$\overrightarrow{PQ'} = \overrightarrow{PQ} - \ell \frac{\overrightarrow{PQ}}{\|\overrightarrow{PQ}\|}, \qquad \overrightarrow{PQ''} = \overrightarrow{PQ} + \ell \frac{\overrightarrow{PQ}}{\|\overrightarrow{PQ}\|}.$$

Again Definition 3.1.3 does not make sense at the point $Q = P$, when P turns out to be a point on the curve.

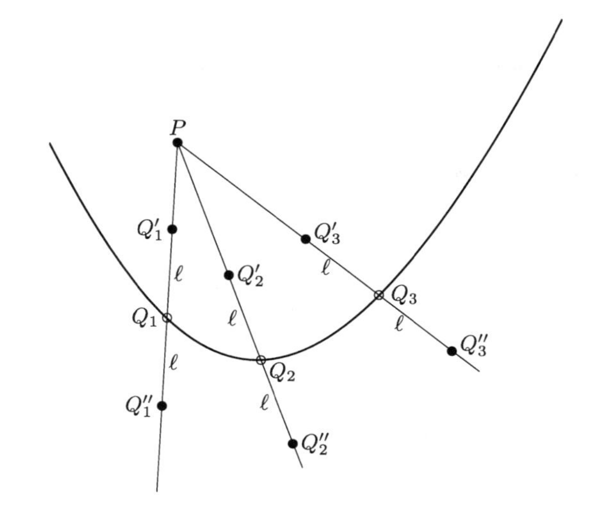

Fig. 3.3

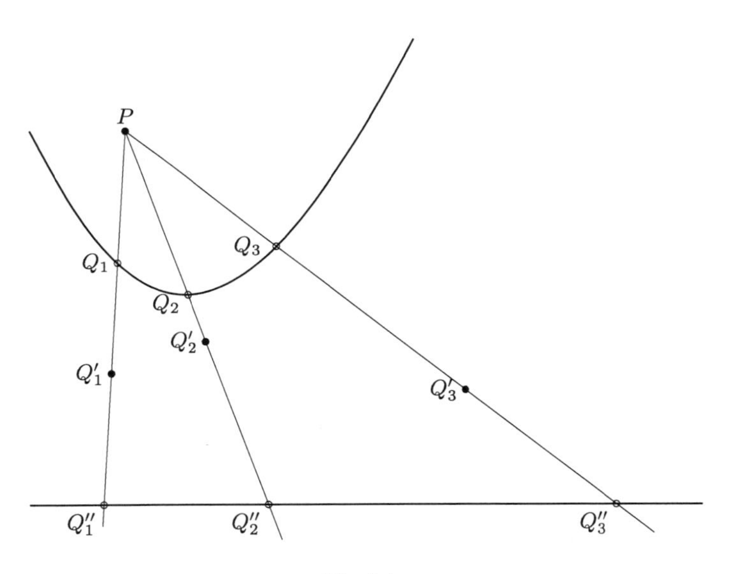

Fig. 3.4

Definition 3.1.4 The *cissoid* of a regular plane curve with respect to a point P and a fixed line d is the locus of those points Q' obtained in the following way. For each point Q of the curve, write Q'' for the intersection of the lines PQ and d. Then Q' is the point on the line PQ given by (see Fig. 3.4)

$$Q' = P + \overrightarrow{QQ''}.$$

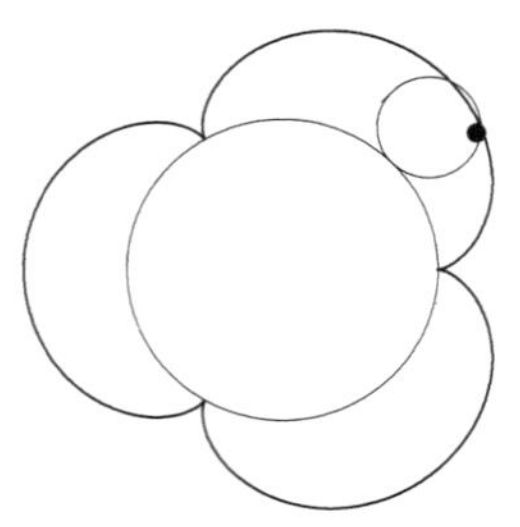

Fig. 3.5

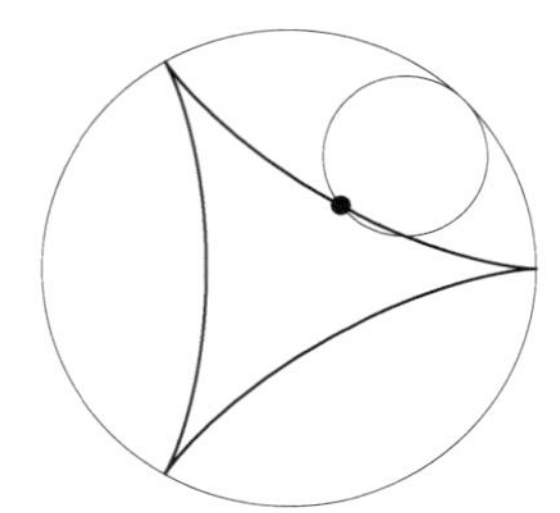

Fig. 3.6

Clearly this definition makes sense only when $P \neq Q$ and when moreover, the line PQ is not parallel to d.

Definition 3.1.5 An *epicycloid* is the trajectory of a point P of a circle of radius r which rolls on another fixed circle of radius R, around the exterior of this other circle (see Fig. 3.5).

Definition 3.1.6 A *hypocycloid* is the trajectory of a point P of a circle of radius r which rolls on another fixed circle of radius R, around the interior of this other circle (see Fig. 3.6).

We shall also refer to the notions of *evolute* and *involutes* of a curve, as studied in Sect. 2.11.

3.2 The Circle

The most basic curve is certainly the circle. With radius R and center the origin, we get:

- Parametric representation

$$f(t) = (R\cos t, R\sin t), \qquad R \neq 0.$$

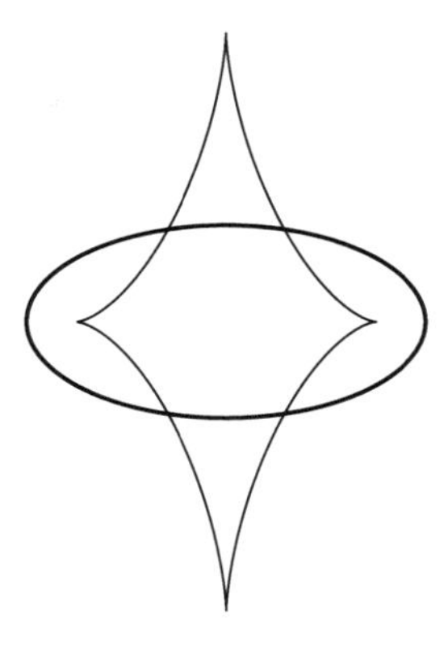

Fig. 3.7 The ellipse and its evolute

- Cartesian equation:

$$x^2 + y^2 = R^2, \quad R \neq 0.$$

- Equation in polar coordinates

$$r = R, \qquad R \in \mathbb{R}, \ R \neq 0.$$

- Intrinsic equation

$$\kappa = \frac{1}{R}, \qquad R \in \mathbb{R}, \ R \neq 0.$$

- Length: $2\pi R$.
- Area: πR^2.

3.3 The Ellipse

The interested reader will find more about the ellipse in Sect. 1.11 of [4], *Trilogy II*.

- Parametric representation

$$f(t) = (a \cos t, b \sin t).$$

- Cartesian equation

$$\left(\frac{x}{a}\right)^2 + \left(\frac{y}{b}\right)^2 = 1.$$

- The area is πab.
- The evolute is the curve with equation

$$(ax)^{\frac{2}{3}} + (by)^{\frac{2}{3}} = \left(a^2 - b^2\right)^{\frac{2}{3}} \quad \text{(see Fig. 3.7)}.$$

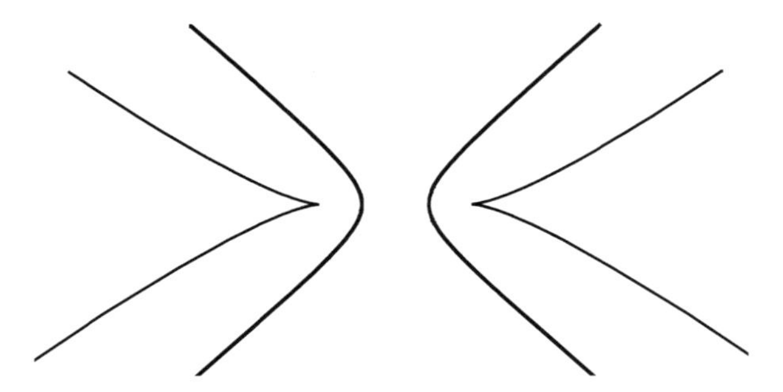

Fig. 3.8 The hyperbola and its evolute

- The ellipse is the locus of those points of the plane whose sum $2R$ of distances to two fixed points $F_1 = (k, 0)$ and $F_2 = (-0, k)$ (the foci) is constant. More precisely, $R = a$ and $k = \sqrt{a^2 - b^2}$, with the convention $a \geq b$ and of course, $R > k$.
- The ellipse admits the two axes $x = 0$ and $y = 0$ as axes of symmetry and the origin $(0, 0)$ as center of symmetry.
- The ellipse is the section of a circular cone by a plane not containing the vertex of the cone and cutting all the rulings of the same sheet of the cone.

3.4 The Hyperbola

The interested reader will find more about the hyperbola in Sect. 1.12 of [4], *Trilogy II*. Of course the hyperbola should be regarded as the union of two curves or as a curve with two branches, depending of the point of view (see the discussion in Sect. 1.1).

- Parametric representation

$$f(t) = (a \cosh t, b \sinh t)$$

or

$$g(t) = \left(\frac{a}{2}\left(t + \frac{1}{t}\right), \frac{b}{2}\left(t - \frac{1}{t}\right)\right), \quad t \neq 0.$$

- Cartesian equation

$$\left(\frac{x}{a}\right)^2 - \left(\frac{y}{b}\right)^2 = 1.$$

- The evolute is the curve with equation (see Fig. 3.8)

$$(ax)^{\frac{2}{3}} - (by)^{\frac{2}{3}} = \left(a^2 + b^2\right)^{\frac{2}{3}}.$$

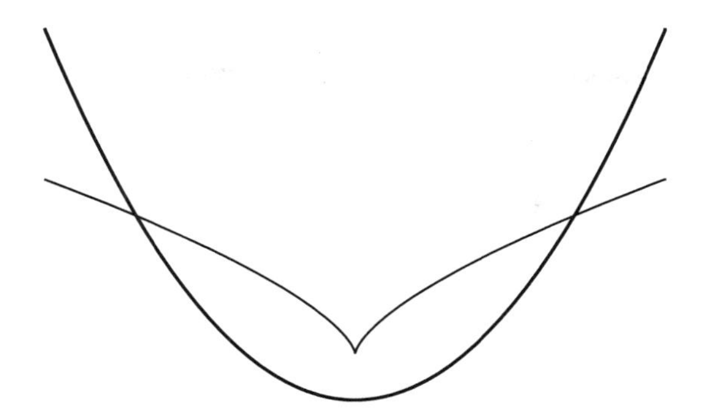

Fig. 3.9 The parabola and its evolute

- The hyperbola is the locus of those points of the plane whose difference $2R$ of distances to two fixed points $F_1 = (k, 0)$ and $F_2 = (-k, 0)$ (the foci) is constant. More precisely, $R = a$ and $k = \sqrt{a^2 + b^2}$ with of course $k > R$.
- The hyperbola admits the two axes $x = 0$ and $y = 0$ as axes of symmetry and the origin $(0, 0)$ as center of symmetry.
- The two lines $y = \pm\sqrt{\frac{b}{a}}x$ are asymptotes of the hyperbola.
- The hyperbola is the section of a circular cone by a plane not containing the vertex of the cone and cutting both sheets of the cone.
- Choosing the asymptotes as (in general, non-orthogonal) coordinate axes, the Cartesian equation takes the form

$$xy = k$$

and a parametric representation is thus

$$f(t) = \left(t, \frac{k}{t}\right), \quad t \neq 0.$$

3.5 The Parabola

The interested reader will find more about the parabola in Sect. 1.13 of [4], *Trilogy II*. The parabola has also already been considered several times in this book.

- Parametric representation:

$$f(t) = (t, at^2).$$

- Cartesian equation:

$$y = ax^2.$$

- The evolute is the semi-cubic parabola with equation (see Fig. 3.9)

$$27ax^2 = 2(2y - a)^3.$$

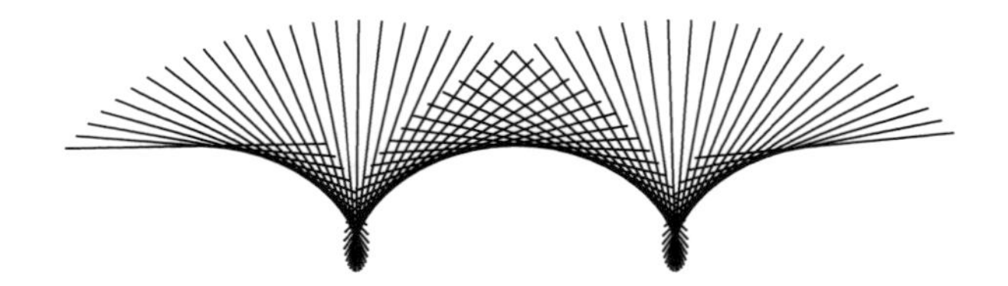

Fig. 3.10 The cycloid as an envelope

- The parabola is the locus of the points in the plane whose distance to a fixed line d (the focal line) is equal to the distance to a fixed point F (the focus). More precisely, d is the line with equation $y = -\frac{1}{4a}x$ and F has the coordinates $(0, \frac{1}{4a})$.
- The axis $x = 0$ is an axis of symmetry of the parabola.
- The parabola is the section of a circular cone by a plane not containing the vertex of the cone and parallel to a ruling.

3.6 The Cycloid

The cycloid has already largely been studied in Sects. 1.9 and 1.12.

- Parametric representation

$$f(t) = R(t - \sin t, 1 - \cos t).$$

- Area of an arch: $3\pi R^2$.
- Length of an arch: $8R$.
- The evolute of a cycloid is another cycloid with the same size.
- The cycloid is the trajectory of a point of a circle of radius R which rolls along the x-axis.
- The cycloid is also the envelope of a diameter of a circle of radius $2R$ which rolls along the x-axis (see Fig. 3.10).
- Turned "upside-down", a half arch of a cycloid

$$g(t) = R(t - \sin t, \cos t - 1), \quad 0 \leq t \leq \pi$$

  is the curve of fastest descent of a particle, between two of its points, in a gravitational field.
- In the same setting, the time taken for the particle to reach the bottom point of the arch is independent of the starting point (see Theorem B.1).

3.7 The Cardioid

The *cardioid*—the *heart-shaped curve*—was considered early in the 18th century, by *La Hire*, who calculated its length.

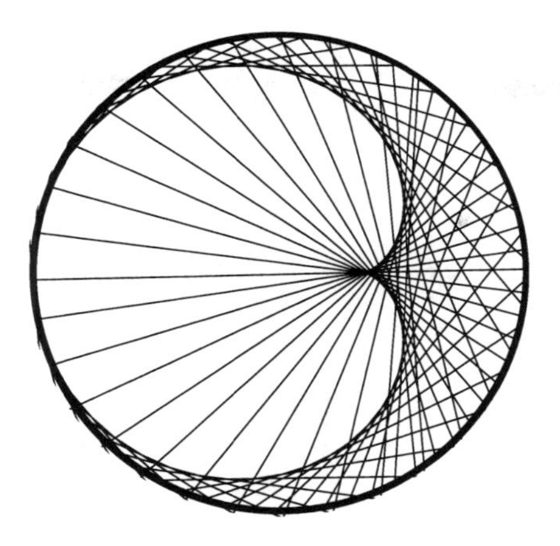

Fig. 3.11 The cardioid as a caustic curve

- The parametric representation of the cardioid is

$$f(t) = R(2\cos t - \cos 2t, 2\sin t - \sin 2t).$$

- The length is $16R$.
- The area is $6\pi R^2$.
- $(R, 0)$ is a cusp point.
- The equation in polar coordinates with the cusp point as origin is

$$r = 2R(1 - \cos\theta).$$

- In this polar equation, the tangents at two points of parameters θ and $\theta + \frac{2\pi}{3}$ are parallel; in particular, there are always three tangents in a given direction.
- All chords of the cardioid passing through the cusp point have the same length $4R$.
- The tangents to the cardioid, at the two extremities of a chord containing the cusp point, are perpendicular.
- The evolute of a cardioid is another cardioid, three times smaller.
- The cardioid is the epicycloid obtained when a circle of radius R rolls around a circle of the same radius R.
- The cardioid is the pedal curve of a circle with respect to a point P of this circle.
- The cardioid is the conchoid of a circle with respect to a point P of this circle and a length ℓ equal to the radius of the circle.
- The cardioid is the caustic of a circle with respect to a point P of this circle (see Fig. 3.11, where the segments are the "reflected rays" with $(-1, 0)$ as emission point).
- The cardioid is the envelope of the family of circles whose center is on a given circle and passing through a fixed point P of this circle.
- The cardioid is the inverse (see Sect. 5.7 in [3], *Trilogy I*) of a parabola with respect to its focus.

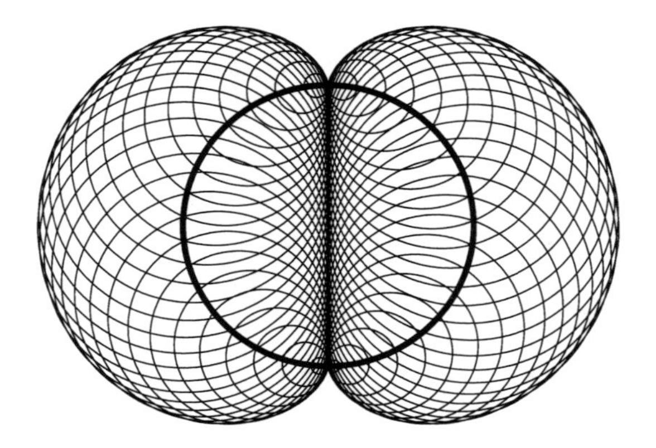

Fig. 3.12 The nephroid as an envelope

3.8 The Nephroid

The name *nephroid* refers to the shape of this *kidney-shaped curve*.

- A parametric representation is given by

$$f(t) = R(3\sin t - \sin 3t, 3\cos t - \cos 3t).$$

- The length is $24R$.
- The area is $12\pi R^2$.
- The evolute is another nephroid, twice smaller.
- The nephroid is the epicycloid obtained when a circle of radius R rolls on a circle of radius $2R$.
- The nephroid is the envelope of the family of those circles whose center is on a fixed circle and which are tangent to a fixed diameter of the fixed circle (see Fig. 3.12).
- The nephroid is the envelope of a fixed diameter of a circle which rolls around a fixed circle of the same diameter.
- The nephroid is the caustic of a cardioid with respect to its cusp point.

3.9 The Astroid

The name *astroid* clearly comes from the *star-shaped* aspect of the curve. This curve was considered by *Leibniz* in 1715.

- A parametric representation is

$$f(t) = R\bigl(\cos^3 t, \sin^3 t\bigr).$$

- The Cartesian equation is

$$x^{\frac{2}{3}} + y^{\frac{2}{3}} = R^{\frac{2}{3}}.$$

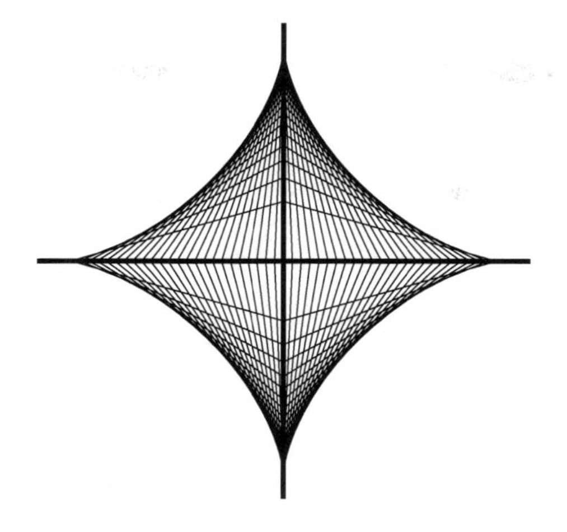

Fig. 3.13 The asteroid as an envelope

- The length is $6R$.
- The area is $6\pi R^2$.
- The evolute is another astroid, twice as large.
- The astroid is the hypocycloid obtained when a circle of radius R rolls inside a circle of radius $4R$.
- The astroid is the hypocycloid obtained when a circle of radius $3R$ rolls inside a circle of radius $4R$.
- The astroid is the envelope of the family of segments of fixed length whose extremities run on two perpendicular axis (see Fig. 3.13).
- The astroid is the envelope of the family of all ellipses

$$\left(\frac{x}{a}\right)^2 + \left(\frac{y}{b}\right)^2 = 1$$

whose sum $k = a + b$ of the lengths of the two axes is constant.

3.10 The Deltoid

The name *deltoid* refers to the Greek letter *Delta*: Δ. The *Deltoid* is also called *Steiner's hypocycloid*.

- A parametric representation is given by

$$f(t) = R(2\cos t + \cos 2t, 2\sin t - \sin 2t).$$

- The length is $16R$.
- The area is $2\pi R^2$.

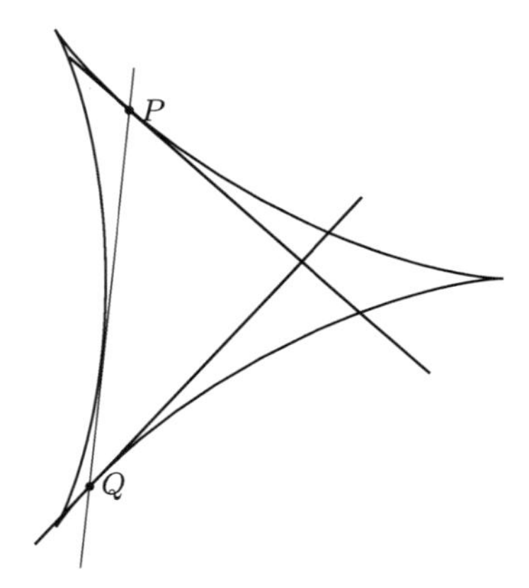

Fig. 3.14 The deltoid

- The evolute is another deltoid, three times bigger.
- The deltoid is the hypocycloid obtained when a circle of radius R rolls inside a circle of radius $3R$.
- The deltoid is the hypocycloid obtained when a circle of radius $2R$ rolls inside a circle of radius $3R$.
- The deltoid is the envelope of a fixed diameter of a circle of radius $2R$ rolling inside a circle of radius $3R$.
- The tangent to the deltoid, terminated at the other two points P and Q of intersection with the deltoid, is of constant length.
- In these circumstances, the tangent at P is perpendicular to the tangent at Q (see Fig. 3.14).

3.11 The Limaçon of Pascal

The *Limaçon of Pascal* (also known as the *snail of Pascal*) is in fact a family of curves which contains as a special case the *cardioid* (see Sect. 3.7). Let us mention that the name *Limaçon of Pascal* does not refer to the famous mathematician and philosopher *Blaise Pascal*, but to his father, *Étienne Pascal*.

- A parametric representation is

$$f(t) = (k\cos t + a\cos 2t, k\sin t + a\sin 2t).$$

  (see Fig. 3.15 which presents—at different scales—the cases $(a = 1, k = 3)$ and $(a = 1, k = 1)$).
- When $k = 2a$, the curve is a cardioid (see Fig. 3.11).
- When $k = a$, the curve is the so-called *trisectrix* (see Sect. 2.3 in [3], *Trilogy I* and the right hand picture in Fig. 3.15).

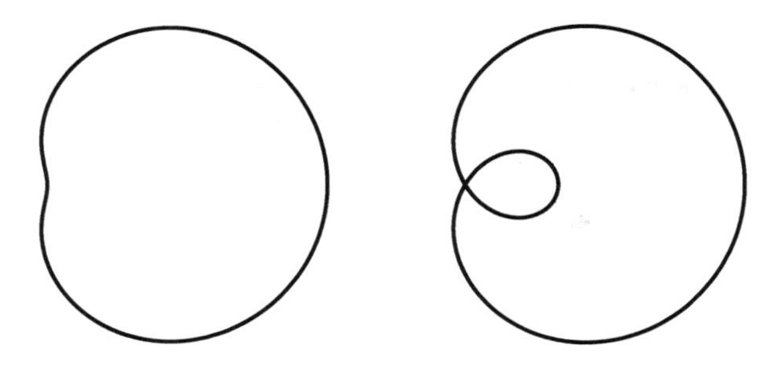

Fig. 3.15 The limaçon of Pascal

- The limaçon is the pedal curve of a circle of radius k with respect to a point whose distance to the center of the circle is $2a$.
- The limaçon is the conchoid of a circle of radius a with respect to a point on this circle, the fixed distance ℓ being k.
- The limaçon is the inverse of a conic (see Sect. 5.7 in [3], *Trilogy I*) with respect to one of its foci.
- When $k \geq 2a$ (see the left picture in Fig. 3.15), the area is $(2a^2 + k^2)\pi$.
- When $k < 2a$ (see the right picture in Fig. 3.15), the area of the inner loop is $a^2(\pi - \frac{3}{2}\sqrt{3})$ while the area between the two loops is $a^2(\pi + 3\sqrt{3})$.
- The polar equation, with origin the cusp or multiple point, is

$$r = k - 2a\cos\theta.$$

3.12 The Lemniscate of Bernoulli

A *lemniscate* is the pedal curve of a hyperbola with respect to its center of symmetry. The *lemniscate of (James) Bernoulli* corresponds to the case where the hyperbola is *rectangular*, that is, with orthogonal asymptotes:

$$\left(\frac{x}{a}\right)^2 - \left(\frac{y}{a}\right)^2 = 1.$$

The *bow* or *ribbon* shape of this curve is at the origin of its name; in Latin, *lemniscus* indicates a type of ribbon.

- A parametric representation is

$$f(t) = a\left(\frac{\sin t}{1 + \cos^2 t}, \frac{\sin t \cdot \cos t}{1 + \cos^2 t}\right).$$

- The Cartesian equation is

$$\left(x^2 + y^2\right)^2 = a^2\left(x^2 - y^2\right).$$

Fig. 3.16 The lemniscate of Bernoulli as an envelope

- The polar equation is

$$r^2 = a^2 \cos 2\theta.$$

- The area is a^2.
- The lemniscate of Bernoulli is the pedal curve of a hyperbola with respect to its center of symmetry.
- The lemniscate of Bernoulli is the inverse of a rectangular hyperbola with respect to its center of symmetry.
- It is the envelope of those circles whose center is on a fixed rectangular hyperbola and passing through the center of symmetry of this hyperbola (see Fig. 3.16).
- The lemniscate of Bernoulli is the cissoid of a circle of radius R with respect to a point at a distance $R\sqrt{2}$ from the center.
- It is also the locus of those points whose product of distances to two fixed points P and Q equals $\frac{1}{4}\|\overrightarrow{PQ}\|$.

3.13 The Conchoid of Nicomedes

The name *conchoid* refers to a *mussel-shell-shaped curve*. According to *Pappus*, *Nicomedes*, in the second century BC, was the first to consider a conchoid. Here we restrict our attention to this *conchoid of Nicomedes*.

- A parametric representation is

$$f(t) = (a + k\cos t, a\tan t + k\sin t).$$

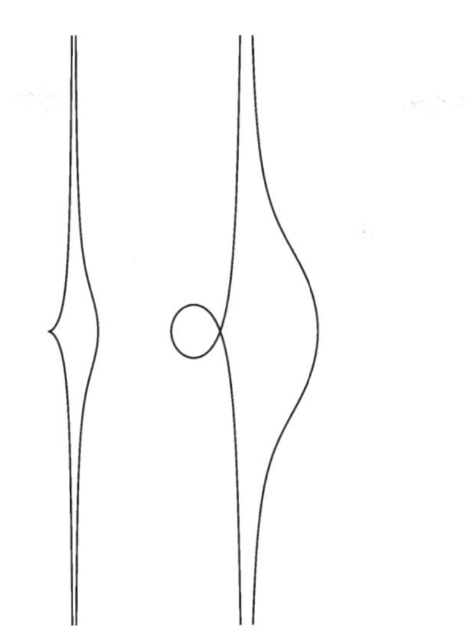

Fig. 3.17 The conchoid of Nicomedes

- The Cartesian equation is

$$k^2x^2 = (a - x)^2\big(x^2 + y^2\big).$$

  (Notice that for $k < a$, this equation includes an isolated point at the origin.)
- The *conchoid of Nicomedes* is the conchoid of a straight line with respect to a point at a distance $a > 0$ from the line; k is the fixed length ℓ of Definition 3.1.3. Depending on the values of the parameters a and k, the curve can take various shapes, with or without a loop (see Fig. 3.17).
- The line $x = a$, which is the line involved in the definition of the conchoid of Nicomedes, is an asymptote.
- The area between each branch of the curve and the asymptote is infinite.

3.14 The Cissoid of Diocles

The *cissoid of Diocles* is one of the numerous curves used to attempt to solve cubical problems. The name *cissoid* suggests an *ivy-leaf-shaped* curve, even if this requires some imagination.

- A parametric representation is

$$f(t) = R\left(\frac{2t^2}{1+t^2}, \frac{2t^3}{1+t^2}\right).$$

- The Cartesian equation is

$$y^2(2R - x) = x^3.$$

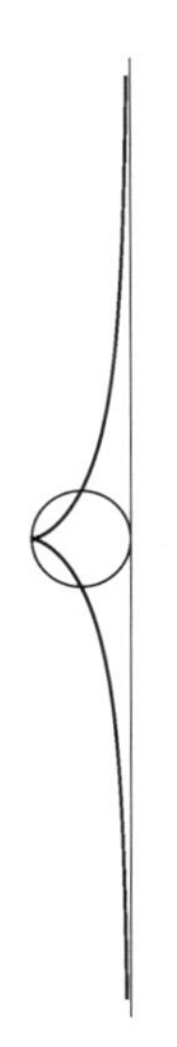

Fig. 3.18 The cissoids of Diocles

- The polar equation is

$$r = 2R \sin\theta \tan\theta.$$

- The area between the curve and its asymptote is equal to $3\pi R^2$.
- The *cissoid of Diocles* is the cissoid of a circle of radius R with respect to a tangent to this circle and the second extremity of the diameter through the tangency point (see Fig. 3.18).
- The line $x = 2R$, which is the tangent involved above, is an asymptote.

3.15 The Right Strophoid

It seems that the study of the right strophoid began during the 17th century, with *Roberval* and *Torricelli*. The origin of the name is not clear: it is perhaps a reference to a *strophos*, the Latin name for a belt with a twist used to hold a sword.

- A parametric representation is

$$f(t) = a\left(\frac{t^2-1}{t^2+1}, t\frac{t^2-1}{t^2+1}\right).$$

- The Cartesian equation is

$$y^2(a-x) = x^2(a+x).$$

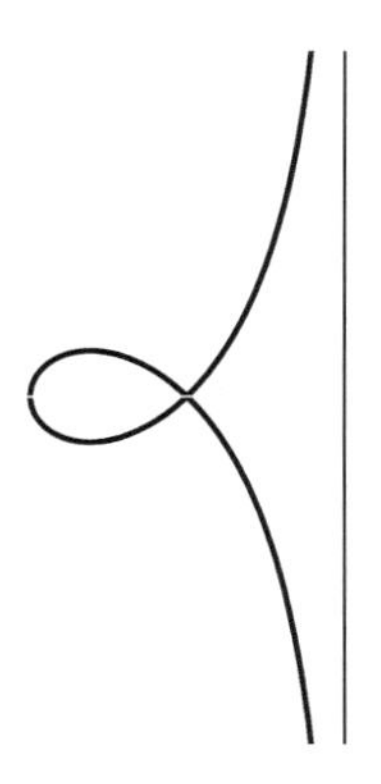

Fig. 3.19 The right strophoid

- The polar equation is

$$r = a\left(\frac{1}{\cos\theta} - 2\cos\theta\right).$$

- The line $x = a$ is an asymptote (see Fig. 3.19).
- The area of the loop is $\frac{1}{2}a^2(4 - \pi)$.
- The area between the curve and its asymptote is $\frac{1}{2}a^2(4 + \pi)$.
- The right strophoid is the cissoid of a circle with respect to one of its diameters and one extremity of the diameter perpendicular to the first diameter.
- The strophoid is the pedal curve of a parabola with respect to the intersection of the directrix and the axis of symmetry of that parabola.
- It is also the inverse of a rectangular hyperbola with respect to one of its vertices.
- The strophoid is its own inverse for an inversion with center the origin and power a^2 (see Sect. 5.7 in [3], *Trilogy I*).
- A secant through the point $A = (-a, 0)$ makes equal angles with the tangents to the strophoid, at the two intersection points.

3.16 The Tractrix

Imagine a horse on a towpath, pulling a boat. The problem of computing the trajectory of the boat, in the absence of any steering, was solved by *Huygens*. Very naturally, he gave the name *tractrix* to the corresponding curve.

- A parametric equation is

$$f(t) = k\left(\log\left(\frac{1 + \sin t}{\cos t}\right) - \sin t, \cos t\right).$$

- The x-axis is an asymptote (see Fig. 3.20).

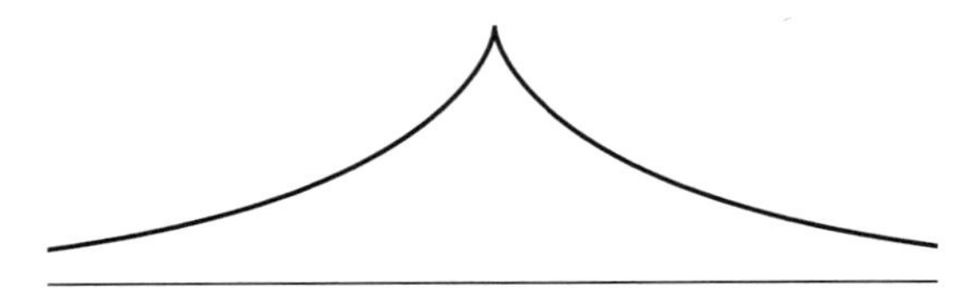

Fig. 3.20 The tractrix

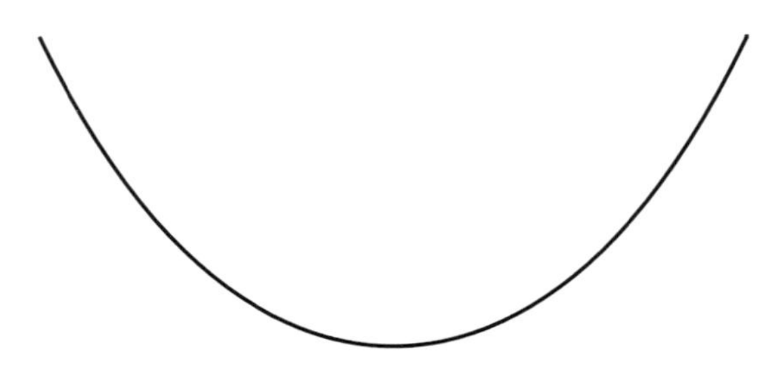

Fig. 3.21 The catenary

- The length of the tangent, from a point of the tractrix to the intersection with the x-axis, is constant.
- The evolute is a catenary.
- The area between the tractrix and its asymptote is $\frac{1}{2}\pi k^2$.
- The tractrix is the trajectory of an object, attached at the end of a string of fixed length, as the other end of the string moves along a straight line.

3.17 The Catenary

Galileo suggested that a rope, attached at its two ends, would hang in the shape of a parabola. This conjecture proved to be false: the correct shape of the rope is the so-called *catenary*, as discovered around 1690. The name refers clearly to a (hanging) chain (a *catena*).

- A parametric representation is

$$f(t) = \left(t, k \cosh \frac{t}{k}\right).$$

- The catenary is the locus of the focus of a parabola which rolls on a straight line.
- The catenary is the form assumed by a flexible chain hanging in a gravitational field (see Fig. 3.21).

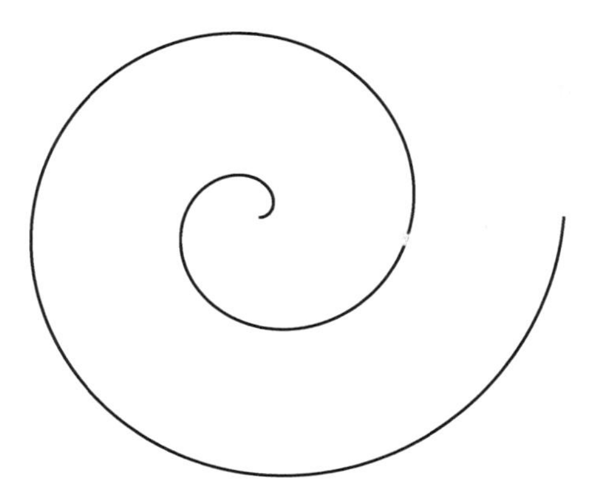

Fig. 3.22 The Archimedean spiral

3.18 The Spiral of Archimedes

The *spiral of Archimedes*—as the name indicates—was first studied in Greek Antiquity (see Sect. 4.5 in [3], *Trilogy I*). *Archimedes* was already able to compute its tangent (see Sect. 1.5).

- A parametric equation of the spiral of *Archimedes* is

$$f(t) = k(t\cos t, t\sin t).$$

  (See Fig. 3.22.)
- The polar equation is $r = kt$.
- The *spiral of Archimedes* is the trajectory of a point which moves at a constant linear speed on a line as this line turns at a constant angular speed around one of its points.

3.19 The Logarithmic Spiral

The *logarithmic spiral* is often introduced as follows. Imagine four runners, initially positioned at the four vertices of a square. They start running at the same time, at the same constant speed, each runner running at each instant in the direction of the position of the next runner. The trajectory of each runner is a special instance of a so-called *logarithmic spiral*, already studied in Sect. 1.7. In this example, if O is the center of the square and A' is the position reached by runner A, considering the square $A'B'C'D'$ we conclude that the angle between the vector OA' and the direction of the trajectory is always 45 degrees.

More generally a *logarithmic spiral*, also called an *equiangular spiral* is a curve such that the position vector $\overrightarrow{OP}$, for all points P of the curve, makes a constant

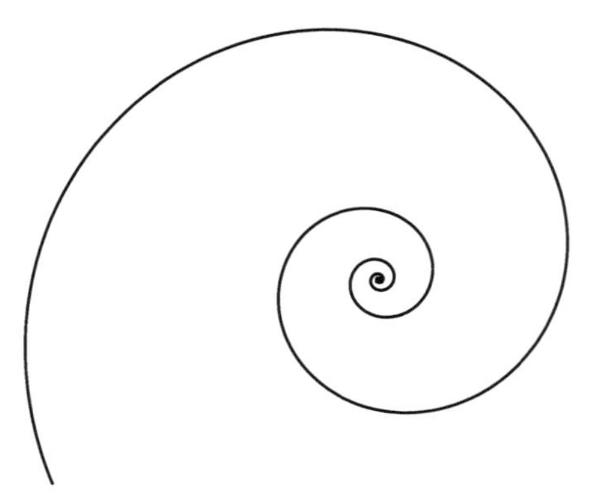

Fig. 3.23 The logarithmic spiral

angle with the tangent to the curve at P. These spiral have striking "stability properties".

- A parametric equation of a logarithmic spiral is

$$f(t) = a\big(e^{bt}\cos t, e^{bt}\sin t\big).$$

  (See Fig. 3.23.)
- The polar equation of a logarithmic spiral is $r = ae^{bt}$.
- The angle between the tangent at a point P of the spiral and the vector $\overrightarrow{OP}$ is constant.
- A secant through the origin meets the curve at points whose distances to the origin are in geometric progression.
- The evolute of a logarithmic spiral is an isometric logarithmic spiral.
- The pedal curve of a logarithmic spiral with respect to the origin is an isometric logarithmic spiral.
- The caustic of a logarithmic spiral with respect to the origin is an isometric logarithmic spiral.
- Every curve homothetic to a logarithmic spiral is an isometric logarithmic spiral.
- The inverse (see Sect. 5.7 in [3], *Trilogy I*) of a logarithmic spiral with respect to the origin (and with arbitrary power) is an isometric logarithmic spiral.

3.20 The Spiral of Cornu

The *spiral of Cornu* is the curve whose curvature is proportional to the distance traveled on the curve.

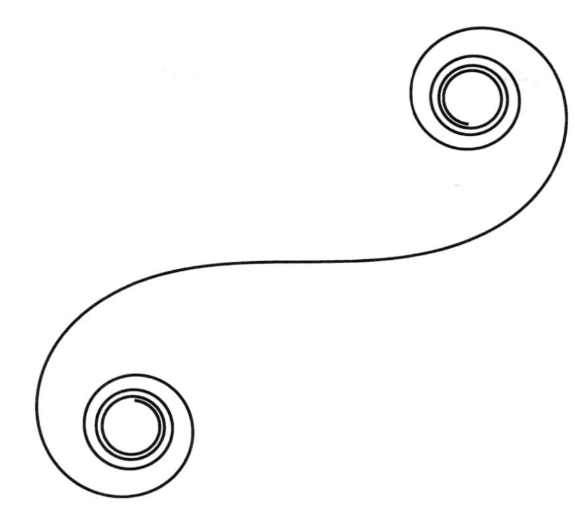

Fig. 3.24 The spiral of Cornu

- Parametric representation:

$$f(t) = \left(\int_0^s \cos \frac{ks^2}{2} ds, \int_0^s \sin \frac{ks^2}{2} ds \right).$$

(See Fig. 3.24.)

- Intrinsic equation:

$$\kappa = ks.$$

Chapter 4
Skew Curves

We now turn our attention to the case of curves in three dimensional real space. Of course many aspects already considered for plane curves extend to the three dimensional case and in fact, to arbitrary finite dimensions. This is so in particular for the notions of parametric representation, tangent, length and normal representation. We therefore treat these questions quite rapidly, since they consist essentially of repeating arguments already developed in Chap. 2.

In fact, we shall focus our attention on the generalization, in dimension 3, of the notion of curvature. The circle is a curve with constant non-zero curvature: in the plane, it is the only such curve. However, in three dimensional space, there are many more such curves: for example every circular helix (a curve having the shape of a bolt; see Fig. 4.1) has non-zero constant curvature. Thus the curvature no longer suffices to characterize a skew curve: another parameter is required: the *torsion*. The plane curves are characterized by having a zero torsion. An intuitive approach to these questions has been developed in Sect. 1.13: we focus here on a systematic treatment of the theory.

After giving precise definitions, we establish efficient formulas to compute the curvature and the torsion. We also study the famous "moving" *Frenet trihedron* attached to each point of the curve. The *Frenet* formulas constitute the basic ingredient for proving our main result concerning skew curves: the existence of *intrinsic equations*, that is, the description of a curve in terms of its curvature and its torsion. In contrast to the case of plane curves, recapturing a parametric equation of the curve from the knowledge of the curvature and the torsion can no longer be done via a simple integration process: in fact the problem reduces to solving a system of differential equations.

All these results can further be generalized in arbitrary finite dimension: we only give the corresponding useful hints in our section devoted to "problems".

4.1 Regular Skew Curves

With the considerations of Sect. 2.1 in mind, we make the following definition:

F. Borceux, *A Differential Approach to Geometry*, DOI 10.1007/978-3-319-01736-5_4,

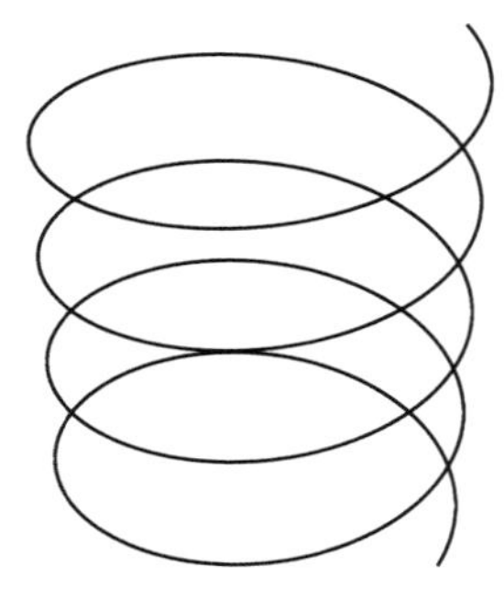

Fig. 4.1 The helix

Definition 4.1.1

1. A *parametric representation* of class $\mathcal{C}^k$ (with $k \in \mathbb{N} \cup \{\infty\}$) of a skew curve is a locally injective function of class $\mathcal{C}^k$

$$f\colon \,]a,b[\, \longrightarrow \mathbb{R}^3, \quad a,b \in \mathbb{R} \cup \{-\infty,+\infty\}.$$

2. Two parametric representations f and g of class $\mathcal{C}^k$ are *equivalent* in class $\mathcal{C}^k$ when there exist inverse bijections φ, φ^{-1} of class $\mathcal{C}^k$ such that

$$f = g \circ \varphi, \qquad g = f \circ \varphi^{-1}.$$

3. A *skew curve* of class $\mathcal{C}^k$ is an equivalence class of parametric representations of class $\mathcal{C}^k$, for the equivalence relation described in 2.
4. The *support* of a skew curve is the image of (any one of) its parametric representations.

Example 4.1.2 The *circular helix* is the curve with parametric representation (see Fig. 4.1)

$$f(t) = (R\cos t, R\sin t, kt), \quad R > 0,\ k \neq 0.$$

Proof The function f is trivially locally injective, since its third component is injective. Notice that the case $k = 0$ would have produced a circle of radius R in the (x,y)-plane. □

Definition 4.1.3

1. A parametric representation $f(t)$ of a skew curve is *regular* when it is of class $\mathcal{C}^1$ and $f'(t) \neq 0$ at each point.
2. This representation is called 2-*regular* when it is of class $\mathcal{C}^2$ and $f'(t)$, $f''(t)$ are linearly independent vectors at each point.

As expected one has:

Proposition 4.1.4 *For a function*

$$f\colon]a,b[\longrightarrow \mathbb{R}^3, \qquad t\mapsto f(t)$$

the following conditions are equivalent:

1. *f is a regular parametric representation of a skew curve;*
2. *f is of class $\mathcal{C}^1$ and $f'(t)\neq 0$ for all $t\in]a,b[$.*

Proof This follows by Lemma 2.2.4. □

Moreover, we have:

Proposition 4.1.5

1. *If a curve of class $\mathcal{C}^k$, $k\geq 1$, admits a regular parametric representation $f(t)$, then all its parametric representations $g(s)$ are regular. Moreover the vectors $f'(t)$ and $g'(s)$, at corresponding points, are proportional.*
2. *If a curve of class $\mathcal{C}^k$, $k\geq 2$, admits a 2-regular parametric representation $f(t)$, then all its parametric representations $g(s)$ are 2-regular. Moreover the vectors $f'(t)$ and $f''(t)$ generate at each point the same vector subspace as the vectors $g'(s)$ and $g''(s)$ at this point.*

Proof We use the notation of Definition 4.1.1. From $f=g\circ\varphi$ we deduce $f'=(g'\circ\varphi)\varphi'$. If $f'\neq 0$, necessarily $g'\neq 0$ and we have the expected proportionality.

Assume now that $k\geq 2$, with f' and f'' linearly independent. We have

$$f''=\big(g''\circ\varphi\big)\big(\varphi'\big)^2+\big(g'\circ\varphi\big)\varphi''.$$

This proves that g'' cannot be a multiple of g', otherwise f'' would be a multiple of g' which is itself a multiple of f', by the first part of the proof. □

By Proposition 4.1.5, it now makes sense to define:

Definition 4.1.6

1. The *tangent* to a regular curve represented by $f(t)$ is the line passing through $f(t)$ and of direction $f'(t)$.
2. The *osculating plane* to a 2-regular curve represented by $f(t)$ is the plane passing through $f(t)$ and whose direction is generated by $f'(t)$ and $f''(t)$.

We have seen in Sect. 2.3 that plane curves can also be described by Cartesian equations. But in three dimensional space, a Cartesian equation

$$F(x,y,z)=0$$

describes in general a surface, not a curve: for example, a quadric (see Sect. 1.14 in [4], *Trilogy II*).

Now a straight line in $\mathbb{R}^3$ can be described by a system of two linear equations: that is, the line can be presented as the intersection of two planes (see Sect. 1.6 in [4], *Trilogy II*). An analogous approach could be developed for curves: a system of two Cartesian equations

$$\begin{cases} F(x, y, z) = 0 \\ G(x, y, z) = 0 \end{cases}$$

can possibly represent a curve as the intersection of two surfaces. We shall not enter into this discussion. Such an approach is in any case most often technically too heavy to allow the development of an elegant theory.

4.2 Normal Representations

We now extend the results of Sects. 2.8 and 2.7. We omit several proofs, identical to those in dimension 2.

Definition 4.2.1 By a *normal* representation of a skew curve is meant a parametric representation $f(s)$ of class $\mathcal{C}^1$ such that $\|f'(s)\| = 1$ at each point.

A normal representation is thus in particular regular. Moreover, we have:

Proposition 4.2.2 *Given a normal representation $f(s)$ of class $\mathcal{C}^2$ of a skew curve:*

1. $(f'|f'') = 0$.
2. *The curve is 2-regular as soon as $f''(s) \neq 0$ at each point.*

Proof Differentiating the equality $(f'|f') = 1$ yields $2(f'|f'') = 0$. Moreover, two perpendicular vectors are linearly independent as soon as they are non-zero. □

But one has more:

Proposition 4.2.3 *Consider two normal representations $f(t)$ and $g(s)$ of the same curve of class $\mathcal{C}^1$.*

1. *The corresponding change of parameter formula has the form*

$$\varphi(t) = \pm t + k, \quad k \in \mathbb{R}.$$

2. *One always has $f' = \pm(g' \circ \varphi)$.*
3. *In class $\mathcal{C}^2$ one has further $f'' = g'' \circ \varphi$.*

Proof Differentiating $f = g \circ \varphi$ yields

$$f' = (g' \circ \varphi)\varphi', \qquad f'' = (g'' \circ \varphi)(\varphi')^2 + (g' \circ \varphi)\varphi''.$$

Since

$$1 = \|f'\| = \|(g' \circ \varphi)\varphi'\| = \|g' \circ \varphi\| \cdot |\varphi'| = |\varphi'|$$

we obtain $\varphi' = \pm 1$, which implies condition 2. By integration we get $\varphi(t) = \pm t + k$, thus condition 1. By differentiation we obtain $\varphi'' = 0$, from which we get condition 3. Notice that the sign must be the same at all points, by continuity of φ. □

Let us now prove the existence of normal representations. With Definition 2.7.2 in mind, we make the following definition.

Definition 4.2.4 Consider a skew curve of class C^1 represented by

$$f\colon]a,b[\longrightarrow \mathbb{R}^3, \qquad t \mapsto f(t).$$

Given $c, d \in]a,b[$, the *length* of the arc of the curve between the points with parameters c and d is the curve integral of the constant function 1 along

$$f\colon [c,d] \longrightarrow \mathbb{R}^3.$$

Of course, we have:

Proposition 4.2.5 *Under the conditions of Definition* 4.2.4, *the length of the arc of the curve is also given by*

$$\int_c^d \|f'\|$$

and it is independent of the chosen parametric representation.

Proof The proof of Proposition 2.7.6 applies as such. □

Example 4.2.6 The length of a turn of the circular helix

$$f(t) = (R\cos t, R\sin t, kt)$$

is equal to $2\pi\sqrt{R^2 + k^2}$.

Proof We have

$$f'(t) = (-R\sin t, R\cos t, k).$$

By Proposition 4.2.5, the length of a turn is

$$\int_0^{2\pi} \sqrt{R^2 + k^2} = 2\pi\sqrt{R^2 + k^2}.$$

□

Proposition 4.2.7 *Consider a regular representation of class C^k of a skew curve*

$$f\colon]a,b[\longrightarrow \mathbb{R}^3, \qquad t \mapsto f(t).$$

Given $t_0 \in]a,b[$, the function

$$\sigma\colon]a,b[\longrightarrow \big]\sigma(a),\sigma(b)\big[, \qquad \sigma(t) = \int_{t_0}^{t} \|f'\|$$

is a change of parameter of class $\mathcal{C}^k$. The corresponding representation

$$\overline{f} = f \circ \sigma^{-1}\colon \big]\sigma(a),\sigma(b)\big[\longrightarrow \mathbb{R}^3, \qquad s \mapsto \overline{f}(s)$$

is normal of class $\mathcal{C}^k$. In particular, when f is 2-regular, so is $\overline{f}$.

Proof The proof of Proposition 2.8.2 applies as such. The last statement follows from Proposition 4.1.5. □

4.3 Curvature

Definition 2.9.1 can be transposed to skew curves since, by Proposition 4.2.3, it is independent of the chosen normal representation:

Definition 4.3.1 Let

$$f\colon]a,b[\longrightarrow \mathbb{R}^3, \qquad s \mapsto f(s)$$

be a normal representation of class $\mathcal{C}^2$ of a skew curve. The *curvature* $\kappa(s)$ at the point with parameter s is the quantity $\|f''(s)\|$.

Our first concern is to establish a formula to easily compute the curvature.

Proposition 4.3.2 *Let*

$$f\colon]a,b[\longrightarrow \mathbb{R}^3, \qquad t \mapsto f(t)$$

be an arbitrary parametric representation of class $\mathcal{C}^2$ of a skew curve. The curvature at the point with parameter t is the quantity

$$\kappa(t) = \frac{\|f'(t) \times f''(t)\|}{\|f'(t)\|^3}.$$

Proof Let us use the notation of Proposition 4.2.7 and consider the normal representation $\overline{f}$. From $\sigma = \int_{t_0}^{t} \|f'\|$ we obtain

$$\sigma' = \|f'\|, \qquad (\sigma^{-1})' = \frac{1}{\sigma' \circ \sigma^{-1}} = \frac{1}{\|f' \circ \sigma^{-1}\|}.$$

From $\overline{f} = f \circ \sigma^{-1}$ we obtain further

$$\overline{f}' = (f' \circ \sigma^{-1})(\sigma^{-1})', \qquad \overline{f}'' = (f'' \circ \sigma^{-1})(\sigma^{-1})'^2 + (f' \circ \sigma^{-1})(\sigma^{-1})''.$$

Using the properties of the cross product (see Sect. 1.7 in [4], *Trilogy II*) together with Proposition 4.2.2, we also have

$$\|\overline{f}' \times \overline{f}''\| = \|\overline{f}'\| \cdot \|\overline{f}''\| \cdot |\sin \measuredangle(\overline{f}', \overline{f}'')| = \|\overline{f}''\| = \kappa.$$

All this yields, again by the properties of the cross product

$$\begin{aligned}\kappa &= \|(f' \circ \sigma^{-1})(\sigma^{-1})' \times (f'' \circ \sigma^{-1})(\sigma^{-1})'^2 + (f' \circ \sigma^{-1})(\sigma^{-1})''\| \\ &= \|(f' \circ \sigma^{-1})(\sigma^{-1})' \times (f'' \circ \sigma^{-1})(\sigma^{-1})'^2\| \\ &= (\sigma^{-1})'^3 \, \|f' \circ \sigma^{-1} \times f'' \circ \sigma^{-1}\| \\ &= \frac{\|f' \circ \sigma^{-1} \times f'' \circ \sigma^{-1}\|}{\|f' \circ \sigma^{-1}\|^3}.\end{aligned}$$

In terms of $t = \sigma^{-1}(s)$, this is precisely the formula of the statement. □

Example 4.3.3 The curvature of the circular helix (see Example 4.1.2)

$$f(t) = (R \cos t, R \sin t, kt)$$

is constant and equal to $\frac{R}{R^2+k^2}$.

Proof We have

$$f'(t) = (-R \sin t, R \cos t, k), \qquad f''(t) = (-R \cos t, -R \sin t, 0)$$

from which, by Proposition 4.3.2

$$\kappa(t) = \frac{\|(kR \sin t, -kR \cos t, R^2)\|}{\|(-R \sin t, R \cos t, k)\|^3} = \frac{\sqrt{k^2R^2 + R^4}}{(\sqrt{R^2 + k^2})^3} = \frac{R}{R^2 + k^2}.$$ □

Example 4.3.4 A regular skew curve of class C^2 has constant curvature 0 if and only if it is a straight line.

Proof Let $f(t)$ be a normal representation of the curve. From $\kappa(t) = 0$ we get $f''(t) = 0$. Integrating twice, we conclude that the three components of f are linear functions. □

4.4 The Frenet Trihedron

We are now going to attach an orthonormal Euclidean basis to each point of a skew curve.

Definition 4.4.1 Consider a normal representation

$$f\colon]a,b[\longrightarrow \mathbb{R}^3, \qquad s \mapsto f(s)$$

of a 2-regular skew curve. The *Frenet trihedron* at the point with parameter s is the orthonormal basis with direct orientation

- whose origin is $f(s)$;
- which is constituted of the three vectors

$$\mathbf{t}(s) = f'(s), \qquad \mathbf{n}(s) = \frac{f''(s)}{\kappa(s)}, \qquad \mathbf{b}(s) = \mathbf{t}(s) \times \mathbf{n}(s).$$

The three vectors are called respectively:

- the *tangent* vector $\mathbf{t}(s)$;
- the *normal* vector $\mathbf{n}(s)$;
- the *binormal* vector $\mathbf{b}(s)$.

The planes generated by two of these vectors have also been given names:

- the *osculating plane*, generated by $\mathbf{t}$ and $\mathbf{n}$ (see Definition 4.1.6);
- the *normal plane*, generated by $\mathbf{n}$ and $\mathbf{b}$;
- the *rectifying plane*, generated by $\mathbf{t}$ and $\mathbf{b}$.

This definition makes perfect sense, by Proposition 4.2.2, Definition 4.3.1 and the properties of the cross product (see Example 3.2.4 in [4], *Trilogy II*).

Notice in particular that since the osculating plane is generated by the vectors $\mathbf{t}$ and $\mathbf{n}$, $\mathbf{b}$ is orthogonal to this osculating plane. Since $\mathbf{b}(s)$ is of constant length 1, its variation is only in direction and thus, measures precisely the variation in direction of the osculating plane. This is what we call the *absolute torsion* of the skew curve.

Definition 4.4.2 Let $f(s)$ be a normal representation of class $\mathcal{C}^3$ of a 2-regular skew curve. The quantity

$$\left|\tau(s)\right| = \left\| \frac{d\mathbf{b}}{ds}(s) \right\|$$

is called the *absolute torsion* of the curve at the point with parameter s.

Let us observe at once that:

Lemma 4.4.3 *The definition of the absolute torsion does not depend on the choice of the normal representation.*

Proof By Proposition 4.2.3, two normal representations give rise to the same binormal vector $\mathbf{b}$, or to opposite vectors $\mathbf{b}$. $\square$

As we did for the curvature of plane curves (see Definition 2.9.8), we shall now provide the torsion with a sign.

Lemma 4.4.4 *Consider a normal representation $f(s)$ of class $\mathcal{C}^2$ of a 2-regular skew curve. Then*

$$\frac{d\mathbf{t}}{ds} = \kappa\,\mathbf{n}.$$

Proof Trivially

$$\frac{d\mathbf{t}}{ds} = f'' = \|f''\|\frac{f''}{\|f''\|} = \kappa\,\mathbf{n}. \qquad \square$$

Lemma 4.4.5 *Consider a normal representation $f(s)$ of a 2-regular skew curve of class $\mathcal{C}^3$. One always has*

$$\frac{d\mathbf{b}}{ds} = \pm\big|\tau(s)\big|\,\mathbf{n}(s).$$

Proof From $(\mathbf{b}|\mathbf{b}) = 1$ we deduce $(\frac{d\mathbf{b}}{ds}|\mathbf{b}) = 0$. Analogously from $(\mathbf{b}|\mathbf{t}) = 0$, we deduce by Lemma 4.4.4

$$0 = \left(\frac{d\mathbf{b}}{ds}\bigg|\mathbf{t}\right) + \left(\mathbf{b}\bigg|\frac{d\mathbf{t}}{ds}\right) = \left(\frac{d\mathbf{b}}{ds}\bigg|\mathbf{t}\right) + (\mathbf{b}|\kappa\,\mathbf{n}) = \left(\frac{d\mathbf{b}}{ds}\bigg|\mathbf{t}\right)$$

since $\mathbf{b}$ is orthogonal to $\mathbf{n}$. But then $\frac{d\mathbf{b}}{ds}$ is orthogonal to both $\mathbf{b}$ and $\mathbf{t}$, thus is parallel to $\mathbf{n}$. Therefore

$$\frac{d\mathbf{b}}{ds}(s) = \pm\left\|\frac{d\mathbf{b}}{ds}(s)\right\|\mathbf{n}(s) = \pm\big|\tau(s)\big|\mathbf{n}(s). \qquad \square$$

Lemma 4.4.5 offers an easy way of providing the torsion with a sign:

Definition 4.4.6 Consider a normal representation $f(s)$ of a 2-regular skew curve of class $\mathcal{C}^3$. The *torsion* at the point with parameter s is the quantity $\tau(s)$ such that

$$\frac{d\mathbf{b}}{ds}(s) = -\tau(s)\mathbf{n}(s).$$

We shall discuss later (see Example 4.5.4) the probably unexpected choice of the sign $-$, instead of $+$, in Definition 4.4.6.

Much more importantly, in striking contrast to the case of the relative curvature for plane curves (see Warning 2.9.9), let us observe that:

Proposition 4.4.7 *Two normal representations of class $\mathcal{C}^3$ of a 2-regular skew curve give rise to the same torsion, with the same sign.*

Proof Let $f(t)$ and $g(s)$ be the two normal representations. We use Proposition 4.2.3 and its notation. When $\varphi(t)=t+k$, both representations yield the same vectors $\mathbf{t}$, $\mathbf{n}$ and $\mathbf{b}$ and the conclusion is immediate.

When $\varphi(t)=-t+k$, the two vectors $\mathbf{t}$ are opposite but the two vectors $\mathbf{n}$ are the same. It follows at once that the two vectors $\mathbf{b}$ are opposite as well. As a consequence

$$\frac{d\mathbf{b}(s)}{ds}(s)=\frac{d(-\mathbf{b}(t))}{dt}\varphi'(t)=-\frac{d\mathbf{b}(t)}{dt}(-1)=\frac{d\mathbf{b}(t)}{dt}.$$

Thus both the derivative of the vectors $\mathbf{b}$ and the vectors $\mathbf{n}$ are the same for both representations, yielding the announced result. □

Again, we will comment on this result in more detail in Sect. 4.5. Let us conclude this section with the celebrated *Frenet formulas*:

Theorem 4.4.8 *Consider a normal representation $f(s)$ of class $\mathcal{C}^3$ of a 2-regular skew curve. The derivatives of the three functions $\mathbf{t}$, $\mathbf{n}$, $\mathbf{b}$ describing the Frenet trihedron are equal to*

$$\frac{d\mathbf{t}}{ds}=\kappa\mathbf{n}$$

$$\frac{d\mathbf{n}}{ds}=-\kappa\mathbf{t}+\tau\mathbf{b}$$

$$\frac{d\mathbf{b}}{ds}=-\tau\mathbf{n}$$

where κ is the curvature and τ is the torsion.

Proof By Lemma 4.4.4 and Definition 4.4.6, it remains to prove the second formula. But from $(\mathbf{n}|\mathbf{n})=1$ we get at once $(\frac{d\mathbf{n}}{ds}|\mathbf{n})=0$, thus $\frac{d\mathbf{n}}{ds}$ is a linear combination of only $\mathbf{t}$ and $\mathbf{b}$. By Proposition 4.6.2 in [4], *Trilogy II*, we thus have

$$\frac{d\mathbf{n}}{ds}=\left(\frac{d\mathbf{n}}{ds}\middle|\mathbf{t}\right)\mathbf{t}+\left(\frac{d\mathbf{n}}{ds}\middle|\mathbf{b}\right)\mathbf{b}.$$

It remains to compute the two coefficients.

From $(\mathbf{n}|\mathbf{t})=0$ we obtain

$$0=\left(\frac{d\mathbf{n}}{ds}\middle|\mathbf{t}\right)+\left(\mathbf{n}\middle|\frac{d\mathbf{t}}{ds}\right)=\left(\frac{d\mathbf{n}}{ds}\middle|\mathbf{t}\right)+(\mathbf{n}|\kappa\mathbf{n})=\left(\frac{d\mathbf{n}}{ds}\middle|\mathbf{t}\right)+\kappa.$$

Therefore

$$\left(\frac{d\mathbf{n}}{ds}\middle|\mathbf{t}\right)=-\kappa.$$

Analogously from $(\mathbf{n}|\mathbf{b}) = 0$ we deduce

$$0 = \left(\frac{d\mathbf{n}}{ds}\Big|\mathbf{b}\right) + \left(\mathbf{n}\Big|\frac{d\mathbf{b}}{ds}\right) = \left(\frac{d\mathbf{n}}{ds}\Big|\mathbf{b}\right) + (\mathbf{n}| - \tau\mathbf{n}) = \left(\frac{d\mathbf{n}}{ds}\Big|\mathbf{b}\right) - \tau.$$

Therefore

$$\left(\frac{d\mathbf{n}}{ds}\Big|\mathbf{b}\right) = \tau.$$

□

4.5 Torsion

The *torsion* of a skew curve has already been defined in Definition 4.4.6. This section is essentially devoted to establishing easy formulas to compute it.

Proposition 4.5.1 *Let f be a normal representation of class $\mathcal{C}^3$ of a 2-regular skew curve. The torsion of this curve is given by the formula*

$$\tau = \frac{(f' \times f''|f''')}{\|f''\|^2}.$$

Proof From $(\mathbf{b}|\mathbf{n}) = 0$ we deduce, via the *Frenet* formulas of Theorem 4.4.8

$$0 = \left(\frac{d\mathbf{b}}{ds}\Big|\mathbf{n}\right) + \left(\mathbf{b}\Big|\frac{d\mathbf{n}}{ds}\right) = (-\tau\mathbf{n}|\mathbf{n}) + \left(\mathbf{b}\Big|\frac{d\mathbf{n}}{ds}\right) = -\tau + \left(\mathbf{b}\Big|\frac{d\mathbf{n}}{ds}\right).$$

It follows that

$$\tau = \left(\mathbf{b}\Big|\frac{d\mathbf{n}}{ds}\right).$$

But by definition of $\mathbf{n}$ (see Definition 4.4.1)

$$\frac{d\mathbf{n}}{ds} = \frac{f'''\kappa - f''\kappa'}{\kappa^2} = \frac{f'''\kappa - \kappa\kappa'\mathbf{n}}{\kappa^2}.$$

Since $\mathbf{n}$ is orthogonal to $\mathbf{b}$, we thus have

$$\tau = \left(\mathbf{b}\Big|\frac{f'''\kappa}{\kappa^2}\right) = \left(f' \times \frac{f''}{\kappa}\Big|\frac{f'''}{\kappa}\right) = \frac{(f' \times f''|f''')}{\|f''\|^2}.$$

□

Let us now consider the case of an arbitrary parametric representation.

Proposition 4.5.2 *Let f be an arbitrary parametric representation of class $\mathcal{C}^3$ of a 2-regular skew curve. The torsion of this curve is given by the formula*

$$\tau = \frac{(f' \times f''|f''')}{\|f' \times f''\|^2}.$$

Proof We use the notation of Proposition 4.2.7. As already observed, from $\sigma(t) = \int_{t_0}^{t} \|f'\|$ we deduce

$$\sigma' = \|f'\|, \qquad (\sigma^{-1})' = \frac{1}{\sigma' \circ \sigma^{-1}} = \frac{1}{\|f' \circ \sigma^{-1}\|}.$$

From $\overline{f} = f \circ \sigma^{-1}$ we deduce further

$$\begin{aligned}
\overline{f}' &= (f' \circ \sigma^{-1})(\sigma^{-1})' \\
\overline{f}'' &= (f'' \circ \sigma^{-1})(\sigma^{-1})'^2 + (f' \circ \sigma^{-1})(\sigma^{-1})'' \\
\overline{f}''' &= (f''' \circ \sigma^{-1})(\sigma^{-1})'^3 + 2(f'' \circ \sigma^{-1})(\sigma^{-1})'(\sigma^{-1})'' \\
&\quad + (f'' \circ \sigma^{-1})(\sigma^{-1})'(\sigma^{-1})'' + (f' \circ \sigma^{-1})(\sigma^{-1})'''.
\end{aligned}$$

By Proposition 4.5.1, the torsion is given by

$$\tau = \frac{(\overline{f}' \times \overline{f}'' | \overline{f}''')}{\|\overline{f}''\|^2} = \frac{(\overline{f}' \times \overline{f}'' | \overline{f}''')}{\kappa^2}.$$

Let us introduce the expressions of $\overline{f}'$, $\overline{f}''$ and $\overline{f}'''$ into this formula.

When computing the cross product $\overline{f}' \times \overline{f}''$, the term involving f' in the expression of $\overline{f}''$ yields a zero component, since f' is parallel to $\overline{f}'$.

In particular $\overline{f}' \times \overline{f}''$ is a vector in the direction of $f' \times f''$, as already observed in Proposition 4.1.5. Therefore f' and f'' are perpendicular to this direction. This implies that the terms involving f' or f'' in the expression of $\overline{f}'''$ also yield a zero component when performing the scalar product with $\overline{f}' \times \overline{f}''$.

Finally, the formula of Proposition 4.5.1 reduces to

$$\tau = \frac{((f' \circ \sigma^{-1})(\sigma^{-1})' \times (f'' \circ \sigma^{-1})(\sigma^{-1})'^2 | (f''' \circ \sigma^{-1})(\sigma^{-1})'^3)}{\kappa^2}.$$

Using further the last formula in the proof of Proposition 4.3.2, we obtain finally

$$\begin{aligned}
\tau &= \frac{\left(\frac{f' \circ \sigma^{-1}}{\|f' \circ \sigma^{-1}\|} \times \frac{f'' \circ \sigma^{-1}}{\|f' \circ \sigma^{-1}\|^2} \Big| \frac{f''' \circ \sigma^{-1}}{\|f' \circ \sigma^{-1}\|^3}\right)}{\frac{\|(f' \circ \sigma^{-1}) \times (f'' \circ \sigma^{-1})\|^2}{\|f' \circ \sigma^{-1}\|^6}} \\
&= \frac{((f' \circ \sigma^{-1}) \times (f'' \circ \sigma^{-1}) | (f''' \circ \sigma^{-1}))}{\|(f' \circ \sigma^{-1}) \times (f'' \circ \sigma^{-1})\|^2}.
\end{aligned}$$

Via the change of parameter $t = \sigma^{-1}(s)$, this is the formula of the statement. □

As expected, we then have:

Proposition 4.5.3 *A 2-regular skew curve of class $\mathcal{C}^3$ has constant zero-torsion if and only if it is a plane curve.*

Proof Up to a change of basis, a plane curve admits a parametric representation of the form

$$f(t) = \big(f_1(t), f_2(t), 0\big).$$

Thus the vectors f' and f'' are in the (x, y)-plane for every value t of the parameter. This implies that the binormal vector $\mathbf{b}(t)$ is a constant vector of length 1 in the direction of the z-axis. Its derivative is thus equal to 0.

Conversely, let us work with a normal representation $f(s)$. If the torsion is constantly 0, we have $\frac{d\mathbf{b}}{ds} = 0$ and thus $\mathbf{b}(s)$ is a constant vector $\mathbf{b}_0$. Differentiating the scalar product $(f(s)|\mathbf{b}_0)$ we obtain

$$\frac{d}{ds}\big(f(s)\big|\mathbf{b}_0\big) = \big(f'(s)\big|\mathbf{b}_0\big) = \big(\mathbf{t}(s)\big|\mathbf{b}_0\big) = 0$$

because $\mathbf{b}_0$ is constant and is orthogonal to $\mathbf{t}$. Integrating this equality thus yields a constant, that is

$$\big(f(s)\big|\mathbf{b}_0\big) = k, \quad k \in \mathbb{R}.$$

It follows at once that the curve $f(s)$ is contained in the plane of equation

$$\big((x, y, z)\big|\mathbf{b}_0\big) = k. \qquad \square$$

Example 4.5.4 The torsion of the circular helix

$$f(t) = (R\cos t, R\sin t, kt), \quad R > 0,\ k \neq 0$$

is equal to $\frac{k}{k^2+R^2}$.

Proof We have

$$\begin{aligned} f' \times f'' &= (-R\sin t, R\cos t, k) \times (-R\cos t, -R\sin t, 0) \\ &= (kR\sin t, -kR\cos t, R^2). \end{aligned}$$

Therefore

$$\big(f' \times f''\big|f'''\big) = \big((kR\sin t, -kR\cos t, R^2)\big|(R\sin t, -R\cos t, 0)\big) = kR^2.$$

The formula of Proposition 4.5.2 then gives

$$\tau = \frac{kR^2}{k^2R^2 + R^4} = \frac{k}{k^2 + R^2}. \qquad \square$$

Example 4.5.4 explains the choice of the sign $-$ in the definition of the torsion (see Definition 4.4.6 and Fig. 4.2).

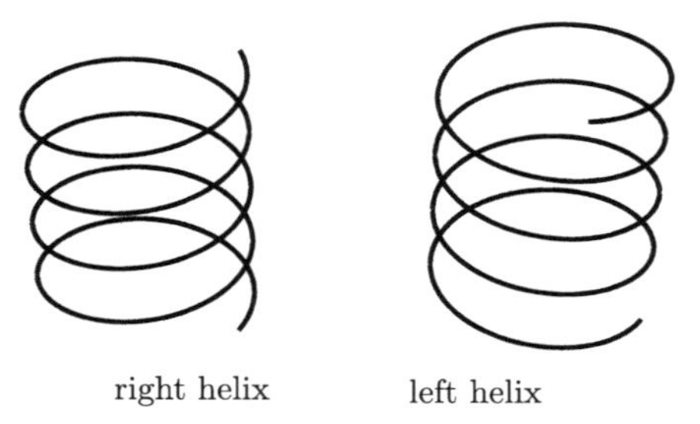

Fig. 4.2

- When $k > 0$, the helix is called a *right helix*, that is, has the shape of a "corkscrew" (or an "ordinary bolt": a bolt with a right thread). The torsion is then strictly positive.
- When $k < 0$, the helix is called a *left helix*, a curve having the shape of a bolt with a left thread. In that case, the torsion is strictly negative.

Scientists are probably big wine drinkers, since they rely so often on the so-called *corkscrew rule*. They definitely want the sign of the torsion to be such that the torsion of a corkscrew (a right helix) is positive. To achieve that, as Example 4.5.4 shows, the sign "−" must be chosen in Definition 4.4.6.

Example 4.5.4 also suggests something else. We have seen that the sign of the torsion is an intrinsic property of the curve, independent of the chosen parametric representation (see Proposition 4.4.7). This is a striking difference with the sign of the relative curvature for plane curves (see Warning 2.9.9), which depends on the representation. How can we intuitively understand this difference? In the plane, there are two ways to "turn": clockwise or counter-clockwise. However, it does not make sense to distinguish between "clockwise circles" and "counter-clockwise circles". On the other hand, in three dimensional space, everybody who has used bolts knows very well the intrinsic difference between "right thread bolts" and "left thread bolts".

4.6 Intrinsic Equations

We now want to prove that a skew curve can "intrinsically" be defined by giving its curvature and its torsion.

First let us prove that a curve is uniquely defined, up to its position in space, by its curvature and its torsion. In the spirit of the comments at the end of Sect. 4.5, observe once more the difference with the case of plane curves: Theorem 2.12.4 refers to an arbitrary isometry while Theorem 4.6.1 concerns a *direct* isometry, that is, just a shifting.

Theorem 4.6.1 *Consider two normal representations*

$$f\colon]a,b[\longrightarrow \mathbb{R}^3, \qquad \widetilde{f}\colon]a,b[\longrightarrow \mathbb{R}^3$$

of class $\mathcal{C}^3$ of two 2-regular skew curves. When these two curves admit the same curvature and the same torsion for each value $s \in]a, b[$ of the parameter, the two curves are the image of each other under a direct isometry.

Proof Let us fix a value $s_0 \in]a, b[$. Up to possible translations of the curves, there is no loss of generality in assuming that $f(s_0) = \widetilde{f}(s_0)$ is the origin of $\mathbb{R}^3$. Up to possible further rotations, there is also no loss of generality in assuming that the two *Frenet* trihedrons at this point coincide with the canonical basis of $\mathbb{R}^3$ (see Definition 4.4.1). We use respectively the notation $\mathbf{t}$, $\mathbf{n}$, $\mathbf{b}$ and $\widetilde{\mathbf{t}}$, $\widetilde{\mathbf{n}}$, $\widetilde{\mathbf{b}}$ to indicate the *Frenet* trihedrons of the two curves.

We shall now use the *Frenet* formulas of Theorem 4.4.8 several times, without further reference. Using the notation $(\ )'$ to indicate a derivative with respect to s, we have, since $\kappa = \widetilde{\kappa}$ and $\tau = \widetilde{\tau}$

$$(\mathbf{t}|\widetilde{\mathbf{t}})' = (\mathbf{t}'|\widetilde{\mathbf{t}}) + (\mathbf{t}|\widetilde{\mathbf{t}}') = (\kappa\mathbf{n}|\widetilde{\mathbf{t}}) + (\mathbf{t}|\widetilde{\kappa}\widetilde{\mathbf{n}}) = \kappa\Big((\mathbf{n}|\widetilde{\mathbf{t}}) + (\mathbf{t}|\widetilde{\mathbf{n}})\Big)$$

$$\begin{aligned}(\mathbf{n}|\widetilde{\mathbf{n}})' &= (\mathbf{n}'|\widetilde{\mathbf{n}}) + (\mathbf{n}|\widetilde{\mathbf{n}}') = (-\kappa\mathbf{t} + \tau\mathbf{b}|\widetilde{\mathbf{n}}) + (\mathbf{n}|-\widetilde{\kappa}\widetilde{\mathbf{t}} + \widetilde{\tau}\widetilde{\mathbf{b}})\\ &= -\kappa\Big((\mathbf{t}|\widetilde{\mathbf{n}}) + (\mathbf{n}|\widetilde{\mathbf{t}})\Big) + \tau\Big((\mathbf{b}|\widetilde{\mathbf{n}}) + (\mathbf{n}|\widetilde{\mathbf{b}})\Big)\end{aligned}$$

$$(\mathbf{b}|\widetilde{\mathbf{b}})' = (\mathbf{b}'|\widetilde{\mathbf{b}}) + (\mathbf{b}|\widetilde{\mathbf{b}}') = (-\tau\mathbf{n}|\widetilde{\mathbf{b}}) + (\mathbf{b}|-\widetilde{\tau}\widetilde{\mathbf{n}}) = -\tau\big((\mathbf{n}|\widetilde{\mathbf{b}}) + (\mathbf{b}|\widetilde{\mathbf{n}})\big).$$

Adding these three equalities we obtain

$$\big((\mathbf{t}|\widetilde{\mathbf{t}}) + (\mathbf{n}|\widetilde{\mathbf{n}}) + (\mathbf{b}|\widetilde{\mathbf{b}})\big)' = 0.$$

Integrating, we thus get a constant

$$(\mathbf{t}|\widetilde{\mathbf{t}}) + (\mathbf{n}|\widetilde{\mathbf{n}}) + (\mathbf{b}|\widetilde{\mathbf{b}}) = k, \quad k \in \mathbb{R}.$$

But at the point with parameter s_0

$$\big(\mathbf{t}(s_0)|\widetilde{\mathbf{t}}(s_0)\big) = (e_1|e_1) = 1$$

and analogously for $\mathbf{n}$ and $\mathbf{b}$. Therefore the constant k is equal to 3. But since each of the vectors $\mathbf{t}(s)$ and $\widetilde{\mathbf{t}}(s)$ has norm 1, the Schwarz inequality (see Proposition 4.2.5 in [4], *Trilogy II*) yields

$$-1 \leq \big(\mathbf{t}(s)|\widetilde{\mathbf{t}}(s)\big) \leq +1.$$

An analogous conclusion holds for $\mathbf{n}$ and $\mathbf{b}$. But for three quantities between -1 and $+1$ to have a sum equal to 3, necessarily all three quantities must be equal to 1. This proves that at all points

$$(\mathbf{t}|\widetilde{\mathbf{t}}) = 1, \qquad (\mathbf{n}|\widetilde{\mathbf{n}}) = 1, \qquad (\mathbf{b}|\widetilde{\mathbf{b}}) = 1.$$

Next considering the formula

$$1 = (\mathbf{t}|\widetilde{\mathbf{t}}) = \|\mathbf{t}\| \cdot \|\widetilde{\mathbf{t}}\| \cdot \cos\measuredangle(\mathbf{t}, \widetilde{\mathbf{t}}) = \cos\measuredangle(\mathbf{t}, \widetilde{\mathbf{t}})$$

we conclude that $\measuredangle(\mathbf{t}, \widetilde{\mathbf{t}}) = 0$ and thus finally, $\mathbf{t} = \widetilde{\mathbf{t}}$. This means $f' = \widetilde{f}'$. Integrating this equality, we find that $f = \widetilde{f} + v$ for some constant vector v. But since we have $f(s_0) = \widetilde{f}(s_0)$, it follows that $v = 0$. Thus $f = \widetilde{f}$. □

We next have to show that there always exists a curve admitting a prescribed curvature function and a prescribed torsion function. Then by Theorem 4.6.1, such a curve will be "unique up to a direct isometry".

Theorem 4.6.2 *Consider a strictly positive function*

$$\kappa\colon]a,b[\longrightarrow \mathbb{R}, \qquad s \mapsto \kappa(s) > 0$$

and an arbitrary function

$$\tau\colon]a,b[\longrightarrow \mathbb{R}, \qquad s \mapsto \tau(s),$$

both of class $\mathcal{C}^1$. In a neighborhood of each point $t \in]a,b[$, there exists a 2-regular normal representation of class $\mathcal{C}^3$ of a skew curve admitting κ and τ as curvature and torsion.

Proof The *Frenet* formulas can be re-written component-wise as

$$\begin{aligned} t_i'(s) &= \kappa(s) n_i(s) \\ n_i'(s) &= -\kappa(s) t_i(s) + \tau(s) b_i(s) \\ b_i'(s) &= -\tau(s) n_i(s) \end{aligned}$$

where $i \in \{1, 2, 3\}$. This is a homogeneous system of nine differential equations with coefficients 1, $\pm\kappa(s)$ and $\pm\tau(s)$ of class $\mathcal{C}^1$ and nine indeterminate functions

$$t_i,\ n_i,\ b_i\colon]a,b[\longrightarrow \mathbb{R}, \quad i = 1, 2, 3.$$

Fix a value $s_0 \in]a,b[$ and impose the initial values

$$\begin{aligned} \big(t_1(s_0), t_2(s_0), t_3(s_0)\big) &= (1, 0, 0) \\ \big(n_1(s_0), n_2(s_0), n_3(s_0)\big) &= (0, 1, 0) \\ \big(b_1(s_0), b_2(s_0), b_3(s_0)\big) &= (0, 0, 1). \end{aligned}$$

The existence (and uniqueness) theorem for a problem of this type (see Proposition B.1.1) proves—in a neighborhood of each point—the existence and uniqueness of nine functions of class $\mathcal{C}^2$ t_i, n_i, b_i satisfying all the requirements above.

Using the solutions of the system, let us then define the following three functions, from $]a,b[$ to $\mathbb{R}^3$:

$$\mathbf{t} = (t_1, t_2, t_3), \qquad \mathbf{n} = (n_1, n_2, n_3), \qquad \mathbf{b} = (b_1, b_2, b_3).$$

The statement that the functions t_i, n_i, b_i are solutions of the system of differential equations can be rephrased as the fact that $\mathbf{t}$, $\mathbf{n}$, $\mathbf{b}$ satisfy the *Frenet* formulas of Theorem 4.4.8. We shall now prove that for each value $s \in]a,b[$, $(\mathbf{t}(s), \mathbf{n}(s), \mathbf{b}(s))$ constitute an orthonormal basis with direct orientation.

Since $\mathbf{t}$, $\mathbf{n}$ and $\mathbf{b}$ satisfy the *Frenet* formulas, we have at once

$$\begin{aligned}
(\mathbf{t}|\mathbf{t})' &= 2\kappa(\mathbf{t}|\mathbf{n})\\
(\mathbf{t}|\mathbf{n})' &= \kappa(\mathbf{n}|\mathbf{n}) - \kappa(\mathbf{t}|\mathbf{t}) + \tau(\mathbf{t}|\mathbf{b})\\
(\mathbf{t}|\mathbf{b})' &= \kappa(\mathbf{n}|\mathbf{b}) - \tau(\mathbf{t}|\mathbf{n})\\
(\mathbf{n}|\mathbf{n})' &= -2\kappa(\mathbf{t}|\mathbf{n}) + 2\tau(\mathbf{b}|\mathbf{n})\\
(\mathbf{n}|\mathbf{b})' &= -\kappa(\mathbf{t}|\mathbf{b}) + \tau(\mathbf{b}|\mathbf{b}) - \tau(\mathbf{n}|\mathbf{n})\\
(\mathbf{b}|\mathbf{b})' &= -2\tau(\mathbf{n}|\mathbf{b}).
\end{aligned}$$

This can be viewed as a new system of differential equations in the following six functions from $]a,b[$ to $\mathbb{R}$:

$$(\mathbf{t}|\mathbf{t}), \qquad (\mathbf{t}|\mathbf{n}), \qquad (\mathbf{t}|\mathbf{b}), \qquad (\mathbf{n}|\mathbf{n}), \qquad (\mathbf{n}|\mathbf{b}), \qquad (\mathbf{b}|\mathbf{b})$$

that is, in all the possible scalar products of two of the functions $\mathbf{t}$, $\mathbf{n}$, $\mathbf{b}$. In addition to admitting these six functions as solutions, the new system of differential equations is such that the following initial conditions are satisfied, simply because of the initial conditions imposed on the functions t_i, n_i, b_i.

$$\begin{aligned}
\big(\mathbf{t}(s_0)\big|\mathbf{t}(s_0)\big) &= 1\\
\big(\mathbf{t}(s_0)\big|\mathbf{n}(s_0)\big) &= 0\\
\big(\mathbf{t}(s_0)\big|\mathbf{b}(s_0)\big) &= 0\\
\big(\mathbf{n}(s_0)\big|\mathbf{n}(s_0)\big) &= 1\\
\big(\mathbf{n}(s_0)\big|\mathbf{b}(s_0)\big) &= 0\\
\big(\mathbf{b}(s_0)\big|\mathbf{b}(s_0)\big) &= 1.
\end{aligned}$$

Again the (existence and) uniqueness theorem for a problem of this type (see Proposition B.1.1) tells us that the six functions indicated constitute the only possible such solution.

But now it is trivial that another possible solution of the system, with the same initial conditions, is given by

$$\begin{aligned}
\big(\mathbf{t}(s)\big|\mathbf{t}(s)\big) &= 1\\
\big(\mathbf{t}(s)\big|\mathbf{n}(s)\big) &= 0\\
\big(\mathbf{t}(s)\big|\mathbf{b}(s)\big) &= 0
\end{aligned}$$

$$\begin{aligned}
\big(\mathbf{n}(s)\big|\mathbf{n}(s)\big) &= 1\\
\big(\mathbf{n}(s)\big|\mathbf{b}(s)\big) &= 0\\
\big(\mathbf{b}(s)\big|\mathbf{b}(s)\big) &= 1.
\end{aligned}$$

By uniqueness of the solution, the three functions $\mathbf{t}$, $\mathbf{n}$ and $\mathbf{b}$ defined above thus satisfy these last equalities.

The rest of the proof is now easy. The definition of the tangent vector $f' = \mathbf{t}$ in the *Frenet* trihedron (see Definition 4.4.1) suggests at once to define

$$f(s) = \left(\int_{s_0}^{s} t_1(s)\,ds, \int_{s_0}^{s} t_2(s)\,ds, \int_{s_0}^{s} t_3(s)\,ds\right).$$

We shall prove that f is the expected curve.

By definition, the function $\mathbf{t}$ is of class $\mathcal{C}^2$. Therefore the function f defined above is of class $\mathcal{C}^3$. The first two derivatives of f are thus

$$f' = \mathbf{t}, \qquad f'' = \mathbf{t}' = \kappa\mathbf{n}.$$

Since $\|f'\| = \|\mathbf{t}\| = 1$, f is a regular parametric representation by Proposition 4.1.4 and thus, is a normal representation. Since $(\mathbf{t}, \mathbf{n}, \mathbf{b})$ constitute at each point an orthonormal basis, f is also 2-regular.

Next the form of f' indicates at once that $\mathbf{t}$ is the tangent vector to f, while the form of f'' indicates further that $\mathbf{n}$ is its normal vector. Since

$$f'' = \mathbf{t}' = \kappa\mathbf{n}$$

we have $\|f''\| = \kappa$ and κ is thus also the curvature of f (see Definition 4.3.1). But then the binormal vector of f is the cross product $\mathbf{t} \times \mathbf{n}$, that is $\mathbf{b}$. Since

$$\frac{d\mathbf{b}}{ds} = -\tau\mathbf{n}$$

we conclude by Definition 4.4.6 that τ is also the torsion of f. □

4.7 Problems

4.7.1 A skew regular curve is said to have *constant slope* when there is a fixed vector v in $\mathbb{R}^3$ such that the tangent vector $\mathbf{t}$ to the curve makes a constant angle $\theta \neq 0$ with the vector v. Prove that a curve with a constant slope admits a normal representation of the form

$$f(s) = \big(f_1(s), f_2(s), s\cos\theta\big).$$

Verify that a circular helix has constant slope.

4.7.2 Prove that a 2-regular curve of class $\mathcal{C}^3$ has constant slope if and only if the ratio $\frac{\tau}{\kappa}$ is constant.

4.7.3 Show that a 2-regular curve of class $\mathcal{C}^3$ is a plane curve if and only if all its osculating planes have a common point of intersection.

4.7.4 Consider two skew curves defined in terms of the same parameter

$$f\colon]a,b[\longrightarrow \mathbb{R}^3, \qquad \widetilde{f}\colon]a,b[\longrightarrow \mathbb{R}^3.$$

Assume that these curves are 2-regular and of class $\mathcal{C}^3$. These curves are called *Bertrand curves* when their normal lines coincide at corresponding points (the normal line is that passing through the point in the direction of the normal vector $\mathbf{n}$). Prove that:

1. the distance between corresponding points is constant;
2. the angle between tangents at corresponding points is constant.

4.7.5 The results of Sects. 4.1 and 4.2 generalize immediately to curves in $\mathbb{R}^n$. A curve in $\mathbb{R}^n$ is called *k-regular* when it is of class $\mathcal{C}^k$ and the first k derivatives of a parametric representation are linearly independent at each point. Consider then a normal representation f of an $(n-1)$-regular curve of class $\mathcal{C}^n$ in $\mathbb{R}^n$. By the *Gram-Schmidt* process (see Theorem 4.6.6 in [3], *Trilogy I*), orthonormalize the sequence of the successive derivatives $f', \ldots, f^{(n-1)}$; call these vectors $\mathbf{t}_1, \ldots, \mathbf{t}_{n-1}$. Complete this sequence with a vector $\mathbf{t}_n$ in order to obtain an orthonormal basis with direct orientation. The $n-1$ *curvatures* are defined by

$$\kappa_i = \left(\frac{d\mathbf{t}_i}{ds} \middle| \mathbf{t}_{i+1} \right) = -\left(\mathbf{t}_i \middle| \frac{d\mathbf{t}_{i+1}}{ds} \right).$$

The *Frenet formulæ* are then

$$\begin{aligned}
\frac{d\mathbf{t}_1}{ds} &= \kappa_1 \mathbf{t}_2 \\
\frac{d\mathbf{t}_i}{ds} &= -\kappa_{i-1}\mathbf{t}_{i-1} + \kappa_i \mathbf{t}_{i+1} \quad (1 < i < n) \\
\frac{d\mathbf{t}_n}{ds} &= -\kappa_{n-1}\mathbf{t}_{n-1}.
\end{aligned}$$

4.8 Exercises

4.8.1 Give a normal parametric representation of the curve defined by

$$f\colon \mathbb{R} \longrightarrow \mathbb{R}^3, \qquad t \mapsto \left(\frac{t^2}{2}, \cos t + t\sin t, \sin t - t\cos t \right).$$

4.8.2 Consider the circular helix (see Fig. 4.1)

$$f: \mathbb{R} \longrightarrow \mathbb{R}^3, \qquad t \mapsto (a\cos t, a\sin t, b), \quad a, b > 0.$$

1. Give the equation of the osculating plane to f at the point $(a, 0, 0)$.
2. Give a parametric representation of the curve obtained as the intersection of this plane π with the cylinder Γ of radius a and axis Oz.

4.8.3 Consider the skew curve represented by

$$f: \mathbb{R} \longrightarrow \mathbb{R}^3, \qquad t \mapsto \left(e^t, e^{-t}, \sqrt{2}t\right).$$

Compute the tangent, the curvature and the torsion for each value of t.

4.8.4 Give a parametric representation of the locus of the orthogonal projections of the point $O = (0, 0, 0)$ on the osculating planes to the helix of Example 4.8.2.

4.8.5 Consider again the helix of Example 4.8.2. At each point, determine the Frenet trihedron and give the equations of the osculating plane, the normal plane and the rectifying plane.

4.8.6 In $\mathbb{R}^2$ consider the *astroid* represented by (see Fig. 2.38)

$$f: \mathbb{R} \longrightarrow \mathbb{R}^2, \qquad t \mapsto \left(2\cos^3 t, 2\sin^3 t\right).$$

View this astroid as a curve in the xy-plane of $\mathbb{R}^3$ and "project" it, parallel to the z-axis, onto the upper half-sphere of radius 2 centered at the origin, to obtain a skew curve $\mathcal{P}$. Give a parametric representation of this curve $\mathcal{P}$.

4.8.7 Suppose that a 2-regular skew curve of class $\mathcal{C}^3$ admits the intrinsic equations

$$\begin{cases} \kappa(s) = f(s) \\ \tau(s) = g(s). \end{cases}$$

Give the intrinsic equations of the curve $\mathcal{C}^*$, the image of $\mathcal{C}$ by:

1. a translation;
2. a central symmetry;
3. an axial symmetry;
4. an orthogonal symmetry with respect to a plane.

Chapter 5
The Local Theory of Surfaces

Our next concern is the theory of surfaces in three dimensional real space. Just as curves are "good deformations" in $\mathbb{R}^2$ or $\mathbb{R}^3$ of a "good piece" of the real line, surfaces are analogous "good deformations" in $\mathbb{R}^3$ of a "good piece" of the real plane. The basic definitions are thus straightforward generalizations of those in the case of curves. This includes the parametric representations, the Cartesian equations and the study of the tangent plane.

However—expectedly or not—the situation for surfaces rapidly turns out to present striking differences with the case of curves. For example, a (good) curve admits a normal representation: the curve can be presented as a deformation of the real line, which respects lengths. Intuitively, you can represent a curve as a piece of iron wire, without any stretching of this wire. But of course you cannot possibly construct a sphere by folding a sheet of metal, without stretching it.

The central topic of this chapter is the study of the *curvature* of a surface at a given point. In each direction at the given point, cut the surface by a plane: you get a curve and the curvature of this curve can be regarded as the curvature of the surface in the given direction. Of course different directions will generally yield different curvatures, as in the case of an ellipsoid. The so-called *normal curvature* is thus, at a given point, a function of the direction. The study of this function, essentially due to *Euler*, is among the main concerns of this chapter.

Following an idea of *Gauss*, we also devote some attention to the case of the *Gaussian curvature*. Intuitively, the *Gaussian curvature* at a given point of the surface measures the "oscillation" of the tangent plane as you move on the surface in the neighborhood of the point. This notion will be further studied in subsequent chapters, especially in Chap. 6 devoted to *Riemannian geometry*, where it plays a central role. In contrast to the *normal curvature*, the *Gaussian curvature* at a given point is just a real number, not a function. The information recaptured in this way is certainly less precise, but it nevertheless gives rise to a "qualitative" idea of the shape of the surface in a neighborhood of the point. As we shall see later, it is "more intrinsic" than the normal curvature.

The reader will have noticed that all properties mentioned in this introduction are properties valid "at a given point of the surface". This justifies the title of the

F. Borceux, *A Differential Approach to Geometry*, DOI 10.1007/978-3-319-01736-5_5,

chapter. Some "global" properties, that is, properties involving the whole surface, not just the neighborhood of a given point, will be studied in Chap. 7.

5.1 Parametric Representation of a Surface

In this chapter, we limit our study to the case of (some) surfaces in $\mathbb{R}^3$; this first approach will be generalized in Chap. 6.

With the theory of curves in mind, we shall define a surface as a continuous and locally injective deformation of a "good piece" of the real plane in $\mathbb{R}^3$. But what is a "good piece $U \subseteq \mathbb{R}^2$? Of course the answer to this question is again a matter of choice, but we shall follow the same arguments as in the case of a curve.

- We want U to be open, in order to avoid "points on the border" (these would otherwise require a special treatment each time continuity or differentiability is concerned).
- To avoid pathologies, we also want U to be "one piece": such open subsets are called *connected* (see Definition A.10.1).

It should be observed that the connected open subsets of the real line are precisely the generalized intervals (see Example A.10.9), so that our choice for surfaces extends the choice that we made for curves.

Therefore we make the following definition:

Definition 5.1.1

1. A *parametric representation* of class $\mathcal{C}^k$ (with $k \in \mathbb{N} \cup \{\infty\}$) of a surface is a locally injective function of class $\mathcal{C}^k$
$$f : U \longrightarrow \mathbb{R}^3, \qquad (u, v) \mapsto \big(f_1(u, v), f_2(u, v), f_3(u, v)\big)$$
where $U \subseteq \mathbb{R}^2$ is a connected open subset of the real plane.
2. Given another parametric representation of class $\mathcal{C}^k$
$$g : V \longrightarrow \mathbb{R}^3, \qquad (r, s) \mapsto \big(g_1(r, s), g_2(r, s), g_3(r, s)\big),$$
the two parametric representations f and g are *equivalent in class* $\mathcal{C}^k$ when there exist inverse bijections φ, φ^{-1} of class $\mathcal{C}^k$ such that
$$f = g \circ \varphi, \qquad g = f \circ \varphi^{-1}.$$
3. A *surface* of class $\mathcal{C}^k$ is an equivalence class of parametric representations of class $\mathcal{C}^k$, with respect to the equivalence relation described in 2.
4. The *support* of a surface is the image of (any one of) its parametric representations.

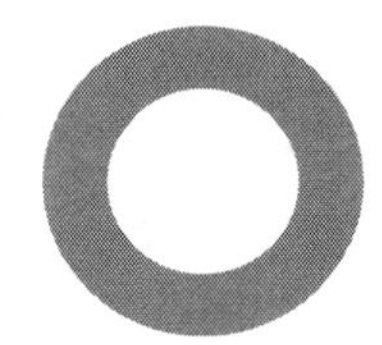

Fig. 5.1

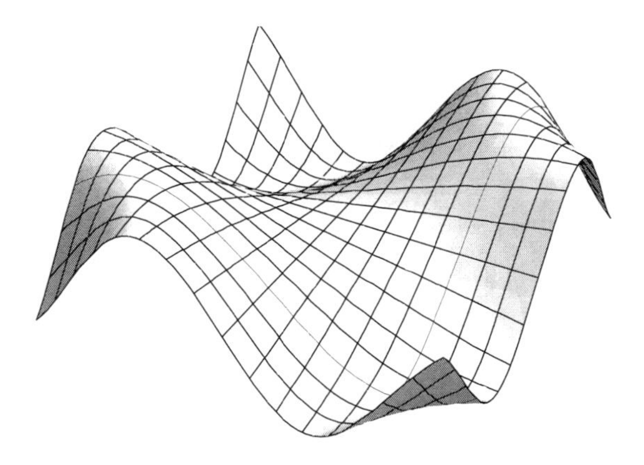

Fig. 5.2

Clearly, *locally injective* means as usual that for every point $P \in U$, there exists a neighborhood $V \subseteq U$ of P on which f is injective.

The fact that Definition 5.1.1 makes perfect sense is a direct transposition of the straightforward observations already made for plane curves in Sect. 2.1.

Notice nevertheless that a *connected open subset of* $\mathbb{R}^2$ can present some peculiarities which cannot possibly appear in the case of an open interval. For example an open "ring" (see Fig. 5.1, the area between two concentric circles) is a connected open subset of $\mathbb{R}^2$ "with a hole in it". Of course, an open interval never has a hole in it, but we can easily imagine a natural deformation of an open ring which provides an interesting surface in $\mathbb{R}^3$, for example a cylinder. So there is certainly no reason to try to avoid such connected open subsets.

Example 5.1.2 Let $g\colon \mathbb{R}^2 \longrightarrow \mathbb{R}$ be a function of class $\mathcal{C}^k$. Then

$$f\colon \mathbb{R}^2 \longrightarrow \mathbb{R}^3, \qquad (x, y) \mapsto \bigl(x, y, g(x, y)\bigr)$$

is a parametric representation of class $\mathcal{C}^k$ of a surface, whose support is the graph of g (see Fig. 5.2 where $f(x, y) = \sin xy$).

Proof The function f is injective and of class $\mathcal{C}^k$. □

Let us now investigate the case of the most interesting quadrics of $\mathbb{R}^3$ (see Sect. 1.14, [4], *Trilogy II*).

Example 5.1.3 The paraboloids are surfaces of class $\mathcal{C}^\infty$.

Proof Going back to Sect. 1.14 in [4], *Trilogy II*, and applying Example 5.1.2, a paraboloid can be presented as the graph of a function $g(x,y)=\frac{x^2}{a^2}\pm\frac{y^2}{b^2}$, with $a,b>0$. □

Now since we insisted on defining surfaces on *connected* open subsets of the real plane, as in the case of the hyperbola (see Sect. 1.2), we must consider separately the possible various sheets of a quadric:

Example 5.1.4 Each sheet of a hyperboloid with two sheets is a surface of class $\mathcal{C}^\infty$.

Proof The two sheets of the hyperboloid

$$\frac{x^2}{a^2}-\frac{y^2}{b^2}-\frac{z^2}{c^2}=1, \quad a,b,c>0$$

are respectively the graphs of the two functions

$$x=\pm g(y,z)$$

where

$$g(y,z)=a\sqrt{1+\frac{y^2}{b^2}+\frac{z^2}{c^2}}.$$

By Example 5.1.2, each of them is therefore a surface of class $\mathcal{C}^\infty$. □

The argument in Example 5.1.2 does not apply to the hyperboloid with one sheet, which is by no means the graph of a function $\mathbb{R}^2 \longrightarrow \mathbb{R}$. Nevertheless:

Example 5.1.5 The hyperboloid with one sheet is a surface of class $\mathcal{C}^\infty$.

Proof We consider the hyperboloid with equation

$$\frac{x^2}{a^2}+\frac{y^2}{b^2}-\frac{z^2}{c^2}=1, \quad a,b,c>0.$$

Cutting this hyperboloid by the plane $z=z_0$, with $z_0\in\mathbb{R}$, we obtain an ellipse whose equation can be re-written as

$$\frac{x^2}{a^2(1+\frac{z_0^2}{c^2})}+\frac{y^2}{b^2(1+\frac{z_0^2}{c^2})}=1.$$

This is an ellipse of radii

$$a\sqrt{1+\frac{z_0^2}{c^2}}, \qquad b\sqrt{1+\frac{z_0^2}{c^2}}.$$

A parametric representation of this ellipse is thus (see Example 2.1.4)

$$f(-,z_0)\colon \mathbb{R}\mapsto\mathbb{R}^2, \qquad \theta\mapsto\left(a\sqrt{1+\frac{z_0^2}{c^2}}\cos\theta, b\sqrt{1+\frac{z_0^2}{c^2}}\sin\theta\right).$$

The function

$$f\colon \mathbb{R}^2\to\mathbb{R}^3, \qquad (\theta,z)\mapsto\left(a\sqrt{1+\frac{z^2}{c^2}}\cos\theta, b\sqrt{1+\frac{z^2}{c^2}}\sin\theta, z\right)$$

is then a parametric representation of class $\mathcal{C}^\infty$ of the hyperboloid with one sheet. This function is trivially of class $\mathcal{C}^\infty$. It is also locally injective because different values of z of course yield different points of $\mathbb{R}^3$, while for $z=z_0$ fixed, the function $f(-,z_0)$ is locally injective by Example 2.1.4. □

However, the most popular of all quadrics—the sphere—is the one which causes troubles! Indeed, one can prove that:

There does not exist a parametric representation of a sphere, in the sense of Definition 5.1.1.

Such a negative result calls at once for a more general notion of "surface", in order to recapture (at least!) the example of the sphere. Proving this negative result elegantly requires some sophisticated tools; proving it with elementary arguments is very long and tedious. We shall not dwell here on this negative result: we shall simply observe why the most natural parametrisation of the sphere does not fulfil the requirements of Definition 5.1.1.

All of us are used to the representation of a part of the surface of the Earth on a geographical map, and a geographical map is by nature a piece of $\mathbb{R}^2$. The two parameters classically used to represent in this way a piece of the Earth (which we shall approximate as a sphere) are the *longitude* and the *latitude*. Let us investigate this example further and observe why it does not provide a parametric representation of the full sphere.

Example 5.1.6 The sphere (and more generally, an ellipsoid) punctured at two opposite poles is a surface of class $\mathcal{C}^\infty$, admitting a parametric representation in terms of the *longitude* and the *latitude*.

Proof Consider first the case of a sphere of radius r, centered at the origin of $\mathbb{R}^3$. View the xy-plane as that of the equator and the x-axis as the origin of the longitudes. A point P of longitude θ and latitude τ thus has the coordinates (see Fig. 5.3)

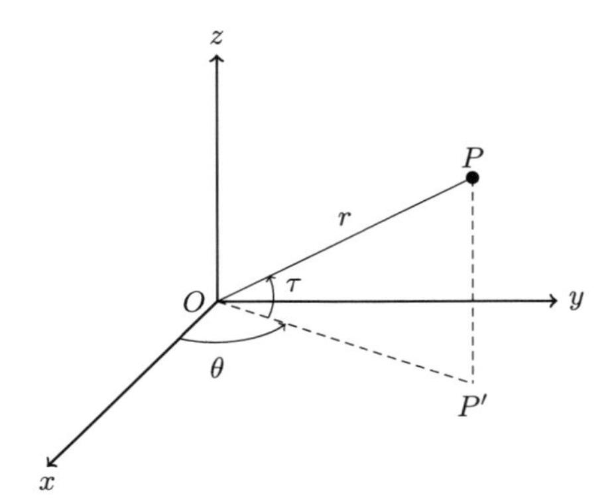

Fig. 5.3

$$f(\theta,\tau) = (r\cos\tau\cos\theta, r\cos\tau\sin\theta, r\sin\tau).$$

Trivially the function

$$f\colon \mathbb{R}^2 \longrightarrow \mathbb{R}^3$$

is of class $\mathcal{C}^\infty$ and describes precisely all the points of the sphere. But this function is not locally injective because all points of the form $(\theta, \frac{\pi}{2})$ are mapped onto the "North pole" $(0,0,r)$, while all points of the form $(\theta, -\frac{\pi}{2})$ are mapped onto the "South pole" $(0,0,-r)$. However, restricting f as a function

$$f\colon \mathbb{R} \times \left]-\frac{\pi}{2}, +\frac{\pi}{2}\right[\longrightarrow \mathbb{R}^3$$

avoids the problem and yields this time a locally injective function, that is, a parametric representation of class $\mathcal{C}^\infty$ of the sphere punctured at its two "poles".

More generally, the function

$$g\colon \mathbb{R} \times \left]-\frac{\pi}{2}, +\frac{\pi}{2}\right[\longrightarrow \mathbb{R}^3, \qquad (\theta,\tau) \mapsto (a\cos\tau\cos\theta, b\cos\tau\sin\theta, c\sin\tau)$$

is such that

$$\left(\frac{x}{a}\right)^2 + \left(\frac{y}{b}\right)^2 + \left(\frac{z}{c}\right)^2 = 1$$

and is a parametric representation of class $\mathcal{C}^\infty$ of an ellipsoid of radii a, b, c punctured at its two poles $(0,0,c)$ and $(0,0,-c)$ (see Sect. 1.14, [4], *Trilogy II*). □

One can even do a little bit better: avoiding just one point of the sphere, not two! Problem 5.17.2 presents a parametric representation of the sphere punctured only at the North pole.

The *torus* (the surface having the shape of the inner tubes of your bicycle) is another celebrated surface:

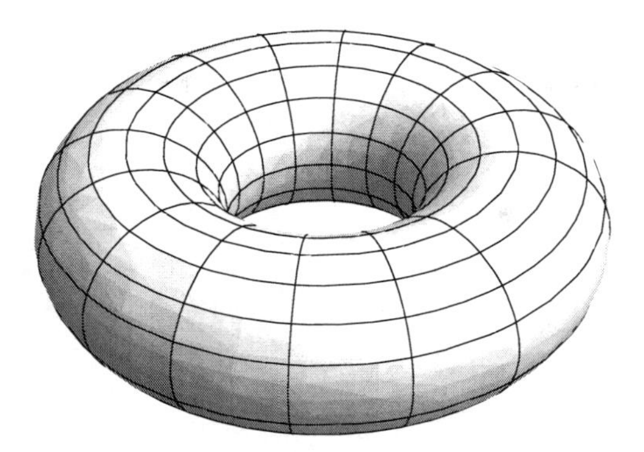

Fig. 5.4 The torus

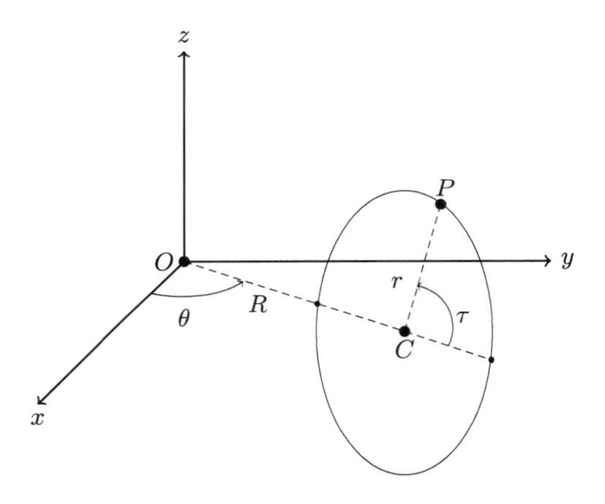

Fig. 5.5

Example 5.1.7 The *torus* (see Fig. 5.4) is a surface of class $\mathcal{C}^\infty$.

Proof The torus can be obtained by letting a circle of radius r, positioned in a vertical plane containing the z-axis, rotate around the z-axis, provided the distance R between the center of the circle and the z-axis is strictly greater than r. In terms of the angles θ and τ as in Fig. 5.5, the coordinates of the corresponding point P of the torus can be obtained by adding the two vectors $\overrightarrow{OC}$ and $\overrightarrow{CP}$. This yields

$$f(\theta,\tau) = (R\cos\theta, R\sin\theta, 0) + (r\cos\tau\cos\theta, r\cos\tau\sin\theta, r\sin\tau).$$

It follows at once that

$$f\colon \mathbb{R}^2 \longrightarrow \mathbb{R}^3$$

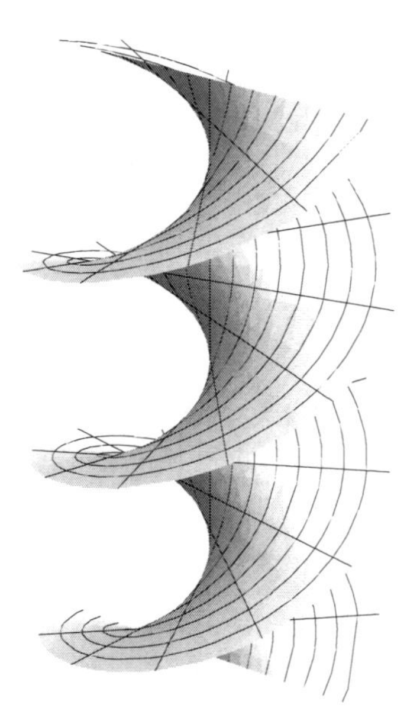

Fig. 5.6 The helicoid

is a parametric representation of class $\mathcal{C}^\infty$ of the torus. Indeed, f is of class $\mathcal{C}^\infty$ and injective on the neighborhood

$$]\theta_0 - \pi, \theta_0 + \pi[\times \left]\tau_0 - \frac{\pi}{2}, \tau_0 + \frac{\pi}{2}\right[$$

of (θ_0, τ_0). □

The *helicoid* is obtained by letting a straight line d rotate uniformly around one of its points P_0, while this point P_0 translates uniformly along a fixed line ℓ perpendicular to d.

Example 5.1.8 The helicoid (see Fig. 5.6) is a surface of class $\mathcal{C}^\infty$.

Proof A parametric representation of the helicoid is given by

$$f(u, v) = (u \cos v, u \sin v, v)$$

where v is the position of the point P_0 on the line ℓ while u is the position of the point $f(u, v)$ on the moving line d. Notice that f is injective. □

The *Möbius strip* is another famous surface: take a rectangle of paper and glue together two opposite sides, but after having twisted the strip of paper by a half turn.

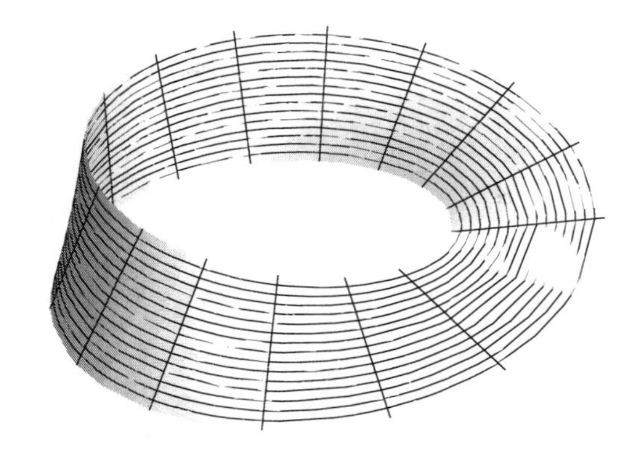

Fig. 5.7 The Möbius string

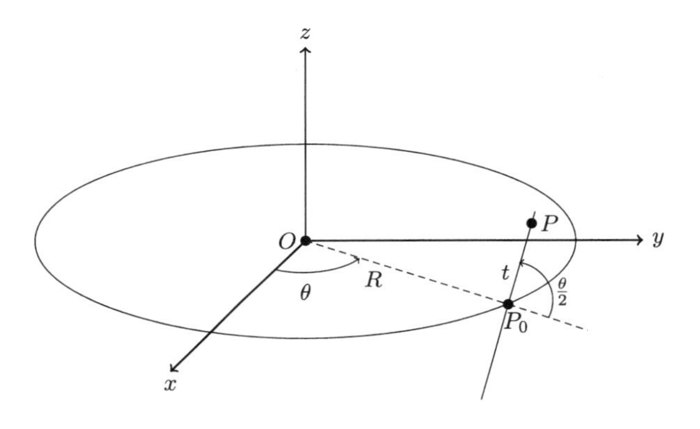

Fig. 5.8

Example 5.1.9 The *Möbius strip* (see Fig. 5.7) is a surface of class $\mathcal{C}^\infty$.

Proof We consider a circle of radius R, centered at the origin in the xy-plane, and a segment of length $2r$, with $r < R$, perpendicular in $\mathbb{R}^3$ to the tangent to the circle. The middle point P_0 of the segment moves uniformly along the circle; as this middle point travels by a full turn along the circle, the segment rotates uniformly by a half turn. This yields a parametric representation of the corresponding *Möbius strip*

$$f: \mathbb{R} \times]-r, +r[\longrightarrow \mathbb{R}^3, \qquad (\theta, t) \mapsto f(\theta, t)$$

$$f(\theta, t) = \left(\left(R + t\cos\frac{\theta}{2}\right)\cos\theta, \left(R + t\cos\frac{\theta}{2}\right)\sin\theta, t\sin\frac{\theta}{2}\right)$$

where θ is the around of rotation arround the circle in the xy-plane and t measures the position on the rotating segment (see Fig. 5.8). The function f is of class $\mathcal{C}^\infty$ and injective on each rectangle $]x, x + 2\pi[\times]-r, +r[$. □

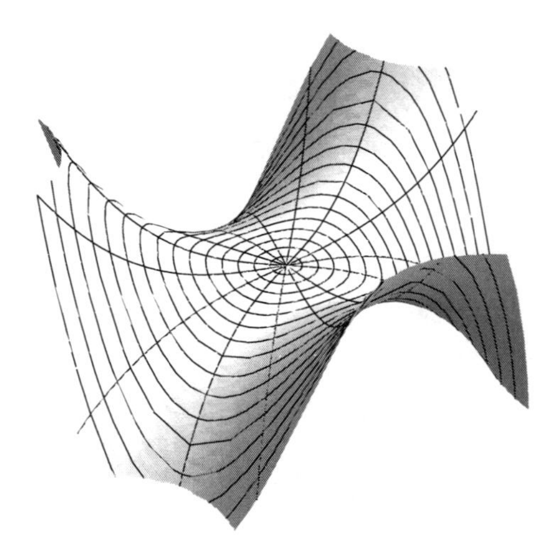

Fig. 5.9 The monkey saddle

Let us conclude this list of examples with a somewhat "peculiar one" (at least, as far as the name is concerned):

Example 5.1.10 The graph of the function

$$f : \mathbb{R}^2 \longrightarrow \mathbb{R}, \qquad (x, y) \mapsto x^3 - 3x^2 y$$

is the so-called *Monkey saddle* (see Fig. 5.9). By Example 5.1.2, this is a surface of class $\mathcal{C}^\infty$.

5.2 Regular Surfaces

We want now to transpose to the case of surfaces the notion of *regularity* of a parametric representation (see Definition 2.2.1).

A parametric representation $f(t)$ of a curve is *regular* when it is of class $\mathcal{C}^1$ and $\frac{df}{dt} \neq 0$ at each point. But a parametric representation $f(u, v)$ of class $\mathcal{C}^1$ of a surface has *two* partial derivatives $\frac{\partial f}{\partial u}$ and $\frac{\partial f}{\partial v}$. Requiring that both partial derivatives are not zero is certainly not the correct generalization of regularity. Everybody familiar with linear algebra knows that the sensible generalization of *a single vector is non-zero* is rather *several vectors are linearly independent*. Therefore we make the following definition:

Definition 5.2.1 Consider a parametric representation $f : U \longrightarrow \mathbb{R}^3$ of class $\mathcal{C}^1$ of a surface.

- The point with parameters $(u_0, v_0) \in U$ is *regular* when $\frac{\partial f}{\partial u}(u_0, v_0)$ and $\frac{\partial f}{\partial v}(u_0, v_0)$ are linearly independent.

- The representation f itself is *regular* when it is regular at each point $(u, v) \in U$.
- A point which is not regular is called *singular*.

Being *regular* is in fact a property of the corresponding surface of class $\mathcal{C}^1$, as our next result proves.

Proposition 5.2.2 *With the notation of Definition* 5.1.1, *consider two parametric representations $f(u, v)$ and $g(r, s)$ of class $\mathcal{C}^1$ of a surface, equivalent in class $\mathcal{C}^1$ via a change of parameters φ.*

- *The point with parameters $(u, v) \in U$ is regular for f if and only if the point with parameters $\varphi(u, v)$ is regular for g.*
- *The parametric representation f is regular if and only if the parametric representation g is regular.*
- *The matrix*

$$\begin{pmatrix} \frac{\partial \varphi_1}{\partial u} & \frac{\partial \varphi_2}{\partial u} \\ \frac{\partial \varphi_1}{\partial v} & \frac{\partial \varphi_2}{\partial v} \end{pmatrix}$$

is regular.

Proof From $f = g \circ \varphi$ we deduce at once

$$\frac{\partial f}{\partial u} = \frac{\partial g}{\partial r}\frac{\partial \varphi_1}{\partial u} + \frac{\partial g}{\partial s}\frac{\partial \varphi_2}{\partial u}, \qquad \frac{\partial f}{\partial v} = \frac{\partial g}{\partial r}\frac{\partial \varphi_1}{\partial v} + \frac{\partial g}{\partial s}\frac{\partial \varphi_2}{\partial v}.$$

If $\varphi(u, v)$ is singular, then $\frac{\partial g}{\partial r}$ and $\frac{\partial g}{\partial s}$ are proportional vectors, thus $\frac{\partial f}{\partial u}$ and $\frac{\partial f}{\partial v}$ are also proportional. So (u, v) is a singular point. The same argument, using φ^{-1}, proves the converse assertion.

The last statement is a well-known fact concerning Jacobian matrices. Indeed the formulas above indicate that

$$\begin{pmatrix} \frac{\partial f}{\partial u} & \frac{\partial f}{\partial u} \end{pmatrix}, \qquad \begin{pmatrix} \frac{\partial g}{\partial r} & \frac{\partial g}{\partial s} \end{pmatrix}$$

are two bases of the same 2-dimensional vector space, with

$$\begin{pmatrix} \frac{\partial \varphi_1}{\partial u} & \frac{\partial \varphi_2}{\partial u} \\ \frac{\partial \varphi_1}{\partial v} & \frac{\partial \varphi_2}{\partial v} \end{pmatrix}$$

as the change of basis matrix. This matrix is thus regular. □

As in the case of curves (see Lemma 2.2.4), we observe that the "regularity condition" immediately implies "local injectivity".

Lemma 5.2.3 *Consider a function of class $\mathcal{C}^1$*

$$f : U \longrightarrow \mathbb{R}^n$$

where U is an open subset of $\mathbb{R}^2$. When the two partial derivatives of f are linearly independent at some point $(u_0, v_0) \in U$, f is injective on a neighborhood of (u_0, v_0).

Proof By the regularity assumption, the matrix

$$\begin{pmatrix} \frac{\partial f_1}{\partial u}(u_0, v_0) & \frac{\partial f_2}{\partial u}(u_0, v_0) & \frac{\partial f_3}{\partial u}(u_0, v_0) \\ \frac{\partial f_1}{\partial v}(u_0, v_0) & \frac{\partial f_2}{\partial v}(u_0, v_0) & \frac{\partial f_3}{\partial v}(u_0, v_0) \end{pmatrix}$$

has rank 2. So at least one 2×2-sub-matrix is regular. Let us say that

$$\begin{pmatrix} \frac{\partial f_1}{\partial u}(u_0, v_0) & \frac{\partial f_2}{\partial u}(u_0, v_0) \\ \frac{\partial f_1}{\partial v}(u_0, v_0) & \frac{\partial f_2}{\partial v}(u_0, v_0) \end{pmatrix}$$

is regular. By the *local inverse theorem* (see Theorem 1.3.1), the function

$$(u, v) \mapsto \big(f_1(u, v), f_2(u, v)\big)$$

is then invertible—thus *a fortiori* injective—on a neighborhood of (u_0, v_0). Therefore f is injective on that same neighborhood. □

This lemma implies at once:

Proposition 5.2.4 *For a function of class $\mathcal{C}^1$*

$$f : U \longrightarrow \mathbb{R}^3$$

with $U \subseteq \mathbb{R}^2$ a connected open subset, the following conditions are equivalent:

1. *f is a regular parametric representation of a surface*;
2. *at each point $(u, v) \in U$, the partial derivatives of f are linearly independent.*

Another very useful related result is:

Proposition 5.2.5 *Consider a parametric representation of class $\mathcal{C}^k$ $(k \geq 1)$*

$$f : U \longrightarrow \mathbb{R}^n$$

of a surface. In a neighborhood of a regular point, the surface coincides with the graph of a function φ of class $\mathcal{C}^k$, expressing one of the three coordinates in terms of the other two.

Proof Going back to the situation in the proof of Lemma 5.2.3, let us write

$$(x, y) \mapsto (u, v) = \big(h_1(x, y), h_2(x, y)\big)$$

for the inverse of the function

$$(u, v) \mapsto (x, y) = \big(f_1(u, v), f_2(u, v)\big)$$

on a neighborhood of (u_0, v_0). This allows us to locally rewrite the parametric representation of the surface as

$$(x, y) \mapsto \big(x, y, f_3\big(h_1(x, y), h_2(x, y)\big)\big).$$

The function φ announced in the statement is thus

$$\varphi(x, y) = f_3\big(h_1(u, v), h_2(u, v)\big).$$

□

5.3 Cartesian Equation

Here we simply generalize the considerations of Sect. 2.3 for plane curves.

Definition 5.3.1 By a *Cartesian equation* of a *Cartesian surface* we mean an equation

$$F(x, y, z) = 0$$

where

- $F: \mathbb{R}^3 \longrightarrow \mathbb{R}$ is a function of class $\mathcal{C}^1$;
- the equation admits infinitely many solutions;
- there are at most finitely many solutions (x, y, z) of the equation where all three partial derivatives of F vanish.

The corresponding *Cartesian surface* is the set of those points which are solutions of the equation $F(x, y, z) = 0$.

We define further:

Definition 5.3.2 Consider a Cartesian equation $F(x, y, z) = 0$ of a Cartesian surface.

- A point of the Cartesian surface is *multiple* when all three partial derivatives of F vanish at this point.
- A point of the Cartesian surface is *simple* when at least one partial derivative of F does not vanish at this point.

As expected, we then have:

Proposition 5.3.3 *Consider a surface of class $\mathcal{C}^1$. In a neighborhood of each regular point, the support of the surface can be described by a Cartesian equation.*

Proof With the notation in the proof of Proposition 5.2.5, the surface can locally be presented as $z = \varphi(x, y)$, up to a permutation of the variables; thus it suffices to put $F(x, y, z) = z - \varphi(x, y)$. □

Proposition 5.3.4 *Let $F(x, y, z) = 0$ be a Cartesian equation of a surface. On a neighborhood of a simple point (x_0, y_0, z_0) of this surface, the Cartesian surface is the support of a regular surface.*

Proof At (x_0, y_0, z_0), one of the partial derivatives of F does not vanish (see Definition 5.3.2); we consider the case $\frac{\partial F}{\partial z} \neq 0$. By the *Implicit Function Theorem* (see Theorem 1.3.5), there exist a neighborhood U of (x_0, y_0) (and there is no loss of generality in assuming that it is open and connected) and a mapping

$$\varphi : U \longrightarrow \mathbb{R}$$

of class $\mathcal{C}^1$, such that

$$\varphi(x_0, y_0) = z_0, \qquad \forall (x, y) \in U \quad F\big(x, y, \varphi(x, y)\big) = 0.$$

The graph of φ (see Example 5.1.2)

$$f : U \longrightarrow \mathbb{R}^3, \qquad x \mapsto \big(x, y, \varphi(x, y)\big)$$

is then a parametric representation of class $\mathcal{C}^1$ of a surface whose support coincides with the Cartesian surface of equation $F(x, y, z) = 0$, in a neighborhood of (x_0, y_0). □

5.4 Curves on a Surface

A "curve on a surface" is the deformation, by the parametric representation of the surface, of a plane curve in the domain of definition of the surface.

Definition 5.4.1 Consider a parametric representation

$$f : U \longrightarrow \mathbb{R}^3, \qquad (u, v) \mapsto f(u, v)$$

of a surface. By a *curve on this surface* is meant a curve represented by $f \circ h$, where

$$h : \,]a, b[\longrightarrow U \subseteq \mathbb{R}^2, \qquad t \mapsto h(t)$$

is a parametric representation of a plane curve. A curve on a regular surface is itself called *regular* if the plane curve represented by h is regular.

Let us observe at once that this definition makes sense:

Lemma 5.4.2 *Under the conditions of Definition* 5.4.1, $f \circ h$ *is a parametric representation of a skew curve. This skew curve is regular as soon as* f *and* h *are regular.*

Proof It is immediate that the local injectivity of f and h forces that of $f \circ h$. Analogously, the regularity of f and h forces that of $f \circ h$: indeed

$$(f \circ h)' = \frac{\partial f}{\partial u} h_1' + \frac{\partial f}{\partial v} h_2'$$

where the two partial derivatives are linearly independent and $(h_1', h_2') \neq (0, 0)$; thus $(f \circ h)' \neq 0$. □

Our first concern about curves on a surface is to exhibit a particular formula for computing their length.

Proposition 5.4.3 *Consider a regular surface represented by*

$$f : U \longrightarrow \mathbb{R}^3, \qquad (u, v) \mapsto f(u, v)$$

and the three functions $U \longrightarrow \mathbb{R}$ *defined by*

$$E = \left(\frac{\partial f}{\partial u} \middle| \frac{\partial f}{\partial u} \right), \qquad F = \left(\frac{\partial f}{\partial u} \middle| \frac{\partial f}{\partial v} \right), \qquad G = \left(\frac{\partial f}{\partial v} \middle| \frac{\partial f}{\partial v} \right).$$

Given a regular curve on this surface

$$h :]a, b[\longrightarrow U \subseteq \mathbb{R}^2, \qquad t \mapsto h(t),$$

the length of an arc of this curve, between the points with parameters t_0, t_1, *is given by*

$$\int_{t_0}^{t_1} \sqrt{\begin{pmatrix} h_1'(t) & h_2'(t) \end{pmatrix} \begin{pmatrix} E(h_1(t), h_2(t)) & F(h_1(t), h_2(t)) \\ F(h_1(t), h_2(t)) & G(h_1(t), h_2(t)) \end{pmatrix} \begin{pmatrix} h_1'(t) \\ h_2'(t) \end{pmatrix}}\, dt.$$

Proof In the formula $\int_{t_0}^{t_1} \|(f \circ h)'\|$ for the length (see Proposition 2.7.5), simply replace $(f \circ h)'$ by its expansion

$$(f \circ h)' = \frac{\partial f}{\partial u} h_1' + \frac{\partial f}{\partial v} h_2'$$

and apply the definition of a norm (see 4.2.4, [4], *Trilogy II*). □

Analogously, we have:

Proposition 5.4.4 *Consider a regular surface represented by*

$$f: U \longrightarrow \mathbb{R}^3, \qquad (u,v) \mapsto f(u,v)$$

and two regular curves on this surface

$$g:]a,b[\longrightarrow U, \qquad t \mapsto g(t), \qquad h:]c,d[\longrightarrow U, \qquad s \mapsto h(s)$$

passing through the same point

$$g(t_0) = (u_0, v_0) = h(t_0).$$

The angle θ between these two curves $f \circ g$ and $f \circ h$ on the surface at the point $f(u_0, v_0)$ (that is, the angle between their tangent vectors) is given by

$$\cos\theta = \frac{\begin{pmatrix} g_1'(t_0) & g_2'(t_0) \end{pmatrix} \begin{pmatrix} E(u_0,v_0) & F(u_0,v_0) \\ F(u_0,v_0) & G(u_0,v_0) \end{pmatrix} \begin{pmatrix} h_1'(s_0) \\ h_2'(s_0) \end{pmatrix}}{\|g'(t_0)\| \cdot \|h'(s_0)\|}$$

where moreover

$$\|g'(t_0)\| = \sqrt{\begin{pmatrix} g_1'(t_0) & g_2'(t_0) \end{pmatrix} \begin{pmatrix} E(u_0,v_0) & F(u_0,v_0) \\ F(u_0,v_0) & G(u_0,v_0) \end{pmatrix} \begin{pmatrix} g_1'(t_0) \\ g_2'(t_0) \end{pmatrix}}$$

$$\|h'(s_0)\| = \sqrt{\begin{pmatrix} h_1'(s_0) & h_2'(s_0) \end{pmatrix} \begin{pmatrix} E(u_0,v_0) & F(u_0,v_0) \\ F(u_0,v_0) & G(u_0,v_0) \end{pmatrix} \begin{pmatrix} h_1'(s_0) \\ h_2'(s_0) \end{pmatrix}}.$$

Proof In the formulæ for the angle and the norm (see Definitions 4.2.6 and 4.2.4 in [4], *Trilogy II*), simply replace $(f \circ g)'$ and $(f \circ h)'$ by their expansions

$$(f \circ g)' = \frac{\partial f}{\partial u} g_1' + \frac{\partial f}{\partial v} g_2', \qquad (f \circ h)' = \frac{\partial f}{\partial u} h_1' + \frac{\partial f}{\partial v} h_2'. \qquad \square$$

The matrix in the statement of Propositions 5.4.3 and 5.4.4 will play a central role in *Riemannian geometry* (see Chap. 6), where it will be called the *metric tensor*. In classical surface theory, it is instead called the *first fundamental form of the surface*.

Definition 5.4.5 Consider a regular surface represented by

$$f: U \longrightarrow \mathbb{R}^3, \qquad (u,v) \mapsto f(u,v).$$

The quadratic form

$$\mathbf{I}: \mathbb{R}^2 \longrightarrow \mathbb{R}, \qquad (\alpha, \beta) \mapsto \begin{pmatrix} \alpha & \beta \end{pmatrix} \begin{pmatrix} E(u,v) & F(u,v) \\ F(u,v) & G(u,v) \end{pmatrix} \begin{pmatrix} \alpha \\ \beta \end{pmatrix}$$

is called the *first fundamental form* of the surface at the point with parameters (u, v), with respect to the parametric representation f.

Proposition 5.4.6 *At each point, the first fundamental form of a regular surface is symmetric, definite and positive. In particular:*

1. $EG - F^2 > 0$;
2. $E > 0, G > 0$;
3. $(\alpha, \beta) \neq (0,0) \implies E\alpha^2 + 2F\alpha\beta + G\beta^2 > 0.$

Proof The matrix

$$\begin{pmatrix} E & F \\ F & G \end{pmatrix} = \begin{pmatrix} \left(\frac{\partial f}{\partial u} \middle| \frac{\partial f}{\partial u}\right) & \left(\frac{\partial f}{\partial u} \middle| \frac{\partial f}{\partial v}\right) \\ \left(\frac{\partial f}{\partial u} \middle| \frac{\partial f}{\partial v}\right) & \left(\frac{\partial f}{\partial v} \middle| \frac{\partial f}{\partial v}\right) \end{pmatrix}$$

is the matrix of the scalar product in the two-dimensional subspace of $\mathbb{R}^3$ having as basis the two partial derivatives of f (see Definition 4.2.1 and Proposition 4.2.2, [4], *Trilogy II*). In particular, its determinant is strictly positive (see Proposition G.3.4, [4], *Trilogy II*). Since the partial derivatives are linearly independent, they are certainly non-zero, proving that $E > 0$, $F > 0$. The (more general) last assertion expresses the positivity of the scalar product. □

Now as far as the *cross product* (see Sect. 1.7, [4], *Trilogy II*) of the partial derivatives is concerned:

Proposition 5.4.7 *Consider a regular surface represented by*

$$f: U \longrightarrow \mathbb{R}^3, \qquad (u,v) \mapsto f(u,v).$$

Then

$$\left\| \frac{\partial f}{\partial u} \times \frac{\partial f}{\partial v} \right\| = \sqrt{EG - F^2}.$$

Proof By Proposition 1.7.2 in [4], *Trilogy II*,

$$\left\| \frac{\partial f}{\partial u} \times \frac{\partial f}{\partial v} \right\| = \left\| \frac{\partial f}{\partial u} \right\| \cdot \left\| \frac{\partial f}{\partial v} \right\| \cdot \left| \sin \measuredangle \left(\frac{\partial f}{\partial u}, \frac{\partial f}{\partial v} \right) \right|.$$

We have already

$$\left\| \frac{\partial f}{\partial u} \right\| = \sqrt{E}, \qquad \left\| \frac{\partial f}{\partial v} \right\| = \sqrt{G}$$

while

$$\left| \sin \measuredangle \left(\frac{\partial f}{\partial u}, \frac{\partial f}{\partial v} \right) \right| = \sqrt{1 - \cos^2 \measuredangle \left(\frac{\partial f}{\partial u}, \frac{\partial f}{\partial v} \right)} = \sqrt{1 - \frac{F^2}{EG}}$$

(see Sect. 4.2 in [4], *Trilogy II*) from which the result follows at once. □

5.5 The Tangent Plane

Of course one could consider three points on a surface, converging to each other, and determine (when this makes sense) the "limit" of the plane passing through these three points. This "limit plane" would be a good candidate for being the *tangent plane*. We prefer the following more useful approach.

Lemma 5.5.1 *Consider a regular parametric representation of a surface*

$$f: U \longrightarrow \mathbb{R}^3, \qquad (u,v) \mapsto f(u,v)$$

and a point $P = f(u_0, v_0)$ *of this surface. All the regular curves* $f \circ h$ *of class* $\mathcal{C}^1$ *on this surface*

$$h:]a,b[\longrightarrow U \subseteq \mathbb{R}^2, \qquad t \mapsto h(t)$$

passing through the point (u_0, v_0), *that is*

$$\exists t_0 \in]a,b[\quad h(t_0) = (u_0, v_0),$$

have a tangent at t_0 *which is contained in the plane passing through* P *and spanned by the two partial derivatives*

$$\frac{\partial f}{\partial u}(u_0, v_0), \qquad \frac{\partial f}{\partial v}(u_0, v_0).$$

Proof Keeping in mind that $h(t_0) = (u_0, v_0)$, simply observe that (see Definition 2.4.2)

$$(f \circ h)'(t_0) = \frac{\partial f}{\partial u}(u_0, v_0)h_1'(t_0) + \frac{\partial f}{\partial v}(u_0, v_0)h_2'(t_0). \qquad \square$$

This plane containing all the tangents to all the curves passing through P on the surface is certainly worthy of the title: the *tangent plane to the surface*.

Definition 5.5.2 The *tangent plane* at a regular point of parameters (u_0, v_0) on a surface of class C^1 represented by

$$f: U \longrightarrow \mathbb{R}^3, \qquad (u,v) \mapsto f(u,v)$$

is the plane passing through $f(u_0, v_0)$ and spanned by the two partial derivatives

$$\frac{\partial f}{\partial u}(u_0, v_0), \qquad \frac{\partial f}{\partial v}(u_0, v_0).$$

As expected:

Lemma 5.5.3 *Two regular parametric representation of class* $\mathcal{C}^1$, *equivalent in class* $\mathcal{C}^1$, *determine the same tangent plane at the corresponding point of the surface.*

Proof Consider two parametric representations $f(u,v)$, $g(r,s)$ equivalent via a change of parameters φ, thus $f = g \circ \varphi$. It follows at once that

$$\frac{\partial f}{\partial u} = \frac{\partial g}{\partial r}\frac{\partial \varphi_1}{\partial u} + \frac{\partial g}{\partial s}\frac{\partial \varphi_2}{\partial u}, \qquad \frac{\partial f}{\partial v} = \frac{\partial g}{\partial r}\frac{\partial \varphi_1}{\partial v} + \frac{\partial g}{\partial s}\frac{\partial \varphi_2}{\partial v}.$$

Thus the partial derivatives of f are linear combinations of the partial derivatives of g. □

Lemma 5.5.3 allows us to rephrase the observation made in the proof of Proposition 5.4.6:

Proposition 5.5.4 *Consider a parametric representation*

$$f : U \longrightarrow \mathbb{R}^3, \qquad (u,v) \mapsto f(u,v)$$

of class C^1 of a surface. At a regular point with parameters (u_0, v_0), the matrix

$$\begin{pmatrix} E(u_0,v_0) & F(u_0,v_0) \\ F(u_0,v_0) & G(u_0,v_0) \end{pmatrix}$$

is the matrix of the scalar product in the tangent plane, expressed with respect to the affine basis

$$\left(f(u_0,v_0); \frac{\partial f}{\partial u}(u_0,v_0), \frac{\partial f}{\partial v}(u_0,v_0) \right).$$

Proposition 5.2.5 can also be specialized:

Proposition 5.5.5 *Consider a parametric representation*

$$f : U \longrightarrow \mathbb{R}^3, \qquad (u,v) \mapsto f(u,v)$$

of class C^k $(k \geq 1)$ of a surface. In a neighborhood of a regular point with parameters (u_0, v_0), with respect to an affine basis

$$\bigl(f(u_0,v_0); e_1, e_2, e_3 \bigr)$$

with the vectors e_1, e_2 in the tangent plane, the surface admits the equation $x_3 = \varphi(x_1, x_2)$, with φ a function of class C^k.

Proof Apply a change of basis and write $g(r,s)$ for the parametric representation of the surface with respect to the basis of the statement; let (r_0, s_0) correspond to (u_0, v_0). The two partial derivatives of g at (r_0, s_0) are in the tangent plane, thus do not have any components along e_3. The linear independence of the two partial derivatives of g at (r_0, s_0)

$$\frac{\partial g}{\partial r}(r_0,s_0) = \left(\frac{\partial g_1}{\partial r}(r_0,s_0), \frac{\partial g_2}{\partial r}(r_0,s_0), 0 \right)$$

$$\frac{\partial g}{\partial s}(r_0, s_0) = \left(\frac{\partial g_1}{\partial s}(r_0, s_0), \frac{\partial g_2}{\partial s}(r_0, s_0), 0\right)$$

thus forces the linear independence of the vectors

$$\left(\frac{\partial g_1}{\partial r}(r_0, s_0), \frac{\partial g_2}{\partial r}(r_0, s_0)\right), \qquad \left(\frac{\partial g_1}{\partial s}(r_0, s_0), \frac{\partial g_2}{\partial s}(r_0, s_0)\right).$$

The result follows as in the proof of Proposition 5.2.5. □

Definition 5.5.2 provides the vectorial equation of the tangent plane at the point of parameters (u_0, v_0):

$$\overrightarrow{x} = f(u_0, v_0) + \alpha \frac{\partial f}{\partial u}(u_0, v_0) + \beta \frac{\partial f}{\partial v}(u_0, v_0)$$

thus the system of parametric equations

$$\begin{cases} x_1 = f_1(u_0, v_0) + \alpha \frac{\partial f_1}{\partial u}(u_0, v_0) + \beta \frac{\partial f_1}{\partial v}(u_0, v_0) \\ x_2 = f_2(u_0, v_0) + \alpha \frac{\partial f_2}{\partial u}(u_0, v_0) + \beta \frac{\partial f_2}{\partial v}(u_0, v_0) \\ x_3 = f_3(u_0, v_0) + \alpha \frac{\partial f_3}{\partial u}(u_0, v_0) + \beta \frac{\partial f_3}{\partial v}(u_0, v_0). \end{cases}$$

Let us now investigate the situation where the surface is described by a Cartesian equation.

Proposition 5.5.6 *Consider a Cartesian equation $F(x, y, z) = 0$ of a surface. At a simple point (x_0, y_0, z_0) of this surface, the tangent plane admits the equation*

$$\frac{\partial F}{\partial x}(x_0, y_0, z_0)(x - x_0) + \frac{\partial F}{\partial y}(x_0, y_0, z_0)(y - y_0) + \frac{\partial F}{\partial z}(x_0, y_0, z_0)(z - z_0) = 0.$$

Proof We freely use the notation and the situation described in the proof of Proposition 5.3.4, thus assuming that

$$\frac{\partial F}{\partial z}(x_0, y_0, z_0) \neq 0.$$

The tangent plane is the one passing through (x_0, y_0, z_0) and spanned by the two partial derivatives (see Definition 5.5.2)

$$\frac{\partial f}{\partial x}(x_0, y_0) = \left(1, 0, \frac{\partial \varphi}{\partial x}(x_0, y_0)\right), \qquad \frac{\partial f}{\partial y}(x_0, y_0) = \left(0, 1, \frac{\partial \varphi}{\partial y}(x_0, y_0)\right).$$

To compute the partial derivatives of φ, it suffices to differentiate with respect to x and y the equality $F(x, y, \varphi(x, y)) = 0$. This gives

$$\frac{\partial F}{\partial x} + \frac{\partial F}{\partial z}\frac{\partial \varphi}{\partial x} = 0, \qquad \frac{\partial F}{\partial y} + \frac{\partial F}{\partial z}\frac{\partial \varphi}{\partial y} = 0$$

and therefore

$$\frac{\partial \varphi}{\partial x} = -\frac{\frac{\partial F}{\partial x}}{\frac{\partial F}{\partial z}}, \qquad \frac{\partial \varphi}{\partial y} = -\frac{\frac{\partial F}{\partial y}}{\frac{\partial F}{\partial z}}.$$

The parametric equations of the tangent plane are thus

$$\begin{cases} x = x_0 + \alpha \\ y = y_0 + \beta \\ z = z_0 - \alpha \dfrac{\frac{\partial F}{\partial x}(x_0, y_0, z_0)}{\frac{\partial F}{\partial z}(x_0, y_0, z_0)} - \beta \dfrac{\frac{\partial F}{\partial y}(x_0, y_0, z_0)}{\frac{\partial F}{\partial z}(x_0, y_0, z_0)}. \end{cases}$$

The first two equations yield at once

$$\alpha = x - x_0, \qquad \beta = y - y_0.$$

Introducing these values into the third equation and multiplying the equality by the denominator yields the equation of the statement. □

Let us introduce another useful notion:

Definition 5.5.7 The *normal vector* at the point with parameters (u_0, v_0) on a regular surface represented by

$$f: U \longrightarrow \mathbb{R}^3, \qquad (u, v) \mapsto f(u, v)$$

is the vector

$$\overrightarrow{n}(u_0, v_0) = \frac{\frac{\partial f}{\partial u}(u_0, v_0) \times \frac{\partial f}{\partial v}(u_0, v_0)}{\|\frac{\partial f}{\partial u}(u_0, v_0) \times \frac{\partial f}{\partial v}(u_0, v_0)\|}$$

of length 1, perpendicular to the tangent plane (see Sect. 1.7 in [4], *Trilogy II*).

To be precise, Definition 5.5.7 only defines "a" normal vector to the surface: indeed there are two vectors of length 1 perpendicular to the tangent plane, one on each side of the tangent plane. The fact of obtaining one normal vector or its opposite depends on the parametric representation. For example if you consider the equivalent parametric representation $g(v, u) = f(u, v)$, you simply interchange the two partial derivatives and obtain the opposite normal vector (see Proposition 1.7.2 in [4], *Trilogy II*). However, a possible change in the orientation of the normal vector is necessarily a global fact:

Proposition 5.5.8 *Consider two regular parametric representations*

$$f: U \longrightarrow \mathbb{R}^3, \qquad g: V \longrightarrow \mathbb{R}^3$$

of a surface, equivalent in class $\mathcal{C}^1$ via a change of parameters φ. Write $\overrightarrow{n}$, $\overrightarrow{\eta}$ for the corresponding normal vectors. Then the two functions

$$\overrightarrow{n}, \ \overrightarrow{\eta} \circ \varphi: U \longrightarrow \mathbb{R}^3$$

- *are either equal at all points,*
- *or are opposite at all points.*

Proof We freely use the notation of Lemma 5.5.3 and the properties of the cross product studied in Sect. 1.7 of [4], *Trilogy II*. In particular since for all vectors $\overrightarrow{x}$, $\overrightarrow{y}$, one has

$$\overrightarrow{x} \times \overrightarrow{x} = \overrightarrow{0}, \qquad \overrightarrow{x} \times \overrightarrow{y} = -(\overrightarrow{y} \times \overrightarrow{x})$$

we obtain

$$\begin{aligned} \frac{\partial f}{\partial u} \times \frac{\partial f}{\partial v} &= \left(\frac{\partial g}{\partial r}\frac{\partial \varphi_1}{\partial u} + \frac{\partial g}{\partial s}\frac{\partial \varphi_2}{\partial u} \right) \times \left(\frac{\partial g}{\partial r}\frac{\partial \varphi_1}{\partial v} + \frac{\partial g}{\partial s}\frac{\partial \varphi_2}{\partial v} \right) \\ &= \left(\frac{\partial g}{\partial r} \times \frac{\partial g}{\partial s} \right) \frac{\partial \varphi_1}{\partial u}\frac{\partial \varphi_2}{\partial v} + \left(\frac{\partial g}{\partial s} \times \frac{\partial g}{\partial r} \right) \frac{\partial \varphi_2}{\partial u}\frac{\partial \varphi_1}{\partial v} \\ &= \left(\frac{\partial g}{\partial r} \times \frac{\partial g}{\partial s} \right) \left(\frac{\partial \varphi_1}{\partial u}\frac{\partial \varphi_2}{\partial v} - \frac{\partial \varphi_2}{\partial u}\frac{\partial \varphi_1}{\partial v} \right). \end{aligned}$$

The proportionality factor is the determinant of the matrix

$$\begin{pmatrix} \frac{\partial \varphi_1}{\partial u} & \frac{\partial \varphi_2}{\partial u} \\ \frac{\partial \varphi_1}{\partial v} & \frac{\partial \varphi_2}{\partial u} \end{pmatrix}.$$

Since φ admits an inverse of class $\mathcal{C}^1$, this matrix is invertible at each point, with precisely the matrix of partial derivatives of φ^{-1} as inverse. The determinant of the matrix of partial derivatives of φ is thus a continuous function of (u, v) which never takes the value 0. Therefore it is of constant sign: positive or negative. □

5.6 Tangent Vector Fields

A *tangent vector field* on a surface consists of, at each point of the surface, a vector in the tangent plane. The most efficient way to describe such a vector is to give its two components in the basis of the tangent plane determined by the partial derivatives of a parametric representation.

Definition 5.6.1 Consider a regular parametric representation

$$f\colon U \longrightarrow \mathbb{R}^3, \qquad (u, v) \mapsto f(u, v)$$

of a surface. By a *tangent vector field* of class $\mathcal{C}^k$ to the surface is meant a function of the form

$$(u, v) \mapsto \alpha(u, v) = \alpha_1(u, v)\frac{\partial f}{\partial u}(u, v) + \alpha_2(u, v)\frac{\partial f}{\partial v}(u, v)$$

where

$$\alpha_i : U \longrightarrow \mathbb{R}, \quad i = 1, 2$$

are functions of class $\mathcal{C}^k$.

Of course every arbitrary choice of the continuous functions α_i yields a tangent vector field. Among these, choosing the constant functions

$$\alpha_1(u, v) = 1, \qquad \alpha_2(u, v) = 0$$

yields the tangent vector field given at each point by the first partial derivative of f, and analogously for the second partial derivative. These two tangent vector fields thus constitute at each point a basis of the tangent plane. We shall need the following result, which generalizes this situation to the case of arbitrary linearly independent tangent vector fields.

Theorem 5.6.2 *Consider a regular parametric representation*

$$f : U \longrightarrow \mathbb{R}^3, \qquad (u, v) \mapsto f(u, v)$$

of class $\mathcal{C}^k$ of a surface and two tangent vector fields α, β of class $\mathcal{C}^k$ on this surface ($k \geq 1$). Suppose that at a given point with parameters (u_0, v_0), the two vectors $\alpha(u_0, v_0)$ and $\beta(u_0, v_0)$ are linearly independent. Then there exists, in a neighborhood of (u_0, v_0), a change of parameters of class $\mathcal{C}^k$ such that the two tangent vector fields of partial derivatives now become oriented along the tangent vector fields α and β.

Proof An elegant proof of this theorem consists of translating it into a sophisticated result on partial differential equations, accepted at once as a known result. To keep this book "as accessible as possible", we choose a more extensive proof based only on the few basic results introduced in Appendix B.

Saying that the vectors $\alpha(u_0, v_0)$ and $\beta(u_0, v_0)$ are linearly independent means that the determinant constituted of their components is non-zero. By continuity, this determinant remains non-zero on a neighborhood of (u_0, v_0).

Moreover since $\alpha(u_0, v_0)$ and $\beta(u_0, v_0)$ are linearly independent, at least one of them is not in the direction of the second partial derivative. There is no loss of generality in assuming that this holds for β. Thus β and $\frac{\partial f}{\partial v}$ are linearly independent at (u_0, v_0) and the argument just given for α and β implies that this remains the case on a neighborhood of (u_0, v_0).

We restrict our attention to a neighborhood V on which α and β, but also β and $\frac{\partial f}{\partial v}$, are linearly independent.

For each fixed point (u_0, v) in this neighborhood V, let us first investigate the existence of a curve on the surface which admits at each point the vector β as tangent vector. We are looking for a curve

$$h^{(v)} :]a, b[\longrightarrow V, \qquad t \mapsto \left(h_1^{(v)}(t), h_2^{(v)}(t)\right)$$

such that $h^{(v)}(0) = (u_0, v)$ and

$$\left(\frac{dh_1^{(v)}}{dt}, \frac{dh_2^{(v)}}{dt}\right) = \left(\beta_1\left(h_1^{(v)}(t), h_2^{(v)}(t)\right), \beta_2\left(h_1^{(v)}(t), h_2^{(v)}(t)\right)\right).$$

By Proposition B.2.1, such a curve of class $\mathcal{C}^k$ exists on a neighborhood of 0. By Proposition B.3.1, on some neighborhood of $(0, v_0)$ in $\mathbb{R}^2$, the functions

$$H_i(t, v) = h_i^{(v)}(t)$$

are themselves of class $\mathcal{C}^k$.

Let us now consider the function

$$H(t, v) = \left(H_1(t, v), H_2(t, v)\right).$$

We have

$$\frac{\partial H}{\partial t} = \left(\frac{dh_1^{(v)}}{dt}, \frac{dh_2^{(v)}}{dt}\right) = \left(\beta_1\left(h^{(v)}\right), \beta_2\left(h^{(v)}\right)\right).$$

On the other hand

$$\frac{\partial H}{\partial v}(0, v) = \frac{\partial (u_0, v)}{\partial v} = (0, 1).$$

The linear independence of β and $\frac{\partial f}{\partial v}$ proves that at each point (u_0, v), the two partial derivatives of H are linearly independent. This is in particular the case at (u_0, v_0) and by the *Local Inverse Theorem* (see Theorem 1.3.1), the function H is invertible in a neighborhood of $(0, v_0)$ and its inverse is still of class $\mathcal{C}^k$.

We have thus exhibited a first change of parameter H. The new parametric representation $g = f \circ H$ is now such that

$$\begin{aligned}
\frac{\partial g}{\partial t}(t, v) &= \frac{\partial f}{\partial u}\left(H(t, v)\right)\frac{\partial H_1}{\partial t}(t, v) + \frac{\partial f}{\partial v}\left(H(t, v)\right)\frac{\partial H_2}{\partial t}(t, v) \\
&= \frac{\partial f}{\partial u}\left(h^{(v)}(t)\right) \cdot \beta_1\left(h^{(v)}(t)\right) + \frac{\partial f}{\partial v}\left(h^{(v)}(t)\right) \cdot \beta_2\left(h^{(v)}(t)\right) \\
&= \beta\left(h^{(v)}(t)\right).
\end{aligned}$$

Let us now proceed with this new parametric representation $g(t, v)$ and the two vector fields $\alpha(t, v)$, $\beta(t, v)$ now expressed in terms of the new parameters.

For each fixed value v, keeping unchanged the various curves $v = k$ with k a constant, we shall now introduce on each of these a change of parameter $t = \psi^{(v)}(s)$ in order to force the conclusion of the theorem. We are thus interested in a global change of parameters of the form

$$(t, v) = \varphi(s, v) = \left(\psi(v, s), v\right).$$

The curves $v = k$, with k a constant, will thus indeed remain the same. Therefore the tangent vectors to these curves will remain in the same direction, that is, in the direction of the vector field β. But now we want the tangent vectors to the curves $s = k$, with k constant, to be oriented along the vector field α. Since

$$\begin{aligned}\frac{\partial(g \circ \varphi)}{\partial v} &= \frac{\partial g}{\partial t}\big(\varphi(s,v)\big) \cdot \frac{\partial \psi}{\partial v}(s,v) + \frac{\partial g}{\partial v}\big(\varphi(s,v)\big) \cdot \frac{\partial v}{\partial v}(s,v) \\ &= \frac{\partial g}{\partial t}\big(\varphi(s,v)\big) \cdot \frac{\partial \psi}{\partial v}(s,v) + \frac{\partial g}{\partial v}\big(\varphi(s,v)\big) \cdot 1\end{aligned}$$

what we want is the vector

$$\left(\frac{\partial \psi}{\partial v}(s,v), 1\right)$$

to be proportional to the vector

$$\big(\alpha_1\big(\varphi(s,v)\big), \alpha_2\big(\varphi(s,v)\big)\big).$$

Notice that $\alpha_2 = 0$ at some point would imply that at this point, α is oriented along $\frac{\partial g}{\partial t} = \beta$. This is never the case, by choice of the neighborhood V. Thus the proportionality above can be achieved by requiring that

$$\frac{\partial \psi}{\partial v}(s,v) = \frac{\alpha_1}{\alpha_2}\big(\psi(s,v), v\big).$$

The local existence of a function ψ of class C^k satisfying these requirements is again attested by Propositions B.2.1 and B.3.1. For each value of s, a function $\psi^{(s)}(v)$ satisfying

$$\frac{\partial \psi^{(s)}}{\partial v}(v) = \frac{\alpha_1}{\alpha_2}\big(\psi^{(s)}(v), v\big)$$

together with the initial condition

$$\psi^{(s)}(v_0) = s$$

exists by Proposition B.2.1. Then Proposition B.3.1 attests that

$$\psi(s,v) = \psi^{(s)}(v)$$

remains of class C^k on a neighborhood of $(0, v_0)$.

It remains to make sure that the function φ is a change of parameters in a neighborhood of $(0, v_0)$. By the *Local Inverse Theorem* (see Theorem 1.3.1), it suffices to check that the partial derivatives of φ are linearly independent at $(0, v_0)$. First,

$$\frac{\partial \varphi}{\partial s}(0, v_0) = \left(\frac{d\psi(s, v_0)}{ds}(0), \frac{dv}{ds}(0)\right) = (1, 0)$$

since $\psi(s, v_0) = s$. On the other hand

$$\frac{\partial \varphi}{\partial v}(0, v_0)(v_0) = \left(\frac{d\psi(0, v)}{dv}(v_0), \frac{dv}{dv}(v_0)\right) = \left(\frac{\alpha_1}{\alpha_2}(\varphi(s, v_0)), 1\right).$$

These two vectors are indeed trivially linearly independent. □

5.7 Orientation of a Surface

In practical life orientation is something we take for granted, we can tell left from right, clockwise from anticlockwise. However, there are surfaces for which it is not possible to define a global orientation in this way.

Imagine that you are living on the *Möbius strip* of Example 5.1.9 and you are driving along the "central circle" along which the segment rotates (see the proof of this Example). This central circle is your road A. At some point, you want to enter the road B determined by the position of the rotating segment at this point. You want to reach a specific point P on this road B and for that, you need—let us say—to turn right. But instead of turning right immediately, you first make a full tour around the central circle. You return to the same point, but "on the other face of the strip"! If you want to enter the road B and reach your original objective P, you now have to turn left! The *Möbius* strip is a *non-orientable* surface. As we have just observed, it is a peculiar surface with only "one face": moving along a face one eventually returns to the same point, but on what you would intuitively call the "other" side. A similar observation can be made concerning the "edge" moving along the "edge" eventually leads you to the opposite end of the segment you started from! The Möbius strip has one face and one edge.

There is an easy way to connect these observations with the notion of the *orientation of a real affine space* studied in Sect. 3.2.

Definition 5.7.1 A surface is *orientable* when it admits a parametric representation of class $\mathcal{C}^1$

$$f : U \longrightarrow \mathbb{R}^3$$

satisfying the following property. When two restrictions of f on connected open subsets $U_1 \subseteq U$, $U_2 \subseteq U$

$$f : U_1 \mapsto \mathbb{R}^3, \qquad f : U_2 \mapsto \mathbb{R}^3$$

constitute equivalent representations of class $\mathcal{C}^1$ of the same piece of the support of the surface, the respective partial derivatives constitute two bases of the tangent plane having the same orientation.

Of course we immediately observe that:

Lemma 5.7.2 *Definition* 5.7.1 *is independent of the choice of the parametric representation of class* $\mathcal{C}^1$.

Proof Under the conditions of Definition 5.7.1, write

$$\varphi\colon U_1 \longrightarrow U_2$$

for the change of parameters involved. Consider another parametric representation of class $\mathcal{C}^1$

$$g\colon V \longrightarrow \mathbb{R}^3$$

equivalent to f in class $\mathcal{C}^1$ via a change of parameter

$$\psi\colon U \longrightarrow V.$$

In terms of g, the change of parameters given by φ becomes

$$\psi \circ \varphi \circ \psi^{-1}\colon V_1 = \varphi(U_1) \longrightarrow \varphi(U_2) = V_2.$$

The change of basis matrix between the two bases of partial derivatives at two points $P \in V_1$ and $\psi\varphi\psi^{-1}(P)$ is thus comprised of the partial derivatives of the two components of this composite. But this matrix of partial derivatives is simply the product of the three matrices constituted respectively of the partial derivatives of ψ, φ and ψ^{-1}. Since the third matrix is the inverse of the first one, the determinant of the composite is the same as that of the second matrix. In particular, these two determinants have the same sign. □

Notice also that:

Proposition 5.7.3 *Every surface admitting an injective parametric representation of class* $\mathcal{C}^1$ *is orientable.*

Proof By injectivity, two distinct restrictions of the parametric representation always describe different pieces of the support of the surface. □

Corollary 5.7.4 *The graph of a function* $g\colon \mathbb{R}^2 \longrightarrow \mathbb{R}$ *of class* $\mathcal{C}^1$ *is an orientable surface.*

Proof As observed in Example 5.1.2, the corresponding parametric representation is injective. □

As expected:

Counterexample 5.7.5 The *Möbius strip* is not orientable.

Proof Going back to the proof of Example 5.1.9, we take

$$U_1 =]a, b[\times]-r, +r[, \qquad U_2 =]a + 2\pi, b + 2\pi[\times]-r, +r[.$$

The change of parameters in Definition 5.7.1 is simply

$$(\theta, t) \mapsto (\theta + 2k\pi, -t).$$

The two partial derivatives of f are

$$\begin{aligned}\frac{\partial f}{\partial \theta} &= \left(-\frac{t}{2}\cos\frac{\theta}{2}\cos\theta - \left(R + t\cos\frac{\theta}{2}\right)\sin\theta,\right.\\ &\qquad \left.-\frac{t}{2}\cos\frac{\theta}{2}\sin\theta + \left(R + t\cos\frac{\theta}{2}\right)\cos\theta, \frac{t}{2}\cos\frac{\theta}{2}\right)\\ \frac{\partial f}{\partial t} &= \left(\cos\frac{\theta}{2}\cos\theta, \cos\frac{\theta}{2}\sin\theta, 0\right).\end{aligned}$$

Observe that $\frac{\partial f}{\partial \theta}$ is unaffected by the change of parameters, while $\frac{\partial f}{\partial t}$ changes its sign. □

On the other hand—for example—

Example 5.7.6 The hyperboloid of one sheet is an orientable surface.

Proof Going back to the proof of Example 5.1.5, the change of parameters in Definition 5.7.1 is simply

$$\mathbb{R}^2 \mapsto \mathbb{R}^2, \qquad (\theta, z) \mapsto (\theta + 2k\pi, z).$$

The two partial derivatives of the parametric representation $f(\theta, z)$ are

$$\begin{aligned}\frac{\partial f}{\partial \theta} &= \left(-a\sqrt{1 + \frac{z^2}{c^2}}\sin\theta, b\sqrt{1 + \frac{z^2}{c^2}}\cos\theta, 0\right)\\ \frac{\partial f}{\partial z} &= \left(\frac{2az\cos\theta}{c^2\sqrt{1 + \frac{z^2}{c^2}}}, \frac{2bz\sin\theta}{c^2\sqrt{1 + \frac{z^2}{c^2}}}, 1\right).\end{aligned}$$

These partial derivatives are unaffected by the change of parameters. □

5.8 Normal Curvature

We arrive now at the central notion of this chapter: the *curvature* of a surface. The idea behind this notion is very simple (see Fig. 5.10) and is due to *Euler*.

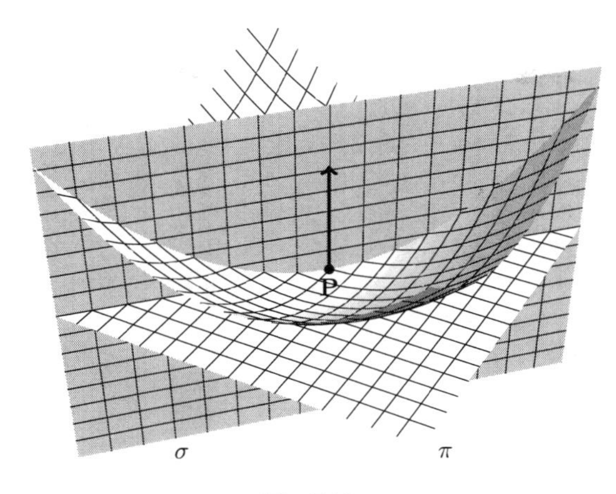

Fig. 5.10

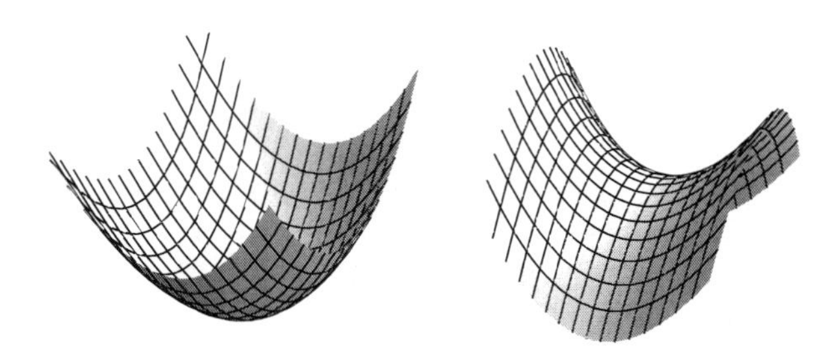

Fig. 5.11

> *Consider the tangent plane π at some point P of the surface. Cut the surface by a plane σ passing through P and perpendicular to the tangent plane π. You obtain a curve whose curvature at P will be called the* curvature of the surface at P in the direction of the plane σ.

Of course cutting the surface by different planes σ through P in different directions will *a priori* yield curves with different curvatures at P. Thus

> *The curvature of a surface at a given point P should be a function of the direction in the tangent plane at this point P.*

But there is still an important piece information to take care of. In some cases—such as an elliptic paraboloid—the curve involved in the "definition" above is always on the same side of the tangent plane; but in other cases—for example a hyperbolic paraboloid—the curve involved is sometimes on one side of the tangent plane, sometimes on the other side (see Fig. 5.11). It is rather clear that

> *Taking care of the position of the section curve with respect to the tangent plane should be achieved by giving a sign to the curvature of the surface in a given direction.*

Having said that, one may be a little bit puzzled by the situation where the section curve, in one or even in all directions, "crosses" the tangent plane, that is, it is on "both sides" of it. The *Monkey saddle* of Fig. 5.9 presents such a situation! How can we choose the sign in this case? As we shall see later (see Example 5.9.2), this is just an apparent problem, because here the normal curvature will turn out to be zero.

The presentation above is very intuitive and—being particularly careful with the necessary assumptions—it can easily be made perfectly rigorous. But it is certainly not very convenient for computing the curvature of the surface! Indeed, let us observe what has to be done.

We start with a parametric representation $f(u,v)$ of the surface and a direction in the tangent plane at the point of parameters (u_0,v_0). This direction is determined by some vector

$$\overrightarrow{w} = \alpha\frac{\partial f}{\partial u}(u_0,v_0) + \beta\frac{\partial f}{\partial v}(u_0,v_0), \qquad (0,0) \neq (\alpha,\beta) \in \mathbb{R}^2.$$

We consider next the normal vector to the surface at the point $f(u_0,v_0)$

$$\overrightarrow{n}(u_0,v_0) = \frac{\partial f}{\partial u}(u_0,v_0) \times \frac{\partial f}{\partial v}(u_0,v_0)$$

and the plane σ passing through $f(u_0,v_0)$ and spanned by the two vectors $\overrightarrow{w}$ and $\overrightarrow{n}$. The plane σ thus admits the vectorial equation, in terms of two parameters r and s,

$$(x,y,z) = f(u_0,v_0) + r\overrightarrow{w} + s\overrightarrow{n}(u_0,v_0).$$

The intersection between the surface and the plane σ is so determined by a system of six equations with four parameters u, v, r, s. It "suffices" then to eliminate three of these parameters to end up with a system of only three equations with one single parameter t

$$x = h_1(t), \qquad y = h_2(t), \qquad z = h_3(t).$$

This should provide a parametric representation $h(t)$ of the curve involved in the problem. The considerations of Sect. 2.9 now allow us to compute the curvature.

This is certainly a challenging technical problem. Fortunately, there is a much more efficient and interesting approach.

The plane curve on the surface, the intersection of the plane σ and the surface as in Fig. 5.10, thus has a parametric representation

$$h\colon\,]a,b[\longrightarrow \mathbb{R}^3.$$

Consider further its *normal* parametric representation (see Sect. 4.2)

$$\overline{h}\colon\,]c,d[\longrightarrow \mathbb{R}^3.$$

The tangent to the curve lies in the tangent plane π to the surface (see Definition 5.5.2) and of course in the plane σ of the curve: it is thus the intersection of the two planes π and σ. The vector $\overline{h}''$ remains in the plane σ of the curve and is perpendicular to the tangent vector $\overline{h}'$ (see Proposition 2.8.3), that is to the intersection of π and σ. Since σ is perpendicular to π, it follows that $\overline{h}''$ is perpendicular to π, thus is in the direction of the normal vector to the surface. Therefore, at the point considered,

$$\overline{h}'' = k\overrightarrow{n}, \qquad k \in \mathbb{R}.$$

But since the (unsigned) curvature κ of the curve is given by $\kappa = \|\overline{h}''\|$ (see Definition 2.9.1), one has further

$$\overline{h}'' = k\overrightarrow{n} = \pm\kappa\overrightarrow{n}.$$

In particular

$$k = (\overline{h}''|\overrightarrow{n}) = \pm\kappa.$$

Considering k instead of κ already takes care of the sign, which will be positive or negative according to whether or not $\overline{h}''$ is on the same side of π as $\overrightarrow{n}$.

The trick to avoid heavy calculations is the following. Instead of working with the specific curve constructed above, consider an arbitrary curve h on the surface, passing through the point P. In that case the quantity

$$(\overline{h}''|\overrightarrow{n})$$

represents—up to the sign—the length of the orthogonal projection of the "curvature vector" $\overline{h}''$ on $\overrightarrow{n}$, while the sign takes care of the orientation of this orthogonal projection, compared with that of the normal vector $\overrightarrow{n}$. Intuitively, this quantity represents *the component of the curvature of the curve in the direction of the normal vector to the surface*. This is what we shall call the *normal curvature* of the curve. The key observation will then be Theorem 5.8.2: all the curves on a surface, at a given point and in a given direction, have the same normal curvature! There is thus no need to compute the precise form of the very specific intersection curve considered at the beginning of this section: any curve in the same direction will yield the same result.

Definition 5.8.1 Consider a regular surface of class $\mathcal{C}^2$ represented by

$$f: U \longrightarrow \mathbb{R}^3, \qquad (u,v) \mapsto f(u,v)$$

and a regular curve of class $\mathcal{C}^2$ on this surface, represented by

$$g:]a,b[\longrightarrow U \subseteq \mathbb{R}^2.$$

Write $\overline{h}(s)$ for the normal representation of the curve $h = f \circ g$ on the surface. The *normal curvature* of this curve (with respect to f) at the point with parameter s_0 is the quantity

$$\kappa_n(s_0) = \big(\overline{h}''(s_0)\big|\overrightarrow{n}\big(\overline{h}(s_0)\big)\big)$$

where $\overrightarrow{n}$ indicates the normal vector to the surface.

Of course choosing another equivalent parametric representation of the surface can possibly change the orientation of the normal vector (see the considerations at the end of Sect. 5.5): thus the *sign* of the normal curvature depends explicitly on the choice of the parametric representation f.

Theorem 5.8.2 *Consider a regular parametric representation of class* $\mathcal{C}^2$

$$f: U \longrightarrow \mathbb{R}^3, \qquad (u,v) \mapsto f(u,v)$$

of a surface. All the regular curves of class $\mathcal{C}^2$

$$g: \left]a,b\right[\longrightarrow U \subseteq \mathbb{R}^2, \qquad t \mapsto g(t)$$

on this surface passing through a fixed point with parameters (u_0, v_0) *and having the same tangent at this point, also have the same normal curvature at this point. For a tangent in the direction*

$$\alpha \frac{\partial f}{\partial u}(u_0, v_0) + \beta \frac{\partial f}{\partial v}(u_0, v_0), \qquad (0,0) \neq (\alpha, \beta) \in \mathbb{R}^2$$

the normal curvature is given by

$$\kappa_n(\alpha, \beta) = \frac{L(u_0,v_0)\alpha^2 + 2M(u_0,v_0)\alpha\beta + N(u_0,v_0)\beta^2}{E(u_0,v_0)\alpha^2 + 2F(u_0,v_0)\alpha\beta + G(u_0,v_0)\beta^2}$$

where E, F, G *are the coefficients of the first fundamental form of the surface (see Definition* 5.4.5 *and Proposition* 5.4.3*) and the three functions*

$$L, M, N: U \longrightarrow \mathbb{R}$$

are defined by

$$L = \left(\frac{\partial^2 f}{\partial u^2} \Big| \overrightarrow{n} \right), \qquad M = \left(\frac{\partial^2 f}{\partial u \partial v} \Big| \overrightarrow{n} \right), \qquad N = \left(\frac{\partial^2 f}{\partial v^2} \Big| \overrightarrow{n} \right).$$

Proof We consider the curve $h(t) = (f \circ g)(t)$ and its normal representation

$$\overline{h}(s) = (h \circ \sigma^{-1})(s)$$

where σ has been defined in Sect. 4.2. We first compute

$$\overline{h}' = \left(h' \circ \sigma^{-1}\right)\left(\sigma^{-1}\right)'$$

$$\overline{h}'' = \left(h'' \circ \sigma^{-1}\right)\left(\sigma^{-1}\right)'^2 + \left(h' \circ \sigma^{-1}\right)\left(\sigma^{-1}\right)''.$$

Since h' is orthogonal to $\overrightarrow{n}$, it follows that

$$\left(\overline{h}'' \big| \overrightarrow{n}\right) = \left(\left(h'' \circ \sigma^{-1}\right)\left(\sigma^{-1}\right)'^2 \big| \overrightarrow{n}\right).$$

Of course, this is an abbreviated notation to indicate the scalar product of $\overline{h}''(s)$ with the normal vector $\overrightarrow{n}(\overline{h}(s))$ to the surface at the point $\overline{h}(s)$.

Next we compute

$$h' = \frac{\partial f}{\partial u}(g_1, g_2)g_1' + \frac{\partial f}{\partial v}(g_1, g_2)g_2'$$

which as usual we abbreviate again as

$$h' = \frac{\partial f}{\partial u}g_1' + \frac{\partial f}{\partial v}g_2'.$$

With analogous abbreviated notation

$$h'' = \left(\frac{\partial^2 f}{\partial u^2}g_1' + \frac{\partial^2 f}{\partial u \partial v}g_2'\right)g_1' + \frac{\partial f}{\partial u}g_1'' + \left(\frac{\partial^2 f}{\partial v \partial u}g_1' + \frac{\partial^2 f}{\partial v^2}g_2'\right)g_2' + \frac{\partial f}{\partial v}g_2''.$$

Again the first partial derivatives of f are in the tangent plane, thus perpendicular to $\overrightarrow{n}$. This proves that

$$(\overline{h}''|\overrightarrow{n}) = \left(\left(\frac{\partial^2 f}{\partial u^2}g_1'^2 + 2\frac{\partial^2 f}{\partial u \partial v}g_1'g_2' + \frac{\partial^2 f}{\partial v^2}g_2'^2\right) \circ \sigma^{-1} \middle| \overrightarrow{n}\right)(\sigma^{-1})'^2.$$

With the notation of the statement, this can be re-written as

$$(\overline{h}''|\overrightarrow{n}) = ((L\,g_1'^2 + 2Mg_1'g_2' + Ng_2'^2) \circ \sigma^{-1})(\sigma^{-1})'^2.$$

But (see the proof of Proposition 4.3.2)

$$(\sigma^{-1})'^2 = \frac{1}{\|h' \circ \sigma^{-1}\|^2} = \frac{1}{(h' \circ \sigma^{-1}|h' \circ \sigma^{-1})}.$$

With the form of h' as calculated above, we obtain

$$(\sigma^{-1})'^2 = \frac{1}{(\frac{\partial f}{\partial u}|\frac{\partial f}{\partial u})g_1'^2 + 2(\frac{\partial f}{\partial u}|\frac{\partial f}{\partial v})g_1'g_2' + (\frac{\partial f}{\partial v}|\frac{\partial f}{\partial v})(g_2')^2} \circ \sigma^{-1}.$$

With the notation of Proposition 5.4.3, this can be re-written as

$$(\sigma^{-1})'^2 = \frac{1}{(E\,g_1'^2 + 2F\,g_1'g_2' + G\,g_2'^2)} \circ \sigma^{-1}.$$

The normal curvature of the curve $h = f \circ g$, in terms of the parameter s, is thus given by

$$\frac{Lg_1'^2 + 2Mg_1'g_2' + Ng_2'^2}{Eg_1'^2 + 2F\,g_1'g_2' + Gg_2'^2} \circ \sigma^{-1}.$$

Since $\sigma^{-1}(s) = t$, this proves the formula of the statement, in terms of the original parameter t.

The rest of the proof is now easy. The numerator and the denominator are homogeneous polynomials of the same degree 2, thus the quotient remains unchanged if we replace h' by a vector proportional to it. This proves that two regular curves with the same tangent line—that is, with proportional tangent vectors—have the same normal curvature. □

Theorem 5.8.2 gives rise to the following definition:

Definition 5.8.3 Consider a regular parametric representation of class $\mathcal{C}^2$

$$f: U \longrightarrow \mathbb{R}^3, \qquad (u, v) \mapsto f(u, v)$$

of a surface. The *normal curvature* of the surface (with respect to f), at the point with parameters (u_0, v_0), in the direction

$$\alpha \frac{\partial f}{\partial u}(u_0, v_0) + \beta \frac{\partial f}{\partial v}(u_0, v_0), \qquad (0, 0) \neq (\alpha, \beta) \in \mathbb{R}^2$$

of the tangent plane, is the normal curvature of any regular curve on the surface, passing through (u_0, v_0) and whose tangent is oriented in the direction (α, β).

Still by Theorem 5.8.2, we thus have:

Proposition 5.8.4 *Consider a regular parametric representation $f(u, v)$ of class $\mathcal{C}^2$ of a surface. The normal curvature in the direction $(\alpha, \beta) \neq (0, 0)$ of the tangent plane at the point with parameters (u_0, v_0) is given by*

$$\kappa_n(\alpha, \beta) = \frac{L(u_0, v_0)\alpha^2 + 2M(u_0, v_0)\alpha\beta + N(u_0, v_0)\beta^2}{E(u_0, v_0)\alpha^2 + 2F(u_0, v_0)\alpha\beta + G(u_0, v_0)\beta^2}$$

where E, F, G, L, M, N are the six functions defined in Proposition 5.4.3 *and Theorem* 5.8.2.

Let us recall that the denominator of the fraction in Proposition 5.8.4 is never zero (see Proposition 5.4.6).

Let us also emphasize the fact that the notion of normal curvature is "almost" independent of the choice of the parametric representation chosen to define it:

Proposition 5.8.5 *Consider two equivalent regular parametric representations of class $\mathcal{C}^2$ of a surface. One of the following possibilities holds:*

- *the two representations yield at each point the same notion of normal curvature;*
- *the two representations yield at each point two notions of normal curvature opposite in sign.*

Proof This follows by Definition 5.8.1 and Proposition 5.5.8. □

Let us conclude with a definition:

Definition 5.8.6 Consider a regular surface of class $\mathcal{C}^2$ represented by

$$f: U \longrightarrow \mathbb{R}^3, \qquad (u, v) \mapsto f(u, v).$$

The quadratic form

$$\mathbf{II}: \mathbb{R}^2 \longrightarrow \mathbb{R}, \qquad (\alpha, \beta) \mapsto (\alpha \quad \beta) \begin{pmatrix} L(u, v) & M(u, v) \\ M(u, v) & N(u, v) \end{pmatrix} \begin{pmatrix} \alpha \\ \beta \end{pmatrix}$$

is called the *second fundamental form* of the surface at the point with parameters (u, v), with respect to the parametric representation f.

We shall prove in Sect. 6.16 that the knowledge of the two fundamental forms of a surface, that is, the knowledge of the six functions E, F, G, L, M, N, entirely determines the surface up to an isometry. This will generalize our Theorem 2.12.4 in the case of plane curves.

5.9 Umbilical Points

As already mentioned, at a given point of a surface, the curvature is *a priori* different in the various directions of the tangent plane. Studying the "curvature function" will be the topic of the next section. The present short section is instead devoted to the special case where the "curvature function" at a given point is constant.

Definition 5.9.1 Consider a regular surface $f(u, v)$ of class $\mathcal{C}^2$. The point with parameters (u_0, v_0) is *umbilical* when at this point, the normal curvature is the same in all directions of the tangent plane.

Proposition 5.9.2 *Consider a regular surface $f(u, v)$ of class $\mathcal{C}^2$. The point with parameters (u_0, v_0) is* umbilical *if and only if at this point, there exists a real number k such that (see Definitions* 5.8.6 *and* 5.4.5)

$$\mathbf{II}(u_0, v_0) = k\mathbf{I}(u_0, v_0).$$

Under these conditions, the real number k is the constant value of the normal curvature at $f(u_0, v_0)$.

The condition of the statement thus means

$$L(u_0, v_0) = kE(u_0, v_0), \qquad M(u_0, v_0) = kF(u_0, v_0),$$
$$N(u_0, v_0) = kG(u_0, v_0).$$

Let us mention that it is common practice to write instead

$$\frac{L(u_0, v_0)}{E(u_0, v_0)} = \frac{M(u_0, v_0)}{F(u_0, v_0)} = \frac{N(u_0, v_0)}{G(u_0, v_0)}$$

where of course k is then the common value of these three fractions. When $F(u_0, v_0) = 0$ the formula must by convention be interpreted as

$$\frac{L(u_0, v_0)}{E(u_0, v_0)} = \frac{N(u_0, v_0)}{G(u_0, v_0)}, \qquad F(u_0, v_0) = 0 = M(u_0, v_0).$$

On the other hand let us recall that in all cases, $E \neq 0$ and $G \neq 0$ (see Proposition 5.4.6).

Proof Assume that we have an umbilical point. With abbreviated notation and writing k for the constant value of the normal curvature, for every pair $(\alpha, \beta) \neq (0, 0)$, but trivially also for $(\alpha, \beta) = (0, 0)$, we have (see Theorem 5.8.2)

$$(L - kE)\alpha^2 + 2(M - kF)\alpha\beta + (N - kG)\beta^2 = 0.$$

We obtain a real polynomial $p(\alpha, \beta)$ taking only the value 0: this is thus the zero polynomial, that is

$$L - kE = 0, \qquad M - kF = 0, \qquad N - kG = 0.$$

The conclusion follows at once.

The converse implication is immediate from the formula in Theorem 5.8.2. □

Our first example of an umbilical point has already been suggested in the introduction to Sect. 5.8.

Example 5.9.3 The origin is an umbilical point of the *Monkey saddle* (see Example 5.1.10), with normal curvature 0.

Proof The parametric representation of the *Monkey saddle* is thus

$$f(x, y) = \big(x, y, x^3 - 3x^2y\big).$$

Its partial derivatives are

$$\frac{\partial f}{\partial x} = \big(1, 0, 3x^2 - 6xy\big), \qquad \frac{\partial f}{\partial y} = \big(0, 1, -3x^2\big)$$

and are linearly independent at each point. The second partial derivatives are

$$\frac{\partial^2 f}{\partial x^2} = (0, 0, 6x - 6y), \qquad \frac{\partial^2 f}{\partial x \partial y} = (0, 0, -6x), \qquad \frac{\partial^2 f}{\partial y^2} = (0, 0, 0).$$

All three second partial derivatives vanish for $(x, y) = (0, 0)$, forcing $L = M = N = 0$ at this point. Thus the normal curvature itself is equal to 0 in all directions at the origin. □

Example 5.9.3 presents the situation of an "isolated" umbilical point. At the opposite extreme we have:

Example 5.9.4 All the points of a sphere of radius R are umbilical, with normal curvature $\pm\frac{1}{R}$.

Proof Cutting a sphere of radius R by a plane orthogonal to a tangent plane yields a "great circle" of radius R, thus with curvature $\frac{1}{R}$ (see Example 2.9.5). So the normal curvature is equal to $\frac{1}{R}$ (or to $-\frac{1}{R}$, depending on the orientation of the normal vector) at all points, in all directions. □

Of course, we have:

Example 5.9.5 All the points of a plane are umbilical with normal curvature 0.

Proof Cutting the plane by another plane orthogonal to it yields a straight line, thus with curvature 0 (see Example 2.9.4). □

It is not always easy "at first glance", to guess where the possible umbilical points of a surface lie. To convince you of that, let us consider the case of an ellipsoid.

Example 5.9.6 The ellipsoid with equation

$$\frac{x^2}{a^2} + \frac{y^2}{b^2} + \frac{z^2}{c^2} = 1, \qquad a > b > c > 0,$$

admits the four umbilical points

$$\left(\pm a\sqrt{\frac{a^2 - b^2}{a^2 - c^2}}, 0, \pm c\sqrt{\frac{b^2 - c^2}{a^2 - c^2}}\right).$$

Proof Notice first that at the two "poles" $(0, 0, \pm c)$ of the ellipsoid, the tangent plane is horizontal. Cutting the ellipsoid by the vertical xz-plane of equation $y = 0$ yields an ellipse with equation

$$\frac{x^2}{a^2} + \frac{z^2}{c^2} = 1$$

whose curvature at the two poles indicated (see Example 2.9.7) is equal to $\frac{c}{a^2}$. Cutting the ellipsoid by the vertical yz-plane analogously yields a curvature $\frac{c}{b^2}$ at the two poles. Since $a > b$, the normal curvatures of the ellipsoid in the directions of the x and y axis are different at the two poles $(0, 0, \pm c)$, thus these poles are certainly not umbilical points. Of course—even if not needed in this proof—an analogous argument holds for the other "poles" $(\pm a, 0, 0)$ and $(0, \pm b, 0)$.

A parametric representation of the ellipsoid punctured at its two poles $(0, 0, \pm c)$ is given by (see Example 5.1.6)

$$f(\theta, \tau) = (a\cos\tau\cos\theta, b\cos\tau\sin\theta, c\sin\tau).$$

It follows at once that

$$\frac{\partial f}{\partial \theta} = (-a\cos\tau\sin\theta, b\cos\tau\cos\theta, 0)$$
$$\frac{\partial f}{\partial \tau} = (-a\sin\tau\cos\theta, -b\sin\tau\sin\theta, c\cos\tau)$$

from which we get in particular the normal vector

$$\vec{n} = \frac{(bc\cos^2\tau\cos\theta, ac\cos^2\tau\sin\theta, ab\sin\tau\cos\tau)}{\|(bc\cos^2\tau\cos\theta, ac\cos^2\tau\sin\theta, ab\sin\tau\cos\tau)\|}.$$

For short, let us write $\eta(\theta, \tau)$ for the norm in the denominator of this fraction. Let us compute further

$$\frac{\partial^2 f}{\partial \theta^2} = (-a\cos\tau\cos\theta, -b\cos\tau\sin\theta, 0)$$
$$\frac{\partial^2 f}{\partial \theta\partial \tau} = (a\sin\tau\sin\theta, -b\sin\tau\cos\theta, 0)$$
$$\frac{\partial^2 f}{\partial \tau^2} = (-a\cos\tau\cos\theta, -b\cos\tau\sin\theta, -c\sin\tau).$$

We therefore obtain

$$E = \cos^2\tau\left(a^2\sin^2\theta + b^2\cos^2\theta\right)$$
$$F = \sin\tau\cos\tau\sin\theta\cos\theta\left(a^2 - b^2\right)$$
$$G = a^2\sin^2\tau\cos^2\theta + b^2\sin^2\tau\sin^2\theta + c^2\cos^2\tau$$

and further

$$L = -\frac{abc\cos^3\tau}{\eta(\theta, \tau)}$$
$$M = 0$$

$$N = -\frac{abc\cos\tau}{\eta(\theta,\tau)}.$$

If $F \neq 0$ at some point, then by Proposition 5.9.2, to have an umbilical point, we must have $L = 0 = N$, because $M = 0$. This means $\cos\tau = 0$, thus $\tau = \pm\frac{\pi}{2}$. This corresponds to the two "poles" $(0, 0, \pm c)$ which are not represented by f (and which in any case, as we have seen above, are not umbilical points). So this case must be excluded and in order to have an umbilical point, it is thus necessary, again by Proposition 5.9.2, to have $F = 0$ as well as the equality $\frac{L}{E} = \frac{N}{G}$, that is

$$\frac{-\frac{abc\cos^3\tau}{\eta(\theta,\tau)}}{\cos^2\tau(a^2\sin^2\theta + b^2\cos^2\theta)} = \frac{-\frac{abc\cos\tau}{\eta(\theta,\tau)}}{a^2\sin^2\tau\cos^2\theta + b^2\sin^2\tau\sin^2\theta + c^2\cos^2\tau}.$$

Since we have already excluded the case $\cos\tau = 0$, this reduces to

$$a^2\sin^2\theta + b^2\cos^2\theta = a^2\sin^2\tau\cos^2\theta + b^2\sin^2\tau\sin^2\theta + c^2\cos^2\tau.$$

The necessary condition $F = 0$ corresponds to the four possibilities

$$\sin\tau = 0, \qquad \cos\tau = 0, \qquad \sin\theta = 0, \qquad \cos\theta = 0.$$

The second possibility has already been excluded. In the case $\sin\tau = 0$, thus $\cos\tau = \pm 1$, the requirement $\frac{L}{E} = \frac{N}{G}$ becomes

$$a^2\sin^2\theta + b^2\cos^2\theta = c^2.$$

This is impossible because

$$a^2\sin^2\theta + b^2\cos^2\theta \geq b^2\sin^2\theta + b^2\cos^2\theta = b^2 > c^2.$$

Analogously, the case $\cos\theta = 0$, thus $\sin\theta = \pm 1$, yields

$$a^2 = b^2\sin^2\tau + c^2\cos^2\tau.$$

Again this is impossible, because

$$b^2\sin^2\tau + c^2\cos^2\tau \leq b^2\sin^2\tau + b^2\cos^2\tau = b^2 < a^2.$$

We are thus left with the single possibility $\sin\theta = 0$, thus $\cos\theta = \pm 1$. In that case the requirement $\frac{L}{E} = \frac{N}{G}$ becomes

$$b^2 = a^2\sin^2\tau + c^2\cos^2\tau.$$

This can be re-written as

$$a^2 - b^2 = (a^2 - c^2)\cos^2\tau.$$

It follows at once that this condition holds for

$$\cos\tau = \pm\sqrt{\frac{a^2-b^2}{a^2-c^2}} \quad \text{from which} \quad \sin\tau = \pm\sqrt{\frac{b^2-c^2}{a^2-c^2}}.$$

These formulas make perfect sense because $a > b > c > 0$. Introducing these values, together with $\sin\theta = 0$ and $\cos\theta = \pm 1$, into the parametric representation f, yields the four umbilical points announced in the statement. □

5.10 Principal Directions

We arrive at the most important theorem of this chapter. To avoid any ambiguity, let us recall that a function

$$f: U \longrightarrow \mathbb{R}, \quad U \subseteq \mathbb{R}^n$$

admits a *global maximum* on U at $P \in U$ when

$$\forall Q \in U \quad f(Q) \leq f(P).$$

The notion of a *local maximum* at P is more subtle:

> *The function f admits a* local maximum *at $P \in U$ when there exists a neighborhood V of P in $\mathbb{R}^n$ which is* entirely contained in U *and such that*
>
> $$\forall Q \in V \quad f(Q) \leq f(P).$$

Of course when U is open, a global maximum at P is at once a local maximum (simply choose $V = U$). Being a *local maximum* is definitely an intrinsic property of the function f at the neighborhood of the point considered, while being a *global maximum* depends heavily on the choice of the domain U.

Theorem 5.10.1 *At a non-umbilical point of a regular surface of class $\mathcal{C}^2$, there exists:*

- *exactly one direction in which the normal curvature reaches a local maximal value;*
- *exactly one direction in which the normal curvature reaches a local minimal value,*

and these two directions are orthogonal.

Proof Let us write

$$f: U \longrightarrow \mathbb{R}^3, \qquad (u,v) \mapsto f(u,v)$$

for a regular parametric representation of class $\mathcal{C}^2$ of the surface. The normal curvature, at a non-umbilical point with parameters (u_0, v_0), in the direction of the vector

$$\alpha\frac{\partial f}{\partial u} + \beta\frac{\partial f}{\partial v}, \qquad (0,0) \neq (\alpha, \beta) \in \mathbb{R}^2$$

is the quantity

$$\kappa_n(\alpha,\beta) = \frac{L(u_0,v_0)\alpha^2 + 2M(u_0,v_0)\alpha\beta + N(u_0,v_0)\beta^2}{E(u_0,v_0)\alpha^2 + 2F(u_0,v_0)\alpha\beta + G(u_0,v_0)\beta^2} = \frac{\mathbf{II}_{(u_0,v_0)}(\alpha,\beta)}{\mathbf{I}_{(u_0,v_0)}(\alpha,\beta)}$$

(see Theorem 5.8.2, Definitions 5.4.5 and 5.8.6). From now on, for short and since the values (u_0, v_0) of the parameters are fixed once and for all, we shall simply write E instead of $E(u_0, v_0)$, and analogously for $F, G, L, M, N, \mathbf{I}, \mathbf{II}$.

The function κ_n is of class $\mathcal{C}^\infty$ in α and β, on $\mathbb{R}^2 \setminus \{(0,0)\}$, as a quotient of two polynomials whose denominator never vanishes (see Proposition 5.4.6). Of course proportional values of the pair (α, β) yield the same value for κ_n. Therefore it is equivalent to study the function κ_n on—for example—a circle of radius 1 centered at $f(u_0, v_0)$, in the tangent plane. But this circle $\mathcal{C}$ is a compact subset of $\mathbb{R}^2$. By a well-known theorem in analysis (see our Corollary A.8.4), the continuous function

$$\kappa_n\colon \mathcal{C} \longrightarrow \mathbb{R}, \qquad (\alpha,\beta) \mapsto \kappa_n(\alpha,\beta)$$

is then bounded and attains its bounds. Since the function is not constant (the point is not umbilical), the function attains a global maximum and a global minimum, which are distinct values.

However, viewing κ_n again as a function defined for all $(\alpha, \beta) \neq (0,0)$, that is, on the open subset $\mathbb{R}^2 \setminus \{(0,0)\} \subseteq \mathbb{R}^2$, a global maximum or minimum is in particular a local one. This already proves that the function κ_n admits at least one local maximal value and at least one local minimal value, these two values being distinct.

But a *necessary* condition for the function κ_n of class $\mathcal{C}^1$ to admit a local minimum or maximum, is the nullity of its partial derivatives at the corresponding point. Thus if the pair $(\alpha_0, \beta_0) \neq (0,0)$ corresponds to a local maximum or a local minimum, then certainly

$$\frac{\partial \kappa_n}{\partial \alpha}(\alpha_0,\beta_0) = 0, \qquad \frac{\partial \kappa_n}{\partial \beta}(\alpha_0,\beta_0) = 0.$$

This can be re-written as

$$\begin{aligned}
&\frac{\frac{\partial \mathbf{II}}{\partial \alpha}(\alpha_0,\beta_0)\mathbf{I}(\alpha_0,\beta_0) - \mathbf{II}(\alpha_0,\beta_0)\frac{\partial \mathbf{I}}{\partial \alpha}(\alpha_0,\beta_0)}{\mathbf{I}^2(\alpha_0,\beta_0)} \\
&\quad = \frac{\frac{\partial \mathbf{II}}{\partial \alpha}(\alpha_0,\beta_0) - \kappa_n(\alpha_0,\beta_0)\frac{\partial \mathbf{I}}{\partial \alpha}(\alpha_0,\beta_0)}{\mathbf{I}(\alpha_0,\beta_0)} = 0 \\
&\frac{\frac{\partial \mathbf{II}}{\partial \beta}(\alpha_0,\beta_0)\mathbf{I}(\alpha_0,\beta_0) - \mathbf{II}(\alpha_0,\beta_0)\frac{\partial \mathbf{I}}{\partial \beta}(\alpha_0,\beta_0)}{\mathbf{I}^2(\alpha_0,\beta_0)}
\end{aligned}$$

$$= \frac{\frac{\partial \mathbf{II}}{\partial \beta}(\alpha_0, \beta_0) - \kappa_n(\alpha_0, \beta_0)\frac{\partial \mathbf{I}}{\partial \beta}(\alpha_0, \beta_0)}{\mathbf{I}(\alpha_0, \beta_0)} = 0.$$

For these fractions to be zero, the numerators must be zero, so that (after division by 2), the nullity of the partial derivatives in the direction (α_0, β_0) reduces to

$$\begin{cases} (L\alpha_0 + M\beta_0) - \kappa_n(\alpha_0, \beta_0)(E\alpha_0 + F\beta_0) = 0 \\ (M\alpha_0 + N\beta_0) - \kappa_n(\alpha_0, \beta_0)(F\alpha_0 + G\beta_0) = 0. \end{cases}$$

This system can be further re-written as

$$\begin{cases} (L - \kappa_n(\alpha_0, \beta_0)E)\alpha_0 + (M - \kappa_n(\alpha_0, \beta_0)F)\beta_0 = 0 \\ (M - \kappa_n(\alpha_0, \beta_0)F)\alpha_0 + (N - \kappa_n(\alpha_0, \beta_0)G)\beta_0 = 0. \end{cases}$$

For a fixed value (α_0, β_0) corresponding to a local maximum or minimum of κ_n, let us then consider the following system in α, β.

$$\begin{cases} (L - \kappa_n(\alpha_0, \beta_0)E)\alpha + (M - \kappa_n(\alpha_0, \beta_0)F)\beta = 0 \\ (M - \kappa_n(\alpha_0, \beta_0)F)\alpha + (N - \kappa_n(\alpha_0, \beta_0)G)\beta = 0. \end{cases}$$

This is a homogeneous system of two linear equations with two unknowns and we know that it admits a non-zero solution, namely, (α_0, β_0). This is possible if and only if the determinant of the system is zero (see any algebra course). Thus

$$\begin{aligned} &\big(L - \kappa_n(\alpha_0, \beta_0)E\big)\big(N - \kappa_n(\alpha_0, \beta_0)G\big) \\ &\quad - \big(M - \kappa_n(\alpha_0, \beta_0)F\big)\big(M - \kappa_n(\alpha_0, \beta_0)F\big) = 0. \end{aligned}$$

This can be re-written as

$$\big(EG - F^2\big)\kappa_n^2(\alpha_0, \beta_0) + (2FM - EN - GL)\kappa_n(\alpha_0, \beta_0) + \big(LN - M^2\big) = 0.$$

This proves in particular that every locally maximal or locally minimal value of $\kappa_n(\alpha, \beta)$ is a root of the equation

$$\big(EG - F^2\big)\kappa^2 + (2FM - EN - GL)\kappa + \big(LN - M^2\big) = 0.$$

Since this is an equation of degree 2, it has at most two roots. This proves that κ_n has at most two locally extremal values. But we know already that κ_n admits at least a locally minimal value and at least a locally maximal value and these two are distinct. Therefore, κ_n admits exactly one locally minimal value κ_1, reached in *at least one* direction (α_1, β_1), and exactly one locally maximal value κ_2, reached in *at least one* direction (α_2, β_2). Furthermore, these values are distinct and are given precisely by the roots of the second degree equation above.

Our next concern is to prove that each locally extremal value of κ_n is reached in *exactly one* direction. We consider the case of κ_1; the case of κ_2 is analogous. As observed earlier in the proof, every direction (α_0, β_0) yielding κ_1 as corresponding normal curvature is a solution of the system

$$\begin{cases}(L - \kappa_1 E)\alpha + (M - \kappa_1 F)\beta = 0\\ (M - \kappa_1 F)\alpha + (N - \kappa_1 G)\beta = 0.\end{cases}$$

But all the solutions of this homogeneous system constitute a sub-vector-space of $\mathbb{R}^2$. This subspace thus has dimension 0, 1 or 2: our concern is to prove that it has precisely dimension 1. Notice that if we do so, we shall get as a by-result that (α, β) is a direction of normal curvature κ_1 *if and only if* it is solution of this system.

We therefore have to exclude the dimensions 0 and 2. First, the subspace of solutions does not have dimension 0, because there exists at least a direction (α_1, β_1) with normal curvature $\kappa_n(\alpha_1, \beta_1) = \kappa_1$. On the other hand the dimension of the subspace of solutions is equal to 2 (the number of unknowns) minus the rank of the matrix of the system. So a subspace of dimension 2 would mean a matrix of rank 0, that is, the zero matrix. This would yield

$$L = \kappa_1 E, \qquad M = \kappa_1 F, \qquad N = \kappa_1 G.$$

This case has to be excluded as well, because by assumption the point is not umbilical (see Proposition 5.9.2 again).

To prove the orthogonality requirement, it "suffices" to solve the equation of the second degree yielding κ_1 and κ_2, and next for each of these values to solve the system of equations yielding the corresponding directions (α_1, β_1) and (α_2, β_2). It remains then (see Proposition 5.5.4) to check that

$$\begin{pmatrix}\alpha_1 & \beta_1\end{pmatrix}\begin{pmatrix}E & F\\ F & G\end{pmatrix}\begin{pmatrix}\alpha_2\\ \beta_2\end{pmatrix} = 0.$$

A more subtle approach will avoid those lengthy calculations.

Let us work in an affine basis with direct orientation

$$(P; e_1, e_2, e_3)$$

where

- P is the point of the surface that we are considering;
- e_1 is of length 1, in the tangent plane at P, in the direction of minimal curvature;
- e_2 is of length 1, in the tangent plane at P, in the direction of maximal curvature;
- e_3 is of length 1, perpendicular to the tangent plane at P.

By Proposition 5.5.5, we know that in a neighborhood of P the surface admits a parametric representation of the form

$$g(r, s) = \big(r, s, \varphi(r, s)\big)$$

with $g(0,0)=P$. The partial derivatives at $(0,0)$ are in the tangent plane at P, thus have a component 0 along e_3. They are thus

$$\frac{\partial g}{\partial r}(0,0)=(1,0,0)=e_1, \qquad \frac{\partial g}{\partial s}(0,0)=(0,1,0)=e_2.$$

The two directions of extremal curvatures thus coincide with those of the partial derivatives of g. The orthogonality of these directions then reduces to

$$\left(\frac{\partial g}{\partial r}(0,0)\middle|\frac{\partial g}{\partial s}(0,0)\right)=0,$$

that is to $F(0,0)=0$. Of course the coefficient F, as well as the other coefficients E, G, L, M, N, are now calculated in terms of the parametric representation g.

In the case of the minimal normal curvature κ_1, reached in the direction $(1,0)$, the system of equations already considered several times above reduces to

$$\begin{cases} L-\kappa_1 E=0 \\ M-\kappa_1 F=0. \end{cases}$$

In the case of the maximal normal curvature κ_2, reached in the direction $(0,1)$

$$\begin{cases} M-\kappa_2 F=0 \\ N-\kappa_2 G=0. \end{cases}$$

This yields in particular

$$\kappa_1 F=M=\kappa_2 F.$$

This indeed forces $F=0$, otherwise one would have $\kappa_1=\kappa_2$, which is not the case because the point is not umbilical.

Notice further (even if it is not needed for this proof) that in this last situation we also have $E(u_0,v_0)=1$ and $G(u_0,v_0)=1$ since e_1, e_2 have been chosen to be of length 1. Therefore

$$L=\kappa_1, \qquad M=0, \qquad N=\kappa_2.$$

□

Let us formalize in a definition the conclusions of Theorem 5.10.1:

Definition 5.10.2 At a non-umbilical point of a regular surface of class $\mathcal{C}^2$, the two orthogonal directions in which the normal curvature reaches an extremal value are called the two *principal directions*. The corresponding extremal values of the normal curvature are called the *principal curvatures*.

The following result is essentially the last part of the proof of Theorem 5.10.1:

Proposition 5.10.3 *Consider a point P of a regular surface of class $\mathcal{C}^2$. Consider an orthonormal basis $(P; e_1, e_2, e_3)$ such that e_1 and e_2 are in the tangent plane* *and—in the case of a non-umbilical point—are oriented along the principal directions. In a neighborhood of P, the surface admits a parametric representation of the form*

$$g\colon U \longrightarrow \mathbb{R}^3, \qquad (r,s) \mapsto \big(r,s,\varphi(r,s)\big)$$

such that

$$E(0,0)=1, \qquad F(0,0)=0, \qquad G(0,0)=1$$
$$L(0,0)=\kappa_1, \qquad M(0,0)=0, \qquad N(0,0)=\kappa_2$$

where κ_i is the normal curvature in the direction of e_i.

Proof The proof in the case of a non-umbilical point, where κ_1 and κ_2 are then the principal curvatures, is precisely the content of the last part of the proof of Theorem 5.10.1.

In the case of an umbilical point, a parametric representation of the form indicated exists locally by Proposition 5.5.5 and since e_1 and e_2 have been chosen to be orthogonal and of length 1, the same argument as in the proof of Theorem 5.10.1 forces

$$E(0,0)=1, \qquad F(0,0)=0, \qquad G(0,0)=1.$$

Since the point is umbilical, Proposition 5.9.2 then forces

$$L(0,0)=\frac{L(0,0)}{E(0,0)}=\kappa(0,0)=\frac{N(0,0)}{G(0,0)}=N(0,0), \qquad M(0,0)=0$$

where $\kappa(0,0)$ is the constant normal curvature. □

Let us emphasize the characterization of the principal curvatures and principal directions established in the proof of Theorem 5.10.1:

Proposition 5.10.4 *At a non-umbilical point of a regular surface of class C^2, the two principal curvatures are the two solutions of the second degree equation*

$$\big(EG-F^2\big)\kappa^2+(2FM-EN-GL)\kappa+\big(LN-M^2\big)=0.$$

When the point is umbilical, the constant normal curvature is a double root of this equation.

Proof The case of a non-umbilical point has been treated in the proof of Theorem 5.10.1. In the umbilical case, by Proposition 5.9.2, the equation of the statement becomes

$$\big(EG-F^2\big)\kappa^2-2k\big(EG-F^2\big)\kappa+k^2\big(EG-F^2\big)=0$$

with k the normal curvature. Since $EG - F^2 \neq 0$ (see Proposition 5.4.6), the equation reduces to

$$\kappa^2 - 2k\kappa + k^2 = 0$$

that is

$$(\kappa - k)^2 = 0.$$

□

Proposition 5.10.5 *At a non-umbilical point of a regular surface of class $\mathcal{C}^2$, a direction (α, β) is principal if and only if there exists a real number κ such that*

$$\begin{cases} (L - \kappa E)\alpha + (M - \kappa F)\beta = 0 \\ (M - \kappa F)\alpha + (N - \kappa G)\beta = 0. \end{cases}$$

Under these conditions, κ is the corresponding principal curvature. When the point is umbilical, every triple (α, β, κ), with $(\alpha, \beta) \neq (0, 0)$, a solution of this system, is such that κ is the constant normal curvature; moreover, for this κ, all directions (α, β) satisfy the system.

Proof Let us first assume that the point is non-umbilical. If (α, β) is a non-zero solution of the system for some real number κ, the determinant of the system must be zero and so κ is a principal curvature by Proposition 5.10.4. The rest has been established in the proof of Theorem 5.10.1.

In the umbilical case, by Proposition 5.9.2, the system becomes

$$\begin{cases} (k - \kappa)E\alpha + (k - \kappa)F\beta = 0 \\ (k - \kappa)F\alpha + (k - \kappa)G\beta = 0 \end{cases}$$

with k the constant normal curvature. If a solution (α, β, κ) exists, with $(\alpha, \beta) \neq (0, 0)$, fixing this value κ, the system in α, β must have a determinant equal to 0, that is

$$(k - \kappa)^2 \big(EG - F^2\big) = 0.$$

Again since $EG - F^2 \neq 0$ by Proposition 5.4.6, we conclude that $\kappa = k$, the constant normal curvature. But then the system becomes the trivial zero system and is satisfied by all pairs (α, β). □

Another interesting characterization of principal directions is given by the following result:

Proposition 5.10.6 *Consider a regular parametric representation $f(u, v)$ of class $\mathcal{C}^2$ of a surface. At a given non-umbilical point $f(u_0, v_0)$, for a direction*

$$\alpha_1 \frac{\partial f}{\partial u}(u_0, v_0) + \alpha_2 \frac{\partial f}{\partial v}(u_0, v_0)$$

of the tangent plane, the following conditions are equivalent:

1. *the direction* (α_1, α_2) *is principal*;
2. *the direction* (α_1, α_2) *satisfies at* (u_0, v_0) *the equation*

$$(EM - LF)\alpha_1^2 + (EN - LG)\alpha_1\alpha_2 + (FN - MG)\alpha_2^2 = 0.$$

At an umbilical point, all directions satisfy this equation.

Proof In the case of a non-umbilical point, let us first re-write the system of Proposition 5.10.5 as

$$\begin{cases} (L\alpha_1 + M\alpha_2)\lambda - (E\alpha_1 + F\alpha_2)\kappa = 0 \\ (M\alpha + N\alpha_2)\lambda - (F\alpha_1 + G\alpha_2)\kappa = 0. \end{cases}$$

For every principal direction (α_1, α_2), this homogeneous system in λ, κ admits the non-zero solution $(1, \kappa)$, with κ the corresponding principal direction. Thus the determinant of the system is zero, which yields precisely the equation in condition 2 of the statement.

Conversely if the equation of condition 2 holds, the determinant of the system above is zero and thus the system admits a non-zero solution. Let us recall that a solution of the system is only defined up to a multiple. If a solution has the form $(1, \kappa)$, the result follows by Proposition 5.10.5. So it suffices to prove that a non-zero solution cannot have the form (0, 1).

Indeed (0, 1) being a solution would imply

$$E\alpha_1 + F\alpha_2 = 0, \qquad F\alpha_1 + G\alpha_2 = 0.$$

Since $E \neq 0$ and $G \neq 0$ (see Proposition 5.4.6), this would imply further

$$\alpha_1 = -\frac{F}{E}\alpha_2 = \frac{F}{E}\frac{F}{G}\alpha_1.$$

Notice that $\alpha_1 = 0$ would imply $G\alpha_2 = 0$, thus $\alpha_2 = 0$; this is impossible since by assumption, $(\alpha_1, \alpha_2) \neq (0, 0)$ as a direction. But $\alpha_1 \neq 0$ then implies $\frac{F^2}{EG} = 1$, that is $EG - F^2 = 0$, which is also excluded by Proposition 5.4.6.

At an umbilical point, Proposition 5.9.2 indicates that the equation of the statement degenerates as

$$0\alpha_1^2 + 0\alpha_1\alpha_2 + 0\alpha_2^2 = 0.$$

It is of course satisfied in every direction. □

5.11 The Case of Quadrics

Let first now investigate umbilical points and principal directions on some quadrics.

Example 5.11.1 Consider the elliptic paraboloid with equation

$$z = \frac{x^2}{a^2} + \frac{y^2}{b^2}$$

where $a, b > 0$.

- The origin is an umbilical point if and only if $a = b$.
- Otherwise, the principal directions at the origin are those of the x- and y-axis with principal curvatures $\frac{2}{a^2}$ and $\frac{2}{b^2}$.

Proof Considering the parametric representation

$$g(x, y) = \left(x, y, \frac{x^2}{a^2} + \frac{y^2}{b^2}\right)$$

we obtain immediately

$$\frac{\partial g}{\partial x} = \left(1, 0, \frac{2x}{a^2}\right), \qquad \frac{\partial g}{\partial y} = \left(0, 1, \frac{2y}{b^2}\right).$$

This yields further

$$\frac{\partial^2 g}{\partial x^2} = \left(0, 0, \frac{2}{a^2}\right), \qquad \frac{\partial^2 g}{\partial x \, \partial y} = (0, 0, 0), \qquad \frac{\partial^2 g}{\partial y^2} = \left(0, 0, \frac{2}{b^2}\right).$$

At the origin, the first partial derivatives are simply $(1, 0, 0)$ and $(0, 1, 0)$, thus the tangent plane is the xy-plane and the normal vector is $(0, 0, 1)$. It follows at once that at the origin

$$E = 1, \qquad F = 0, \qquad G = 1, \qquad L = \frac{2}{a^2}, \qquad M = 0, \qquad N = \frac{2}{b^2}.$$

By Proposition 5.9.2, the origin is an umbilical point precisely when $\frac{2}{a^2} = \frac{2}{b^2}$, that is when $a = b$.

When the origin is a non-umbilical point, by Theorem 5.10.1 the principal curvatures are the roots of the equation

$$\kappa^2 - \left(\frac{2}{a^2} + \frac{2}{b^2}\right)\kappa + \frac{4}{a^2 b^2} = 0.$$

They are thus

$$\kappa_1 = \frac{2}{a^2}, \qquad \kappa_2 = \frac{2}{b^2}$$

and are both positive. We treat the case of κ_1: the other case is analogous. The corresponding principal direction is given by the solutions of the system (see again

Proposition 5.10.5)

$$\begin{cases} 0x + 0y = 0 \\ 0x + \left(\frac{2}{b^2} - \frac{2}{a^2}\right)y = 0. \end{cases}$$

These are precisely all the points $(x, 0)$. □

Example 5.11.2 Consider the hyperbolic paraboloid with equation

$$z = \frac{x^2}{a^2} - \frac{y^2}{b^2}$$

with $a, b > 0$.

- The origin is never an umbilical point.
- The principal directions at the origin are those of the x- and y-axis with principal curvatures $\frac{2}{a^2}$ and $-\frac{2}{b^2}$.

Proof Exactly the same arguments as in Example 5.11.1 now yield

$$E = 1, \qquad F = 0, \qquad G = 1, \qquad L = \frac{2}{a^2}, \qquad M = 0, \qquad N = -\frac{2}{b^2}$$

and lead to the conclusions of the statement. □

Example 5.11.3 At all points, the parabolic cylinder with equation $z = \frac{x^2}{a^2}$ admits the direction of the y-axis as a principal direction with principal curvature 0. It does not have any umbilical points. At the origin, the principal curvature in the direction of the x-axis is equal to $\frac{2}{a^2}$.

Proof The parametric representation

$$f(x, y) = \left(x, y, \frac{x^2}{a^2}\right)$$

admits as first partial derivatives

$$\frac{\partial f}{\partial x} = \left(1, 0, \frac{2x}{a^2}\right), \qquad \frac{\partial f}{\partial y} = (0, 1, 0).$$

It follows that

$$E = 1 + \frac{4x^2}{a^4}, \qquad F = 0, \qquad G = 1$$

while the normal vector is

$$\vec{n}(x, y) = \frac{1}{\sqrt{1 + \frac{4x^2}{a^4}}}\left(-\frac{2x}{a^2}, 0, 1\right).$$

The second partial derivatives of f are

$$\frac{\partial^2 f}{\partial x^2} = \left(0, 0, \frac{2}{a^2}\right), \qquad \frac{\partial^2 f}{\partial x \partial y} = (0, 0, 0), \qquad \frac{\partial^2 f}{\partial y^2} = (0, 0, 0).$$

It follows that

$$L = \frac{2}{a^2\sqrt{1 + \frac{4x^2}{a^4}}}, \qquad M = 0, \qquad N = 0.$$

The normal curvature at the point $f(x, y)$ in the direction (α, β) of the tangent plane is thus (see Proposition 5.8.4)

$$\kappa_n(\alpha, \beta) = \frac{\frac{2}{a^2\sqrt{1+\frac{4x^2}{a^4}}}\alpha^2}{(1 + \frac{4x^2}{a^4})\alpha^2 + \beta^2}.$$

This function is always positive and takes the value 0 when $\alpha = 0$. This value 0 is thus minimal. Thus the direction $(0, 1)$, that is the direction $\frac{\partial f}{\partial y} = (0, 1, 0)$ of the y-axis, is always a principal direction with principal curvature 0. Furthermore, when $\alpha \neq 0$, $\kappa_n(\alpha, \beta) > 0$, proving the absence of umbilical points.

At the origin, the formula above becomes simply

$$\kappa_n(\alpha, \beta) = \frac{\frac{2}{a^2}\alpha^2}{\alpha^2 + \beta^2} = \frac{2}{a^2} \frac{\alpha^2}{\alpha^2 + \beta^2}.$$

The maximal value $\frac{2}{a^2}$ is reached for $\beta = 0$, i.e. in the direction of the x-axis. □

5.12 Approximation by a Quadric

Theorem 5.10.1 is surprising and one is even tempted to think that it cannot be true, as suggested by our Problem 5.17.6. Indeed, one can easily imagine "counterexamples" to Theorem 5.10.1, such as the surface pictured in Fig. 5.12! This gives the clear impression of admitting two directions of maximal curvature and two directions of minimal curvature.

It is certainly quite easy to find surfaces which violate the conclusions of Theorem 5.10.1; but what about the assumptions of this theorem? For example, just by looking at a picture, we can often discern the difference between the classes $\mathcal{C}^0$ and $\mathcal{C}^1$; but not really between the classes $\mathcal{C}^1$ and $\mathcal{C}^2$. Nor can we visually recognize the difference between the class $\mathcal{C}^1$ and the regularity, and it is certainly not easy to find umbilical points on a picture (see Example 5.9.6)!

This section intends to "demystify" the question, by showing that the whole situation can be reduced to that at a "vertex" of a quadric, as studied in Sect. 5.11. In that case, Theorem 5.10.1 certainly becomes "less amazing".

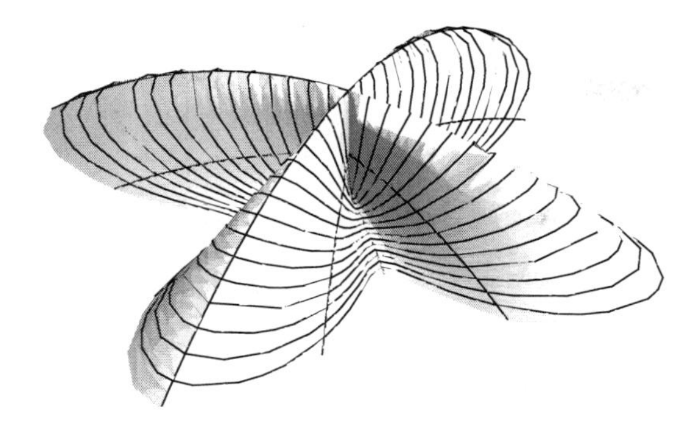

Fig. 5.12

Theorem 5.12.1 *In a neighborhood of a regular point of a surface of class* $\mathcal{C}^2$, *there exists a second order approximation of the surface by a quadric admitting at this point the same normal curvature as the surface.*

Proof Let P be a regular point of the surface. We apply Proposition 5.5.5 and work in an orthonormal basis with origin P, with direct orientation, whose first two vectors e_1, e_2 are in the tangent plane: e_1 and e_2 are oriented along the principal directions at P in the case of a non-umbilical point and are arbitrary otherwise. In a neighborhood of P, the surface admits a parametric representation of the form $g(x,y) = (x, y, \varphi(x,y))$ with $g(0,0) = P = (0,0,0)$, thus in particular $\varphi(0,0) = 0$. Notice that in the case of a non-umbilical point, this is precisely the basis considered in the last part of the proof of Theorem 5.10.1.

The partial derivatives of g at $(0,0)$ are in the tangent plane at P, thus have a component 0 along e_3. In other words

$$\frac{\partial g}{\partial x}(0,0) = \left(1, 0, \frac{\partial \varphi}{\partial x}(0,0)\right) = (1,0,0) = e_1$$

$$\frac{\partial g}{\partial y}(0,0) = \left(0, 1, \frac{\partial \varphi}{\partial y}(0,0)\right) = (0,1,0) = e_2.$$

In particular, since (e_1, e_2) is an orthonormal basis of the tangent plane at P

$$E(0,0) = 1, \qquad F(0,0) = 0, \qquad G(0,0) = 1.$$

Since the normal vector at P is $e_3 = (0,0,1)$, computing the second partial derivatives of g and performing the scalar product with e_3 yields at once

$$L(0,0) = \frac{\partial^2 \varphi}{\partial x^2}(0,0), \qquad M(0,0) = \frac{\partial^2 \varphi}{\partial x \partial y}(0,0), \qquad N(0,0) = \frac{\partial^2 \varphi}{\partial y^2}(0,0).$$

Notice further that when the point is umbilical, then $M(0,0) = 0$ because $F(0,0) = 0$ (see Proposition 5.9.2). When the point is non-umbilical, we have observed at the end of the proof of Theorem 5.10.1 that we also have $M(0,0)=0$. So in all cases, $M(0,0)=0$.

Again the proof of Theorem 5.10.1 tells us that in the case of a non-umbilical point

$$L(0,0)=\kappa_1(0,0), \qquad N(0,0)=\kappa_2(0,0)$$

where κ_i is the normal curvature in the direction e_i. In the case of an umbilical point, Proposition 5.9.2 indicates that

$$L(0,0)=\kappa(0,0)\,E(0,0)=\kappa(0,0), \qquad N(0,0)=\kappa(0,0)\,G(0,0)=\kappa(0,0).$$

Thus in all cases, $L(0,0)$ and $N(0,0)$ are the normal curvatures κ_i at P in the directions e_i.

Since $\varphi(0,0)=0$ and the first partial derivatives of φ vanish at $(0,0)$, the Taylor development of degree 2 of the function φ at $(0,0)$ reduces to

$$\frac{1}{2}\left(\frac{\partial^2\varphi}{\partial x^2}(0,0)x^2+2\frac{\partial^2\varphi}{\partial x\,\partial y}(0,0)xy+\frac{\partial^2\varphi}{\partial y^2}(0,0)y^2\right).$$

Considering the values of L, M, N calculated above, this can be re-written as

$$\frac{1}{2}\big(\kappa_1(0,0)x^2+\kappa_2(0,0)y^2\big).$$

The quadric with equation

$$z=\frac{1}{2}\big(\kappa_1(0,0)x^2+\kappa_2(0,0)y^2\big)$$

is thus a second order approximation of the surface in a neighborhood of P; it admits the parametric representation

$$h(x,y)=\left(x,y,\frac{1}{2}\big(\kappa_1x^2+\kappa_2(0,0)y^2\big)\right).$$

The first partial derivatives of h at $(0,0)$ are then

$$\frac{\partial h}{\partial x}(0,0)=(1,0,0), \qquad \frac{\partial h}{\partial y}(0,0)=(0,1,0).$$

These coincide with the partial derivatives of g at $(0,0)$, thus both surfaces have the same normal vector at P, but also the same coefficients E, F, G as the first fundamental form at $(x,y)=(0,0)$ (see Definition 5.4.5).

The second partial derivatives of h are

$$\frac{\partial^2 h}{\partial x^2}(0,0)=\big(0,0,\kappa_1(0,0)\big),$$

$$\frac{\partial^2 h}{\partial x\,\partial y}(0,0) = (0,0,0),$$

$$\frac{\partial^2 h}{\partial y^2}(0,0) = \bigl(0,0,\kappa_2(0,0)\bigr).$$

Since both surfaces have the same normal vector $(0,0,1)$ at P, these formulas imply that they also have the same coefficients

$$L = \kappa_1, \qquad M = 0, \qquad N = \kappa_2$$

of the second fundamental form at $(x,y) = (0,0)$ (see Definition 5.8.6).

Finally, since both surfaces admit at P the same coefficients for the two fundamental forms, they admit in particular the same normal curvature at P (see Definition 5.8.3). □

Corollary 5.12.2 *The quadric of Theorem* 5.12.1 *is a plane, a parabolic cylinder or a paraboloid.*

Proof With respect to the orthonormal basis used in the proof of Theorem 5.12.1, the quadric admits the equation

$$z = \frac{\kappa_1(0,0)}{2}x^2 + \frac{\kappa_2(0,0)}{2}y^2.$$

This is a plane when $\kappa_1(0,0) = 0 = \kappa_2(0,0)$, a parabolic cylinder when exactly one of the two coefficients $\kappa_1(0,0)$, $\kappa_2(0,0)$ is non-zero and a paraboloid when both coefficients are non-zero (see Sect. 1.14 in [4], *Trilogy II*). □

Our Theorem 5.10.1 is thus somehow "demystified": in all cases, the situation reduces to that of a plane or to Examples 5.11.1, 5.11.2 and 5.11.3.

5.13 The Rodrigues Formula

This short section is devoted to proving a celebrated result on principal directions:

Theorem 5.13.1 (Rodrigues Formula) *Given a regular parametric representation $f(u,v)$ of class $\mathcal{C}^2$ of a surface, consider the function $\overrightarrow{n}$: "the normal vector to the surface". Consider further the differentials of f and $\overrightarrow{n}$ at a non-umbilical point with parameters (u_0, v_0):*

$$df(\alpha,\beta) = \frac{\partial f}{\partial u}(u_0,v_0)\alpha + \frac{\partial f}{\partial v}(u_0,v_0)\beta$$

$$d\overrightarrow{n}(\alpha,\beta) = \frac{\partial \overrightarrow{n}}{\partial u}(u_0,v_0)\alpha + \frac{\partial \overrightarrow{n}}{\partial v}(u_0,v_0)\beta.$$

The direction

$$\alpha\frac{\partial f}{\partial u}(u_0, v_0) + \beta\frac{\partial f}{\partial v}(u_0, v_0)$$

is principal if and only if there exists a scalar κ such that

$$d\overrightarrow{n}(\alpha, \beta) = -\kappa\, df(\alpha, \beta).$$

Under these conditions, κ is the corresponding principal curvature.

Proof Let (α_0, β_0) be a principal direction. Differentiating the equality

$$\left(\overrightarrow{n}\,\middle|\,\frac{\partial f}{\partial u}\right) = 0$$

with respect to u, we obtain

$$\left(\frac{\partial \overrightarrow{n}}{\partial u}\,\middle|\,\frac{\partial f}{\partial u}\right) = -\left(\overrightarrow{n}\,\middle|\,\frac{\partial^2 f}{\partial u^2}\right).$$

Analogously

$$\left(\frac{\partial \overrightarrow{n}}{\partial v}\,\middle|\,\frac{\partial f}{\partial u}\right) = -\left(\overrightarrow{n}\,\middle|\,\frac{\partial^2 f}{\partial u\,\partial v}\right).$$

This implies

$$\begin{aligned}
&\left(d\overrightarrow{n}(\alpha, \beta) + \kappa_n(\alpha_0, \beta_0) df(\alpha_0, \beta_0)\,\middle|\,\frac{\partial f}{\partial u}\right)\\
&\quad = \left(\frac{\partial n}{\partial u}\alpha + \frac{\partial n}{\partial v}\beta + \kappa_n(\alpha_0, \beta_0)\frac{\partial f}{\partial u}\alpha + \kappa_n(\alpha_0, \beta_0)\frac{\partial f}{\partial v}\beta\,\middle|\,\frac{\partial f}{\partial u}\right)\\
&\quad = -\left(\overrightarrow{n}\,\middle|\,\frac{\partial^2 f}{\partial u^2}\right)\alpha - \left(\overrightarrow{n}\,\middle|\,\frac{\partial^2 f}{\partial u \partial v}\right)\beta\\
&\qquad + \kappa_n(\alpha_0, \beta_0)\left(\frac{\partial f}{\partial u}\,\middle|\,\frac{\partial f}{\partial u}\right)\alpha + \kappa_n(\alpha_0, \beta_0)\left(\frac{\partial f}{\partial v}\,\middle|\,\frac{\partial f}{\partial u}\right)\beta\\
&\quad = -(L\alpha + M\beta) + \kappa_n(\alpha_0, \beta_0)(E\alpha + F\beta)\\
&\quad = -\frac{1}{2}\frac{\partial \mathbf{II}}{\partial \alpha}(\alpha_0, \beta_0) + \kappa_n(\alpha_0, \beta_0)\frac{1}{2}\frac{\partial \mathbf{I}}{\partial \alpha}(\alpha_0, \beta_0)\\
&\quad = -\frac{1}{2}\frac{\partial(\mathbf{II} - \kappa_n \mathbf{I})}{\partial \alpha}(\alpha_0, \beta_0)\\
&\quad = 0
\end{aligned}$$

where the last but one equality holds because in the principal direction (α_0, β_0) we have

$$\frac{\partial \kappa_n}{\partial \alpha}(\alpha_0, \beta_0) = 0$$

as observed in the proof of Theorem 5.10.1. The last equality follows at once from Definition 5.8.3, which implies $\mathbf{II} - \kappa_n \mathbf{I} = 0$.

We have thus proved that

$$d\overrightarrow{n}(\alpha_0, \beta_0) + \kappa_n(\alpha_0, \beta_0) df(\alpha_0, \beta_0)$$

is orthogonal to $\frac{\partial f}{\partial u}$. Analogously, it is perpendicular to $\frac{\partial f}{\partial v}$ and hence orthogonal to the tangent plane. But differentiating the equality $(\overrightarrow{n} \mid \overrightarrow{n}) = 0$ with respect to u and v shows that the partial derivatives of $\overrightarrow{n}$ are perpendicular to $\overrightarrow{n}$, thus parallel to the tangent plane. Therefore $d\overrightarrow{n}$ is parallel to the tangent plane. Of course by definition, df is in the tangent plane as well. Therefore, in a principal direction, $d\overrightarrow{n} + \kappa_n df$ is both perpendicular and parallel to the tangent plane: it is thus equal to zero.

Conversely, assume that

$$d\overrightarrow{n}(\alpha_0, \beta_0) + \kappa\, df(\alpha_0, \beta_0) = 0$$

for some direction (α_0, β_0) and some scalar k. This trivially implies

$$\left(d\overrightarrow{n}(\alpha_0, \beta_0) + k\, df(\alpha_0, \beta_0) \middle| \frac{\partial f}{\partial u} \right) = 0.$$

This equality can be re-written as

$$\left(\left(\frac{\partial \overrightarrow{n}}{\partial u} \middle| \frac{\partial f}{\partial u} \right) + k \left(\frac{\partial f}{\partial u} \middle| \frac{\partial f}{\partial u} \right) \right) \alpha_0 + \left(\left(\frac{\partial \overrightarrow{n}}{\partial v} \middle| \frac{\partial f}{\partial u} \right) + k \left(\frac{\partial f}{\partial v} \middle| \frac{\partial f}{\partial u} \right) \right) \beta_0.$$

As already observed above, this can be further written as

$$(-L + kE)\alpha_0 + (-M + kF)\beta_0 = 0.$$

An analogous computation using this time $\frac{\partial f}{\partial v}$ finally yields the system of Proposition 5.10.5. Therefore (α, β) is a principal direction with principal curvature κ. □

5.14 Lines of Curvature

The curves on a surface "oriented along the principal directions" are of some interest.

Definition 5.14.1 A regular curve on a regular surface of class C^2 is a *line of curvature* when at each non-umbilical point, the tangent to the curve is oriented in a principal direction of the surface.

The following theorem underlines the interest of the lines of curvature:

Theorem 5.14.2 *Consider a regular surface of class $\mathcal{C}^k$ ($k \geq 3$). In a neighborhood of a non-umbilical point P, there exists a parametric representation $g(r,s)$ of class $\mathcal{C}^{k-1}$ of the surface such that all curves*

$$r = k, \qquad s = l, \quad \text{with } k, l \text{ constants}$$

are lines of curvature, while moreover $F = 0$ and $M = 0$ at each point.

Proof Consider a regular parametric representation

$$f : U \longrightarrow \mathbb{R}^3, \qquad (u,v) \mapsto f(u,v)$$

of class $\mathcal{C}^k$ of the surface and a non-umbilical point with parameters (u_0, v_0). Saying that the point is non-umbilical is equivalent to saying that the two roots of the second degree equation in Proposition 5.10.4 are distinct. This can be rephrased as the fact that the discriminant

$$(2FM - EN - GL)^2 - 4\big(EG - F^2\big)\big(LN - M^2\big)$$

is non-zero. By continuity, this discriminant remains non-zero on a neighborhood V of (u_0, v_0), thus all corresponding points remain non-umbilical.

For each non-umbilical point with parameters $(u,v) \in V$, let us write

$$\big(\alpha_1(u,v), \alpha_2(u,v)\big), \qquad \big(\beta_1(u,v), \beta_2(u,v)\big)$$

for the coordinates of the two principal directions with respect to the basis of partial derivatives. Expressing α and β via the classical formulas for the solutions of the second degree equation in Proposition 5.10.6, we obtain four functions α_i, β_i of class $\mathcal{C}^{k-1}$. We have thus obtained two tangent vector fields (see Definition 5.6.1) which are linearly independent at each point, since all points considered are non-umbilical. The result then follows by Theorem 5.6.2.

Since the partial derivatives are oriented at each point along the principal directions, they are orthogonal (see Theorem 5.10.1), thus $F = 0$ by definition (see Proposition 5.4.3). The first principal curvature is in the direction $(1,0)$, thus the first principal curvature is equal to $\frac{L}{E}$ (see Theorem 5.8.2); analogously the second principal curvature is $\frac{G}{N}$. Therefore by Definition 5.16.1, the total curvature is equal to $\frac{LN}{EG}$; since $F = 0$, comparing with Proposition 5.16.3 yields $M = 0$. □

5.15 Gauss' Approach to Total Curvature

This very informal section focuses on the ideas which led *Gauss* to introduce the notion of the *total curvature* of a surface, today usually called the *Gaussian curvature*.

The formal study of the *Gaussian curvature* will be the topic of our next section. Readers only interested in this formal study can skip the present "historical" section.

As we have done several times already, for this "historical" approach, we shall not pay particular attention to the necessary assumptions for our arguments to be valid. We just want—in those good cases where everything makes sense—to end up with a notion which we can easily formalize rigorously.

Consider a parametric representation of a regular surface of class $\mathcal{C}^2$

$$f\colon U \mapsto \mathbb{R}^3, \qquad (u,v) \mapsto f(u,v).$$

Since we shall work on the neighborhood of a point $P = f(u_0, v_0)$, by local injectivity, there is no loss of generality in assuming that f is injective. The normal vectors $\overrightarrow{n}(u,v)$ to the surface are of length 1, thus belong to the sphere $\mathcal{S}$ of radius 1 centered at the origin of $\mathbb{R}^3$. We thus consider the mapping

$$\overrightarrow{n} : U \longrightarrow \mathcal{S}, \qquad (u,v) \mapsto \overrightarrow{n}(u,v).$$

Consider now a neighborhood $(u_0, v_0) \in V \subseteq U$ and the corresponding piece $f(V)$ of the surface around $P = f(u_0, v_0)$. When the surface is "very slightly curved" in the neighborhood of P, the vector $\overrightarrow{n}(u,v)$ does not vary much and therefore the area of the piece $\overrightarrow{n}(V)$ of the sphere is rather small. At the opposite extreme, when the surface is "very highly curved" in the neighborhood of P, the vector $\overrightarrow{n}(u,v)$ varies a lot and so the piece $\overrightarrow{n}(V)$ of the sphere has a rather large area. Thus, the quotient

$$\frac{\text{area of } \overrightarrow{n}(V)}{\text{area of } f(V)}$$

somehow measures the "global curvature of the surface in the neighborhood V", provided these areas exist, of course!

With this idea in mind, *Gauss* "defines" the *total curvature* of the surface at the point $P = f(u_0, v_0)$ as being

$$\kappa_\tau(u_0, v_0) = \lim_V \frac{\text{area of } \overrightarrow{n}(V)}{\text{area of } f(V)}$$

where the limit is taken in some sense over "smaller and smaller neighborhoods of (u_0, v_0), converging to the point (u_0, v_0)". Of course if such an approach makes sense, it should be independent of the parametric representation chosen.

Let us thus work with the orthonormal basis and the parametric representation

$$g(x,y) = \bigl(x, y, \varphi(x,y)\bigr)$$

already considered in the proof of Theorem 5.12.1. Our point of the surface is now the origin of the orthonormal basis.

If the limit formula above makes sense, the limit can equivalently be calculated on a converging subfamily of neighborhoods of $(0,0)$ and we choose

$$V_{(a,b)} =]-a,+a[\times]-b,+b[, \quad a,b \in \mathbb{R},\ a,b > 0.$$

In good cases, the limit

$$\lim_{(a,b)\to(0,0)} \frac{\text{area of } \vec{n}\,(V_{(a,b)})}{\text{area of } g(V_{(a,b)})}$$

of this quotient must remain unchanged if we replace the numerator and the denominator by approximations of them at the first order. We shall therefore replace them by their orthogonal projections on the corresponding tangent plane to the surface or to the sphere, at the point considered.

For the denominator, the problem is easy: the projection of $g(x, y)$ on the tangent plane at the origin is simply $(x, y, 0)$. The corresponding area is thus simply $4ab$.

For the numerator, we have

$$\frac{\partial g}{\partial x} = \left(1, 0, \frac{\partial \varphi}{\partial x}\right), \qquad \frac{\partial g}{\partial y} = \left(0, 1, \frac{\partial \varphi}{\partial y}\right)$$

from which we get

$$\vec{n} = \frac{(-\frac{\partial \varphi}{\partial x}, -\frac{\partial \varphi}{\partial y}, 1)}{\sqrt{(\frac{\partial \varphi}{\partial x})^2 + (\frac{\partial \varphi}{\partial y})^2 + 1}}.$$

The normal vector at $g(0, 0)$ is simply $\vec{n}\,(0, 0) = (0, 0, 1)$, thus the corresponding tangent plane to the sphere is $z = 1$. Therefore the projection of $\vec{n}\,(x, y)$ on the tangent plane to the sphere for $(x, y) = (0, 0)$, translated further on the xy-plane (which does not change the area) is thus

$$p(x, y) = \frac{(-\frac{\partial \varphi}{\partial x}, \frac{\partial \varphi}{\partial y}, 0)}{\sqrt{(\frac{\partial \varphi}{\partial x})^2 + (\frac{\partial \varphi}{\partial y})^2 + 1}}.$$

A well-known formula in analysis tells us that the corresponding area is (up to its sign)

$$\int_{-a}^{+a} \int_{-b}^{+b} \left\| \frac{\partial p}{\partial x} \times \frac{\partial p}{\partial y} \right\| dx\, dy.$$

We now have to go through the lengthy calculation of the norm involved in this formula.

Writing for short

$$\sqrt{\bullet} = \sqrt{\left(\frac{\partial \varphi}{\partial x}\right)^2 + \left(\frac{\partial \varphi}{\partial y}\right)^2 + 1}$$

we first compute that

$$\frac{\partial p}{\partial x} = \frac{1}{(\sqrt{\bullet})^2}\left(-\frac{\partial^2 \varphi}{\partial x^2}\sqrt{\bullet} + \frac{\partial \varphi}{\partial x}\frac{\partial \sqrt{\bullet}}{\partial x}, -\frac{\partial^2 \varphi}{\partial x \partial y}\sqrt{\bullet} + \frac{\partial \varphi}{\partial y}\frac{\partial \sqrt{\bullet}}{\partial x}, 0\right)$$

$$\frac{\partial p}{\partial y} = \frac{1}{(\sqrt{\bullet})^2}\left(-\frac{\partial^2 \varphi}{\partial x \partial y}\sqrt{\bullet} + \frac{\partial \varphi}{\partial x}\frac{\partial \sqrt{\bullet}}{\partial y}, -\frac{\partial^2 \varphi}{\partial y^2}\sqrt{\bullet} + \frac{\partial \varphi}{\partial y}\frac{\partial \sqrt{\bullet}}{\partial y}, 0\right).$$

It follows that

$$\frac{\partial p}{\partial x} \times \frac{\partial p}{\partial y} = \frac{1}{(\sqrt{\bullet})^4}(0,0,z)$$

where, after simplification,

$$\begin{aligned} z = {} & \frac{\partial^2 \varphi}{\partial x^2}\frac{\partial^2 \varphi}{\partial y^2}(\sqrt{\bullet})^2 - \frac{\partial^2 \varphi}{\partial x^2}\frac{\partial \varphi}{\partial y}\sqrt{\bullet}\frac{\partial \sqrt{\bullet}}{\partial y} - \frac{\partial \varphi}{\partial x}\frac{\partial^2 \varphi}{\partial y^2}\frac{\partial \sqrt{\bullet}}{\partial x}\sqrt{\bullet} \\ & - \left(\frac{\partial^2 \varphi}{\partial x \partial y}\right)^2 (\sqrt{\bullet})^2 + \frac{\partial \varphi}{\partial x}\frac{\partial^2 \varphi}{\partial x \partial y}\frac{\partial \sqrt{\bullet}}{\partial y}\sqrt{\bullet}. \end{aligned}$$

We compute further that

$$\begin{aligned} \frac{\partial \sqrt{\bullet}}{\partial x} &= \frac{1}{\sqrt{\bullet}}\left(\frac{\partial \varphi}{\partial x}\frac{\partial^2 \varphi}{\partial x^2} + \frac{\partial \varphi}{\partial y}\frac{\partial^2 \varphi}{\partial x \partial y}\right) \\ \frac{\partial \sqrt{\bullet}}{\partial y} &= \frac{1}{\sqrt{\bullet}}\left(\frac{\partial \varphi}{\partial y}\frac{\partial^2 \varphi}{\partial y^2} + \frac{\partial \varphi}{\partial x}\frac{\partial^2 \varphi}{\partial x \partial y}\right). \end{aligned}$$

Introducing these values into the expression above, a straightforward calculation eventually yields

$$\left\| \frac{\partial p}{\partial x} \times \frac{\partial p}{\partial y} \right\| = \frac{\frac{\partial^2 \varphi}{\partial x^2}\frac{\partial^2 \varphi}{\partial y^2} - (\frac{\partial^2 \varphi}{\partial x \partial y})^2}{(\sqrt{\bullet})^4}.$$

We thus have to compute the limit

$$\lim_{(a,b)\to(0,0)} \frac{\int_{-a}^{+a}\int_{-b}^{+b} \frac{\frac{\partial^2 \varphi}{\partial x^2}\frac{\partial^2 \varphi}{\partial y^2} - (\frac{\partial^2 \varphi}{\partial x \partial y})^2}{((\frac{\partial \varphi}{\partial x})^2 + (\frac{\partial \varphi}{\partial y})^2 + 1)^2}\, dx\, dy}{4ab}.$$

But in good cases, the limit of a quotient does not change if we differentiate both the numerator and the denominator. Let us recall further that (still in good cases)

$$\frac{d \int_0^a h(t)\, dt}{da} = h(a).$$

Therefore, differentiating with respect to both a and b,

$$\lim_{(a,b)\to(0,0)} \frac{\int_0^a \int_0^b \frac{\frac{\partial^2 \varphi}{\partial x^2}\frac{\partial^2 \varphi}{\partial y^2} - (\frac{\partial^2 \varphi}{\partial x \partial y})^2}{((\frac{\partial \varphi}{\partial x})^2 + (\frac{\partial \varphi}{\partial y})^2 + 1)^2}\, dx\, dy}{4ab}$$

$$= \lim_{(a,b)\to(0,0)} \frac{\frac{\frac{\partial^2\varphi}{\partial x^2}\frac{\partial^2\varphi}{\partial y^2}-(\frac{\partial^2\varphi}{\partial x\partial y})^2}{((\frac{\partial\varphi}{\partial x})^2+(\frac{\partial\varphi}{\partial y})^2+1)^2}(a,b)}{4}$$

$$= \frac{1}{4}\frac{\frac{\partial^2\varphi}{\partial x^2}\frac{\partial^2\varphi}{\partial y^2}-(\frac{\partial^2\varphi}{\partial x\partial y})^2}{((\frac{\partial\varphi}{\partial x})^2+(\frac{\partial\varphi}{\partial y})^2+1)^2}(0,0).$$

An analogous computation on the other quarters of $V_{(a,b)}$

$$[-a,0]\times[0,b], \qquad [0,a]\times[-b,0], \qquad [-a,0]\times[-b,0]$$

yields the same conclusion. Thus eventually, adding the results,

$$\kappa_\tau(0,0) = \frac{\frac{\partial^2\varphi}{\partial x^2}\cdot\frac{\partial^2 g}{\partial y^2}-(\frac{\partial^2\varphi}{\partial x\partial y})^2}{((\frac{\partial\varphi}{\partial x})^2+(\frac{\partial\varphi}{\partial y})^2+1)^2}(0,0).$$

In the proof of Theorem 5.12.1, we observed that

$$\frac{\partial\varphi}{\partial x}(0,0)=0, \qquad \frac{\partial\varphi}{\partial y}(0,0)=0,$$

while

$$\frac{\partial^2\varphi}{\partial x^2}(0,0)=L(0,0)=\kappa_1(0,0), \qquad \frac{\partial^2\varphi}{\partial y^2}(0,0)=N(0,0)=\kappa_2(0,0)$$

where κ_1 and κ_2 indicate the two principal curvatures (twice the constant normal curvature in the case of an umbilical point). Moreover

$$\frac{\partial^2\varphi}{\partial x\partial y}(0,0)=M(0,0)=0.$$

Therefore the formula above reduces simply to

$$\kappa_\tau(0,0)=\kappa_1(0,0)\cdot\kappa_2(0,0).$$

We conclude therefore that

The total curvature at the point with parameters (u_0, v_0) is the product of the two principal curvatures at this point.

5.16 Gaussian Curvature

With the considerations of Sect. 5.15 in mind, we make the following definition:

Definition 5.16.1 Consider a regular surface of class $\mathcal{C}^2$. The *Gaussian curvature* (also called the *total curvature*) at a given point is:

- the product of the two principal curvatures, if the point is non-umbilical;
- the square of the constant normal curvature, if the point is umbilical.

Let us first observe that:

Lemma 5.16.2 *The value of the Gaussian curvature is independent of the regular parametric representation of class $\mathcal{C}^2$ used to compute it.*

Proof By Proposition 5.8.5, both factors of the product in Definition 5.16.1 are equal or both are opposite in sign. □

Let us next prove a formula for computing the *Gaussian curvature*:

Proposition 5.16.3 *Consider a regular surface of class $\mathcal{C}^2$. The* Gaussian curvature *is given by the formula*

$$\kappa_\tau = \frac{LN - M^2}{EG - F^2}$$

where the six coefficients E, F, G, L, M, N are computed using any regular parametric representation of class $\mathcal{C}^2$ of the surface.

Proof We know that at a non-umbilical point, the two principal curvatures are the two solutions of the equation in Proposition 5.10.4. The product of the two roots is thus indeed given by the formula of the statement.

In the case of an umbilical point with constant normal curvature k, Proposition 5.9.2 tells us that

$$L = kE, \qquad M = kF, \qquad N = kG$$

which again implies the formula of the statement. □

Example 5.16.4 A plane has Gaussian curvature 0 at each point.

Proof This follows by Example 5.9.5. □

Example 5.16.5 The sphere of radius R has Gaussian curvature $\frac{1}{R^2}$ at each point.

Proof This follows by Example 5.9.4. □

Examples 5.16.4 and 5.16.7 present well-known situations where the Gaussian curvature is constant: zero for the plane, positive for the sphere. These results are very predictable since these surfaces have exactly the same shape in a neighbourhood of any point. But the converse is not true: having the same Gaussian curvature

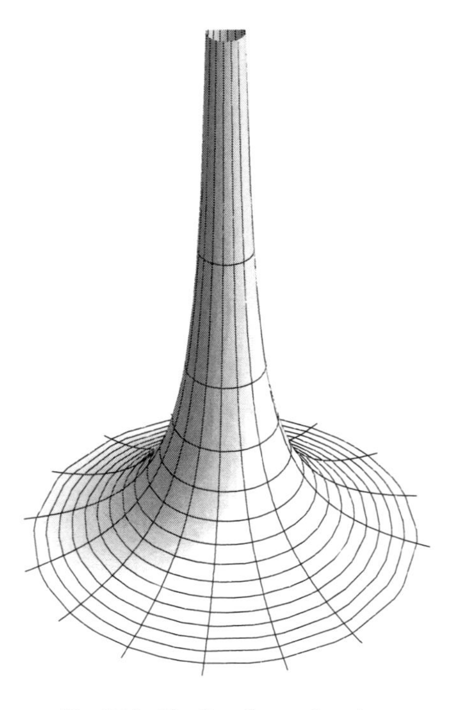

Fig. 5.13 The (hemi)-pseudo-sphere

at all points does not imply having the same shape in a neighborhood of all points. For example:

Example 5.16.6 The parabolic cylinder has constant Gaussian curvature 0.

Proof This follows by Example 5.11.3, since at each point one of the principal curvatures is equal to zero. □

The case of a constant negative Gaussian curvature is probably less popular. An example is provided by the *pseudo-sphere*: the surface obtained by rotating the *tractrix* (see Example 2.5.4) around its asymptote. More precisely, this surface should better be called the *hemi-pseudo-sphere* (see Problem 5.17.15 for the definition of the "full" *pseudo-sphere*).

Example 5.16.7 The *hemi-pseudo-sphere* of "pseudo-radius" R is the surface represented by

$$f\colon]0,R[\times]0,\infty[\to\mathbb{R}^3,\qquad (r,\theta)\mapsto\left(r\cos\theta, r\sin\theta, \int_r^R\sqrt{\frac{R^2}{t^2}-1}\,dt\right)$$

(see Fig. 5.13). It has constant Gaussian curvature $-\frac{1}{R^2}$.

Proof We first compute

$$\frac{\partial f}{\partial r} = \left(\cos\theta, \sin\theta, -\sqrt{\frac{R^2}{r^2} - 1}\right)$$
$$\frac{\partial f}{\partial \theta} = (-r\sin\theta, r\cos\theta, 0)$$

from which

$$E = \frac{R^2}{r^2}, \qquad F = 0, \qquad G = r^2.$$

We thus also have

$$\vec{n} = \frac{r}{R}\left(\cos\theta\sqrt{\frac{R^2}{r^2} - 1}, \sin\theta\sqrt{\frac{R^2}{r^2} - 1}, 1\right)$$

and we compute further

$$\frac{\partial^2 f}{\partial r^2} = \left(0, 0, \frac{R^2}{r^2\sqrt{\frac{R^2}{r^2} - 1}}\right)$$
$$\frac{\partial^2 f}{\partial r \partial \theta} = (-\sin\theta, \cos\theta, 0)$$
$$\frac{\partial^2 f}{\partial \theta^2} = (-r\cos\theta, -r\sin\theta, 0).$$

This yields

$$L = \frac{R}{r^2\sqrt{\frac{R^2}{r^2} - 1}}, \qquad M = 0, \qquad N = -\frac{r^2\sqrt{\frac{R^2}{r^2} - 1}}{R}.$$

Applying Proposition 5.16.3, we conclude that at each point the Gaussian curvature is equal to

$$\kappa_\tau = \frac{LN - M^2}{EG - F^2} = -\frac{1}{R^2}.$$ □

A "good" surface, at a given point, has the same normal curvature as a quadric (see Theorem 5.12.1). Knowing the nature of this quadric thus gives a rough idea of the shape of the surface in a neighborhood of the point. Let us formalize this as follows:

Theorem 5.16.8 *Consider a regular surface of class $\mathcal{C}^2$. At a given point, the quadric locally approximating the surface as in Theorem* 5.12.1 *is:*

- *an elliptic paraboloid when the Gaussian curvature is strictly positive;*
- *a hyperbolic paraboloid when the Gaussian curvature is strictly negative;*
- *a parabolic cylinder when the Gaussian curvature is zero and the point is not umbilical;*
- *a plane when the Gaussian curvature is zero and the point is umbilical.*

Proof The surface and the quadric have the same normal curvature at the given point, by Proposition 5.12.1. They thus also have the same Gaussian curvature at this point, by Definition 5.16.1. But by Corollary 5.12.2, the quadric can only take one of the following forms:

- a plane, whose Gaussian curvature is zero by Example 5.9.5 and all of whose points are umbilical by Example 5.9.5;
- a parabolic cylinder, whose Gaussian curvature is zero as attested by Example 5.11.3 and all of whose points are non-umbilical by Example 5.11.3;
- an elliptic paraboloid, whose Gaussian curvature is strictly positive at the point considered (see Example 5.11.1), and in fact also at all points (see Problem 5.17.13);
- a hyperbolic paraboloid, whose Gaussian curvature is strictly negative at the point considered (see Example 5.11.2), and in fact at all points (see Problem 5.17.13).

□

In view of Theorem 5.16.8, the following is common terminology:

Definition 5.16.9 A point of a regular surface of class $\mathcal{C}^2$ is:

1. *elliptic*, when the Gaussian curvature is strictly positive;
2. *hyperbolic*, when the Gaussian curvature is strictly negative;
3. *parabolic*, when the Gaussian curvature is zero, but the point is not umbilical;
4. *planar*, when the Gaussian curvature is zero and the point is umbilical.

Warning 5.16.10 *The sole consideration of Gaussian curvature does not allow the detection of umbilical points.*

Proof Examples 5.16.4 and 5.16.6 show that the plane and the parabolic cylinder have the same Gaussian curvature, namely, 0 at each point. But all the points of the plane are umbilical (see Example 5.9.5) while the parabolic cylinder does not have any umbilical points (see Example 5.11.3). □

Thus in terms of just the Gaussian curvature at a given point, it is impossible to distinguish between cases 3 and 4. Furthermore, in case 1, it is again impossible, just in terms of the Gaussian curvature at the point, to decide if the point is umbilical or not. An umbilical point can never be detected by just considering the Gaussian curvature at this point.

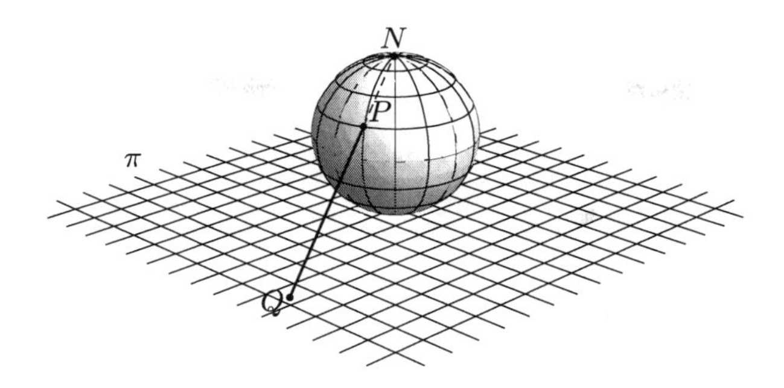

Fig. 5.14 The pseudo-sphere

On the other hand a hyperbolic point is never umbilical, since the two normal curvatures are non-zero and of opposite signs, thus certainly distinct.

Finally, let us mention that various results in Chap. 7 will show that more precise conclusions on the form of the surface at a given point can sometimes be inferred from the consideration of the Gaussian curvature, provided that one considers this Gaussian curvature *on a whole neighborhood of the point*, not just at the point itself. But this is definitely not a general fact, as the proof of Warning 5.16.10 shows.

5.17 Problems

5.17.1 Show that the following can be presented as surfaces of class $\mathcal{C}^\infty$: a plane, an elliptic cylinder, a parabolic cylinder, a sheet of a hyperbolic cylinder, a half cone punctured at its vertex.

5.17.2 The stereographic projection (see Fig. 5.14 and Sect. 5.9 in [3], *Trilogy I*) is a mapping of the sphere $\mathcal{S}$ punctured at its North pole N onto the tangent plane π at the South pole S. It maps a point P of the sphere to the intersection Q of π with the line joining N and P. This is a bijection and its inverse provides an injective parametric representation of the sphere punctured at the North pole.

5.17.3 Prove that all quadrics (or sheets of them) are orientable surfaces.

5.17.4 Prove that a regular surface is orientable precisely when it is possible to make a choice of the normal vector at each point of the support $\mathcal{S} \subseteq \mathbb{R}^3$, in such a way to obtain a continuous function $\overrightarrow{n} : \mathcal{S} \longrightarrow \mathbb{R}^3$.

5.17.5 Consider a regular curve of class $\mathcal{C}^2$ on a regular surface of class $\mathcal{C}^2$. At a given point of the curve, one defines its *Darboux* orthonormal trihedron, comprising the tangent vector $\overrightarrow{t}$ to the curve, the normal vector $\overrightarrow{n}$ to the surface and their cross product $\overrightarrow{g}$. Study the change of coordinates matrix between the *Frenet* trihedron of

Fig. 5.15

the curve and its *Darboux* trihedron. Express this change of coordinates in terms of the curvature, the torsion and the normal curvature of the curve.

5.17.6 Consider, in the xy-plane, the curve with polar equation

$$R = \cos 4\theta + 2$$

pictured in Fig. 5.15. In each plane containing the z-axis, consider a parabola with vertex $(0, 0, 1)$ cutting the xy-plane on the given curve. When the plane through the z-axis turns, the curvature of the various parabolas at the point $(0, 0, 1)$ clearly passes twice through a local maximum and twice through a local minimum, while the corresponding directions form angles of 45 degrees. Verify that the corresponding surface (pictured in Fig. 5.12) is not a counterexample to Theorem 5.10.1 because it is not of class $\mathcal{C}^2$. Next replace the usual parabolas by "quartic parabolas" (curves of equations $y = kx^4$). The problem of differentiability vanishes, but the point $(0, 0, 1)$ is now umbilical.

5.17.7 The surface of Fig. 5.16 is the one admitting the equation

$$z = x^4 + y^4 - 4x^2y^2.$$

The situation at the origin does not contradict Theorem 5.10.1, because that point is umbilical.

5.17.8 Consider the ellipsoid with equation

$$\frac{x^2}{a^2} + \frac{y^2}{b^2} + \frac{z^2}{c^2} = 1$$

where a, b, c are distinct and non-zero. Prove that at each intersection with a coordinate axis, the principal directions are in the directions of the other two axes.

5.17.9 At a regular point P of a surface of class $\mathcal{C}^2$, consider in the tangent plane at P a line through P making an angle θ with the first principal direction. Prove that

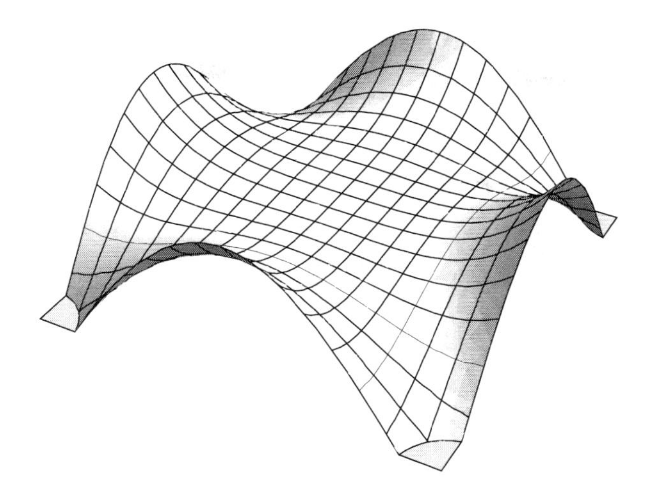

Fig. 5.16

the normal curvature of the surface in this direction is equal to

$$\kappa_n(\theta) = \kappa_1 \cos^2\theta + \kappa_2 \sin^2\theta$$

where κ_1, κ_2 are respectively the first and the second principal curvatures.

5.17.10 Consider a regular parametric representation $f(u,v)$ of class $\mathcal{C}^2$ of a surface. Prove that at a non-umbilical point $f(u_0,v_0)$, the curves $f(u_0,v)$ and $f(u,v_0)$ are oriented in the principal directions if and only if $F(u_0,v_0)=0$ and $M(u_0,v_0)=0$. In that case, the two principal curvatures are given by

$$\kappa_1 = \frac{L(u_0,v_0)}{E(u_0,v_0)}, \qquad \kappa_2 = \frac{N(u_0,v_0)}{G(u_0,v_0)}.$$

5.17.11 In the spirit of Theorem 5.14.2, now using a system of partial differential equations, prove that in a neighborhood of a non-umbilical point of a regular surface of class $\mathcal{C}^3$ represented by $f(u,v)$, there exists a change of parameters φ

$$u = \varphi_1(r,s), \qquad v = \varphi_2(r,s)$$

for which the equivalent parametric representation $g = f \circ \varphi$ is such that all curves $g(r_0,s)$ and $g(r,s_0)$ are lines of curvature. In particular, the conclusions of Problem 5.17.10 apply at each point.

5.17.12 A regular curve on a regular surface of class $\mathcal{C}^2$ is called an *asymptotic line* when at each point, its tangent is oriented in a direction with normal curvature zero. Prove that in a neighborhood of a non-umbilical point, there exist on the surface two asymptotic lines passing through that point.

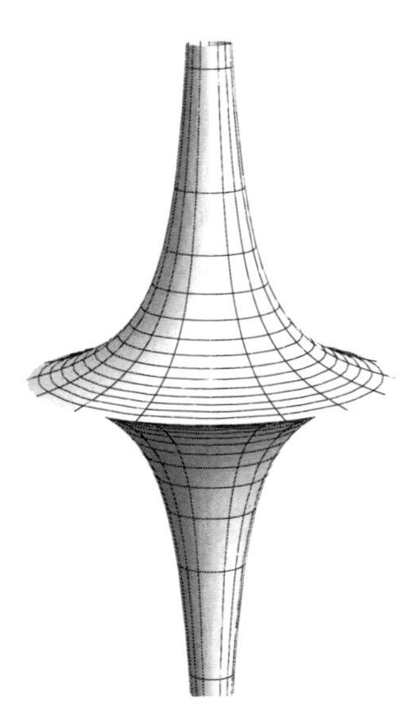

Fig. 5.17 The pseudo-sphere

5.17.13 Prove that the second degree approximation as in Theorem 5.12.1 is at each point:

- an elliptic paraboloid in the case of the ellipsoid, the hyperboloid with two sheets and the elliptic paraboloid;
- a hyperbolic paraboloid in the case of the hyperboloid with one sheet and the hyperbolic paraboloid;
- a parabolic cylinder in the case of a cylinder or a cone.

5.17.14 Consider the torus of Example 5.1.7. Prove that the point with parameters (θ, τ) is:

- elliptic, when $-\frac{\pi}{2} < \tau < \frac{\pi}{2}$;
- hyperbolic when $\frac{\pi}{2} < \tau < \frac{3\pi}{2}$;
- parabolic when $\tau = \pm\frac{\pi}{2}$.

5.17.15 The *pseudo-sphere* (see Fig. 5.17) is the surface obtained by gluing together, along the circle of radius R centered at the origin in the xy-plane, the *hemi-pseudo-sphere* of Example 5.16.7 and its symmetric copy with respect to the xy-plane. In terms of the function g of Example 2.5.4, extended continuously on $[0, \infty[$ by defining $g(0) = R$, the pseudo-sphere thus admits the parametric representation of class $\mathcal{C}^0$

$$f : \mathbb{R}^2 \longrightarrow \mathbb{R}^3, \qquad (z, \theta) \mapsto \big(g(|z|)\cos\theta, g(|z|)\sin\theta, z\big).$$

The *pseudo-sphere* is clearly not bounded: nevertheless it has the same area $4\pi R^2$ and the same volume $\frac{4}{3}\pi R^3$ as the ordinary sphere of radius R. Together with the fact that the Gaussian curvature is constant and equal to $-\frac{1}{R^2}$, this fully justifies the name *pseudo-sphere*.

5.18 Exercises

5.18.1 Sketch the shape of the surfaces represented by

1. $f: \mathbb{R}^2 \longrightarrow \mathbb{R}^3$, $(u, v) \mapsto (u \cos v, u \sin v, v)$;
2. $f: \mathbb{R}^2 \longrightarrow \mathbb{R}^3$, $(u, v) \mapsto (u \cos u, u \sin u, v)$;
3. $f: \mathbb{R}^2 \longrightarrow \mathbb{R}^3$, $(u, v) \mapsto (u \cos u, v, u)$;
4. $f: \mathbb{R}^2 \longrightarrow \mathbb{R}^3$, $(u, v) \mapsto (u \cos v, u \sin v, u)$.

Determine if these parametric representations are regular.

5.18.2 Give a parametric representation of the surface obtained by letting the curve $f(t) = (e^t, 0, t)$ rotate

1. around the x-axis;
2. around the z-axis.

5.18.3 In $\mathbb{R}^3$ consider the lines

$$d \equiv \begin{cases} x = y \\ z = 1 \end{cases} \quad \text{and} \quad d' \equiv \begin{cases} y = 0 \\ z = 0. \end{cases}$$

For each point $P \in d$, let d_P be the line through P perpendicular to d'. Give a parametric representation of the surface comprising all these lines d_P, as P runs along d.

5.18.4 Determine if the curve represented by

$$g: \mathbb{R} \longrightarrow \mathbb{R}^3, \qquad t \mapsto (t \cos t, t \sin t, t)$$

lies on the surface whose a parametric representation is:

1. $f(u, v) = (u \cos v, u \sin v, v)$;
2. $f(u, v) = (u \cos u, v, u)$.

5.18.5 Give a parametric representation, a normal vector and the equation of the tangent plane to the hyperbolic paraboloid with equation $z = x^2 - y^2$ at the point $(1, 1, 0)$ (see Fig. 1.14.8 in [4], *Trilogy II*).

5.18.6 Calculate the tangent plane at each regular point of the cone with equation

$$z^2 = a^2(x^2 + y^2), \quad a > 0.$$

5.18.7 Calculate the first fundamental form of the helicoid (see Fig. 5.6)

$$f(u,v)=(u\cos v, u\sin v, kv), \quad k>0$$

1. at the origin;
2. at the point with parameters $(1,0)$;
3. at the point with parameters $(2,\frac{\pi}{2})$.

5.18.8 Calculate the length of the arc of the curve

$$h(t)=\left(\int_{\frac{\pi}{4}}^{t}\frac{1}{\sin\tau}\,d\tau, t\right)$$

drawn on the sphere

$$f(\theta,\varphi)=(\cos\theta\sin\varphi, \sin\theta\sin\varphi, \cos\varphi)$$

as t runs from 0 to π.

5.18.9 Calculate the second fundamental form of the helicoid at the origin (see Example 5.1.8).

5.18.10 Calculate, at an arbitrary point, the normal curvature of the helix (see Example 4.1.2)

$$h(t)=(a\cos t, a\sin t, bt)$$

on the helicoid (see Example 5.1.8)

$$f(u,v)=(u\cos v, u\sin v, bv).$$

5.18.11 Let $\mathcal{S}$ be a surface of class $\mathcal{C}^2$ whose intersection with some plane π is a circle of radius R. Suppose that at a given point P of this circle, the tangent plane to the surface is perpendicular to the plane π. What is the normal curvature of the circle at P?

5.18.12 In $\mathbb{R}^3$, consider the curve $\mathcal{C}$ obtained as the intersection of the paraboloid $\mathcal{P}$ with equation $x^2+y^2=z$ and the plane $z=1$. Consider further the point $(1,0,1)$ on this curve $\mathcal{C}$. At this point P, compute the normal curvature of the paraboloid $\mathcal{P}$ in the direction of the tangent to $\mathcal{C}$.

5.18.13 Consider the two spheres

- $\mathcal{S}_1$, with center $(0,0,0)$ and radius 3;
- $\mathcal{S}_2$, with center $(0,0,4)$ and radius 5.

Compute the normal curvature of the intersection $\mathcal{S}_1\cap\mathcal{S}_2$ as a curve on $\mathcal{S}_1$.

5.18.14 Calculate the normal curvature of the hyperbolic paraboloid $z = x^2 - y^2$ at the origin, in all directions.

5.18.15 Let $\mathcal{S}$ be the surface of $\mathbb{R}^3$ obtained by letting the curve $c(t) = (0, t^2, t)$ rotate around the y-axis. Consider the mapping

$$f\colon]0, \infty[\times \mathbb{R} \longrightarrow \mathbb{R}^3, \qquad (\theta, \varphi) \mapsto (\theta \sin\varphi, \theta^2, \theta\cos\varphi).$$

1. Prove that f is a regular parametric representation of class $\mathcal{C}^\infty$.
2. Prove that $\mathcal{S}$ is the support of f.
3. Prove that the curve $\mathcal{C}$ represented by

$$h\colon]0, \infty[\longrightarrow \mathbb{R}^3, \qquad t \mapsto (t\cos t, t^2, t\sin t)$$

   lies on the surface $\mathcal{S}$.
4. Compute the normal curvature of $\mathcal{C}$ on $\mathcal{S}$ at an arbitrary point.

5.18.16 Consider the surface represented by $f(u, v) = (u, v, uv)$. At the point with parameters $(1, 2)$ compute the principal directions

1. with respect to the basis of partial derivatives;
2. with respect to the canonical basis of $\mathbb{R}^3$.

What is the nature of this surface?

5.18.17 Consider the helicoid (see Fig. 5.6) represented by

$$f(u, v) = (u\cos v, u\sin v, bv).$$

1. Compute the normal curvature at an arbitrary point, in an arbitrary direction.
2. Compute at each point the principal curvatures.
3. Find the possible umbilical points.
4. Compute the Gaussian curvature at an arbitrary point.

5.18.18 Let us recall that the *hyperbolic cosine* and the *hyperbolic sine* are the functions defined by

$$\cosh x = \frac{e^x + e^{-x}}{2}, \qquad \sinh x = \frac{e^x - e^{-x}}{2};$$

these functions satisfy the equality

$$\cosh^2 x - \sinh^2 x = 1.$$

Consider the surface $\mathcal{S}$ represented by

$$f\colon \mathbb{R}^2 \longrightarrow \mathbb{R}^3, \qquad (u, v) \mapsto (\cosh u \cos v, \cosh u \sin v, \sinh u).$$

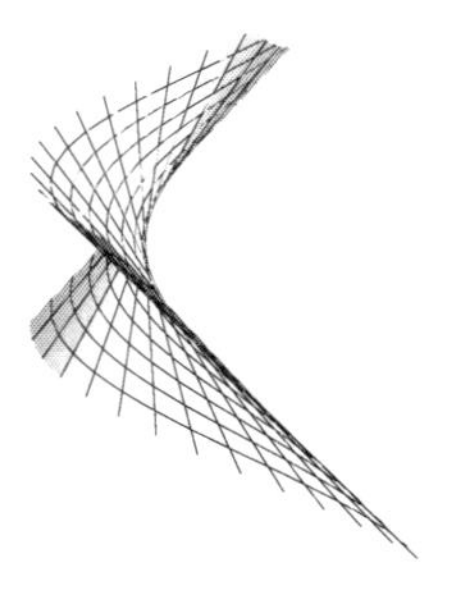

Fig. 5.18

1. Give the normal vector and an equation of the tangent plane at the point $(\frac{\sqrt{2}}{2}, \frac{\sqrt{2}}{2}, 0)$.
2. Let $\mathcal{C}_0$ be the curve on $\mathcal{S}$ determined by $u = u_0$. Compute the normal curvature of $\mathcal{C}_0$ on $\mathcal{S}$ and prove that it is constant along $\mathcal{C}_0$.
3. Give a Cartesian equation of $\mathcal{S}$. What is the nature of this surface?
4. Show that $\mathcal{S}$ is the surface obtained by letting a plane curve $\mathcal{C}$ rotate around an axis situated in the plane of $\mathcal{C}$. What is this curve $\mathcal{C}$?

5.18.19 Let $\mathcal{C}$ be a 2-regular skew curve of class $\mathcal{C}^3$, whose curvature and torsion never vanish. The *tangent surface* to this curve is the surface $\mathcal{S}$ comprising all the tangents to the curve (see Fig. 5.18).

1. Given a normal parametric representation $f(s)$ of the curve $\mathcal{C}$, give a parametric representation of the corresponding tangent surface $\mathcal{S}$.
2. Show that, among the points of the tangent surface $\mathcal{S}$, those situated on the curve $\mathcal{C}$ are singular and the other points are regular.
3. Show that the osculating plane to the curve $\mathcal{C}$ at a point with parameter s_0 is also the tangent plane to the surface $\mathcal{S}$ at all regular points of the surface $\mathcal{S}$ lying on the tangent to the curve $\mathcal{C}$ at this point $f(s_0)$.
4. Prove that all regular points of the surface $\mathcal{S}$ are parabolic.

5.18.20 Let $f: \mathbb{R} \longrightarrow \mathbb{R}$ be a function of class $\mathcal{C}^\infty$ such that $f(x)f''(x) > 0$ for all $x \in \mathbb{R}$. In the yz-plane of $\mathbb{R}^3$, consider the curve $\mathcal{C}$ with equation $y = f(z)$. Let $\mathcal{S}$ be the surface obtained by letting the curve $\mathcal{C}$ rotate around the z-axis. Compute, at each point, the Gaussian curvature of $\mathcal{S}$.

5.18.21 Let $\mathcal{S}$ be the surface admitting the parametric representation

$$f(t, u) = \left(\cos t \cos u, \cos t \sin u, \frac{1}{2} \sin t\right).$$

Let $\mathcal{C}$ be the curve obtained as the intersection of $\mathcal{S}$ with the plane π with equation $x = y$. Consider a regular point P of the surface $\mathcal{S}$ situated on the curve $\mathcal{C}$. Prove that at P, the tangent plane to $\mathcal{S}$ is orthogonal to the plane π. What can you say about the normal curvature of $\mathcal{C}$?

Chapter 6
Towards Riemannian Geometry

As the title indicates, the purpose of this chapter is *not* to develop *Riemannian geometry* as such. The idea is to first re-visit the theory of surfaces in $\mathbb{R}^3$, as studied in Chap. 5, adopting a different point of view and different notation. One of the objectives of this somehow unusual exercise—besides proving some important new theorems—is to provide a good intuition for the basic notions and techniques involved in general *Riemannian geometry*.

The idea of *Riemannian geometry* is to consider a surface as a *universe in itself*, not as a part of a "bigger universe", for example as a part of $\mathbb{R}^3$. Thus *Riemannian geometry* is interested in the study of those properties of the surface which can be established by measures performed directly "within" the surface, without any reference to the possible surrounding space. The "key" to performing measures on the surface will be the consideration of its first fundamental form (see Sect. 5.4), called the *metric tensor* in *Riemannian geometry*.

In Chap. 5, our main concern regarding surfaces in $\mathbb{R}^3$ has been the study of their curvature. The *normal curvature* of a given curve on the surface is the orthogonal projection of its "curvature" vector on the normal vector to the surface (see Sect. 5.8). This *normal curvature* cannot be determined by measures performed on the surface, but the *geodesic curvature* can: the geodesic curvature is "the other component" of the curvature vector, that is, the length of the orthogonal projection of the curvature vector on the tangent plane to the surface. When you have the impression of "moving without turning" on a given surface—like when you follow a great circle on the surface of the Earth—you are in fact following a curve with zero geodesic curvature. Such a curve is called a *geodesic*: we pay special attention to the study of these geodesics.

In Sect. 5.16 we also studied the *Gaussian curvature*, which provides less precise information than the *normal curvature*. But in *Riemannian geometry*—where the *normal curvature* no longer makes sense—the *Gaussian curvature* assumes its full importance: the *Gaussian curvature* can be determined by measures performed on the surface itself. This is the famous *Theorema Egregium* of *Gauss*.

We characterize—in terms of the coefficients E, F, G, L, M, N of their two fundamental quadratic forms—those Riemann surfaces which arise from a surface

F. Borceux, *A Differential Approach to Geometry*, DOI 10.1007/978-3-319-01736-5_6,

embedded in $\mathbb{R}^3$, as studied in Chap. 5. As an example of a Riemann surface not obtained from a surface in $\mathbb{R}^3$, we describe the so-called *Poincaré half plane*, which is a model of non-Euclidean geometry.

We conclude with a first discussion of *what a tensor is* and a precise definition of a *Riemann surface*. This last definition refers explicitly to topological notions: the reader not familiar with them is invited to consult Appendix A.

Let us close this introduction with an observation which, in this chapter, will play an important role in supporting our intuition. Consider a surface represented by

$$f: U \mapsto \mathbb{R}^3, \qquad (u, v) \mapsto f(u, v)$$

and suppose that f is injective, not just locally injective. Consider a point $P = f(u, v)$, for $(u, v) \in U$. The injectivity of f allows us to speak equivalently of the point P of the surface or the point with parameters (u, v) on the surface. In this case, the points of the open subset U describe precisely the points of the surface: f is a bijection between these two sets of points.

6.1 What Is Riemannian Geometry?

Turning through the pages of Chap. 5, we find many pictures of surfaces, as if we had taken photographs of these surfaces. But of course when you take a photograph of a surface, you do not put the lens of the camera *on* the surface itself: you stay *outside* the surface, sufficiently far, at some point from which you have a good view of the shape of the surface. Doing this, you study your surface *from the outside*, taking full advantage of the fact that the surface is embedded in $\mathbb{R}^3$ and that you are able to move in $\mathbb{R}^3$, outside the surface.

Let us proceed to a completely different example. We are three dimensional beings living in a three-dimensional world. We are interested in studying the world in which we are living. Of course if we are interested in only studying our solar system, we can take $\mathbb{R}^3$ as a reliable mathematical model of our universe and use the rules of classical mechanics to study the trajectories of the planets. But we know that if we are interested in cosmology and the theory of the expansion of the universe, the very "static" model $\mathbb{R}^3$ is no longer appropriate to the question. Physicists perform a lot of experiments to study our universe: they use large telescopes to capture very remote information. But these telescopes are *inside* our universe and take pictures of things that are *inside* our universe. This time—we have no other choice—we study our universe *from the inside*. From this study *inside the universe itself* physicists try to determine—for example—the possible *curvature* of our universe.

The topic of *Riemannian geometry* is precisely this: the study of a universe from the inside, from measures taken inside that universe. In this book we shall focus on *Riemann surfaces*, that is, two dimensional universes. Thus we imagine that we are very clever two-dimensional beings, living in a two-dimensional universe and knowing a lot of geometry. We do our best to study our universe *from the inside*, since of course there is no way for us to escape and look at it from the outside.

Our first challenge is to *mathematically model* this idea. For that, we shall rely on our study of surfaces embedded in $\mathbb{R}^3$ in order to guess what it can possibly mean to study these surfaces *from the inside*. The above discussion suggests a first answer:

> A Riemannian property *of a surface in* $\mathbb{R}^3$ *is a property which can be established by measures performed on the support of the surface, without any reference to its parametric representation.*

Of course a two-dimensional being living on a surface of $\mathbb{R}^3$ is able to measure the length of an arc of a curve on this surface, or the angle between two curves on the surface. These operations can trivially be done *inside the surface*, without any need to escape from the surface.

But given a regular curve

$$c: I \longrightarrow U \subseteq \mathbb{R}^2, \qquad t \mapsto \bigl(c_1(t), c_2(t)\bigr)$$

on a regular surface

$$f: U \longrightarrow \mathbb{R}^3, \qquad (u,v) \mapsto f(u,v),$$

we have seen (see Proposition 5.4.3) how to calculate its length:

$$\int \|(f\circ c)'\| = \int \sqrt{\begin{pmatrix} c_1' & c_2' \end{pmatrix} \begin{pmatrix} E & F \\ F & G \end{pmatrix} \begin{pmatrix} c_1' \\ c_2' \end{pmatrix}}$$

where the three functions $E(u,v)$, $F(u,v)$, $G(u,v)$ are the coefficients of the first fundamental form of the surface. The same matrix also allows us to calculate the angle between two curves on the surface (see Proposition 5.4.4). The matrix

$$\begin{pmatrix} E & F \\ F & G \end{pmatrix}$$

thus allows us to compute lengths and angles on the surface: it is intuitively the "mathematical measuring tape" on the surface.

However, we must stress the following: knowledge of this "mathematical measuring tape" is (somehow) *equivalent* to being able to perform measures *inside* the surface. That is, only from measures performed *inside* the surface, you can infer the values of the three functions $E(u,v)$, $F(u,v)$, $G(u,v)$. In this statement, the "somehow" restriction is the fact that to reach this goal, you have to perform infinitely many measures, because there are infinitely many points on the surface.

Indeed consider the curve $v = v_0$

$$u \mapsto f(u, v_0)$$

for some fixed value v_0. The length on this curve from an origin u_0 to the point with parameter u is thus

$$\ell(u) = \int_{u_0}^{u} \sqrt{\begin{pmatrix} 1 & 0 \end{pmatrix} \begin{pmatrix} E(u,v_0) & F(u,v_0) \\ F(u,v_0) & G(u,v_0) \end{pmatrix} \begin{pmatrix} 1 \\ 0 \end{pmatrix}}\, du = \int_{u_0}^{u} \sqrt{E(u,v_0)}\, du.$$

The two dimensional being living on the surface can thus *measure* the value $\ell(u)$ for any value of the parameter u, and so "somehow" determine the function $\ell(u)$. If he makes the additional effort to attend a first calculus course, he will be able to compute the derivative

$$\ell'(u) = \sqrt{E(u, v_0)}$$

of that function and thus eventually, get the value of $E(u, v_0)$. An analogous argument holds for $G(u_0, v)$. Notice further that the angle θ between the two curves $u = u_0$ and $v = v_0$ is given by

$$\cos\theta = \frac{F(u_0, v_0)}{\sqrt{G(u_0, v_0)}\sqrt{E(u_0, v_0)}}.$$

Since $E(u_0, v_0)$ and $G(u_0, v_0)$ are already known, the two-dimensional being gets the value of $F(u_0, v_0)$ from the measure of the angle θ.

All this suggests reformulating the above statement as follows:

> *The* Riemannian geometry *of a surface*
>
> $$f: U \longrightarrow \mathbb{R}^3, \qquad (u, v) \mapsto f(u, v)$$
>
> *embedded in* $\mathbb{R}^3$ *is the study of those properties of the surface which can be inferred from the sole knowledge of the three functions*
>
> $$E, F, G: U \longrightarrow \mathbb{R}.$$

Now working with three symbols E, F, G and two parameters u, v remains technically quite tractable. But imagine that you are no longer interested in "two dimensional universes" (surfaces), but in "three dimensional universes", such as the universe in which we are living! Instead of two parameters, you now have to handle three parameters; analogously, as we shall see in Definition 6.17.6, the corresponding "mathematical measuring tape" will become a 3×3-matrix. If you are interested—for example—in studying *relativity*, you will have to handle a fourth dimension, "time". Thus four parameters and a 4×4-matrix. In such higher dimensions, one has to use "notation with indices" in order to cope with all the quantities involved! We shall do this in the case of surfaces and introduce the classical notation of Riemannian geometry.

The Riemannian notation for the first fundamental form is:

$$\begin{pmatrix} E(u, v) & F(u, v) \\ F(u, v) & G(u, v) \end{pmatrix} = \begin{pmatrix} g_{11}(x^1, x^2) & g_{12}(x^1, x^2) \\ g_{21}(x^1, x^2) & g_{22}(x^1, x^2) \end{pmatrix}.$$

Having changed the notation, we shall also change the terminology.

Definition 6.1.1 Consider a regular parametric representation of a surface

$$f: U \longrightarrow \mathbb{R}^3, \qquad (x^1, x^2) \mapsto f(x^1, x^2).$$

The matrix of functions

$$g_{ij}\colon U \longrightarrow \mathbb{R}, \qquad (x^1, x^2) \mapsto g_{ij}(x^1, x^2), \quad 1 \le i, j \le 2$$

defined by

$$\begin{pmatrix} g_{11} & g_{12} \\ g_{21} & g_{22} \end{pmatrix} = \begin{pmatrix} \left(\frac{\partial f}{\partial x^1} \middle| \frac{\partial f}{\partial x^1}\right) & \left(\frac{\partial f}{\partial x^1} \middle| \frac{\partial f}{\partial x^2}\right) \\ \left(\frac{\partial f}{\partial x^2} \middle| \frac{\partial f}{\partial x^1}\right) & \left(\frac{\partial f}{\partial x^2} \middle| \frac{\partial f}{\partial x^2}\right) \end{pmatrix}$$

is called the *metric tensor* of the surface.

The "magic word" *tensor* suddenly appears! The reason for such a terminology will be "explained" in Sect. 6.12. For the time being, this is just a point of terminology which does not conceal any hidden properties and so formally, does not require any justification.

Using symbols like g_{ij} to indicate the various elements of the "tensor"—which after all is just a matrix—sounds perfectly reasonable, as does using symbols (x_1, x_2) to indicate the two parameters. The use of upper indices x^1 and x^2 might seem like an invitation for confusion at this point, but we shall come back to this in Sect. 6.12. For the time being, we just decide to use the unusual notation (x^1, x^2).

Even if we do not yet know the reason for using these "upper indices", let us at least be consistent. Given a curve

$$c\colon]a, b[\longrightarrow U$$

on the surface, we should now write

$$c(t) = \bigl(c^1(t), c^2(t)\bigr)$$

for the two components of the function c.

We have thus described the "challenge" of Riemannian geometry, as far as surfaces embedded in $\mathbb{R}^3$ are concerned, and we have introduced the classical notation and terminology of Riemannian geometry. But to help us guess which properties have a good chance to be Riemannian, we shall add a "slogan".

Consider again our friendly and clever two-dimensional being living on the surface. This two-dimensional being should have full knowledge of what happens "at the level of the surface" but no knowledge at all of what happens "outside the surface". From a quantity that lives in the "outside world $\mathbb{R}^3$", the two-dimensional being should only see its "shadow on the surface", its "component at the level of the surface", that is, its "orthogonal projection at the level of the surface". Let us write this quantity of $\mathbb{R}^3$ in terms of the basis comprising the two partial derivatives of the parametric representation and the normal vector to the surface. What happens along the normal to the surface, that is, what projects as "zero" on the surface, is the part of the information that the two-dimensional being cannot possibly access. So the rest of the information, that is, the components along the partial derivatives, should probably be accessible to our two-dimensional being. Let us take this as a slogan for discovering Riemannian properties.

Slogan:
The component of a geometric quantity along the normal vector to the surface is not Riemannian, but its components along the tangent plane should be Riemannian.

This is of course just a "slogan", not a precise mathematical statement!

6.2 The Metric Tensor

Let us first recall (Proposition 5.5.4) that at each point of a regular surface

$$f : U \longrightarrow \mathbb{R}^3, \qquad (x^1, x^2) \mapsto f(x^1, x^2)$$

the matrix

$$\begin{pmatrix} g_{11}(x_0^1, x_0^2) & g_{12}(x_0^1, x_0^2) \\ g_{21}(x_0^1, x_0^2) & g_{22}(x_0^1, x_0^2) \end{pmatrix} = \begin{pmatrix} E(x_0^1, x_0^2) & F(x_0^1, x_0^2) \\ F(x_0^1, x_0^2) & G(x_0^1, x_0^2) \end{pmatrix}$$

is that of the scalar product in the tangent plane at $f(x_0^1, x_0^2)$, with respect to the affine basis

$$\left(f(x_0^1, x_0^2); \frac{\partial f}{\partial x^1}(x_0^1, x_0^2), \frac{\partial f}{\partial x^2}(x_0^1, x_0^2) \right).$$

This matrix is thus symmetric, definite and positive (Proposition 5.4.6).

We are now ready to give a first (restricted) definition of a *Riemann surface*, a definition which no longer refers to any parametric representation:

Definition 6.2.1 A *Riemann patch* of class $\mathcal{C}^k$ consists of:

1. a connected open subset $U \subseteq \mathbb{R}^2$;
2. four functions of class $\mathcal{C}^k$

$$g_{i,j} : U \longrightarrow \mathbb{R}, \qquad (x^1, x^2) \mapsto g_{ij}(x^1, x^2), \quad 1 \le i, j \le 2$$

so that at each point $(x^1, x^2) \in U$, the matrix

$$\begin{pmatrix} g_{11}(x^1, x^2) & g_{12}(x^1, x^2) \\ g_{21}(x^1, x^2) & g_{22}(x^1, x^2) \end{pmatrix}$$

is symmetric definite positive. The matrix of functions

$$(g_{ij})_{ij}$$

is called the *metric tensor* of the Riemann patch.

The observant reader will have noticed that if we start with a regular parametric representation f of class $\mathcal{C}^k$ of a surface in $\mathbb{R}^3$, the corresponding metric tensor as in Definition 6.2.1 is only of class $\mathcal{C}^{k-1}$. This is the reason why some authors declare a Riemann patch to be of class $\mathcal{C}^{k+1}$ when the functions g_{ij} are of class $\mathcal{C}^k$. This is just a matter of taste!

The term *Riemann patch* instead of *Riemann surface* underlines the fact that in this chapter, we shall again essentially work "locally". The more general notion of *Riemann surface* is investigated in Sect. 6.17. We can now express will full precision the concern of Riemannian geometry:

> *Local Riemannian geometry is the study of the properties of a Riemann patch.*

Of course what has been explained above suggests that we should think of the metric tensor intuitively as being that of a hypothetical surface in $\mathbb{R}^3$. This can indeed support our intuition but this is perhaps not the best way to look at a *Riemann patch*.

Let us go back to the example of the sphere (or part of it) in terms of the "longitude" and "latitude", as in Example 5.1.6:

$$f: U \longrightarrow \mathbb{R}^3, \qquad (\theta, \tau) \mapsto (\cos\tau\cos\theta, \cos\tau\sin\theta, \sin\tau).$$

Assume that we have restricted our attention to an open subset U on which f is injective. Think of the sphere as being the Earth. The open subset $U \subseteq \mathbb{R}^2$ is then the *geographical map* of the corresponding piece of the Earth. The two coordinates of a point of $U \subseteq \mathbb{R}^2$ are the longitude and the latitude of the corresponding point of the Earth. But how can you—for example—determine the distance between two points of the Earth, simply by inspecting your map? Certainly not by measuring the distance on the map using your ruler! Indeed on the map, the further one moves away from the equator, the more distorted the distances on the map become. Of course we can determine the longitude and the latitude of the two points on the map and use our knowledge of spherical trigonometry to calculate the corresponding distance on the surface of the Earth. However, to do this, one has to know that the Earth is approximately a sphere in the surrounding universe $\mathbb{R}^3$: an attitude which does not make sense in Riemannian geometry.

What would be better would be to have an elastic ruler which is able to adjust itself to the correct length, depending on where it has been placed on the map. We are in luck, such an elastic ruler exists: it is the *metric tensor*. The metric tensor is at each point the matrix of a scalar product, but a scalar product which varies from point to point, compensating for the distortion of the map.

It is better to imagine that the map is of some unknown planet, the shape of which is totally unknown to us. From only the longitude and the latitude of points on this planet, as given by the map, we cannot draw any conclusions since we have no idea of the shape of the planet and thus of the distortion of the map together with the "elastic ruler" (*the metric tensor*), to "calculate" from the map the actual distances on the planet (we shall make this precise in Definitions 6.3.2 and 6.3.3). This probably gives a clearer intuitive way to think about *Riemann patches*.

Let us return to our formal definition of a Riemann patch.

Proposition 6.2.2 *Given a Riemann patch as in Definition* 6.2.1, *the metric tensor is at each point* (x^1, x^2) *an invertible matrix with strictly positive determinant. Moreover at each point,* $g_{11} > 0$ *and* $g_{22} > 0$.

Proof By Proposition G.3.4 in [4], *Trilogy II*, the determinant of the matrix is strictly positive, thus the matrix is invertible. On the other hand

$$\begin{pmatrix} 1 & 0 \end{pmatrix} \begin{pmatrix} g_{11} & g_{12} \\ g_{21} & g_{22} \end{pmatrix} \begin{pmatrix} 1 \\ 0 \end{pmatrix} = g_{11}$$

thus this quantity is strictly positive by positivity and definiteness. An analogous argument holds for g_{22}. □

Definition 6.2.3 Given a Riemann patch as in Definition 6.2.1, the *inverse metric tensor*

$$\begin{pmatrix} g^{11}(x^1, x^2) & g^{12}(x^1, x^2) \\ g^{21}(x^1, x^2) & g^{22}(x^1, x^2) \end{pmatrix} = \begin{pmatrix} g_{11}(x^1, x^2) & g_{12}(x^1, x^2) \\ g_{21}(x^1, x^2) & g_{22}(x^1, x^2) \end{pmatrix}^{-1}$$

is at each point (x^1, x^2) the inverse of the metric tensor.

The matrix $(g^{ij})_{ij}$ has again received the label *tensor* and the indices have now been put "upstairs". Once more, this is for the moment simply a matter of terminology and notation. We will comment further on this in Sect. 6.12.

Proposition 6.2.4 *Given a Riemann patch of class* $\mathcal{C}^k$, *the coefficients* g^{ij} *of the inverse metric tensor are still functions of class* $\mathcal{C}^k$.

Proof From any algebra course, we know that the inverse metric tensor is equal to

$$\begin{pmatrix} g^{11} & g^{12} \\ g^{21} & g^{22} \end{pmatrix} = \begin{pmatrix} \frac{g_{22}}{g_{11}g_{22}-g_{12}g_{21}} & \frac{-g_{12}}{g_{11}g_{22}-g_{12}g_{21}} \\ \frac{-g_{21}}{g_{11}g_{22}-g_{12}g_{21}} & \frac{g_{11}}{g_{11}g_{22}-g_{12}g_{21}} \end{pmatrix}.$$

This forces the conclusion because the denominator is never zero (Proposition 6.2.2) while by Definition 6.2.1, the functions g_{ij} are of class $\mathcal{C}^k$. □

Let us conclude this section with a useful point of notation. Since at each point of a Riemann patch, the metric tensor is a 2×2 symmetric definite positive matrix (Definition 6.2.1), it is the matrix of a scalar product in $\mathbb{R}^2$. Let us introduce a notation for this scalar product.

Notation 6.2.5 Given a Riemann patch

$$g_{ij} : U \longrightarrow \mathbb{R}, \quad 1 \le i, j \le 2$$

and a point $(x^1, x^2) \in U$, we shall write

$$\big((a,b)\big|(c,d)\big)_{(x^1,x^2)} = \begin{pmatrix} a & b \end{pmatrix} \begin{pmatrix} g_{11}(x^1,x^2) & g_{12}(x^1,x^2) \\ g_{21}(x^1,x^2) & g_{22}(x^1,x^2) \end{pmatrix} \begin{pmatrix} c \\ d \end{pmatrix}$$

for the corresponding scalar product on $\mathbb{R}^2$, and by analogy

$$\big\|(a,b)\big\|_{(x^1,x^2)} = \sqrt{\big((a,b)\big|(a,b)\big)_{(x^1,x^2)}}.$$

6.3 Curves on a Riemann Patch

The notion of a curve on a Riemann patch is the most obvious one.

Definition 6.3.1 A *curve* on a Riemann patch

$$g_{ij}\colon U \longrightarrow \mathbb{R}, \quad 1 \le i, j \le 2$$

is simply a plane curve in the sense of Sect. 2.1, admitting a parametric representation

$$c\colon]a,b[\longrightarrow U \subseteq \mathbb{R}^2, \qquad t \mapsto \big(c^1(t), c^2(t)\big).$$

The curve on the Riemann patch is *regular* when the plane curve represented by c is regular.

Using Notation 6.2.5 and in view of Propositions 5.4.3 and 5.4.4, we define:

Definition 6.3.2 Consider a regular curve

$$c\colon]a,b[\longrightarrow U, \quad t \mapsto \big(c^1(t), c^2(t)\big)$$

on a Riemann patch

$$g_{ij}\colon U \longrightarrow \mathbb{R}, \quad 1 \le i, j \le 2.$$

Given $a < k < l < b$, the *length* of the arc of the curve between the points with parameters k and l is defined as being

$$\mathsf{Length}_k^l(c) = \int_k^l \big\|c'(t)\big\|_{c(t)}.$$

Definition 6.3.3 Consider two regular curves

$$c\colon]a,b[\longrightarrow U, \qquad t \mapsto \big(c^1(t), c^2(t)\big), \qquad d\colon]k,l[\longrightarrow U,$$
$$s \mapsto \big(d^1(s), d^2(s)\big)$$

on a Riemann patch

$$g_{ij}\colon U \longrightarrow \mathbb{R}, \quad 1 \le i, j \le 2.$$

When these two curves have a common point

$$c(t_0) = \left(x_0^1, x_0^2\right) = d(s_0)$$

the *angle* between these two curves at their common point is the real number $\theta \in [0, 2\pi[$ such that

$$\cos\theta = \frac{(c'(t_0)|d'(s_0))_{(x_0^1,x_0^2)}}{\|c'(t_0)\|_{(x_0^1,x_0^2)} \cdot \|d'(s_0)\|_{(x_0^1,x_0^2)}}.$$

Let us now, for a curve in a Riemann patch, investigate the existence of a *normal* representation.

Definition 6.3.4 Consider a Riemann patch

$$g_{ij}\colon U \longrightarrow \mathbb{R}, \quad 1 \le i, j \le 2$$

and a regular curve

$$c\colon]a, b[\longrightarrow U, \qquad s \mapsto \left(c^1(s), c^2(s)\right)$$

in it. The parametric representation c is said to be a *normal representation* when at each point

$$\left\|c'(s)\right\|_{c(s)} = 1.$$

One should be well aware that in Definition 6.3.4, c is *not* a normal representation of the ordinary plane curve in U of which it is a parametric representation. In the case of a surface of $\mathbb{R}^3$ represented by f, the condition in Definition 6.3.4 requires in fact that $f \circ c$ be a normal representation of the corresponding skew curve.

The existence of normal representations in the case of a Riemann patch can be established just as in the case of plane or skew curves (see Propositions 2.8.2 and 4.2.7).

Proposition 6.3.5 *Consider a Riemann patch*

$$g_{ij}\colon U \longrightarrow \mathbb{R}, \quad 1 \le i, j \le 2$$

and a regular curve

$$c\colon]a, b[\longrightarrow U, \qquad t \mapsto \left(c^1(t), c^2(t)\right).$$

Fixing a point $t_0 \in]a, b[$, the function

$$\sigma(t) = \int_{t_0}^{t} \left\|c'\right\|_c$$

is a change of parameter of class $\mathcal{C}^1$ making $c \circ \sigma^{-1}$ a normal representation.

Proof The derivative of σ is simply $\sigma' = \|c'\|_c$. This derivative is strictly positive at each point because the matrix $(g_{ij})_{ij}$ is symmetric definite positive (see Definition 6.2.1) and by regularity, the vector c' is never zero. Therefore σ'—thus also σ—is still of class $\mathcal{C}^1$, since so are the g_{ij} and c, as well as the square root function "away from zero". Since its derivative is always strictly positive, σ is a strictly increasing function. Therefore σ admits an inverse σ^{-1}, still of class $\mathcal{C}^1$, whose first derivative is given by

$$(\sigma^{-1})' = \frac{1}{\sigma' \circ \sigma^{-1}} = \frac{1}{\|c'\|_c \circ \sigma^{-1}} = \frac{1}{\|c' \circ \sigma^{-1}\|_{c\circ\sigma^{-1}}}.$$

Let us write $\overline{c} = c \circ \sigma^{-1}$. We must prove that $\|\overline{c}'\|_{\overline{c}} = 1$ (see Definition 6.3.4). But

$$\overline{c}' = (c' \circ \sigma^{-1})(\sigma^{-1})' = \frac{c' \circ \sigma^{-1}}{\|c' \circ \sigma^{-1}\|_{\overline{c}}}$$

which forces at once the conclusion. □

Of course, we have:

Proposition 6.3.6 *Consider a Riemann patch*

$$g_{ij} : U \longrightarrow \mathbb{R}, \quad 1 \le i, j \le 2$$

and a regular curve

$$c : \,]a, b[\longrightarrow U, \qquad t \mapsto (c^1(t), c^2(t))$$

given in normal representation. Then for $a < k < l < b$

$$\mathsf{Length}_k^l(c) = l - k.$$

Proof In Definition 6.3.2, the integral is that of the constant function 1. □

6.4 Vector Fields Along a Curve

Following the "slogan" at the end of Sect. 6.1, what happens in the tangent plane to a surface should be a Riemannian notion. In the theory of skew curves (see Chap. 3) we have considered several vectors attached to each point of a curve: its successive derivatives, the tangent vector, the normal vector, the binormal vector, and so on. When the curve is drawn on a surface, our "slogan" suggests that only the components of these vectors in the tangent plane should be relevant in Riemannian geometry. Therefore we make the following definition.

Definition 6.4.1 Let us consider a curve c on a surface f

$$]a,b[\xrightarrow{c} U \xrightarrow{f} \mathbb{R}^3,$$

both being regular and of class $\mathcal{C}^k$. A *vector field* of class $\mathcal{C}^k$ along the curve, tangent to the surface, is a function of class $\mathcal{C}^k$

$$\xi \colon]a,b[\longrightarrow \mathbb{R}^3, \qquad t \mapsto \xi(t)$$

where for each $t \in]a,b[$, $\xi(t)$ belongs to the direction of the tangent plane to the surface at the point $(f \circ c)(t)$ (see Definition 2.4.1 in [4], *Trilogy II*).

Of course, working as usual in the affine basis of the partial derivatives in each tangent plane, we can re-write (with upper indices)

$$\xi(t) = \xi^1(t)\frac{\partial f}{\partial x^1}(f \circ c)(t) + \xi^2(t)\frac{\partial f}{\partial x^2}(f \circ c)(t).$$

The knowledge of the vector field ξ is of course equivalent to the knowledge of its two components ξ^1 and ξ^2. This suggests at once that being a tangent vector field can easily be made a Riemannian notion:

Definition 6.4.2 Consider a Riemann patch of class $\mathcal{C}^k$

$$g_{ij} \colon U \to \mathbb{R}, \quad 1 \leq i,j \leq 2$$

and a regular curve of class $\mathcal{C}^k$ in it

$$c \colon]a,b[\longrightarrow U.$$

A *vector field* ξ of class $\mathcal{C}^k$ along this curve consists of giving two functions of class $\mathcal{C}^k$

$$\xi^1, \xi^2 \colon]a,b[\longrightarrow \mathbb{R}.$$

Of course in Definition 6.4.2 we *intuitively* think of the two functions ξ^1, ξ^2 as being the two components of a vector in the tangent plane (see Definition 6.4.1), even if in the case of a Riemann patch, no such tangent plane is *a priori* defined. Let us conclude this section with a very natural definition:

Example 6.4.3 Consider a Riemann patch of class $\mathcal{C}^k$

$$g_{ij} \colon U \to \mathbb{R}, \quad 1 \leq i,j \leq 2$$

and a regular curve of class $\mathcal{C}^k$

$$c \colon]a,b[\longrightarrow U, \qquad s \mapsto \big(c^1(s), c^2(s)\big).$$

The vector field τ of class $\mathcal{C}^{k-1}$ with components

$$\tau^1 = \frac{(c^1)'}{\|c'\|_c}, \qquad \tau^2 = \frac{(c^2)'}{\|c'\|_c} : \,]a,b[\longrightarrow \mathbb{R}$$

is called "the" *tangent vector field to the curve*; it is such that $\|\tau\|_c = 1$. When c is given in normal representation, one has further $\tau = c'$.

Proof At each point, the vector field τ is the vector $c' \in \mathbb{R}^2$ divided by its norm for the scalar product $(-|-)_c$ (see Notation 6.3.4). The result follows by Definition 6.3.4. □

In Example 6.4.3, it is clear that equivalent parametric representations of the same curve can possibly give corresponding *tangent vector fields* opposite in sign.

Definition 6.4.4 Consider a Riemann patch

$$g_{ij} : U \longrightarrow \mathbb{R}, \quad 1 \leq i, j \leq 2$$

and a regular curve

$$c : \,]a,b[\longrightarrow U, \qquad t \mapsto \big(c^1(t), c^2(t)\big).$$

With Notation 6.2.5:

1. The *norm* of a vector field ξ along c is the positive real valued function

$$\|\xi\|(t) = \big\|\xi(t)\big\|_{c(t)}.$$

2. Two vector fields ξ and χ along c are *orthogonal* when at each point

$$\big(\xi(t)\big|\chi(t)\big)_{c(t)} = 0.$$

6.5 The Normal Vector Field to a Curve

We are now interested in transposing, to the context of a Riemann patch, the notion of the *normal vector* to a curve in the sense of the *Frenet trihedron* (see Definition 4.4.1).

Proposition 6.5.1 *Consider a Riemann patch of class $\mathcal{C}^k$*

$$g_{ij} : U \to \mathbb{R}, \quad 1 \leq i, j \leq 2$$

and a regular curve of class $\mathcal{C}^k$

$$c : \,]a,b[\longrightarrow U, \qquad t \mapsto \big(c^1(t), c^2(t)\big).$$

There exists a vector field of class $\mathcal{C}^{k-1}$

$$\eta^1, \eta^2 \colon]a, b[$$

along c *with the properties*:

1. η *is orthogonal to the tangent vector field of* c;
2. $\|\eta\|_c = 1$;
3. *the basis* (c', η) *has at each point direct orientation* (*see Sect.* 3.2 *in* [4], Trilogy II).

This vector field is called "the" normal vector field *to the curve*.

Proof To get a vector field μ satisfying the orthogonality condition of the statement, at each point we must have (see Definition 6.4.4)

$$\begin{pmatrix} \mu^1 & \mu^2 \end{pmatrix} \begin{pmatrix} g_{11} & g_{12} \\ g_{21} & g_{22} \end{pmatrix} \begin{pmatrix} (c^1)' \\ (c^2)' \end{pmatrix} = 0$$

or in other words

$$\begin{pmatrix} \mu^1 & \mu^2 \end{pmatrix} \begin{pmatrix} g_{11}(c^1)' + g_{12}(c^2)' \\ g_{21}(c^1)' + g_{22}(c^2)' \end{pmatrix} = 0.$$

It suffices to put

$$\mu^1 = g_{21}(c^1)' + g_{22}(c^2)', \qquad \mu^2 = -\big(g_{11}(c^1)' + g_{12}(c^2)'\big)$$

or—of course—the opposite choice

$$\mu^1 = -\big(g_{21}(c^1)' + g_{22}(c^2)'\big), \qquad \mu^2 = g_{11}(c^1)' + g_{12}(c^2)'.$$

This can be re-written as

$$\begin{pmatrix} \mu^1 \\ \mu^2 \end{pmatrix} = \begin{pmatrix} g_{21} & g_{22} \\ -g_{11} & -g_{12} \end{pmatrix} \begin{pmatrix} (c^1)' \\ (c^2)' \end{pmatrix}.$$

In this formula, the square matrix is regular at each point (it has the same determinant as the metric tensor) and c' is non-zero at each point, by regularity of c. Thus μ is non-zero at each point and the expected normal vector field is

$$\eta^1 = \frac{\mu^1}{\|\mu\|_c}, \qquad \eta^2 = \frac{\mu^2}{\|\mu\|_c} \colon]a, b[\longrightarrow \mathbb{R}.$$

Observe that the two-fold possibility in the choice of the vector field μ yields two bases (c', η) with opposite orientations: it remains to choose the basis with direct orientation. $\square$

6.6 The Christoffel Symbols

Chapter 5 has provided evidence that all important properties of the surface can be expressed in terms of the six functions E, F, G, L, M, N defined in Proposition 5.4.3 and Theorem 5.8.2. As emphasized in Sect. 6.2, the functions E, F, G constitute the *metric tensor* of the surface. But what about the functions L, M, N. Are they *Riemannian quantities*, quantities that we can determine by measures performed on the surface? The answer is definitely "No":

Counterexample 6.6.1 The coefficients L, M, N of the second fundamental form of a surface cannot be deduced from the sole knowledge of the coefficients E, F, G of the first fundamental form.

Proof At each point of the plane with parametric representation

$$f(x, y) = (x, y, 0)$$

we have trivially

$$E = 1, \qquad F = 0, \qquad G = 1, \qquad L = 0, \qquad M = 0, \qquad N = 0.$$

At each point of the circular cylinder (see Sect. 1.14 in [4], *Trilogy II*) with parametric representation

$$g(\theta, z) = (\cos\theta, \sin\theta, z)$$

we have

$$\frac{\partial g}{\partial \theta} = (-\sin\theta, \cos\theta, 0), \qquad \frac{\partial g}{\partial z} = (0, 0, 1)$$

from which again

$$E = 1, \qquad F = 0, \qquad G = 1.$$

On the other hand

$$\vec{n} = (\cos\theta, \sin\theta, 0)$$

and

$$\frac{\partial g^2}{\partial \theta^2} = (-\cos\theta, -\sin\theta, 0), \qquad \frac{\partial g^2}{\partial \theta\, \partial z} = (0, 0, 0), \qquad \frac{\partial g^2}{\partial z^2} = (0, 0, 0)$$

from which

$$L = -1, \qquad M = 0, \qquad N = 0.$$

The two surfaces have the same functions E, F, G, but not the same functions L, M, N. □

So L, M, N are *not* Riemannian quantities. Let us recall that they are obtained from the second partial derivatives of the parametric representation, by performing the scalar product with the normal vector $\overrightarrow{n}$ to the surface (see Theorem 5.8.2). Applying the "slogan" at the end of Sect. 6.1 to the case of the second partial derivatives of the parametric representation, the following definition sounds sensible:

Definition 6.6.2 Consider a regular parametric representation

$$f : U \longrightarrow \mathbb{R}^3, \qquad (x^1, x^2) \mapsto f(x^1, x^2)$$

of class $\mathcal{C}^k$ of a surface.

1. The *Christoffel symbols* of the first kind are the functions

$$\Gamma_{ijk} = \left(\frac{\partial^2 f}{\partial x^i \partial x^j} \middle| \frac{\partial f}{\partial x^k} \right), \qquad 1 \le i, j, k \le 2.$$

2. The Christoffel symbols of the second kind are the quantities Γ_{ij}^k, the components of the second partial derivatives of f with respect to the basis comprising the first partial derivatives and the normal to the surface:

$$\frac{\partial^2 f}{\partial x^i \partial x^j} = \Gamma_{ij}^1 \frac{\partial f}{\partial x^1} + \Gamma_{ij}^2 \frac{\partial f}{\partial x^2} + h_{ij} \overrightarrow{n}.$$

The observant reader will have noticed the use of the word "symbols", not "tensor"; and the presence of upper and lower indices! Again, we shall comment upon this later.

Proposition 6.6.3 *Under the conditions of Definition* 6.6.2,

$$\begin{pmatrix} h_{11} & h_{12} \\ h_{21} & h_{22} \end{pmatrix} = \begin{pmatrix} L & M \\ M & N \end{pmatrix}.$$

Proof Simply take the scalar product of

$$\frac{\partial f^2}{\partial x^i \partial x^j} = \Gamma_{ij}^1 \frac{\partial f}{\partial x^1} + \Gamma_{ij}^2 \frac{\partial f}{\partial x^2} + h_{ij} \overrightarrow{n}$$

with $\overrightarrow{n}$, keeping in mind that

$$\left(\frac{\partial f}{\partial x^i} \middle| \overrightarrow{n} \right) = 0, \qquad \left(\overrightarrow{n} \middle| \overrightarrow{n} \right) = 1.$$

□

From now on, we shall use the notation h_{ij} instead of L, M, N. Let us also make the following easy observations:

Proposition 6.6.4 *Under the conditions of our Definition* 6.6.2, *the Christoffel symbols are functions of class* $\mathcal{C}^{k-2}$ *with the following properties*:

$$\Gamma_{ijk} = \Gamma_{jik},$$
$$\Gamma_{ij}^{k} = \Gamma_{ji}^{k},$$
$$\Gamma_{ijk} = \sum_{l} g_{lk}\Gamma_{ij}^{l},$$
$$\Gamma_{ij}^{k} = \sum_{l} g^{kl}\Gamma_{ijl}.$$

Proof The first two equalities hold because

$$\frac{\partial^2 f}{\partial x^i \partial x^j} = \frac{\partial^2 f}{\partial x^j \partial x^i}.$$

The third equality is obtained by expanding the scalar product

$$\left(\Gamma_{ij}^{1}\frac{\partial f}{\partial x^1} + \Gamma_{ij}^{2}\frac{\partial f}{\partial x^2} + h_{ij}\vec{n} \,\middle|\, \frac{\partial f}{\partial x^k}\right)$$

keeping in mind that $(\vec{n} \,|\, \frac{\partial f}{\partial x^k}) = 0$.

This third equality can be re-written in matrix form as

$$\begin{pmatrix}\Gamma_{ij1} & \Gamma_{ij2}\end{pmatrix} = \begin{pmatrix}\Gamma_{ij}^{1} & \Gamma_{ij}^{2}\end{pmatrix}\begin{pmatrix} g_{11} & g_{12} \\ g_{21} & g_{22}\end{pmatrix}.$$

Multiplying both sides by the inverse metric tensor, we obtain the fourth formula.

By Definition 6.6.2, the Christoffel symbols of the first kind are functions of class $\mathcal{C}^{k-2}$. By Proposition 6.2.4 and the fourth equality in the statement, the same conclusion holds for the symbols of the second kind. □

The key observation is now:

Proposition 6.6.5 *Under the conditions of Definition* 6.6.2, *the Christoffel symbols of the first kind are also equal to*

$$\Gamma_{ijk} = \frac{1}{2}\left(\frac{\partial g_{jk}}{\partial x^i} + \frac{\partial g_{ki}}{\partial x^j} - \frac{\partial g_{ij}}{\partial x^k}\right).$$

Proof First (see Lemma 1.11.3)

$$\begin{aligned}\frac{\partial g_{ij}}{\partial x^k} &= \frac{\partial}{\partial x^k}\left(\frac{\partial f}{\partial x^i}\,\middle|\,\frac{\partial f}{\partial x^j}\right) \\ &= \left(\frac{\partial^2 f}{\partial x^i \partial x^k}\,\middle|\,\frac{\partial f}{\partial x^j}\right) + \left(\frac{\partial f}{\partial x^i}\,\middle|\,\frac{\partial^2 f}{\partial x^j \partial x^k}\right) \\ &= \Gamma_{ikj} + \Gamma_{jki}.\end{aligned}$$

Therefore

$$\frac{\partial g_{jk}}{\partial x^i} + \frac{\partial g_{ki}}{\partial x^j} - \frac{\partial g_{ij}}{\partial x^k} = \Gamma_{jik} + \Gamma_{kij} + \Gamma_{ijk} + \Gamma_{kji} - \Gamma_{ikj} - \Gamma_{jki} = 2\Gamma_{ijk}$$

by the first formula in Proposition 6.6.4. □

Proposition 6.6.6 *Under the conditions of Definition* 6.6.2, *the Christoffel symbols of the first and second kind can be expressed as functions of the coefficients of the metric tensor.*

Proof Proposition 6.6.5 proves the result for the symbols of the first kind. But the inverse metric tensor can itself be expressed in terms of the metric tensor (see Definition 6.2.3 or the proof of Proposition 6.2.4 for an explicit formula). By the fourth equality in Proposition 6.6.5, the result for the symbols of the second kind follows immediately. □

To stress the fact that the Christoffel symbols are Riemannian quantities, let us conclude this section with a definition inspired by Proposition 6.6.6:

Definition 6.6.7 Given a Riemann patch of class $\mathcal{C}^1$ as in Definition 6.2.1, the *Christoffel symbols of the first kind* are by definition the quantities

$$\Gamma_{ijk} = \frac{1}{2}\left(\frac{\partial g_{jk}}{\partial x^i} + \frac{\partial g_{ki}}{\partial x^j} - \frac{\partial g_{ij}}{\partial x^k}\right), \quad 1 \leq i, j, k \leq 2$$

while the *Christoffel symbols of the second kind* are the quantities

$$\Gamma_{ij}^k = \sum_l g^{kl}\Gamma_{ijl}, \quad 1 \leq i, j, k, l \leq 2.$$

Proposition 6.6.4 carries over to this generalized context.

Proposition 6.6.8 *Consider a Riemann patch of class* $\mathcal{C}^k$

$$g_{ij} : U \longrightarrow \mathbb{R}, \quad 1 \leq i, j \leq 2.$$

The Christoffel symbols are functions of class $\mathcal{C}^{k-1}$ *satisfying the following properties*:

$$\Gamma_{ijk} = \Gamma_{jik},$$

$$\Gamma_{ij}^k = \Gamma_{ji}^k,$$

$$\Gamma_{ijk} = \sum_l g_{lk}\Gamma_{ij}^l,$$

$$\Gamma_{ij}^k = \sum_l g^{kl}\Gamma_{ijl}.$$

Proof The last condition is just Definition 6.6.7. This same definition forces at once the first condition, because the metric tensor is symmetric. This immediately implies the second condition. Finally Definition 6.6.7 can be expressed as the matrix formula

$$\begin{pmatrix} \Gamma_{ij}^1 \\ \Gamma_{ij}^2 \end{pmatrix} \begin{pmatrix} g^{11} & g^{12} \\ g^{21} & g^{22} \end{pmatrix} \begin{pmatrix} \Gamma_{ij1} \\ \Gamma_{ij2} \end{pmatrix}.$$

Multiplying both sides by the metric tensor yields condition 4 in the statement. □

6.7 Covariant Derivative

Our experience of doing mathematics tells us how important the derivative of a function can be. Going back to Definition 6.4.1, we therefore want to consider the derivative of the function ξ describing a vector field. However, since we are working in Riemannian geometry, our "slogan" of Sect. 6.6 suggests that we should focus on the component of this derivative in the tangent plane.

Definition 6.7.1 Consider a regular curve c on a regular surface f

$$]a,b[\xrightarrow{c} U \xrightarrow{f} \mathbb{R}^3.$$

Consider further a vector field of class $\mathcal{C}^1$ along this curve, tangent to the surface

$$\xi :]a,b[\longrightarrow \mathbb{R}^3, \qquad t \mapsto \xi(t).$$

The *covariant derivative* of this vector field is the vector field

$$\frac{\nabla \xi}{dt} :]a,b[\longrightarrow \mathbb{R}^3$$

defined at each point as the orthogonal projection of the derivative of ξ on the direction of the tangent plane to the surface.

Our job is now to explicitly calculate this covariant derivative.

Proposition 6.7.2 *In the situation described in Definition* 6.7.1, *when the surface is of class $\mathcal{C}^2$, the covariant derivative of the vector field ξ is equal to*

$$\frac{\nabla \xi}{dt} = \sum_{k=1}^{2} \left(\frac{d\xi^k}{dt} + \sum_{i,j=1}^{2} \xi^i \frac{dc^j}{dt} \Gamma_{ij}^k (c^1, c^2) \right) \frac{\partial f}{\partial x^k} (c^1, c^2).$$

Proof Let us first differentiate the function

$$\xi(t) = \sum_{i=1}^{2} \xi^i(t) \frac{\partial f}{\partial x^i} \left(c^1(t), c^2(t)\right).$$

We obtain, keeping in mind Definition 6.6.2 and writing $\overrightarrow{n}$ for the normal vector to the surface:

$$\begin{aligned}
\frac{d\xi}{dt} &= \sum_i \left(\frac{d\xi^i}{dt} \frac{\partial f}{\partial x^i} + \xi^i \left(\sum_j \frac{\partial^2 f}{\partial x^i \partial x^j} \frac{dc^j}{dt} \right) \right) \\
&= \sum_i \left(\frac{d\xi^i}{dt} \frac{\partial f}{\partial x^i} + \xi^i \left(\sum_j \left(\Gamma_{ij}^1 \frac{\partial f}{\partial x^1} + \Gamma_{ij}^2 \frac{\partial f}{\partial x^2} + h_{ij} \overrightarrow{n} \right) \frac{dc^j}{dt} \right) \right) \\
&= \left(\frac{d\xi^1}{dt} + \sum_i \xi^i \left(\sum_j \Gamma_{ij}^1 \frac{dc^j}{dt} \right) \right) \frac{\partial f}{\partial x^1} \\
&\quad + \left(\frac{d\xi^2}{dt} + \sum_i \xi^i \left(\sum_j \Gamma_{ij}^2 \frac{dc^j}{dt} \right) \right) \frac{\partial f}{\partial x^2} \\
&\quad + \left(\sum_{ij} \xi^i \frac{dc^j}{dt} h_{ij} \right) \overrightarrow{n}
\end{aligned}$$

where for short, we have used the abbreviated notation

$$\Gamma_{ij}^l = \Gamma_{ij}^l(c^1, c^2), \qquad h_{ij} = h_{ij}(c^1, c^2)$$

and analogously for the partial derivatives of f. The orthogonal projection on the direction of the tangent plane is constituted of the first two lines of this last expression, which is indeed the formula in the statement. □

The observant reader will have noticed that Definition 6.7.1 of the covariant derivative makes perfect sense in class $\mathcal{C}^1$, while its expression given in Proposition 6.7.2 requires the class $\mathcal{C}^2$ because of the presence of the Christoffel symbols (see Definition 6.6.2).

The Christoffel symbols are Riemannian quantities (see Definition 6.6.7), thus by Proposition 6.7.2, so is the covariant derivative:

Definition 6.7.3 Consider a Riemann patch of class $\mathcal{C}^1$

$$g_{ij} \colon U \to \mathbb{R}, \quad 1 \le i, j \le 2$$

and a regular curve

$$c \colon]a, b[\longrightarrow U.$$

The *covariant derivative* of a tangent vector field ξ of class $\mathcal{C}^1$ along this curve is the tangent vector field $\frac{\nabla \xi}{dt}$ whose two components are

$$\frac{d\xi^k}{dt} + \sum_{i,j} \xi^i \frac{dc^j}{dt} \Gamma_{ij}^k(c^1, c^2), \quad 1 \le k \le 2.$$

The covariant derivative inherits the classical properties of an "ordinary" derivative. For example:

Proposition 6.7.4 *Consider a Riemann patch of class $\mathcal{C}^1$*

$$g_{ij}: U \longrightarrow \mathbb{R}, \quad 1 \leq i, j \leq 2$$

and a regular curve

$$c:]a, b[\longrightarrow U, \qquad t \mapsto \big(c^1(t), c^2)\big)$$

in it. Consider two tangent vector fields ξ and χ of class $\mathcal{C}^1$ along this curve, as well as an additional function of class $\mathcal{C}^1$

$$\alpha:]a, b[\longrightarrow \mathbb{R}, \qquad t \mapsto \alpha(t)$$

and a change of parameter of class $\mathcal{C}^1$

$$\varphi:]r, s[\longrightarrow]a, b[, \qquad s \mapsto \varphi(s).$$

The following properties hold:

1. $\frac{\nabla(\xi+\chi)}{dt} = \frac{\nabla\xi}{dt} + \frac{\nabla\chi}{dt}$;
2. $\frac{\nabla(\alpha\cdot\xi)}{dt} = \frac{d\alpha}{dt}\xi + \alpha\frac{\nabla\xi}{dt}$;
3. $\frac{\nabla(\xi\circ\varphi)}{ds} = (\frac{\nabla\xi}{dt} \circ \varphi) \cdot \varphi'$;
4. $\frac{d(\xi|\chi)_c}{dt} = (\frac{\nabla\xi}{dt}|\xi)_c + (\xi|\frac{\nabla\chi}{dt})_c$.

Proof Condition 1 of the statement is trivial. Condition 2 is immediate: the components of $\frac{\nabla(\alpha\cdot\xi)}{dt}$ are

$$\frac{d(\alpha \cdot \xi^k)}{dt} + \sum_{ij} \big(\alpha \cdot \xi^k\big)^i \frac{dc^j}{dt} \Gamma_{ij}^k = \frac{d\alpha}{dt}\xi^k + \alpha\frac{d\xi^k}{dt} + \sum_{ij} \alpha\xi^i \frac{dc^j}{dt}\Gamma_{ij}^k$$

which is the second formula of the statement. Condition 3 is proved in exactly the same straightforward way: the components of $\frac{\nabla(\xi\circ\varphi)}{ds}$ are

$$\begin{aligned}
&\frac{d(\xi^k \circ \varphi)}{ds} + \sum_{ij} (\xi^i \circ \varphi) \frac{d(c^j \circ \varphi)}{ds} \Gamma_{ij}^k \\
&= \left(\frac{d\xi^k}{dt} \circ \varphi\right)\varphi' + \sum_{ij} (\xi^i \circ \varphi) \left(\frac{dc^j}{dt} \circ \varphi\right)\varphi' \Gamma_{ij}^k
\end{aligned}$$

which is again the announced statement.

Proving the fourth formula in the statement is a more involved task. First let us observe that the components of $\frac{d(\xi|\chi)_c}{dt}$ are

$$\begin{aligned}
&\frac{d}{dt}\Big(\sum_{kl} \xi^k(t)\chi^l(t)g_{kl}\big(c(t)\big)\Big)\\
&\quad=\sum_{kl}\frac{d\xi^k(t)}{dt}\chi^l(t)g_{kl}\big(c(t)\big)+\sum_{kl}\xi^k(t)\frac{d\chi^l(t)}{dt}g_{kl}\big(c(t)\big)\\
&\quad\quad+\sum_{kl}\xi^k(t)\chi^l(t)\Big(\sum_m \frac{\partial g_{kl}}{\partial x^m}\big(c(t)\big)\frac{dc^m(t)}{dt}\Big).
\end{aligned}$$

On the other hand the components of $(\frac{\nabla\xi}{dt}|\chi)+(\xi|\frac{\nabla\xi}{dt})$ are

$$\begin{aligned}
&\sum_{kl}\Big(\frac{d\xi^k(t)}{dt}+\sum_{ij}\xi^i(t)\frac{dc^j(t)}{dt}\Gamma_{ij}^k\big(c(t)\big)\Big)\chi^l(t)g_{kl}\big(c(t)\big)\\
&\quad+\sum_{kl}\xi^k(t)\Big(\frac{d\chi(t)}{dt}+\sum_{ij}\chi^i(t)\frac{dc^j(t)}{dt}\Gamma_{ij}^l\big(c(t)\big)\Big)g_{kl}\big(c(t)\big).
\end{aligned}$$

Comparing both expressions, it remains to prove that

$$\sum_{klm}\xi^k\chi^l\frac{\partial g_{kl}}{\partial x^m}\frac{dc^m}{dt}=\sum_{klij}\xi^i\chi^l\frac{dc^j}{dt}\gamma_{ij}^k g_{kl}+\sum_{klij}\xi^k\chi^i\frac{dc^j}{dt}\gamma_{ij}^l g_{kl}.$$

Using Proposition 6.6.8 and Definition 6.6.7, we obtain

$$\begin{aligned}
\sum_{klm}\xi^k\chi^l\frac{dc^m}{dt}\frac{\partial g_{kl}}{\partial x^m}&=\sum_{lij}\xi^i\chi^l\frac{dc^j}{dt}\Gamma_{ijl}+\sum_{kij}\xi^k\chi^i\frac{dc^j}{dt}\Gamma_{ijk}\\
&=\sum_{lij}\xi^i\chi^l\frac{dc^j}{dt}\frac{1}{2}\Big(\frac{\partial g_{jl}}{\partial x^i}+\frac{\partial g_{li}}{\partial x^j}-\frac{\partial g_{ij}}{\partial x^l}\Big)\\
&\quad+\sum_{kij}\xi^k\chi^i\frac{dc^j}{dt}\frac{1}{2}\Big(\frac{\partial g_{jk}}{\partial x^i}+\frac{\partial g_{ki}}{\partial x^j}-\frac{\partial g_{ij}}{\partial x^k}\Big)\\
&=\sum_{klm}\xi^k\chi^l\frac{dc^m}{dt}\frac{1}{2}\Big(\frac{\partial g_{ml}}{\partial x^k}+\frac{\partial g_{lk}}{\partial x^m}-\frac{\partial g_{km}}{\partial x^l}\Big)\\
&\quad+\sum_{klm}\xi^k\chi^l\frac{dc^m}{dt}\frac{1}{2}\Big(\frac{\partial g_{mk}}{\partial x^l}+\frac{\partial g_{kl}}{\partial x^m}-\frac{\partial g_{lm}}{\partial x^k}\Big)\\
&=\sum_{klm}\xi^k\chi^l\frac{dc^m}{dt}\frac{\partial g_{kl}}{\partial x^m}
\end{aligned}$$

which is the expected equality concluding the proof. □

Corollary 6.7.5 *Consider a Riemann patch of class $\mathcal{C}^2$*

$$g_{ij}: U \to \mathbb{R}, \quad 1 \le i, j \le 2$$

and a regular curve of class $\mathcal{C}^2$

$$c:]a,b[\longrightarrow U, \qquad s \mapsto \big(c^1(s), c^2(s)\big)$$

given in normal representation. The tangent vector field c' to the curve and its covariant derivative $\frac{\nabla c'}{ds}$ are orthogonal vector fields (see Definition 6.4.4).

Proof By Definition 6.3.4, $(c'|c')_c = 1$. By Proposition 6.7.4.4, this implies $2(\frac{\nabla c'}{ds}|c')_c = 0$. □

Under the conditions of Corollary 6.7.5, when the covariant derivative of c' is non-zero at each point, the *normal vector field* of Proposition 6.5.1 is given by

$$\eta = \pm \frac{\frac{\nabla c'}{ds}}{\|\frac{\nabla c'}{ds}\|}.$$

Let us conclude this section by noticing that the notion of covariant derivative provides the notion of *covariant partial derivative*:

Definition 6.7.6 Consider a Riemann patch $(U, (g_{ij})_{ij})$ of class $\mathcal{C}^k$ $(k \ge 1)$.

1. A 2-dimensional tangent vector field ξ of class $\mathcal{C}^k$ on this Riemann patch consists of two functions of class $\mathcal{C}^k$

$$\xi^1, \xi^2 : U \longrightarrow \mathbb{R}.$$

2. The covariant partial derivatives of this vector field ξ at a point (x_0^1, x_0^2) are:
   - $\frac{\nabla \xi}{\partial x^1}(x_0^1, x_0^2)$, the covariant partial derivative at x_0^1 of the vector field $\xi(x^1, x_0^2)$ along the curve $x^2 = x_0^2$;
   - $\frac{\nabla \xi}{\partial x^2}(x_0^1, x_0^2)$, the covariant partial derivative at x_0^2 of the vector field $\xi(x_0^1, x^2)$ along the curve $x^1 = x_0^1$.

As expected, one has:

Proposition 6.7.7 *Consider*:

- *a Riemann patch $(U, (g_{ij})_{ij})$ of class $\mathcal{C}^1$ $(k \ge 1)$;*
- *a 2-dimensional tangent vector field $\xi = (\xi^1, \xi^2)$ of class $\mathcal{C}^1$;*
- *a regular curve represented by $c:]a,b[\longrightarrow U$.*

Under these conditions, writing $t \in]a,b[$ for the parameter,

$$\frac{\nabla\xi(c(t))}{dt} = \frac{\nabla\xi}{\partial x^1}\big(c(t)\big)\frac{dc^1}{dt} + \frac{\nabla\xi}{\partial x^2}\big(c(t)\big)\frac{dc^2}{dt}.$$

Proof Since $\frac{\nabla\xi}{\partial x^1}$ is computed along a curve $h(x^1) = (x^1, x_0^2)$, one has

$$\frac{dh^1}{dx^1} = 1, \qquad \frac{dh^2}{dx^1} = 0$$

and analogously for the other partial derivative. Thus by Definition 6.7.3 $\frac{\nabla\xi}{\partial x^j}$ has for components

$$\frac{\partial\xi^k}{\partial x^j} + \sum_i \xi^i \Gamma_{ij}^k.$$

On the other hand, still by Definition 6.7.3, the vector field $\xi(c(t))$ along c has a covariant derivative whose components are given by

$$\begin{aligned}
&\frac{d\xi^k}{dt}\big(c(t)\big) + \sum_{ij}\xi^i\big(c(t)\big)\frac{dc^j}{dt}\Gamma_i{}j^k\big(c(t)\big) \\
&= \sum_j \frac{\partial\xi^k}{\partial x^j}\big(c(t)\big)\frac{dc^j}{dt}(t) + \sum_{ij}\xi^i\big(c(t)\big)\frac{dc^j}{dt}\Gamma_i{}j^k\big(c(t)\big) \\
&= \sum_j \left(\frac{\partial\xi^k}{\partial x^j}\big(c(t)\big) + \sum_i \xi^i\big(c(t)\big)\Gamma_i{}j^k\big(c(t)\big)\right)\frac{dc^j}{dt}(t) \\
&= \sum_j \frac{\nabla\xi^k}{\partial x^j}\big(c(t)\big)\frac{dc^j}{dt}(t).
\end{aligned}$$

This proves the announced formula. □

6.8 Parallel Transport

In the plane $\mathbb{R}^2$, we know at once how to "transport" a fixed vector $\overrightarrow{v}$ along a curve represented by $c(t)$ (see Fig. 6.1): at each point of the curve, simply consider the point

$$P(t) = c(t) + \overrightarrow{v}$$

that is the point $P(t)$ such that $\overrightarrow{c(t)\,P(t)} = \overrightarrow{v}$ (see Definition 2.1.1 in [4], *Trilogy II*).

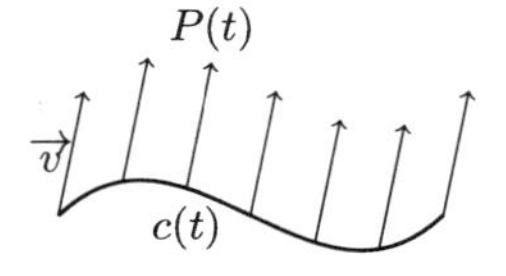

Fig. 6.1

This construction in $\mathbb{R}^2$ thus yields a "constant vector field" along c

$$\overrightarrow{v} : \left]a, b\right[\longrightarrow \mathbb{R}^2, \qquad t \mapsto \overrightarrow{v}.$$

But saying that this function is constant is equivalent to saying that its derivative is equal to zero. The corresponding Riemannian notion is now clear:

Definition 6.8.1 Consider a Riemann patch of class $\mathcal{C}^1$

$$g_{ij} : U \longrightarrow \mathbb{R}, \quad 1 \leq i, j \leq 2,$$

and a regular curve

$$c \colon \left]a, b\right[\longrightarrow U, \qquad t \mapsto \big(c^1(t), c^2(t)\big)$$

in it. A vector field ξ of class $\mathcal{C}^1$ along c is said to be *parallel* when its covariant derivative is everywhere zero.

Let us observe that

Lemma 6.8.2 *Being a* parallel *vector field along a curve is independent of the regular parametric representation chosen for the curve.*

Proof Let φ be a change of parameters of class $\mathcal{C}^1$ for the curve. Differentiating the equality $\varphi \circ \varphi^1 = \mathsf{id}$, we get

$$\left(\varphi' \circ \varphi^{-1}\right) \cdot \left(\varphi^{-1}\right)' = 1$$

proving that φ' is never zero. The conclusion then follows immediately from Proposition 6.7.4.3. □

Proposition 6.8.3 *Consider a Riemann patch of class* $\mathcal{C}^1$

$$g_{ij} : U \longrightarrow \mathbb{R}, \quad 1 \leq i, j \leq 2,$$

and a regular curve in it:

$$c \colon \left]a, b\right[\longrightarrow U, \qquad t \mapsto \big(c^1(t), c^2(t)\big).$$

1. *A parallel vector field ξ of class $\mathcal{C}^1$ along c has a constant norm.*
2. *A parallel vector field ξ of class $\mathcal{C}^1$ along c is orthogonal to its covariant derivative $\frac{\nabla\xi}{dt}$.*
3. *Two non-zero parallel vector fields ξ, χ of class $\mathcal{C}^1$ along c make a constant angle.*

Proof By Proposition 6.7.4

$$\frac{d(\xi|\chi)_c}{dt} = \left(\frac{\nabla\xi}{dt}\middle|\chi\right)_c + \left(\xi\middle|\frac{\nabla\chi}{dt}\right)_c = (0|\chi)_c + (\xi|0)_c = 0.$$

This proves that the scalar product $(\xi|\chi)_c$ is constant. Putting $\xi = \chi$ we conclude that $\|\xi\|_c$ and $\|\chi\|_c$ are constant. Together with the scalar product being constant, this proves that the angle is constant as well (see Notation 6.2.5).

But when $\|\xi\|_c^2 = (\xi|\xi)_c$ is constant, its derivative is zero and by Proposition 6.7.4.4, this yields $2(\frac{\nabla\xi}{dt}|\xi)_c = 0$, thus the orthogonality of ξ and $\frac{\nabla\xi}{dt}$. □

The existence of parallel vector fields is attested by the following theorem:

Theorem 6.8.4 *Consider a Riemann patch of class $\mathcal{C}^k$ $(k \geq 2)$*

$$g_{ij}\colon U \longrightarrow \mathbb{R}, \qquad 1 \leq i, j \leq 2,$$

and a regular curve of class $\mathcal{C}^k$ in it:

$$c\colon]a, b[\longrightarrow U, \qquad t \mapsto \big(c^1(t), c^2(t)\big).$$

Given a vector $\overrightarrow{v} \in \mathbb{R}^2$ and a point $t_0 \in]a, b[$, there exists a sub-interval $]r, s[\subseteq]a, b[$ still containing t_0 and a unique parallel vector field ξ of class $\mathcal{C}^k$ along c

$$\xi^1, \xi^2\colon]r, s[\longrightarrow \mathbb{R}$$

such that $\xi(t_0) = \overrightarrow{v}$. For each value $t \in]r, s[$, the vector $\xi(t)$ is called the parallel transport of $\overrightarrow{v}$ along c.

Proof This is an immediate consequence of the theorem for the existence and uniqueness of a solution of the system of differential equations (see Proposition B.1.1)

$$\frac{d\xi^k}{dt}(t) + \sum_{ij} \xi^i(t)\frac{dc^j}{dt}(t)\Gamma_{ij}^k\big(c^1(t), c^2(t)\big) = 0, \quad 1 \leq k \leq 2$$

together with the initial conditions

$$\xi^1(t_0) = v^1, \qquad \xi^2\big(t^0\big) = v^2$$

(see Definition 6.7.3). Observe that all the coefficients of the differential equations are indeed of class $\mathcal{C}^{k-1}$ (see Proposition 6.6.8). □

6.9 Geodesic Curvature

Let us now switch to the study of the curvature of a curve in a Riemann patch.

Let us first recall the situation studied in Sect. 5.8. Given a curve on a surface

$$]a, b[\xrightarrow{c} U \xrightarrow{f} \mathbb{R}^3,$$

we write $h = f \circ c$ for the corresponding skew curve and $\overline{h}$ for its normal representation. The *normal curvature* (up to its sign) is the length of the orthogonal projection of the "curvature vector" $\overline{h}''$ on the normal vector $\overrightarrow{n}$ to the surface. Following our "slogan" at the end of Sect. 6.1, this *normal curvature* is probably not a Riemannian notion. Indeed we have the following:

Counterexample 6.9.1 The normal curvature of a surface cannot be deduced from the sole knowledge of the three coefficients E, F, G.

Proof In Counterexample 6.6.1, the two surfaces have the same coefficients E, F, G but not the same normal curvature. Indeed by Theorem 5.8.2, the normal curvature of the cylinder is equal to -1 in the direction $(1, 0)$ while in the case of the plane, the normal curvature is equal to 0 in all directions. □

However, as our "slogan" of Sect. 6.1 suggests, in the discussion above, the orthogonal projection of the "curvature vector" $\overline{h}''$ on the tangent plane *should be* a Riemannian notion. That projection—called the *geodesic curvature* of the curve—is intuitively *what the two-dimensional being living on the surface sees of the curvature of the curve* (see Sect. 6.1).

Definition 6.9.2 Consider a curve on a surface

$$]a, b[\xrightarrow{c} U \xrightarrow{f} \mathbb{R}^3,$$

both being regular and of class $\mathcal{C}^2$. Write $\overline{f \circ c}$ for the normal representation of the corresponding skew curve. The *geodesic curvature* of the curve on the surface is the length of the orthogonal projection of the vector $\overline{f \circ c}''$ on the tangent plane to the surface.

We thus get at once:

Proposition 6.9.3 *Consider a curve on a surface, both being regular and of class $\mathcal{C}^2$. Then at each point of this curve*

$$\kappa^2 = \kappa_n^2 + \kappa_g^2$$

where

- *κ indicates the curvature of the curve;*

- κ_n *indicates the normal curvature of the curve;*
- κ_g *indicates the geodesic curvature of the curve.*

Proof This follows by Pythagoras' Theorem (see Theorem 4.3.5 in [4], *Trilogy II*) and Definitions 5.8.1 and 6.9.2. □

Definition 6.9.2 can easily be rephrased:

Proposition 6.9.4 *Consider a curve on a surface*

$$]a,b[\xrightarrow{c} U \xrightarrow{f} \mathbb{R}^3,$$

both being regular and of class $\mathcal{C}^2$. Write $\overline{f \circ c}$ for the normal representation of the corresponding skew curve. The geodesic curvature *of the curve on the surface is the norm of the covariant derivative of the tangent vector field $\overline{f \circ h}'$.*

Proof This follows by Definitions 6.9.2 and 6.7.1. □

By Proposition 6.9.4, the *geodesic curvature* is thus a Riemannian notion. Therefore we make the following definition:

Definition 6.9.5 Consider a Riemann patch of class $\mathcal{C}^2$

$$g_{ij}: U \longrightarrow \mathbb{R}, \quad 1 \le i, j \le 2$$

and a regular curve of class $\mathcal{C}^2$

$$c:]a,b[\longrightarrow U, \qquad s \mapsto \big(c^1(s), c^2(s)\big)$$

given in normal representation. The *geodesic curvature* of that curve is—with Notation 6.2.5—the norm of the covariant derivative of its tangent vector field:

$$\kappa_g = \left\| \frac{\nabla c'}{ds} \right\|_c.$$

Of course one can refine Definition 6.9.5 and provide the geodesic curvature with a sign, as we did for plane curves (see Definition 2.9.8). For that purpose, let us make the following observation:

Proposition 6.9.6 *Consider a Riemann patch of class $\mathcal{C}^2$*

$$g_{ij}: U \longrightarrow \mathbb{R}, \quad 1 \le i, j \le 2$$

and a regular curve of class $\mathcal{C}^2$

$$c:]a,b[\longrightarrow U, \qquad s \mapsto \big(c^1(s), c^2(s)\big)$$

given in normal representation. The geodesic curvature is also equal to

$$\kappa_g = \left| \left(\frac{\nabla c'}{ds} \middle| \eta \right)_c \right|$$

where η is the normal vector field to the curve (see Proposition 6.5.1).

Proof At a point where $\frac{\nabla c'}{ds}(s) = (0,0)$, both the geodesic curvature and the scalar product of the statement are equal to zero. Otherwise we have

$$\left(\frac{\nabla c}{ds}(s) \middle| \eta(s) \right)_{c(s)} = \left\| \frac{\nabla c'}{ds}(s) \right\|_{c(s)} \cdot \|\eta(s)\|_{c(s)} \cdot \cos\theta(s)$$

where $\theta(s)$ is the angle between $\frac{\nabla c'}{ds}(s)$ and $\eta(s)$. By Proposition 6.5.1, $\eta(s)$ is of length 1. But by Proposition 6.7.5, since c is given in normal representation, $\frac{\nabla c}{ds}(s)$ is proportional to $\eta(s)$, thus $\cos\theta(s) = \pm 1$. Therefore

$$\left(\frac{\nabla c'}{ds}(s) \middle| \eta(s) \right)_{c(s)} = \pm \left\| \frac{\nabla c'}{ds}(s) \right\|_{c(s)}$$

which forces the conclusion. □

Definition 6.9.7 Consider a Riemann patch of class $\mathcal{C}^2$

$$g_{ij} : U \longrightarrow \mathbb{R}, \quad 1 \leq i, j \leq 2$$

and a regular curve of class $\mathcal{C}^2$

$$c : \,]a,b[\, \longrightarrow U, \qquad s \mapsto \left(c^1(s), c^2(s)\right)$$

given in normal representation. The *relative geodesic curvature* is the quantity

$$\kappa_g = \left(\frac{\nabla c'}{ds} \middle| \eta \right)_c$$

where η is the normal vector field to the curve (see Proposition 6.5.1).

Clearly, the sign of the geodesic curvature as in Definition 6.9.7 is not an intrinsic property of the curve: for example, it is reversed when considering the equivalent normal parametric representation $\widetilde{c}(\widetilde{s})$ obtained via the change of parameter $\widetilde{s} = -s$.

Of course the following proposition is particularly useful:

Proposition 6.9.8 *Consider a Riemann patch of class* $\mathcal{C}^2$

$$g_{ij} : U \longrightarrow \mathbb{R}, \quad 1 \leq i, j \leq 2$$

and a regular curve of class $\mathcal{C}^2$

$$c\colon\,]a,b[\longrightarrow U,\qquad t\mapsto\big(c^1(t),c^2(t)\big)$$

given in arbitrary representation. The geodesic curvature of c *is equal to*

$$\kappa_g=-\frac{(\frac{\nabla c'}{dt}|\eta)_c}{\|c'\|_c^2}$$

where η *is the normal vector field of the curve* (*see Proposition* 6.5.1).

Proof Let us freely use the notation and the results in the proof of Proposition 6.3.5: we thus write $s=\sigma(t)$ and $\overline{c}(s)=(c\circ\sigma^{-1})(s)$ for the normal representation of the curve. Analogously, we write $\overline{\eta}(s)$ for the normal vector expressed as a function of the parameter s (see Proposition 6.5.1). Thus, by the proof of Proposition 6.3.5, we already know that

$$\big(\sigma^{-1}\big)'=\frac{1}{\|(c'\circ\sigma^{-1})(s)\|_{(c\circ\sigma^{-1})(s)}}=\frac{1}{\|c'(t)\|_{c(t)}}.$$

By Proposition 6.9.6, and using Proposition 6.7.4, the normal curvature in terms of the parameter s is then given by

$$\begin{aligned}
-\kappa_g&=\left(\frac{\nabla(c\circ\sigma^{-1})'}{ds}\middle|\eta\circ\sigma^{-1}\right)_{c\circ\sigma^{-1}}\\
&=\left(\frac{\nabla(c'\circ\sigma^{-1})\cdot(\sigma^{-1})'}{ds}\middle|\eta\right)_{c\circ\sigma^{-1}}\\
&=\left(\left(\frac{\nabla c'}{dt}\circ\sigma^{-1}\right)\cdot\big((\sigma-1)'\big)^2+\big(c'\circ\sigma^{-1}\big)\cdot\big(\sigma^{-1}\big)''\middle|\eta\right)_{c\circ\sigma^{-1}}\\
&=\left(\left(\frac{\nabla c'}{dt}\circ\sigma^{-1}\right)\cdot\big((\sigma-1)'\big)^2\middle|\eta\right)_{c\circ\sigma^{-1}}\\
&=\frac{((\frac{\nabla c'}{dt}\circ\sigma^{-1})|\eta)_{c\circ\sigma^{-1}}}{\|c'\circ\sigma^{-1}\|^2_{c\circ\sigma^{-1}}}
\end{aligned}$$

where the last but one equality holds because c' is orthogonal to η.

Putting $\sigma^{-1}(s)=t$ in these equalities, we get the formula of the statement. □

6.10 Geodesics

Imagine that you traveling on the Earth, around the equator. To achieve this, you have to proceed "straight on", without ever turning left or right. But nevertheless, by doing this you travel along a circle, because the equator *is* a circle. The point is that the "curvature vector" of this circle—the second derivative of a normal representation (see Example 2.9.5)—is oriented towards the center of the circle, and in the case of the equator, the center of the circle is also the center of the Earth. The "curvature vector" is thus perpendicular to the tangent plane to the Earth and so its orthogonal projection on that tangent plane is zero. The *geodesic curvature* of the equator is zero and this is the reason why you have the false impression of not turning at all when you proceed along the equator.

Definition 6.10.1 A *geodesic* in a Riemann patch of class $\mathcal{C}^2$ is a regular curve of class $\mathcal{C}^2$ whose geodesic curvature is zero at each point.

Notice at once that

Proposition 6.10.2 *In a Riemann patch of class $\mathcal{C}^2$, a regular curve of class $\mathcal{C}^2$ is a geodesic if and only if its tangent vector field is a parallel vector field.*

Proof By Lemma 6.8.2, there is no loss of generality in assuming that the curve is given in normal representation. By Definitions 6.10.1 and 6.9.5, being a geodesic is then equivalent to $\|\frac{\nabla c'}{ds}\| = 0$, which is the condition for being a parallel vector field (see Definition 6.8.1). □

The results that we already have yield at once a characterization of the geodesics:

Theorem 6.10.3 *Consider a Riemann patch of class $\mathcal{C}^2$*

$$g_{ij}\colon U \longrightarrow \mathbb{R}, \quad 1 \leq i, j \leq 2$$

and a regular curve of class $\mathcal{C}^2$

$$c\colon]a,b[\longrightarrow U, \qquad s \mapsto \left(c^1(s), c^2(s)\right)$$

given in normal representation. That curve is a geodesic if and only if

$$\frac{d^2c^k}{ds^2} + \sum_{ij} \frac{dc^i}{ds}\frac{dc^j}{ds}\Gamma_{ij}^k\left(c^1, c^2\right) = 0, \quad 1 \leq k \leq 2.$$

Proof By Definition 6.9.5, we must prove that $\|\frac{\nabla c'}{ds}\| = 0$, which is of course equivalent to $\frac{\nabla c'}{ds} = 0$, since at each point of U, the norm is that given by a scalar product (see Definition 6.4.4 and Notation 6.2.5). The result follows by Definition 6.7.3, putting $\xi = c'$. □

Example 6.10.4 The geodesics of a sphere are the great circles.

Proof The argument concerning the equator, at the beginning of this section, works for every great circle, proving that these are geodesics of the sphere.

Conversely, consider a geodesic on a sphere. There is no loss of generality in assuming that the center of the sphere is the origin of $\mathbb{R}^3$. Given a normal representation h of that geodesic viewed as a skew curve, we have h' in the tangent plane to the sphere (Lemma 5.5.1) and h'' perpendicular to that tangent plane (Definition 6.9.2). Therefore h'' is oriented along the radius of the sphere and the osculating plane (Definition 4.1.6) to the curve passes through the center of the sphere. But, since the center of the sphere is the origin of $\mathbb{R}^3$, h'' is also proportional to h. Let us write

$$h''(s) = \alpha(s)h(s).$$

By Proposition 4.5.1, the torsion of the geodesic is equal to

$$\tau = \frac{(h' \times h'' | h''')}{\|h''\|^2} = \frac{(h' \times h'' | \alpha' h + \alpha h')}{\|h''\|^2} = 0$$

because $h' \times h''$ is orthogonal to h', but also to h which is proportional to h''. So the torsion of the curve is equal to zero and by Proposition 4.5.3, the geodesic is a plane curve. The plane of this curve is thus also its osculating plane, which passes through the center of the sphere. So the geodesic lies on the intersection of the sphere with a plane through the center of the sphere. Therefore the geodesic is (a piece of) a great circle. □

Example 6.10.5 A straight line contained in a surface is always a geodesic.

Proof A straight line has a zero curvature vector (see Example 2.9.4). □

Example 6.10.6 The geodesics of the plane are the straight lines.

Proof The straight lines are geodesics by Example 6.10.5. Now as a surface, the plane is its own tangent plane at each point. But given a curve in the plane, its curvature vector is already in the plane, thus coincides with its orthogonal projection on the tangent plane. Therefore the curve is a geodesic if and only if its curvature vector is zero at each point. The result follows by Example 2.12.7. □

Example 6.10.7 The geodesics of the circular cylinder

$$g(\theta, z) = (\cos\theta, \sin\theta, z)$$

are:

1. for each fixed value θ_0, the rulings

$$z \mapsto (\cos\theta_0, \sin\theta_0, z);$$

2. for each fixed value z_0, the circular sections

$$\theta \mapsto (\cos\theta, \sin\theta, z_0);$$

3. for all values $r \neq 0$, $s \in \mathbb{R}$, the circular helices (see Example 4.5.4)

$$\theta \mapsto (\cos\theta, \sin\theta, r\theta + s).$$

Proof Going back to the proof of Example 6.6.1, we observe at once that the second partial derivatives of g are orthogonal to the first partial derivatives. Therefore the Christoffel symbols of the first kind are all equal to zero (Definition 6.6.2). By the fourth formula in Proposition 6.6.4, the Christoffel symbols of the second kind are all zero as well. This trivializes the equations in Theorem 6.10.3: a curve on the cylinder

$$c\colon]a,b[\longrightarrow \mathbb{R}, \qquad s \mapsto \big(c^1(s), c^2(s)\big)$$

such that $g \circ c$ is in normal representation is a geodesic when

$$\frac{d^2c^1}{ds^2} = 0, \qquad \frac{d^2c^2}{ds^2} = 0.$$

Integrating twice, we conclude that c^1 and c^2 are polynomials of degree 1. The geodesics are thus obtained as the deformations by g of the plane curves

$$s \mapsto (as + b, cs + d).$$

The case $a = 0 = c$ is excluded, since it is not a curve. When $a = 0$, $c \neq 0$, we obtain the ruling corresponding to $\theta_0 = b$. When $a \neq 0$, $c = 0$ we obtain the circular section corresponding to $z_0 = d$. When $a \neq 0 \neq c$, the change of parameter $t = as + b$ yields in the plane the parametric representation

$$t \mapsto \left(t, c\frac{t-b}{a} + d\right) = \left(t, \frac{c}{a}t - \frac{cb}{a} + d\right).$$

Putting

$$r = \frac{c}{a}, \qquad s = d - \frac{cb}{a}$$

this curve yields on the cylinder the circular helix of the statement. □

In fact, all surfaces admit geodesics, not just these obvious examples which tend to be the ones we immediately think of. Indeed:

Proposition 6.10.8 *Consider a Riemann patch of class $\mathcal{C}^k$, with $k \geq 3$*

$$g_{ij}\colon U \longrightarrow \mathbb{R}, \quad 1 \leq i, j \leq 2.$$

For each point of $(x_0^1, x_0^2) \in U$ *and every direction* $(\alpha, \beta) \neq (0,0)$, *there exists in a neighborhood of this point a unique geodesic of class* $\mathcal{C}^k$

$$c\colon]a,b[\longrightarrow U, \quad a < 0 < b$$

such that

$$c(0) = (x_0^1, x_0^2), \qquad c'(0) = (\alpha, \beta).$$

Proof We are looking for two functions c^1, c^2 of class $\mathcal{C}^k$ which are solutions of the second order differential equations in Theorem 6.10.3 and satisfy the initial conditions of the statement. Since all coefficients of the differential equations are of class $\mathcal{C}^{k-1}$, such a solution exists and is unique (see Proposition B.2.1). □

6.11 The Riemann Tensor

Both the normal curvature and the Gaussian curvature of a surface are expressed in terms of the six coefficients E, F, G, L, M, N (see Propositions 5.8.4 and 5.16.3). We have seen in Counterexample 6.6.1 that the three functions L, M, N are not Riemannian quantities and, in Counterexample 6.9.1, that the normal curvature is not a Riemannian notion. This might suggest that the Gaussian curvature is also not a Riemannian quantity. Perhaps unexpectedly, it is!

A very striking result, due to *Gauss* himself, is that the Gaussian curvature can be expressed as a function of E, F, G. So the Gaussian curvature *is* a Riemannian notion, while the normal curvature is not. To prove this, in view of the formula

$$\kappa_\tau = \frac{LN - M^2}{EG - F^2}$$

of Proposition 5.16.3, it suffices of course to prove that the quantity $LN - M^2$ can be expressed as a function of E, F, G. For this, let us switch back to the notation h_{ij} and g_{ij} of Definitions 6.6.2 and 6.1.1.

Definition 6.11.1 Consider a regular parametric representation of class $\mathcal{C}^3$ of a surface:

$$f: U \mapsto \mathbb{R}^3, \qquad (x^1, x^2) \mapsto f(x^1, x^2).$$

The *Riemann tensor* of this surface consists of the family of functions

$$R_{ijkl} = h_{jl}h_{ki} - h_{jk}h_{li}, \quad 1 \leq i, j, k, l \leq 2$$

where

$$\begin{pmatrix} h_{11} & h_{12} \\ h_{21} & h_{22} \end{pmatrix} = \begin{pmatrix} L & M \\ M & N \end{pmatrix}$$

are the coefficients of the second fundamental form of the surface (see Theorem 5.8.2).

Notice once more the appearance of the term *tensor*.

Lemma 6.11.2 *Under the conditions of Definition* 6.11.1, *all the components* R_{ijkl} *of the Riemann tensor are equal to one of the following quantities*:

$$LN - M^2, \qquad 0, \qquad -(LN - M^2).$$

Thus, knowing the metric tensor, the knowledge of the Riemann tensor is equivalent to the knowledge of the Gaussian curvature.

Proof Simply observe that

$$R_{1212} = R_{2121} = LN - M^2, \qquad R_{1221} = R_{2112} = -(LN - M^2)$$

while all other components are zero. □

Theorem 6.11.3 (Theorema Egregium, Gauss) *Under the conditions of Definition* 6.11.1, *the Riemann tensor is equal to*

$$R_{ijkl} = \frac{\partial \Gamma_{jli}}{\partial x^k} - \frac{\partial \Gamma_{jki}}{\partial x^l} + \sum_\alpha \left(\Gamma_{jk}^\alpha \Gamma_{li\alpha} - \Gamma_{jl}^\alpha \Gamma_{ki\alpha}\right).$$

In particular, the Riemann tensor can be expressed as a function of the sole coefficients of the metric tensor.

Proof Of course the last sentence in the statement will follow at once from the formula in the statement, since we already know the corresponding result for the Christoffel symbols (see Proposition 6.6.6). Let us therefore prove this formula.

Since the normal vector $\overrightarrow{n}$ has length 1, we can write equivalently

$$R_{ijkl} = (h_{jl}\overrightarrow{n}|h_{ki}\overrightarrow{n}) - (h_{jk}\overrightarrow{n}|h_{li}\overrightarrow{n}).$$

But by Definition 6.6.2

$$h_{ij}\overrightarrow{n} = \frac{\partial^2 f}{\partial x^i \partial x^j} - \Gamma_{ij}^1 \frac{\partial f}{\partial x^1} - \Gamma_{ij}^2 \frac{\partial f}{\partial x^2}.$$

Let us then replace $h_{ij}\overrightarrow{n}$ by the quantity on the right hand side, keeping in mind Definition 6.1.1 of the coefficients of the metric tensor and Definition 6.6.2 of the Christoffel symbols.

$$R_{ijkl} = \left(\frac{\partial^2 f}{\partial x^j \partial x^l} - \Gamma_{jl}^1 \frac{\partial f}{\partial x^1} - \Gamma_{jl}^2 \frac{\partial f}{\partial x^2} \middle| \frac{\partial^2 f}{\partial x^k \partial x^i} - \Gamma_{ki}^1 \frac{\partial f}{\partial x^1} - \Gamma_{ki}^2 \frac{\partial f}{\partial x^2}\right)$$

$$
\begin{aligned}
&- \left(\frac{\partial^2 f}{\partial x^j \partial x^k} - \Gamma_{jk}^1 \frac{\partial f}{\partial x^1} - \Gamma_{jk}^2 \frac{\partial f}{\partial x^2} \middle| \frac{\partial^2 f}{\partial x^l \partial x^i} - \Gamma_{li}^1 \frac{\partial f}{\partial x^1} - \Gamma_{li}^2 \frac{\partial f}{\partial x^2} \right) \\
&= \left(\frac{\partial^2 f}{\partial x^j \partial x^l} \middle| \frac{\partial^2 f}{\partial x^k \partial x^i} \right) - \left(\frac{\partial^2 f}{\partial x^j \partial x^k} \middle| \frac{\partial^2 f}{\partial x^l \partial x^i} \right) \\
&\quad - \Gamma_{jl1}\Gamma_{ki}^1 - \Gamma_{jl2}\Gamma_{ki}^2 - \Gamma_{ki1}\Gamma_{jl}^1 - \Gamma_{ki2}\Gamma_{jl}^2 \\
&\quad + \Gamma_{jk1}\Gamma_{li}^1 + \Gamma_{jk2}\Gamma_{li}^2 + \Gamma_{li1}\Gamma_{jk}^1 + \Gamma_{li2}\Gamma_{jk}^2 \\
&\quad + \Gamma_{jl}^1\Gamma_{ki}^1 g_{11} + \Gamma_{jl}^1\Gamma_{ki}^2 g_{12} + \Gamma_{jl}^2\Gamma_{ki}^1 g_{21} + \Gamma_{jl}^2\Gamma_{ki}^2 g_{22} \\
&\quad - \Gamma_{jk}^1\Gamma_{li}^1 g_{11} - \Gamma_{jk}^1\Gamma_{l2}^1 g_{12} - \Gamma_{jk}^2\Gamma_{li}^1 g_{21} - \Gamma_{jk}^2\Gamma_{li}^2 g_{22}.
\end{aligned}
$$

Let us now use the third formula in Proposition 6.6.4 to simplify this last expression. This formula allows us to combine the first and the third terms in the fourth line to obtain

$$\left(g_{11}\Gamma_{jl}^1 + g_{21}\Gamma_{jl}^2\right)\Gamma_{ki}^1 = \Gamma_{jl1}\Gamma_{ki}^1.$$

That quantity is then exactly the opposite of the first term in the second line. The same process allows us to simplify the second and fourth terms in the fourth line with the second term in the second line. Next, we can apply this process again to the last line and the first two terms in the third line. Eventually, the last four lines reduce to

$$-\Gamma_{ki1}\Gamma_{jl}^1 - \Gamma_{ki2}\Gamma_{jl}^2 + \Gamma_{li1}\Gamma_{jk}^1 + \Gamma_{li2}\Gamma_{jk}^2.$$

This is exactly the sum in α in the formula of the statement.

To conclude, it remains to check that

$$\frac{\partial \Gamma_{jli}}{\partial x^k} - \frac{\partial \Gamma_{jki}}{\partial x^l} = \left(\frac{\partial^2 f}{\partial x^j \partial x^l} \middle| \frac{\partial^2 f}{\partial x^k \partial x^i} \right) - \left(\frac{\partial^2 f}{\partial x^j \partial x^k} \middle| \frac{\partial^2 f}{\partial x^l \partial x^i} \right).$$

Indeed

$$
\begin{aligned}
\frac{\partial \Gamma_{jli}}{\partial x^k} - \frac{\partial \Gamma_{jki}}{\partial x^l} &= \frac{\partial}{\partial x^k}\left(\frac{\partial^2 f}{\partial x^j \partial x^l} \middle| \frac{\partial f}{\partial x^i} \right) - \frac{\partial}{\partial x^l}\left(\frac{\partial^2 f}{\partial x^j \partial x^k} \middle| \frac{\partial f}{\partial x^i} \right) \\
&= \left(\frac{\partial^3 f}{\partial x^j \partial x^l \partial x^k} \middle| \frac{\partial f}{\partial x^i} \right) + \left(\frac{\partial^2 f}{\partial x^j \partial x^l} \middle| \frac{\partial^2 f}{\partial x^i \partial x^k} \right) \\
&\quad - \left(\frac{\partial^3 f}{\partial x^j \partial x^k \partial x^l} \middle| \frac{\partial f}{\partial x^i} \right) - \left(\frac{\partial^2 f}{\partial x^j \partial x^k} \middle| \frac{\partial^2 f}{\partial x^i \partial x^l} \right) \\
&= \left(\frac{\partial^2 f}{\partial x^j \partial x^l} \middle| \frac{\partial^2 f}{\partial x^i \partial x^k} \right) - \left(\frac{\partial^2 f}{\partial x^j \partial x^k} \middle| \frac{\partial^2 f}{\partial x^i \partial x^l} \right)
\end{aligned}
$$

by the well-known property of commutation of partial derivatives. □

As you might now expect, we conclude this section with a corresponding definition:

Definition 6.11.4 Given a Riemann patch of class $\mathcal{C}^2$, the *Riemann tensor* is defined as being the family of functions

$$R_{ijkl} = \frac{\partial \Gamma_{jli}}{\partial x^k} - \frac{\partial \Gamma_{jki}}{\partial x^l} + \sum_\alpha \left(\Gamma_{jk}^\alpha \Gamma_{li\alpha} - \Gamma_{jl}^\alpha \Gamma_{ki\alpha}\right).$$

(See Definition 6.6.7.)

It is worth adding a comment.

Definition 6.11.5 Given a Riemann patch of class $\mathcal{C}^2$, the quantity

$$\kappa_\tau = \frac{R_{1212}}{g_{11}g_{22} - g_{21}g_{12}}$$

is called the *Gaussian curvature* of the Riemann patch.

This terminology is clearly inspired by Lemma 6.11.2 and its proof. This notion of Gaussian curvature makes perfect sense in the "restricted" context of our Definition 6.2.1, simply because Lemma 6.11.2 remains valid in this context (see Problem 6.18.1). However, the possibility of reducing the information given by the metric tensor to a single quantity κ_τ is a very specific peculiarity of the *Riemann patches of dimension* 2. This notion of *Gaussian curvature* does not extend to higher dimensional Riemann patches, as defined in Definition 6.17.6: in higher dimensions, the correct notion to consider is the full *Riemann tensor*.

6.12 What Is a Tensor?

The time has come to discuss the magic word *tensor*. A family of functions receives this "honorary label" when it transforms "elegantly" along a change of parameters. In fact, a formal, general and elegant theory of *tensors* must rely on a good *multilinear algebra* course; but this is beyond the scope of this book.

In Sect. 6.1 we have exhibited the *Riemann patch* corresponding to a specific parametric representation of a surface in $\mathbb{R}^3$, and we know very well that a given surface admits many equivalent parametric representations. However, up to now, we have not paid attention to the question: *What are equivalent Riemann patches?*

Consider a regular surface of class $\mathcal{C}^3$ admitting two equivalent parametric representations

$$f: U \longrightarrow \mathbb{R}^3; \qquad (x^1, x^2) \mapsto f(x^1, x^2)$$
$$\widetilde{f}: U \longrightarrow \mathbb{R}^3; \qquad (\widetilde{x}^1, \widetilde{x}^2) \mapsto \widetilde{f}(\widetilde{x}^1, \widetilde{x}^2).$$

To be able to handle the corresponding change of parameters in our arguments, we have to fix a notation for it. Up to now, we have always used a notation like

$$(\widetilde{x}^1,\widetilde{x}^2)=\varphi(x^1,x^2)=(\varphi^1(x^1,x^2),\varphi^2(x^1,x^2)).$$

Of course if you have many changes of parameters to handle, using various notations such as φ, ψ, θ, τ and so on rapidly becomes unwieldy. Riemannian geometry uses a very standard and efficient notation for a change of parameters:

$$(\widetilde{x}^1,\widetilde{x}^2)=(\widetilde{x}^1(x^1,x^2),\widetilde{x}^2(x^1,x^2)).$$

Of course such a notation is a little ambiguous, since it uses the same symbol for the coordinates $\widetilde{x}^i$ and for the functions $\widetilde{x}^i$. However, in practice no confusion occurs. In fact, this notation significantly clarifies the language. When you have several changes of coordinates, the notation $\widetilde{x}^i(x^1,x^2)$ reminds you at once of both systems of coordinates involved in the question, while a notation such as $\varphi^i(x^1,x^2)$ recalls only one of them.

Proposition 6.12.1 *Consider a regular surface of class $\mathcal{C}^3$ admitting the equivalent parametric representations*

$$\begin{aligned} f\colon U &\longrightarrow \mathbb{R}^3; && (x^1,x^2)\mapsto f(x^1,x^2)\\ \widetilde{f}\colon U &\longrightarrow \mathbb{R}^3; && (\widetilde{x}^1,\widetilde{x}^2)\mapsto \widetilde{f}(\widetilde{x}^1,\widetilde{x}^2). \end{aligned}$$

Write further

$$(g_{ij})_{i,j} \quad \text{and} \quad (\widetilde{g}_{ij})_{i,j}$$

for the corresponding metric tensors. Under these conditions

$$\widetilde{g}_{ij}=\sum_{k,l} g_{kl}\frac{\partial x^k}{\partial\widetilde{x}^i}\frac{\partial x^l}{\partial\widetilde{x}^j}.$$

Proof With the notation just explained for the changes of coordinates, we have

$$\widetilde{f}(\widetilde{x}^1,\widetilde{x}^2)=f(x^1(\widetilde{x}^1,\widetilde{x}^2),x^2(\widetilde{x}^1,\widetilde{x}^2)).$$

It follows that

$$\frac{\partial\widetilde{f}}{\partial\widetilde{x}^i}=\frac{\partial f}{\partial x^1}\frac{\partial x^1}{\partial\widetilde{x}^i}+\frac{\partial f}{\partial x^2}\frac{\partial x^2}{\partial\widetilde{x}^i}=\sum_k\frac{\partial f}{\partial x^k}\frac{\partial x^k}{\partial\widetilde{x}^i}.$$

This implies

$$\left(\frac{\partial\widetilde{f}}{\partial\widetilde{x}^i}\middle|\frac{\partial\widetilde{f}}{\partial\widetilde{x}^j}\right)=\sum_{k,l}\left(\frac{\partial f}{\partial x^k}\middle|\frac{\partial f}{\partial x^l}\right)\frac{\partial x^k}{\partial\widetilde{x}^i}\frac{\partial x^l}{\partial\widetilde{x}^j}$$

that is

$$\widetilde{g}_{ij} = \sum_{k,l} g_{kl} \frac{\partial x^k}{\partial \widetilde{x}^i} \frac{\partial x^l}{\partial \widetilde{x}^j}$$

which is the formula of the statement. □

This elegant formula is what one calls the transformation formula for a tensor which is *twice covariant*. Forgetting about this new jargon "covariant" for the time being, let us repeat the same for the inverse metric tensor (see Definition 6.2.3).

Proposition 6.12.2 *Consider a regular surface of class $\mathcal{C}^3$ admitting the equivalent parametric representations*

$$f : U \longrightarrow \mathbb{R}^3; \qquad (x^1, x^2) \mapsto f(x^1, x^2)$$
$$\widetilde{f} : U \longrightarrow \mathbb{R}^3; \qquad (\widetilde{x}^1, \widetilde{x}^2) \mapsto \widetilde{f}(\widetilde{x}^1, \widetilde{x}^2).$$

Write further

$$\left(g^{ij}\right)_{i,j} \quad \text{and} \quad \left(\widetilde{g}^{ij}\right)_{i,j}$$

for the corresponding inverse metric tensors. Under these conditions

$$\widetilde{g}^{ij} = \sum_{k,l} g^{kl} \frac{\partial \widetilde{x}^i}{\partial x^k} \frac{\partial \widetilde{x}^j}{\partial x^l}.$$

Proof As already observed in the proof of Proposition 6.12.1:

$$\frac{\partial f}{\partial \widetilde{x}^i} = \sum_k \frac{\partial f}{\partial x^k} \frac{\partial x^k}{\partial \widetilde{x}^i}.$$

The matrix

$$B = \begin{pmatrix} \frac{\partial x^1}{\partial \widetilde{x}^1} & \frac{\partial x^1}{\partial \widetilde{x}^2} \\ \frac{\partial x^2}{\partial \widetilde{x}^1} & \frac{\partial x^2}{\partial \widetilde{x}^2} \end{pmatrix}$$

is thus the change of coordinates matrix between the two bases of partial derivatives in the tangent plane (see Sect. 2.20 in [4], *Trilogy II*).

For the needs of this proof, let us write T for the matrix given by the metric tensor. The formula of Proposition 6.12.1 becomes simply

$$\widetilde{T} = B^t T B.$$

Taking the inverses of both sides, we get

$$\widetilde{T}^{-1} = B^{-1} T^{-1} \left(B^t\right)^{-1} = B^{-1} T^{-1} \left(B^{-1}\right)^t$$

since $(B^t)^{-1} = (B^{-1})^t$. But the same argument as above shows that

$$B^{-1} = \begin{pmatrix} \frac{\partial \widetilde{x}^1}{\partial x^1} & \frac{\partial \widetilde{x}^1}{\partial x^2} \\ \frac{\partial \widetilde{x}^2}{\partial x^1} & \frac{\partial \widetilde{x}^2}{\partial x^2} \end{pmatrix}.$$

Therefore the transformation formula for the inverse metric tensor is

$$\widetilde{g}^{ij} = \sum_{k,l} g^{kl} \frac{\partial \widetilde{x}^i}{\partial x^k} \frac{\partial \widetilde{x}^j}{\partial x^l}$$

as announced in the statement. □

Compare now the two formulas in Propositions 6.12.1 and 6.12.2. They are very similar of course, but nevertheless with a major difference! It will be convenient for us to call $(\widetilde{x}^1, \widetilde{x}^2)$ the "new" coordinates and (x^1, x^2) the "old" coordinates.

- In the case of the metric tensor, the coefficients in the change of parameters formula are the derivatives of the "old" coordinates with respect to the "new" coordinates. One says that the tensor is *covariant in both variables* or simply, *twice covariant*. One uses *lower indices* to indicate the covariant indices of a tensor.
- In the case of the inverse metric tensor, the coefficients in the change of parameters formula are the derivatives of the "new" coordinates with respect to the "old" coordinates. One says that the tensor is *contravariant in both variables* or simply, *twice contravariant*. One uses *upper indices* to indicate the contravariant indices of a tensor.

This already clarifies some points of notation and terminology. However, this still does not tell us what a tensor is. As mentioned earlier, in order to give an elegant definition we would need some multi-linear algebra. Nevertheless, as far as surfaces in $\mathbb{R}^3$ are concerned, we can at least take as our definition a famous criterion characterizing the tensors of Riemannian geometry. For simplicity, we state the definition in the particular case of a tensor two times covariant and three times contravariant, but the generalization is obvious.

Definition 6.12.3 Suppose that for each parametric representation of class $\mathcal{C}^3$ of a given surface of $\mathbb{R}^3$ you have a corresponding family of continuous functions

$$T_{ij}^{klm} : U \longrightarrow \mathbb{R}$$

with two lower indices and three upper indices. These families of continuous functions are said to constitute a *tensor covariant in the indices* i, j *and contravariant in the indices* k, l, m when, given any two equivalent parametric representations f, $\widetilde{f}$—and with obvious notation—these functions transform into each other via the formulas

$$\widetilde{T}_{ij}^{klm} = \sum_{r,s,t,u,v} T_{rs}^{tuv} \frac{\partial x^r}{\partial \widetilde{x}^i} \frac{\partial x^s}{\partial \widetilde{x}^j} \frac{\partial \widetilde{x}^k}{\partial x^t} \frac{\partial \widetilde{x}^l}{\partial x^u} \frac{\partial \widetilde{x}^m}{\partial x^v}.$$

Of course, an analogous definition holds for a tensor α times covariant and β times contravariant, for any two integers α, β.

You should now have a clear idea why some quantities are designated as *tensors* and others are not. For example, the *Riemann tensor* of Theorem 6.11.3 is a tensor four times covariant (Problem 6.18.2) while the Christoffel symbols do not constitute a tensor (Problem 6.18.4). This also indicates why some indices are put upside and others downside.

One should be able to guess now why we use upper indices to indicate the coordinates of a point or the coordinates of a tangent vector field.

Proposition 6.12.4 *Consider a regular curve c on a regular surface of class $\mathcal{C}^3$ in $\mathbb{R}^3$. In a change of parameters and with obvious notation, a vector field ξ along the curve c, tangent to the surface, transforms via the formula*

$$\widetilde{\xi}^k = \sum_i \xi^i \frac{\partial \widetilde{x}^k}{\partial x^i}.$$

Proof One has

$$\begin{aligned}
\xi &= \xi^1 \frac{\partial f}{\partial x^1} + \xi^2 \frac{\partial f}{\partial x^2} \\
&= \xi^1 \Bigl(\sum_k \frac{\partial \widetilde{f}}{\partial \widetilde{x}^k} \frac{\partial \widetilde{x}^k}{\partial x^1}\Bigr) + \xi^2 \Bigl(\sum_k \frac{\partial \widetilde{f}}{\partial \widetilde{x}^k} \frac{\partial \widetilde{x}^k}{\partial x^2}\Bigr) \\
&= \Bigl(\sum_i \xi^i \frac{\partial \widetilde{x}^1}{\partial x^i}\Bigr) \frac{\partial \widetilde{f}}{\partial \widetilde{x}^1} + \Bigl(\sum_i \xi^i \frac{\partial \widetilde{x}^2}{\partial x^i}\Bigr) \frac{\partial \widetilde{f}}{\partial \widetilde{x}^2}
\end{aligned}$$

and this proves the formula of the statement. □

Of course a vector field along a curve is not a tensor in the sense of Definition 6.12.3, because it is not defined on the whole subset U. Nevertheless, its transformation law along the curve is exactly that of a tensor one time contravariant. This explains the use of upper indices.

In particular, the components of the tangent vector field to the curve c itself should be written with upper indices: $c' = ((c^1)', (c^2)')$. But then of course, the components of c should use upper indices as well $c = (c^1, c^2)$. To be consistent, when writing the parametric equations of the curve c

$$\begin{cases} x^1 = c^1(t) \\ x^2 = c^2(t) \end{cases}$$

we should use upper indices as well for the two coordinates x^1 and x^2.

Let us conclude this long discussion on tensors by giving the answer to the question raised at the beginning of this section: *What are equivalent Riemann patches?*

Keeping in mind that for a surface of class $\mathcal{C}^{k+1}$ in $\mathbb{R}^3$ the coefficients g_{ij} of the metric tensor are functions of class $\mathcal{C}^k$ (see Definition 6.1.1), we make the following definition:

Definition 6.12.5 Two Riemann patches of class $\mathcal{C}^k$

$$g_{ij}\colon U \longrightarrow \mathbb{R}, \qquad (x^1,x^2) \mapsto g(x^1,x^2)$$
$$\widetilde{g}_{ij}\colon U \longrightarrow \mathbb{R}, \qquad (\widetilde{x}^1,\widetilde{x}^2) \mapsto \widetilde{g}(\widetilde{x}^1,\widetilde{x}^2)$$

are *equivalent* in class $\mathcal{C}^k$ when there exists a change of parameters of class $\mathcal{C}^{k+1}$ (that is, a bijection of class $\mathcal{C}^{k+1}$ with inverse of class $\mathcal{C}^{k+1}$)

$$\varphi\colon U \longrightarrow \widetilde{U}, \qquad (x^1,x^2) \mapsto \big(\widetilde{x}^1(x^1,x^2), \widetilde{x}^2(x^1,x^2)\big)$$

such that

$$g_{ij} = \sum_{k,l} \widetilde{g}_{kl} \frac{\partial \widetilde{x}^k}{\partial x^i} \frac{\partial \widetilde{x}^l}{\partial x^j}.$$

As expected:

Proposition 6.12.6 *A change of parameters φ as in Definition 6.12.5 is a Riemannian isometry, that is, respects lengths and angles in the sense of the Riemannian metric.*

Proof Consider a curve

$$c\colon \,]a,b[\, \to U, \qquad t \mapsto c(t).$$

Under the conditions of Definition 6.12.5, the length of an arc of the curve in $\widetilde{U}$ represented by $\varphi \circ c$ is given by

$$\int_{t_0}^{t_1} \sqrt{\sum_{kl} \widetilde{g}_{kl} \frac{d(\widetilde{x}^k \circ c)}{dt} \frac{d(\widetilde{x}^l \circ c)}{dt}}\, dt = \int_{t_0}^{t_1} \sqrt{\sum_{ijkl} \widetilde{g}_{kl} \frac{\partial \widetilde{x}^k}{\partial x^i} \frac{dc^i}{dt} \frac{\partial \widetilde{x}^l}{\partial x^j} \frac{dc^j}{dt}}\, dt$$
$$= \int_{t_0}^{t_1} \sqrt{\sum_{ij} g_{ij} \frac{dc^i}{dt} \frac{dc^j}{dt}}\, dt$$

and this last formula expresses precisely the length of the curve c in U.

The proof concerning the preservation of angles is perfectly analogous. □

Notice that already for a Riemann patch of class $\mathcal{C}^0$, the form of the change of parameters requires that it be of class $\mathcal{C}^1$. This is another way to justify the "jump" of one unit in the classes of differentiability.

We are almost done. But you are still entitled to ask an intriguing question. If the Christoffel symbols are not tensors, how do we decide to use upper or lower indices? There is another convention in Riemannian geometry: a convention which, deliberately, has not been used in this chapter, and which requires an appropriate choice of position of the indices.

Convention 6.12.7 (Abbreviated Notation) In Riemannian geometry, when in a given term of a formula, the same index appears once as an upper index and once as a lower index, it is understood that a sum is taken over all the possible values of this index.

For example, following this convention, the formula giving the components of the covariant derivative of a tangent vector field (see Definition 6.7.3)

$$\frac{d\xi^k}{dt} + \sum_{i,j} \xi^i \frac{dc^j}{dt} \Gamma_{ij}^k, \qquad 1 \le k \le 2$$

is generally simply written as

$$\frac{d\xi^k}{dt} + \xi^i \frac{dc^j}{dt} \Gamma_{ij}^k, \qquad 1 \le k \le 2$$

because both indices i and j appear once as an upper index and once as a lower index in the "second" term. Notice that the index k appears twice as an upper index and moreover in two different terms: thus no sum is to be taken on this index. It is easy to see why we did not use this convention in this first approach of Riemannian geometry.

6.13 Systems of Geodesic Coordinates

Once again, let us support our intuition with the case of the Earth, regarded as a sphere. The most traditional system of coordinates is in terms of the *latitude* and the *longitude*. Consider the corresponding "geographical map" as in Example 5.1.6

$$f(\tau, \theta) = (\cos\tau \cos\theta, \cos\tau \sin\theta, \sin\tau)$$

where τ is the latitude and θ is the longitude.

- The *equator* is really the "base curve" of the whole system of coordinates: the curve given by $\tau = 0$; this is a great circle on the sphere, that is, a *geodesic* (see Example 6.10.4). Observe that $f(0, \theta)$ is a normal representation of the equator, because the radius of the sphere has been chosen to be equal to 1.
- The curves $\theta = k$ on the sphere, with k constant, are the *meridians*: they are great circles, thus *geodesics*, and moreover they are *orthogonal to the equator*. Observe that $f(\tau, \theta_0)$ is again a normal representation of the meridian with fixed longitude θ_0, again because the radius of the sphere is equal to 1.

- The curves $\tau = k$ on the sphere, with k constant, are the so-called *parallels*; they are not great circles (except for the equator), thus they are not geodesics; but they are *orthogonal to all the meridians*.

This is thus a very particular system of coordinates of which we can expect many properties and advantages. One calls such a system a *system of geodesic coordinates*.

A system of geodesic coordinates exists in a neighborhood of each point of a "good" surface. Let us establish this result in the general context of a Riemann patch.

Theorem 6.13.1 *Consider a regular curve $\mathcal{C}$ passing through a point P in a Riemann patch. Assume that both the Riemann patch and the curve are of class $\mathcal{C}^m$, with $m \geq 2$. There exists a connected open neighborhood of P such that the Riemann patch, restricted to this neighborhood, is equivalent in class $\mathcal{C}^{m-1}$ to a Riemann patch*

$$g_{ij} : U \longrightarrow \mathbb{R}, \qquad (x^1, x^2) \mapsto g_{ij}(x^1, x^2), \quad 1 \leq i, j \leq 2$$

with the following properties:

1. *the point P has coordinates* $(0, 0)$;
2. *the curve $\mathcal{C}$ is the curve $x^1 = 0$ and is now given in normal representation*;
3. *the curves $x^2 = k$, with k constant, are geodesics in normal representation*;
4. *the curves $x^1 = l$, with l constant, are orthogonal to the curves $x^2 = k$, with k constant*;
5. *at all points of U*

$$g_{11} = 1, \qquad g_{21} = 0 = g_{12}, \qquad g_{22} > 0$$

and also

$$g^{11} = 1, \qquad g^{21} = 0 = g^{12}, \qquad g^{22} > 0;$$

6. *at all points of U*

$$\Gamma_{211} = \Gamma_{121} = \Gamma_{112} = \Gamma_{111} = 0; \qquad \Gamma_{11}^2 = \Gamma_{11}^1 = \Gamma_{21}^1 = \Gamma_{12}^1 = 0$$

while

$$\Gamma_{222} = \frac{1}{2}\frac{\partial g_{22}}{\partial x^2}, \qquad \Gamma_{212} = \Gamma_{122} = \frac{1}{2}\frac{\partial g_{22}}{\partial x^1}, \qquad \Gamma_{221} = -\frac{1}{2}\frac{\partial g_{22}}{\partial x^1},$$

and

$$\Gamma_{22}^2 = \frac{1}{2g_{22}}\frac{\partial g_{22}}{\partial x^2}, \qquad \Gamma_{21}^2 = \Gamma_{12}^2 = \frac{1}{2g_{22}}\frac{\partial g_{22}}{\partial x^1}, \qquad \Gamma_{22}^1 = -\frac{1}{2}\frac{\partial g_{22}}{\partial x^1};$$

7. *when moreover the original curve c is a geodesic*

$$g_{22}(0, x^2) = 1, \qquad \frac{\partial g_{22}}{\partial x^1}(0, x^2) = 0$$

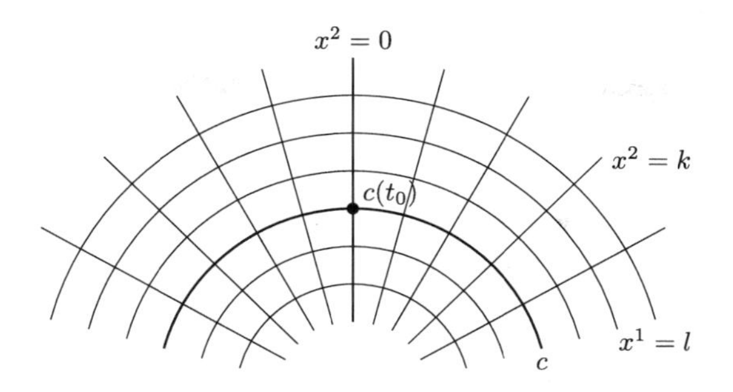

Fig. 6.2

and

$$\Gamma_{ij}^{k}(0, x^2) = 0, \qquad \Gamma_{ijk}(0, x^2) = 0, \quad 1 \le i, j, k \le 2.$$

A system of coordinates satisfying conditions 1 *to* 6 *is called a* geodesic system of coordinates. *When moreover it satisfies condition* 7, *it is called a* Fermi system of geodesic coordinates.

Proof Let us write

$$\widetilde{g}_{ij} : \widetilde{U} \longrightarrow \mathbb{R}, \qquad (\widetilde{x}^1, \widetilde{x}^2) \longrightarrow \widetilde{g}_{ij}(\widetilde{x}^1, \widetilde{x}^2)$$

for the original Riemann patch and

$$c \colon]a, b[\longrightarrow \widetilde{U}, \qquad t \mapsto c(t)$$

for the given curve $\mathcal{C}$. Let us write further $P = c(t_0)$. Follow the construction above on Fig. 6.2.

By Proposition 6.3.5, there is no loss of generality in assuming that the curve $\mathcal{C}$ is given in normal representation with P as origin, thus $P = c(0)$. Under these conditions $c'(t)$ becomes a vector of norm 1 (see Definition 6.3.4).

For each value $t \in]a, b[$, we consider the normal vector $\eta(t)$ to the curve (see Proposition 6.5.1), which is thus a vector of norm 1 orthogonal to $c'(t)$. By Proposition 6.10.8, in a neighborhood of $c(t)$, there exists in the Riemann patch a unique geodesic $h_t(s)$ of class $\mathcal{C}^m$ through $c(t)$ in the direction $\eta(t)$, satisfying an initial condition that we choose to be $h_t(0) = c(t)$. We are interested in the function

$$\varphi(s, t) = h_t(s)$$

which we want to become the expected change of parameters of class $\mathcal{C}^{m-2}$ in a neighborhood of $c(0)$. Since the coefficients of the equations in Proposition 6.10.8 are of class $\mathcal{C}^{m-1}$ and c is of class $\mathcal{C}^m$, the function φ is indeed defined, and of class $\mathcal{C}^m$, on a neighborhood of $(0, 0)$ (see Proposition B.3.2). But to be a good change of parameters, the inverse of φ should also be of class $\mathcal{C}^m$.

Let us compute the partial derivatives of the function φ at the point $(0,0)$:

$$\frac{\partial\varphi}{\partial s}(0,0)=\frac{h_0(s)}{ds}=\eta(0),\qquad \frac{\partial\varphi}{\partial t}(0,0)=\frac{dh_t(0)}{dt}(0,0)=\frac{c(t)}{dt}(0)=c'(0).$$

By regularity of c and Proposition 6.5.1, $c'(0)$ and $\eta(0)$ are perpendicular and of length 1 with respect to the scalar product $(-|-)_{c(0)}$, thus linearly independent. By the *Local Inverse Theorem* (see Theorem 1.3.1), the function φ is thus invertible on some neighborhood U' of $(0,0)$, with an inverse which is still of class $\mathcal{C}^m$. There is no loss of generality in choosing U' open and connected. In this way φ becomes a homeomorphism

$$\varphi\colon U'\longrightarrow\varphi(U').$$

We simply define $U=\varphi(U')$ and use the notation (x^1,x^2) instead of (s,t). Thus our change of parameters φ is now

$$U'\longrightarrow U,\qquad (\widetilde{x}^1,\widetilde{x}^2)\mapsto(s,r)=\big(x^1(\widetilde{x}^1,\widetilde{x}^1),x^2(\widetilde{x}^1,\widetilde{x}^2)\big).$$

Of course there is no difficulty in providing U with the structure of a Riemann patch equivalent to that given by the $\widetilde{g}_{ij}$ on $\widetilde{U}$. With Definition 6.12.5 in mind, simply define

$$g_{ij}=\sum_{k,l}\widetilde{g}_{kl}\frac{\partial\widetilde{x}^k}{\partial x^i}\frac{\partial\widetilde{x}^l}{\partial x^j}.$$

With the notation of Proposition 6.12.3, this definition can be re-written as

$$T=B^t\widetilde{T}B.$$

As observed in the proof of Proposition 6.12.3, the matrix B is that of a change of basis, while $\widetilde{T}$ is at each point the matrix of a scalar product in $\mathbb{R}^2$ (see Notation 6.2.5). By Corollary G.1.4 in [4], *Trilogy II*, T is then at each point the matrix of the same scalar product expressed in another base: it is thus a symmetric definite positive matrix. Therefore the g_{ij} on U constitute a Riemann patch equivalent in class $\mathcal{C}^{m-1}$ to that of the $\widetilde{g}_{ij}$ on U'.

By construction, the curve $x^1=0$ is the curve $h_t(0)$, that is the original curve $c(t)$.

Also by construction, the curves $x^2=k$, with k constant, are the curves $h_k(s)$, which are geodesics given in normal representation.

Next, we prove that $g_{11}=1$. With Notation 6.2.5,

$$\begin{aligned}g_{11}(s,t)&=\sum_{k,l}\widetilde{g}_{kl}\big(h_t(s)\big)\frac{\partial\widetilde{x}^k}{\partial x^1}(s,t)\frac{\partial\widetilde{x}^l}{\partial x^1}(s,t)\\&=\sum_{k,l}\widetilde{g}_{kl}\big(h_t(s)\big)\big(h_t^k\big)'(s)\big(h_t^l\big)'(s)\end{aligned}$$

$$= \left\| h_t'(s) \right\|_{h_t(s)}$$
$$= 1$$

because each $h_t(s)$ is in normal representation (see Definition 6.3.4).

Next, we turn our attention to the Christoffel symbols. The curve $x^2 = k$ is represented by

$$x^1 \mapsto \overline{h}_k(x^1) = (x^1, k).$$

Differentiating with respect to $s = x^1$, we obtain

$$\frac{\partial \overline{h}^2}{\partial x^1} = 0, \qquad \frac{\partial^2 \overline{h}^2}{\partial (x^1)^2} = 0, \qquad \frac{\partial \overline{h}^1}{\partial x^1} = 1, \qquad \frac{\partial^2 \overline{h}^1}{\partial (x^1)^2} = 0.$$

Since this curve $x^2 = k$ is a geodesic in normal representation, it satisfies the system of differential equations of Theorem 6.10.3. The observations that we have just made show that this system reduces simply to its terms in $(i, j) = (1, 1)$, that is,

$$\Gamma_{11}^2(x^1, k) = 0, \qquad \Gamma_{11}^1(x^1, k) = 0.$$

Since this holds for every value k, this proves condition 5 of the statement.

Now the case of $g_{12} = g_{21}$. By Definition 6.6.7, we have at all points

$$0 = \Gamma_{11}^1 = \sum_l g^{1l} \left(\frac{\partial g_{1l}}{\partial x^1} + \frac{\partial g_{l1}}{\partial x^1} - \frac{\partial g_{11}}{\partial x^l} \right).$$

Keeping in mind that $g_{11} = 1$ while $g_{12} = g_{21}$, which also forces $g^{12} = g^{21}$ (the inverse of a symmetric matrix is symmetric), this equality reduces to

$$2g^{21} \frac{\partial g_{21}}{\partial x^1} = 0.$$

Introducing the value of g^{21} (see the proof of Proposition 6.2.4) into this equality, we obtain

$$2 \frac{-g_{21}}{g_{22}g_{11} - g_{21}g_{11}} \frac{\partial g_{21}}{\partial x^1} = 0.$$

We know that $g_{11}g_{22} - g_{12}g_{21} \neq 0$ (Proposition 6.2.1); the equality is thus equivalent to

$$2g_{21} \frac{\partial g_{21}}{\partial x^1} = 0.$$

But this can be re-written as

$$\frac{\partial (g_{21})^2}{\partial x^1} = 0.$$

This proves that $g_{21}(x^1, x^2)$ is a constant function of x^1: thus to conclude that $g_{21} = 0$, it suffices thus to prove that $g_{21}(0, x^2) = 0$. By definition of g_{21} and using the values of the partial derivatives of the change of parameters φ, we indeed obtain

$$\begin{aligned} g_{21}(0, x^2) &= \sum_{k,l} \widetilde{g}_{kl}\big(c(x^2)\big) \frac{\partial \widetilde{x}^k}{\partial x^2}(0, x^2) \frac{\partial \widetilde{x}^l}{\partial x^1}(0, x^2) \\ &= \sum_{k,l} \widetilde{g}_{kl}\big(c(t)\big)\big(c^k\big)'(t)\eta^l(t) \\ &= \big(c'(t)\big|\eta(t)\big)_{c(t)} \\ &= 0. \end{aligned}$$

So $g_{21}(0, x^2) = 0$ and as we have seen, this implies $g_{21} = g_{12} = 0$. Since we know already that $g_{11} = 1$, this forces $g_{22} > 0$ by positivity of the metric tensor (see Definition 6.2.1).

The metric tensor is thus a diagonal matrix; therefore its inverse (see Definition 6.2.3) is obtained by taking the inverses of the diagonal elements and thus

$$g^{11} = 1, \qquad g^{12} = 0 = g^{21}, \qquad g^{22} = \frac{1}{g_{22}}.$$

Since g_{11}, g_{12} and g_{21} are constant, their partial derivatives are zero. Considering the definition of the Christoffel symbols of the first kind (see Definition 6.6.7), only the partial derivatives of g_{22} remain: this gives at once the formulas of the statement concerning the symbols Γ_{ijk} and as an immediate consequence, the formulas concerning the symbols Γ_{ij}^k.

Saying that the curves $x^1 = l$, $x^2 = k$, are orthogonal means

$$\begin{pmatrix} 1 & 0 \end{pmatrix} \begin{pmatrix} g_{11} & g_{12} \\ g_{21} & g_{22} \end{pmatrix} \begin{pmatrix} 0 \\ 1 \end{pmatrix} = 0$$

which is trivially the case since $g_{12} = 0 = g_{21}$. This concludes the proof in the case of an arbitrary base curve c.

Let us now suppose that this curve c is itself a geodesic. In terms of the coordinates (x^1, x^2), we have $t = x^2$ and the curve c is simply $c(x^2) = (0, x^2)$. By Theorem 6.10.3 we have

$$\frac{d^2c^k}{dx^2}(x^2) + \sum_{ij} \frac{dc^i}{dx^2}(x^2) \frac{dc^j}{dx^2}(x^2) \Gamma_{ij}^k(0, x^2) = 0, \quad 1 \le k \le 2.$$

But since $c(x^2) = (0, x^2)$, this reduces to

$$\Gamma_{22}^2(0, x^2) = 0, \qquad \Gamma_{22}^1(0, x^2) = 0.$$

By Proposition 6.7.4.4

$$\|\eta(t)\|_{c(t)} = 1 \quad \Longrightarrow \quad 2\left(\frac{\nabla\eta}{dt}\middle|\eta\right)_c = 0.$$

On the other hand by Proposition 6.7.4

$$0 = (\eta|c')_c \quad \Longrightarrow \quad 0 = \left(\frac{\nabla\eta}{dt}\middle|c'\right)_c + \left(\eta\middle|\frac{\nabla c'}{dt}\right)_c = \left(\frac{\nabla\eta}{dt}\middle|c'\right)_c;$$

the last equality holds because $c(t)$ is a geodesic in normal representation: this implies that c' is a parallel vector field (see Proposition 6.10.2), thus $\frac{\nabla c'}{dt} = 0$ by Definition 6.8.1. But, still by Proposition 6.5.1 and normality of the representation, c' and η constitute at each point an orthonormal basis of $\mathbb{R}^2$ for the scalar product $(-|-)_c$. The orthogonality of $\frac{\nabla\eta}{dt}$ to both c' and η implies

$$\frac{\nabla\eta}{dt}(t) = 0$$

for all values of $t = x^2$. By Definition 6.7.3 we have

$$\frac{d\eta^k}{dx^2}(x^2) + \sum_{ij} \eta^i(x^2)\frac{dc^j}{dx^2}(x^2)\Gamma_{ij}^k(0, x^2), \quad 1 \le k \le 2.$$

But in terms of the coordinates (x^1, x^2), $\eta(x^2) = (1, 0)$ while $c(x^2) = (0, x^2)$. Therefore the two equalities reduce to

$$\Gamma_{12}^1(0, x^2) = 0, \qquad \Gamma_{12}^2(0, x^2) = 0.$$

Of course this also forces

$$\Gamma_{21}^1(0, x^2) = 0, \qquad \Gamma_{21}^2(0, x^2) = 0$$

by Proposition 6.6.8. The third condition in this same proposition also shows that $\Gamma_{ijk}(0, x^2) = 0$ for all indices. □

Corollary 6.13.2 *Consider a Riemann patch of class C^m, with $m \ge 2$*

$$g_{ij}\colon V \longrightarrow \mathbb{R}, \quad 1 \le i, j \le 2.$$

Suppose that:

1. *the curves $x^2 = k$ are geodesics in normal representation;*
2. *each of these curves cuts the curve $x^1 = 0$ orthogonally.*

Under these conditions, (x^1, x^2) is already a system of geodesic coordinates.

Proof Simply observe that in the proof of Theorem 6.13.1, the change of parameters φ is the identity and therefore, is trivially valid on the whole of V. □

At the beginning of Sect. 6.10, we introduced geodesics via the intuition that they are the curves on the surface along which you have the impression of travelling in a straight line without ever turning left or right. As a consequence of the existence of systems of geodesic coordinates, let us now prove a precise result which reinforces the intuition that geodesics are "the best substitute for straight lines" on a surface.

Theorem 6.13.3 *Locally, in a Riemann patch of class $\mathcal{C}^2$, a geodesic is the shortest regular curve joining two of its points.*

Proof We consider a Riemann patch

$$g_{ij}: U \longrightarrow \mathbb{R}, \quad 1 \leq i, j \leq 2$$

and a geodesic

$$h:]a, b[\longrightarrow V, \qquad s \mapsto \left(h^1(s), h^2(s)\right).$$

There is no loss of generality in assuming that h is at once given in normal representation, with $0 \in]a, b[$.

By Proposition 6.10.8, let us consider the geodesic

$$c:]p, q[\longrightarrow V, \qquad t \mapsto \left(c^1(t), c^2(t)\right)$$

such that

$$c(0) = h(0), \qquad c'(0) = \eta(0)$$

where η is the normal vector field of h (see Proposition 6.5.1). By Theorem 6.13.1, there exists in a neighborhood $U \subseteq V$ of $c(0)$ a Fermi system of geodesic coordinates admitting the curve c as base curve. There is no loss of generality in assuming that we are working in this system of coordinates. The curve h is a geodesic perpendicular to the base curve c at $c(0) = h(0)$: it is thus the geodesic $x^2 = 0$.

Consider now an arbitrary regular curve

$$f:]m, n[\longrightarrow U, \qquad u \mapsto \left(f^1(u), f^2(u)\right)$$

joining two points

$$f(u_1) = h(s_1) = (s_1, 0), \qquad f(u_2) = h(s_2) = (s_2, 0)$$

of the geodesic h. Let us compute its length (see Definition 6.12.3); in view of Theorem 6.13.1

$$\begin{aligned}
\mathsf{Length}_{u_1}^{u_2}(f) &= \int_{u_1}^{u_2} \left\| f'(u) \right\|_{f(u)} du \\
&= \int_{u_1}^{u_2} \sqrt{\sum_{ij} (f^i)'(u) \cdot (f^j)'(u) \cdot g_{ij}\big(f^1(u), f^2(u)\big)}\, du \\
&= \int_{u_1}^{u^2} \sqrt{\big((f^1)'(u)\big)^2 + \big((f^i)'(u)\big)^1 g_{22}\big(f^1(u), f^2(u)\big)}\, du \\
&\geq \int_{u_1}^{u^2} \left|(f^1)'(u)\right| du \\
&\geq \left| f^1(u_2) - f^1(u_1) \right| \\
&= |s_2 - s_1| \\
&= \mathsf{Length}_{s_1}^{s_2}(h)
\end{aligned}$$

where the last equality holds by Proposition 6.3.6. □

6.14 Curvature in Geodesic Coordinates

Let us now investigate some simplifications of formulas when working in a system of geodesic coordinates.

Proposition 6.14.1 *Consider a Riemann patch of class $\mathcal{C}^2$*

$$g_{ij} : U \longrightarrow \mathbb{R}, \quad 1 \leq i, j \leq 2$$

and assume that we are working in a system of geodesic coordinates. Under these conditions the length of an arc of the curves $x^1 = k$, with k a constant, is given by $\int \sqrt{g_{22}}$.

Proof These curves admit the parametric representation $c_k(x^2) = (k, x^2)$. Therefore $c_k' = (0, 1)$ and (see Definition 6.3.2 and Theorem 6.13.1)

$$\int \left\| c_k' \right\|_c = \int \sqrt{\begin{pmatrix} 0 & 1 \end{pmatrix} \begin{pmatrix} 1 & 0 \\ 0 & g_{22} \end{pmatrix} \begin{pmatrix} 0 \\ 1 \end{pmatrix}} = \int \sqrt{g_{22}}. \qquad \square$$

Proposition 6.14.2 *Consider a Riemann patch of class $\mathcal{C}^2$*

$$g_{ij} : U \longrightarrow \mathbb{R}, \quad 1 \leq i, j \leq 2$$

and assume that we are working in a system of geodesic coordinates. Under these conditions the relative geodesic curvature of the curves $x^1 = k$, with k a constant, is *given by*

$$\kappa_g = \frac{1}{2g_{22}} \frac{\partial g_{22}}{\partial x^1}.$$

Proof Proposition 6.14.1 explains how to pass to normal representation for these curves $x^1 = k$, but we shall instead Proposition 6.9.8 which allows us to work at once with the representation $h(x^2) = (k, x^2)$. Notice that $h' = (0, 1)$, from which $\|h'\|_h = \sqrt{g_{22}}$ at each point.

Let us first compute the covariant derivative of $h'(x^2) = (0, 1)$ along $h(x^2) = (k, x^2)$. The formula in Definition 6.7.3 reduces to

$$\frac{\nabla h'}{dx^2} = \left(\Gamma_{22}^2, \Gamma_{22}^1\right) = \left(\frac{1}{2g_{22}} \frac{\partial g_{22}}{\partial x^2}, -\frac{1}{2} \frac{\partial g_{22}}{\partial x^1}\right).$$

The curves $x^2 = l$ (with l a constant) are in normal representation and orthogonal to the curve $x^1 = k$ (Theorem 6.13.1), that is, the curve h. The tangent vector field to the curves $x^2 = l$ is thus of constant length 1 (Definition 6.3.4) and orthogonal to h'. Therefore, up to the sign, it is the normal vector field η to h. The "minus sign" must be chosen since (h', η) must have direct orientation. But the curve $x^2 = l$ admits the parametric representation $x^1 \mapsto (x^1, l)$; therefore its tangent vector field is $(1, 0)$ and the normal vector field η to d is given by $\eta = (-1, 0)$.

By Proposition 6.9.8, the geodesic curvature is then given by

$$\kappa_g = \frac{(\frac{\nabla h'}{dx^2}|\eta)}{\|h'\|_d^2} = \frac{((-\frac{1}{2}\frac{\partial g_{22}}{\partial x^1}, \frac{1}{2}g_{22}\frac{\partial g_{22}}{\partial x^2})|(-1,0))_c}{g_{22}} = \frac{1}{2g_{22}} \frac{\partial g_{22}}{\partial x^1}. \qquad \square$$

Our next concern is to exhibit a formula, in geodesic coordinates, for the geodesic curvature of an arbitrary regular curve. It is based on the so-called *Liouville formula*, which does not require the full strength of geodesic coordinates and—after all—is more elegant under these weaker assumptions.

Theorem 6.14.3 (Liouville Formula) *Consider a Riemann patch*

$$g_{ij} : U \longrightarrow \mathbb{R}, \quad i, j = 1, 2$$

of class C^3 and a regular curve in this patch

$$c\colon]a, b[\longrightarrow U, \qquad s \mapsto \left(c^1(s), c^2(s)\right)$$

given in normal representation. Suppose that at each point of the Riemann patch, the curves $x^1 = l$, with l constant, are orthogonal to the curves $x^2 = k$, with k constant. At a point with coordinates (x_0^1, x_0^2), let us write

- *$\kappa_1(x_0^1, x_0^2)$ for the relative geodesic curvature at (x_0^1, x_0^2) of the curve $x^1 \mapsto (x^1, x_0^2)$;*

- $\kappa_2(x_0^1, x_0^2)$ *for the relative geodesic curvature at* (x_0^1, x_0^2) *of the curve* $x^2 \mapsto (x_0^1, x^2)$;
- $\theta(s_0)$ *for the angle at* $c(s_0)$ *between the curve* c *and the curve* $x^1 \mapsto (x^1, c^2(s_0))$.

Under these conditions, the geodesic curvature of the curve c *is given by*

$$\kappa_g = \frac{d\theta}{ds} + \kappa_1 \cos\theta + \kappa_2 \sin\theta.$$

Proof To keep the notation as "light" as possible, let us make the convention that every norm, length, angle, scalar product or orthogonal condition met in this proof has to be understood in the Riemann sense, that is, with respect to the metric tensor. Except when absolutely necessary, we thus avoid repeating the notation introduced in Notation 6.2.5.

Let us fix once and for all a value s_0 of the parameter. We consider the two changes of parameters

$$\widetilde{x}^1 = \widetilde{x}^1(x^1), \qquad \widetilde{x}^2 = \widetilde{x}^2(x^2)$$

putting the two curves

$$\widetilde{x}^1 \mapsto (\widetilde{x}^1, c^2(s_0)), \qquad \widetilde{x}^2 \mapsto (c^1(s_0), \widetilde{x}^2)$$

in normal representations (see Proposition 6.3.5). This yields a mapping

$$(x^1, x^2) \mapsto (\widetilde{x}^1(x^1), \widetilde{x}^2(x^2))$$

which is bijective, of class $\mathcal{C}^3$, with an inverse of class $\mathcal{C}^3$, since this is the case for each of the two components of this function. This is thus a very special change of parameters of class $\mathcal{C}^3$, acting component-wise. Notice that the equivalent Riemann patch is then simply given by (see Definition 6.12.5)

$$\widetilde{g}_{ij}(\widetilde{x}^1, \widetilde{x}^2) = \sum_{kl} g_{kl} \frac{\partial x^k}{\partial \widetilde{x}^i} \frac{\partial x^l}{\partial \widetilde{x}^j} = g_{ij} \frac{\partial x^i}{\partial \widetilde{x}^i} \frac{\partial x^j}{\partial \widetilde{x}^j}$$

because

$$\frac{\partial x^1}{\partial \widetilde{x}^2} = 0, \qquad \frac{\partial x^2}{\partial \widetilde{x}^1} = 0.$$

Let us write

$$\widetilde{f}(\widetilde{x}^1, \widetilde{x}^2) = f(x^1(\widetilde{x}^1), x^2(\widetilde{x}^2))$$

for the parametric representation of the surface in terms of the new parameters $\widetilde{x}^1$, $\widetilde{x}^2$. With respect to this new system of coordinates, the curve c becomes

$$\widetilde{c}(s) = (\widetilde{x}^1(c^1(s)), \widetilde{x}^2(c^2(s)))$$

and is still in normal representation, since the parameter s is still the length s on the curve. The curves $x^1 = l$ are transformed into the curves $\widetilde{x}^1 = \widetilde{x}^1(l)$ (that is, $\widetilde{x}^1$ is equal to a constant) and analogously for the curves $x^2 = k$. Therefore the angle $\theta(s)$ and the geodesic curvatures κ_1 and κ_2 are the same in the new system of coordinates as in the original system. Consequently, it suffices to prove the formula of the statement in the new system of coordinates. Or in other words, there is no loss of generality in assuming that the two curves

$$x^1 \mapsto \big(x^1, c^2(s_0)\big), \qquad x^2 \mapsto \big(c^1(s_0), x^2\big)$$

are in normal representation. This is what we shall do from now on.

With that convention, let us now consider the two "2-dimensional" vector fields (see Definition 6.7.6) given by the normed tangent vectors to the curves $x^2 = k$, $x^1 = l$:

$$e^1\big(x^1, x^2\big) = \frac{(1,0)}{\|(1,0)\|_{(x^1,x^2)}}, \qquad e^2\big(x^1, x^2\big) = \frac{(0,1)}{\|(0,1)\|_{(x^1,x^2)}}.$$

Since the curves $x^2 = k$, $x^1 = l$ are orthogonal at each point, $e^1(x^1, x^2)$ and $e^2(x^1, x^2)$ constitute at each point an orthonormal basis for the metric tensor. Notice that since the two curves $x^1 = c^1(s_0)$ and $x^2 = c^2(s_0)$ are in normal representation

$$e^1\big(c(s_0)\big) = (1,0), \qquad e^2\big(c(s_0)\big) = (0,1).$$

On the other hand the curve c is in normal representation and at each point (e^1, e^2) is an orthonormal basis. Thus the normed vector $c'(s)$ has at each point the form

$$c'(s) = \cos\theta(s)\, e^1\big(c(s)\big) + \sin\theta(s)\, e^2\big(c(s)\big).$$

Let us compute the covariant derivative along the curve $c(s)$ of both sides of this equality, using freely the results of Sect. 6.7.

$$\begin{aligned}
\frac{\nabla c'}{ds}(s) &= -\sin\theta(s)\frac{d\theta}{ds}(s)e^1\big(c(s)\big) \\
&\quad + \cos\theta(s)\left(\frac{\nabla e^1}{\partial x^1}\big(c(s)\big)\frac{dc^1}{ds}(s) + \frac{\nabla e^1}{\partial x^2}\big(c(s)\big)\frac{dc^2}{ds}(s)\right) \\
&\quad + \cos\theta(s)\frac{d\theta}{ds}(s)e^2\big(c(s)\big) \\
&\quad + \sin\theta(s)\left(\frac{\nabla e^2}{\partial x^1}\big(c(s)\big)\frac{dc^1}{ds}(s) + \frac{\nabla e^2}{\partial x^2}\big(c(s)\big)\frac{dc^2}{ds}(s)\right) \\
&= -\sin\theta(s)\,\frac{d\theta}{ds}(s)e^1\big(c(s)\big) \\
&\quad + \cos\theta(s)\left(\frac{\nabla e^1}{\partial x^1}\big(c(s)\big)\cos\theta(s) + \frac{\nabla e^1}{\partial x^2}\big(c(s)\big)\sin\theta(s)\right)
\end{aligned}$$

$$
\begin{aligned}
&+\cos\theta(s)\frac{d\theta}{ds}(s)e^2\big(c(s)\big)\\
&+\sin\theta(s)\left(\frac{\nabla e^2}{\partial x^1}\big(c(s)\big)\cos\theta(s)+\frac{\nabla e^2}{\partial x^2}\big(c(s)\big)\sin\theta(s)\right)\\
=&\cos^2\theta(s)\frac{\nabla e^1}{dx^1}\big(c(s)\big)+\sin^2\theta(s)\frac{\nabla e^2}{dx^2}\big(c(s)\big)\\
&+\left(\frac{\nabla e^1}{dx^2}\big(c(s)\big)+\frac{\nabla e^2}{dx^1}\big(c(s)\big)\right)\cos\theta(s)\sin\theta(s)\\
&+\big(-\sin\theta(s)e^1\big(c(s)\big)+\cos\theta(s)e^2\big(c(s)\big)\big)\frac{d\theta}{ds}(s)\\
=&\cos^2\theta(s)\frac{\nabla e^1}{dx^1}\big(c(s)\big)+\sin^2\theta(s)\frac{\nabla e^2}{dx^2}\big(c(s)\big)\\
&+\left(\frac{\nabla e^1}{dx^2}\big(c(s)\big)+\frac{\nabla e^2}{dx^1}\big(c(s)\big)\right)\cos\theta(s)\sin\theta(s)+\eta(s)\frac{d\theta}{ds}(s)
\end{aligned}
$$

where $\eta(s)$ is the normal vector to the curve c (see Proposition 6.5.1).

Let us now compute the relative geodesic curvature of the two curves

$$p(x^1)=\big(x^1,c^2(s_0)\big),\qquad q(x^2)=\big(c^1(s_0),x^2\big)$$

which are thus in normal representation. We have

$$p'(x^1)=e_1\big(x^1,c^2(s_0)\big),\qquad q'(x^2)=e_2\big(c^1(s_0),x^2\big)$$

which implies that the corresponding normal vectors are

$$e_2\big(x^1,c^2(s_0)\big),\qquad -e_2\big(c^1(s_0),x^2\big).$$

The relative geodesic curvatures κ_1 and κ_2 involved in the statement are thus given by

$$
\begin{aligned}
\kappa_1\big(x^1,c^2(s_0)\big)&=\left(\frac{\nabla e^1}{\partial x^1}\big(x^1,c^2(s_0)\big)\Big|e^2\big(x^1,c^2(s_0)\big)\right),\\
\kappa_2\big(c^1(s_0),x^2\big)&=-\left(\frac{\nabla e^2}{\partial x^2}\big(c^1(s_0),x^2\big)\Big|e^1\big(c^1(s_0),x^2\big)\right).
\end{aligned}
$$

Covariantly differentiating the equalities

$$\big(e^1\big(x^1,c^2(s_0)\big)\big|e^2\big(x^1,c^2(s_0)\big)\big)=0,\qquad \big(e^1\big(c^1(s_0),x^2\big)\big|e^2\big(c^1(s_0),x^2\big)\big)=0$$

(see Proposition 6.7.4), the κ_1 and κ_2 can also be re-written as

$$\kappa_1\big(x^1,c^2(s_0)\big)=-\left(e^1\big(x^1,c^2(s_0)\big)\Big|\frac{\nabla e^2}{\partial x^1}\big(x^1,c^2(s_0)\big)\right),$$

$$\kappa_2\big(c^1(s_0),x^2\big)=\left(e^2\big(c^1(s_0),x^2\big)\Big|\frac{\nabla e^1}{\partial x^2}\big(c^1(s_0),x^2\big)\right).$$

Notice also that covariantly differentiating the equality

$$\big(e^i\big(x^1,c^2(s_0)\big)\big|e^i\big(x^1,c^2(s_0)\big)\big)=1$$

yields

$$\left(\frac{\nabla e^i}{\partial x^1}\big(x^1,c^2(s_0)\big)\Big|e^i\big(x^1,c^2(s_0)\big)\right)=0.$$

Analogously we obtain

$$\left(\frac{\nabla e^i}{\partial x^2}\big(c^1(s_0),x^2\big)\Big|e^i\big(c^1(s_0),x^2\big)\right)=0.$$

Using these various equalities and keeping in mind that

$$\eta\big(c(s_0)\big)=-\sin\theta\big(c(s_0)\big)e^1\big(c(s_0)\big)+\cos\theta\big(c(s_0)\big)e^2\big(c(s_0)\big)$$

we compute further, *at the point* $c(s_0)$ and with obvious abbreviated notation, that

$$\left(\frac{\nabla e^1}{\partial x^1}\Big|\eta\right)=-\sin\theta\left(\frac{\nabla e^1}{\partial x^1}\Big|e^1\right)+\cos\theta\left(\frac{\nabla e^1}{\partial x^1}\Big|e^2\right)=\cos\theta\kappa_1$$

$$\left(\frac{\nabla e^2}{\partial x^2}\Big|\eta\right)=-\sin\theta\left(\frac{\nabla e^2}{\partial x^2}\Big|e^1\right)+\cos\theta\left(\frac{\nabla e^2}{\partial x^2}\Big|e^2\right)=\sin\theta\kappa_2$$

$$\left(\frac{\nabla e^1}{\partial x^2}\Big|\eta\right)=-\sin\theta\left(\frac{\nabla e^1}{\partial x^2}\Big|e^1\right)+\cos\theta\left(\frac{\nabla e^1}{\partial x^2}\Big|e^2\right)=\cos\theta\kappa_2$$

$$\left(\frac{\nabla e^2}{\partial x^1}\Big|\eta\right)=-\sin\theta\left(\frac{\nabla e^2}{\partial x^1}\Big|e^1\right)+\cos\theta\left(\frac{\nabla e^2}{\partial x^1}\Big|e^2\right)=\sin\theta\kappa_1.$$

Now the relative geodesic curvature of the curve c is given by (see Definition 6.9.7)

$$\kappa_g=\left(\frac{\nabla c'}{ds}\Big|\eta\right)_c.$$

Introducing into this formula the various quantities calculated above, we obtain *still at the point* $c(s_0)$ and still with abbreviated notation,

$$\begin{aligned}\kappa_g&=\left(\frac{\nabla c'}{ds}\Big|\eta\right)\\&=\cos^2\theta\left(\frac{\nabla e^1}{\partial x^1}\Big|\eta\right)+\sin^2\theta\left(\frac{\nabla e^2}{\partial x^2}\Big|\eta\right)\end{aligned}$$

$$
\begin{aligned}
&+\sin\theta\cos\theta\left(\frac{\nabla e^1}{\partial x^2}\Big|\eta\right)+\sin\theta\cos\theta\left(\frac{\nabla e^2}{\partial x^1}\Big|\eta\right)\\
&+\frac{d\theta}{ds}(\eta|\eta)\\
&=\cos^3\theta\kappa_1+\sin^3\theta\kappa_2+\sin\theta\cos^2\theta\kappa_2+\sin^2\theta\cos\theta\kappa_1+\frac{d\theta}{ds}\\
&=\frac{d\theta}{ds}+\kappa_1\big(\cos^2\theta+\sin^2\theta\big)+\kappa_2\big(\sin^2\theta+\cos^2\theta\big)\\
&=\frac{d\theta}{ds}+\kappa_1\cos\theta+\kappa_2\in\theta.
\end{aligned}
$$

This concludes the proof. □

Corollary 6.14.4 *Consider a Riemann patch of class $\mathcal{C}^3$*

$$g_{ij}\colon U\longrightarrow\mathbb{R},\quad i,j=1,2$$

given in geodesic coordinates and a regular curve in it

$$c\colon\,]a,b[\,\longrightarrow U,\qquad s\mapsto\big(c^1(s),c^2(s)\big)$$

given in normal representation. Write $\theta(s_0)$ for the angle at $c(s_0)$ between the curve c and the curve $x^1\mapsto(x^1,c^2(s_0))$. Under these conditions, the geodesic curvature of the curve c at the point with parameter s_0 is given by

$$\kappa_g(s_0)=\frac{d\theta}{ds}(s_0)+\frac{1}{2\sqrt{g_{22}}}\frac{\partial g_{22}}{\partial x^1}\frac{dc^2}{ds}.$$

Proof The curves $x^2=l$ are geodesics by Theorem 6.13.1 thus, with the notation of Theorem 6.14.3, $\kappa_1=0$ (see Definition 6.10.1). On the other hand, Proposition 6.14.2 gives the value of κ_2. Then, still following Theorem 6.14.3

$$\kappa_g=\frac{d\theta}{ds}+\frac{1}{2g_{22}}\frac{\partial g_{22}}{\partial x^1}\sin\theta.$$

It remains to compute $\sin\theta$. But

$$\sin\theta=\cos\left(\frac{\pi}{2}-\theta\right)$$

that is at each point, the cosine of the angle between the curve c and the curves $x^1=c^1(s_0)$. Therefore

$$\sin\theta=\frac{((\frac{dc^1}{ds},\frac{dc^2}{ds})|(0,1))}{\|c'(s)\|\cdot\|(0,1)\|}=\frac{\frac{dc^2}{ds}g_{22}}{\sqrt{g_{22}}}=\frac{dc^2}{ds}\sqrt{g_{22}}.$$

Thus finally

$$\kappa_g = \frac{d\theta}{ds} + \frac{1}{2g_{22}} \frac{\partial g_{22}}{\partial x^1} \frac{dc^2}{ds} \sqrt{g_{22}} = \frac{d\theta}{ds} + \frac{1}{2\sqrt{g_{22}}} \frac{\partial g_{22}}{\partial x^1} \frac{dc^2}{ds}. \qquad \square$$

Let us conclude this section with the case of the Gaussian curvature (see Definition 5.16.1).

Proposition 6.14.5 *In a geodesic system of coordinates, the Gaussian curvature of a regular surface of class $\mathcal{C}^3$ in $\mathbb{R}^3$ is given by*

$$\kappa_\tau = -\frac{1}{\sqrt{g_{22}}} \frac{\partial^2 \sqrt{g_{22}}}{\partial (x^1)^2}.$$

Proof In view of condition 5 in Theorem 6.13.1, the formula in Theorem 6.11.3 reduces to

$$R_{1212} = -\frac{\partial \Gamma_{122}}{\partial x^1} + \Gamma_{12}^1 \Gamma_{212}.$$

Applying Proposition 6.6.8 and Theorem 6.13.1 again

$$\Gamma_{212} = \Gamma_{122} = \frac{1}{2} \frac{\partial g_{22}}{\partial x^1}$$

$$\Gamma_{21}^2 = \Gamma_{12}^2 = g^{22} \Gamma_{212} = \frac{1}{g_{22}} \frac{1}{2} \frac{\partial g_{22}}{\partial x^1}.$$

These various observations, together with Lemma 6.11.2, show that

$$\kappa_\tau = \frac{R_{1212}}{g_{11}} = \frac{1}{4} \frac{1}{(g_{22})^2} \left(\frac{\partial g_{22}}{\partial x^1} \right)^2 - \frac{1}{g_{22}} \frac{1}{2} \frac{\partial^2 g_{22}}{\partial (x^1)^2}.$$

On the other hand

$$\frac{\partial \sqrt{g_{22}}}{\partial x^1} = \frac{1}{2} \frac{1}{\sqrt{g_{22}}} \frac{\partial g_{22}}{\partial x^1}$$

$$\frac{\partial^2 \sqrt{g_{22}}}{\partial^2 (x^1)^2} = \frac{-1}{4} \frac{1}{g_{22}\sqrt{g_{22}}} \left(\frac{\partial g_{22}}{\partial x^1} \right)^2 + \frac{1}{2} \frac{1}{\sqrt{g_{22}}} \frac{\partial^2 g_{22}}{\partial (x^1)^2}.$$

Dividing by $\sqrt{g_{22}}$ and changing the sign indeed yields the formula that we have obtained for κ_τ. $\square$

6.15 The Poincaré Half Plane

Our basic example of a Riemann patch is that induced by a surface of $\mathbb{R}^3$ (see Sect. 6.1). Up to now, we have not provide any other examples. Let us fill this gap by describing the so-called *Poincaré half plane*: a Riemann patch which was introduced in order to provide a model of *non-Euclidean geometry*. *Non-Euclidean geometries* have been given full attention in Chap. 7 of [3], *Trilogy I*. Therefore we shall only very briefly remark upon them later in this section.

Definition 6.15.1 Let U be the "upper half plane" in $\mathbb{R}^2$, that is

$$U = \{(x^1, x^2) | x^2 > 0\}.$$

The *Poincaré half plane* is the Riemann patch of class $\mathcal{C}^\infty$

$$g_{ij} : U \longrightarrow \mathbb{R}, \quad 1 \le i, j \le 2$$

given by

$$\begin{pmatrix} g_{11}(x^1, x^2) & g_{12}(x^1, x^2) \\ g_{21}(x^1, x^2) & g_{22}(x^1, x^2) \end{pmatrix} = \begin{pmatrix} \frac{1}{(x^2)^2} & 0 \\ 0 & \frac{1}{(x^2)^2} \end{pmatrix}.$$

Trivially, the matrix $(g_{ij})_{ij}$ is symmetric definite positive, since $x^2 > 0$ at all points of U.

Proposition 6.15.2 *In the Poincaré half plane*:

1. $g^{11} = g^{22} = x^2$ *while* $g^{12} = g^{21} = 0$;
2. $\Gamma_{111} = \Gamma_{122} = \Gamma_{212} = \Gamma_{221} = 0$ *while* $\Gamma_{112} = \frac{1}{(x^2)^3}$ *and* $\Gamma_{121} = \Gamma_{211} = \Gamma_{222} = -\frac{1}{(x^2)^3}$;
3. $\Gamma_{11}^1 = \Gamma_{12}^2 = \Gamma_{21}^2 = \Gamma_{22}^1 = 0$ *while* $\Gamma_{11}^2 = \frac{1}{x^2}$ *and* $\Gamma_{12}^1 = \Gamma_{21}^1 = \Gamma_{22}^2 = -\frac{1}{x^2}$.

Proof This is just a routine application of the formulas in Definitions 6.6.7 and 6.11.4. ☐

Corollary 6.15.3 *The* Poincaré half plane *is such that*

$$\kappa_\tau = \frac{R_{1212}}{g_{11}g_{22} - g_{12}g_{21}} = -1$$

at all points.

Proof This follows by Proposition 6.15.2 and Definition 6.11.4. ☐

Let us recall that the quantity in Corollary 6.15.3, in the case of a 2-dimensional Riemann patch, is sometimes called the *Gaussian curvature* of the Riemann patch (see the comment at the end of Sect. 6.11). The *Poincaré half plane* thus has a constant negative Gaussian curvature equal to -1, just as the *pseudo-sphere* of pseudo-radius 1 (see Example 5.16.7). Notice that nevertheless, the metric tensor of the pseudo-sphere does not have the same form as that of the Riemann patch (see the proof of Example 5.16.7).

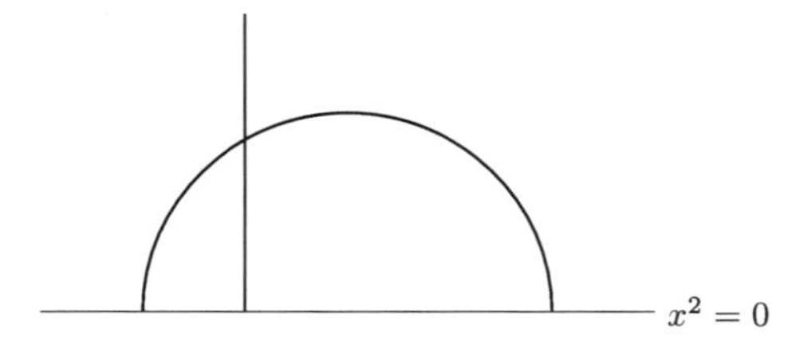

Fig. 6.3

Proposition 6.15.4 *The Riemannian angles of the Poincaré half plane coincide with the Euclidean angles.*

Proof At a given point (x^1, x^2), consider the two vectors

$$v = (v^1, v^2), \qquad w = (w^1, w^2).$$

Their Riemannian angle θ is such that

$$\begin{aligned}\cos\theta &= \frac{\frac{1}{(x^2)^2}v_1w_1 + \frac{1}{(x^2)^2}v_2w_2}{\sqrt{\frac{1}{(x^2)^2}(v_1)^2 + \frac{1}{(x^2)^2}(v_2)^2}\sqrt{\frac{1}{(x^2)^2}(w_1)^2 + \frac{1}{(x^2)^2}(w_2)^2}}\\ &= \frac{v^1w^1 + v^2w^2}{\sqrt{(v^1)^2 + (v^2)^2}\sqrt{(w^1)^2 + (w^2)^2}}\end{aligned}$$

and this last formula is precisely the value of the cosine of the Euclidean angle. □

Let us now turn our attention to the geodesics:

Proposition 6.15.5 *The geodesics of the Poincaré half plane are*:

1. *the parallels to the x^2-axis*;
2. *the half circles with center on the x^1-axis.*

Of course in this statement "parallel" and "circle" should be understood in the sense of the ordinary Euclidean geometry of $\mathbb{R}^2$ (see Fig. 6.3).

Proof In view of Proposition 6.15.2, the conditions of Theorem 6.10.3 for being a geodesic c in normal representation reduce to

$$\begin{cases} \frac{d^2c^1}{ds^2}(s) - \frac{dc^1}{ds}(s)\frac{dc^2}{ds}(s)\frac{1}{x^2(c(s))} - \frac{dc^1}{ds}(s)\frac{dc^2}{ds}(s)\frac{1}{x^2(c(s))} = 0\\ \frac{d^2c^2}{ds^2}(s) + \frac{dc^1}{ds}(s)\frac{dc^1}{ds}(s)\frac{1}{x^2(c(s))} - \frac{dc^2}{ds}(s)\frac{dc^2}{ds}(s)\frac{1}{x^2(c(s))} = 0\end{cases}$$

that is

$$\begin{cases} \frac{d^2c^1}{ds^2} - 2\frac{dc^1}{ds}\frac{dc^2}{ds}\frac{1}{c^2} = 0 \\ \frac{d^2c^2}{ds^2} + (\frac{dc^1}{ds})^2\frac{1}{c^2} - (\frac{dc^2}{ds})^2\frac{1}{c^2} = 0. \end{cases}$$

From now on, let us use instead the more concise notation

$$\begin{cases} (c^1)'' - 2\frac{(c^1)'(c^2)'}{c^2} = 0 \\ (c^2)'' + \frac{(c^1)'^2 - (c^2)'^2}{c^2} = 0. \end{cases}$$

Integrating a system of differential equations is not such an easy task. But we shall nevertheless take it easy, since the statement suggests at once the answer! We shall prove that the curves given in the statement are geodesics and we shall prove further that they exhaust all the possibilities.

The parallels to the x^2-axis are the curves $c(t) = (k, t)$ with k a constant. The change of parameter to pass in normal representation (see Proposition 6.3.5) is

$$\sigma(t) \int_{t_0}^{t} \sqrt{\begin{pmatrix} 0 & 1 \end{pmatrix} \begin{pmatrix} \frac{1}{t^2} & 0 \\ 0 & \frac{1}{t^2} \end{pmatrix} \begin{pmatrix} 0 \\ 1 \end{pmatrix}}\, dt = \int_{t_0}^{t} \frac{1}{t}\, dt = \log t$$

when choosing $t_0 = 1$. Therefore $\sigma^{-1}(s) = e^s$ and we obtain as normal representation

$$\overline{c}(s) = (k, e^s), \qquad \overline{c}'(s) = (0, e^s), \qquad \overline{c}''(s) = (0, e^s)$$

and it is immediate that these data satisfy the above system of differential equations for being a geodesic.

The upper half circle of center $(\alpha, 0)$ and radius R admits the parametric representation

$$c(t) = (\alpha + R\cos t, R\sin t), \quad 0 < t < \pi.$$

Notice in particular that $\sin t > 0$ for all t. The change of parameter for passing to normal representation is this time

$$\begin{aligned} \sigma(t) &= \int_{t_0}^{t} \sqrt{\begin{pmatrix} -R\sin t & R\cos t \end{pmatrix} \begin{pmatrix} \frac{1}{R^2\sin^2 t} & 0 \\ 0 & \frac{1}{R^2\sin^2 t} \end{pmatrix} \begin{pmatrix} -R\sin t \\ R\cos t \end{pmatrix}}\, dt \\ &= \int_{t_0}^{t} \sqrt{1 + \frac{\cos^2 t}{\sin^2 t}}\, dt \\ &= \int_{t_0}^{t} \frac{1}{\sin t}\, dt. \end{aligned}$$

Unfortunately, the explicit form of this last integral is in terms of hyperbolic functions, so we shall avoid calculating it and calculate instead its inverse function. This

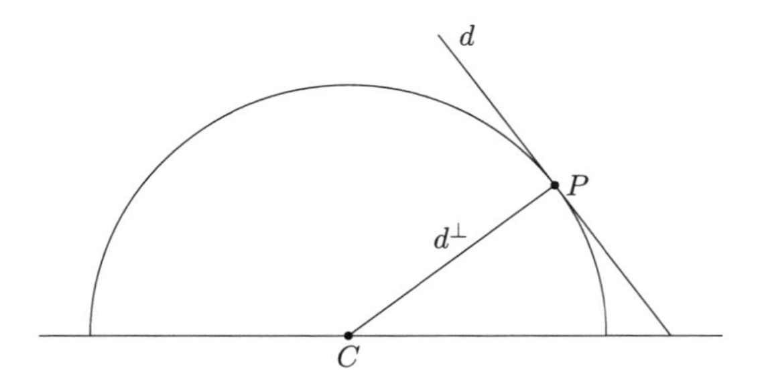

Fig. 6.4

is not really a problem because the differential equations for being a geodesic refer only to the derivatives of the normal representation $\overline{c}$. Therefore, as already observed several times in this book, only the derivatives of σ and σ^{-1} are needed explicitly. We have

$$\sigma'(t) = \frac{1}{\sin t}, \qquad (\sigma^{-1})'(s) = \frac{1}{\sigma'(\sigma^{-1}(s))} = \sin(\sigma^{-1}(s)).$$

The normal representation and its derivatives are thus

$$\begin{aligned}
\overline{c}(s) &= \left(\alpha + R\cos(\sigma^{-1}(s)), R\sin(\sigma^{-1}(s))\right) \\
\overline{c}'(s) &= \left(-R\sin^2(\sigma^{-1}(s)), R\cos(\sigma^{-1}(s))\sin(\sigma^{-1}(s))\right) \\
\overline{c}''(s) &= \left(-2R\sin^2(\sigma^{-1}(s))\cos(\sigma^{-1}(s)),\right. \\
&\qquad \left. -R\sin^3(\sigma^{-1}(s)) + R\cos^2(\sigma^{-1}(s))\sin(\sigma^{-1}(s))\right).
\end{aligned}$$

It is now trivial that these data satisfy the above differential equations for being a geodesic.

So: all the curves mentioned in the statement are geodesics. Are they the only ones? Let us recall that by Proposition 6.10.8, at each point of U, in each direction, there exists a unique geodesic. If we prove that at each point, in each direction, there is already a geodesic of the statement, these geodesics will thus exhaust all the possibilities and the proof will be complete. Of course in the "vertical" direction, we always have the corresponding parallel to the x^2-axis. Consider next a point $P \in U$ and a line d passing through P in a non-vertical direction (see Fig. 6.4). The line $d^\perp$, passing through P and orthogonal to d, intersects the x^1-axis at some point C. The circle with center C passing through P has a tangent at P perpendicular to its radius CP: this is precisely the given line d. So the half circle is a geodesic having at P the direction d. □

Axiomatic geometries were the topic of [3], *Trilogy I*. Roughly speaking, for the reader who is not familiar with these theories, let us say that a *Euclidean plane* consists of:

- a set Π—called *the plane*—whose elements are called *points*;
- a choice of subsets of Π, called *lines*;
- various binary or ternary relations involving points and lines;
- axioms to be satisfied by those data.

Among the relations, some express "geometric configurations" (a point is on a line, a point is between two other points) and others express "congruences": congruence of two segments or congruence of two angles (see [3], *Trilogy I*, for the definition of these notions).

Among the axioms, one has expected statements such as

Through two distinct points passes exactly one line.

But one also has the famous *parallel postulate* (two lines are called *parallel* when their intersection is empty):

Given a point P not on a line d, there passes through P exactly one line parallel to d.

The full list of axioms on these geometric data force Π to be isomorphic to the Euclidean space $\mathbb{R}^2$, with its usual lines. See for example *Hilbert's* axiomatization of the Euclidean plane in Chap. 8 of [3], *Trilogy I*.

Dropping the "parallel axiom" from Hilbert's axiomatization of the plane yields what is called *absolute geometry*. Let us stress that in absolute geometry, the *existence* of a parallel is already a theorem (see Corollary 8.3.36 in [3], *Trilogy I*): a parallel can be constructed via two perpendiculars. Thus the parallel axiom is not needed to prove the existence of a parallel and therefore passing from absolute geometry to Euclidean geometry requires only to state as an axiom the *uniqueness* of the parallel:

Given a point P not on a line d, there passes through P at most one line parallel to d.

Non-Euclidean geometry is obtained when adding instead to absolute geometry the *negation* of this uniqueness requirement:

Given a point P not on a line d, there pass through P several lines parallel to d.

It can then be proved that the number of possible parallel lines is necessarily infinite.

The *Poincaré* half plane is a model of non-Euclidean geometry obtained by choosing as *lines* the *geodesics*. The rest of this section is devoted to proving this result.

Proposition 6.15.6 *Through two distinct points of the* Poincaré half plane *passes exactly one geodesic.*

Proof We freely refer to Theorem 6.16.2. If two points A and B are on the same vertical line, this vertical line is a geodesic joining them. Of course no other vertical line contains these two points A, B and no half circle contains them either: indeed,

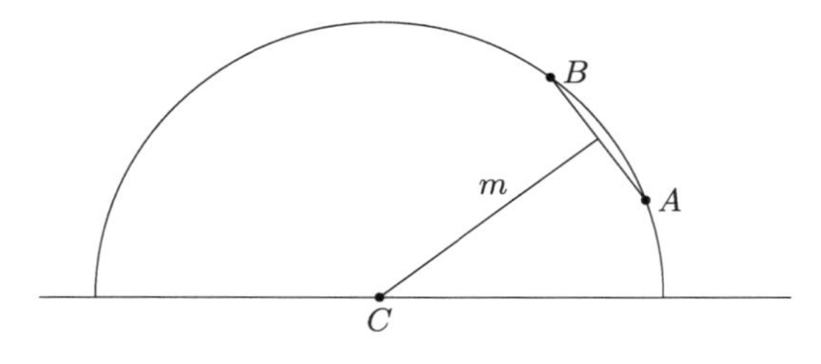

Fig. 6.5

a half circle never contains two points on the same vertical. So the vertical line through A and B is the unique geodesic joining these two points.

If two points A and B are not on the same vertical line, we must prove the existence of a unique half circle joining them. The necessary and sufficient condition for a circle to pass through A and B is that its center C lies on the median perpendicular m of the segment AB (see Fig. 6.5 and Proposition 8.4.10 in [3], *Trilogy I*; the *median perpendicular* is the perpendicular to AB at its middle point). But when the half circle is a geodesic, the center C of the circle must also be on the x^1-axis, thus finally it must be at the intersection of the median m and the x^1-axis. Since A and B are not on the same vertical, the median m is not horizontal, thus it indeed meets the x^1-axis at some unique point C. The half circle with center C passing through A and B is then the expected unique geodesic. □

Now the "parallel axiom":

Proposition 6.15.7 *In the* Poincaré half plane, *given a point P not on a geodesic d, there pass through P infinitely many geodesics not intersecting d. All these geodesics are contained between two "limit" geodesics.*

Proof We freely refer to Theorem 6.16.2. Consider first as geodesic a vertical line d (see Fig. 6.6). Consider also a point $P \notin d$, thus on the left or the right of d; the same argument applies in both cases. Write Q for the intersection in $\mathbb{R}^2$ of d and the x^1-axis. Consider the median m of the segment PQ which—since PQ is not vertical—meets the x^1-axis at some point C in $\mathbb{R}^2$. The half circle with center C passing through P thus also "passes" through Q: this is a first geodesic c_1 which does not meet d in the *Poincaré half plane*. Of course the vertical line through P is another geodesic c_2 not intersecting d. But there are infinitely many others! Every half circle passing through P and whose center is "on the other side of C with respect to Q", intersects the x^1-axis at some point situated between C and Q: thus it does not meet d. In other words, "all geodesics situated between c_1 and c_2 do not intersect the geodesic d".

Next consider as geodesic d a half circle "cutting" in $\mathbb{R}^2$ the x^1-axis at two points A and B. Consider further a point P not on this geodesic d. The point P can thus be "under" or "above" the half circle, but the proof applies to both cases (see Figs. 6.7 and 6.8). Again we consider the median of the segment PA and its intersection C_A

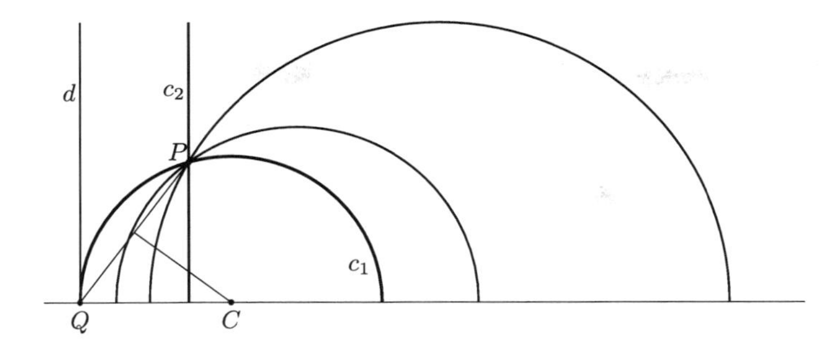

Fig. 6.6

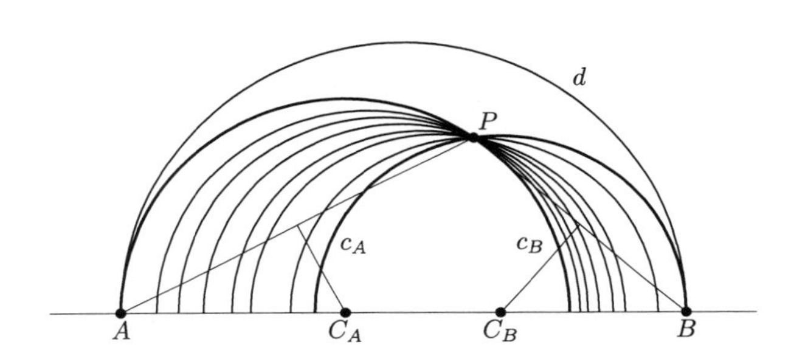

Fig. 6.7

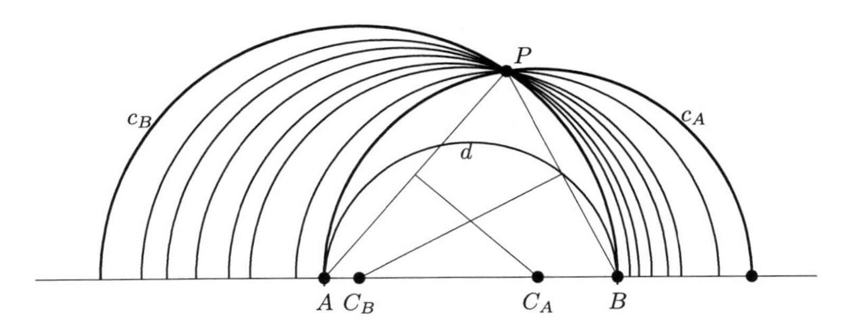

Fig. 6.8

with the x^1-axis. In $\mathbb{R}^2$, the half circle with center C_A passing through P "cuts" the x^1-axis at A; this is thus a first geodesic c_A not intersecting the geodesic d in the *Poincaré half plane*. Analogously the median of PB cuts the x^1-axis at a point C_B and the half circle with center C_B passing through P is a second geodesic c_B not intersecting d. Of course all half circles passing through P "between c_A and c_B" are geodesics not intersecting d. □

Next, we observe that

Lemma 6.15.8 *In the Poincaré half-plane, every normal representation of a geodesic induces a bijection between the real line and the geodesic.*

Proof Consider first a parallel to the x^2-axis:

$$c(x^2) = (\alpha, x^2), \quad x^2 > 0$$

and choose x_0^2 as origin for computing lengths along the geodesic. Since $c'(x^2) = (0, 1)$, we obtain (see Proposition 6.3.5)

$$\begin{aligned}\sigma(x^2) &= \int_{x_0^2}^{x^2} \|(0,1)\|_{c(x^2)} \\ &= \int_{x_0^2}^{x^2} \frac{dx^2}{x^2} \\ &= \log x^2 - \log x_0^2.\end{aligned}$$

As x^2 tends to zero, this quantity tends to $-\infty$ and when x^2 tends to infinity, it tends to $+\infty$. This yields the announced bijection.

Consider next as geodesic a half circle centered on the x^1-axis.

$$c(t) = (\alpha + R\cos t, R\sin t), \quad 0 < t < \pi$$

and fix t_0 as origin for computing lengths along this geodesic. Since

$$c'(t) = (-R\sin t, R\cos t),$$

we obtain (see Proposition 6.3.5 again)

$$\begin{aligned}\sigma(x^2) &= \int_{t_0}^{t} \|(-R\sin t, R\cos t)\|_{c(t)} \\ &= \int_{t_0}^{t} \frac{dt}{\sin t} \\ &= \log\left(\tan\frac{t}{2}\right) - \log\left(\tan\frac{t_0}{2}\right).\end{aligned}$$

As t tends to 0 this quantity tends to $-\infty$ and as t tends to π, it tends to $+\infty$. Again this yields the announced bijection. □

Notice that the bijection in Lemma 6.15.8 allows us to transpose onto every geodesic the natural order of the real line and in particular, the relation "*a point is between two other points*". It thus makes perfect sense to speak of segments and half lines (see Sect. 8.2 in [3], *Trilogy I*). It therefore also makes sense to consider the congruence (in the sense of the Riemannian metric) between two segments or two angles. Notice that:

Corollary 6.15.9 *In the Poincaré half plane, every half-line has infinite length.*

Proof This follows by Lemma 6.15.8. □

To be able to check the validity of all congruence axioms for non-Euclidean geometry (see Sect. 8.3 in [3], *Trilogy II*), we need to exhibit some Riemannian isometries of the Poincaré half plane onto itself: that is, bijections which respect the Riemannian metric.

Proposition 6.15.10 *Every Euclidean horizontal translation is a Riemannian isometry of the Poincaré half plane.*

Proof Such a translation is a bijection having another such translation as inverse. It has the form

$$\tau\left(x^1, x^2\right) = \left(x^1 + \alpha, x^2\right).$$

The matrix of partial derivatives of τ is the identity matrix. The value of the Riemann tensor is the same at (x^1, x^2) and $\tau(x^1, x^2)$, since it depends only on the coordinate x^2. The conditions of Definition 6.12.5 are then trivially satisfied and by Corollary 6.12.6, τ is an isometry. □

Proposition 6.15.11 *Every Euclidean symmetry with respect to a vertical axis is a Riemannian isometry of the Poincaré half plane.*

Proof It suffices to prove the result when the axis of symmetry is the x^2-axis. Indeed, first translating the axis of symmetry onto the x^2-axis, performing the orthogonal symmetry around the x^2-axis and translating the axis back, yields the symmetry indicated. We know already by Proposition 6.15.10 that horizontal translations are Riemannian isometries.

The orthogonal symmetry around the x^2-axis is its own inverse and has the form

$$\sigma\left(x^1, x^2\right) = \left(-x^1, x^2\right).$$

The matrix of partial derivatives of σ thus has the form

$$\begin{pmatrix} -1 & 0 \\ 0 & 1 \end{pmatrix}.$$

The value of the Riemann tensor is the same at (x^1, x^2) and $\sigma(x^1, x^2)$, since it depends only on the coordinate x^2. The conditions of Definition 6.12.5 are then trivially satisfied and by Corollary 6.12.6, σ is an isometry. □

However, the crucial fact is (see Sect. 5.7 in [3], *Trilogy I*, for the theory of inversions):

Proposition 6.15.12 *The Euclidean inversions with center on the x^1-axis are Riemannian isometries.*

Proof It suffices to prove the result when the center of inversion is the point $(0, 0)$. Indeed, first translating the center of inversion to $(0, 0)$, performing the inversion and translating the center of inversion back to its original position, yields the inversion indicated. We know already by Proposition 6.15.10 that horizontal translations are Riemannian isometries.

The inversion with center $(0, 0)$ and power R^2 is its own inverse and is defined everywhere on the Poincaré half plane, since $(0, 0)$ is not a point of the Poincaré half plane. With the notation of Definition 6.12.5, it has the form

$$\iota(x^1, x^2) = \big(\widetilde{x}^1(x^1, x^2), \widetilde{x}^2(x^1, x^2)\big) = \frac{R^2}{(x^1)^2 + (x^2)^2}(x^1, x^2).$$

The matrix of partial derivatives of ι is thus

$$\frac{R^2}{((x^1)^2 + (x^2)^2)^2}\begin{pmatrix} (x^2)^2 - (x^1)^2 & -2x^1x^2 \\ -2x^1x^2 & (x^1)^2 - (x^2)^2 \end{pmatrix}$$

while, with the notation of Corollary 6.12.6,

$$(\widetilde{g}_{ij})_{ij} = \frac{((x^1)^2 + (x^2)^2)^2}{R^4(x^2)^2}\begin{pmatrix} 1 & 0 \\ 0 & 1 \end{pmatrix}.$$

It is then immediate to compute that

$$\left(\sum_{kl} \widetilde{g}_{kl} \frac{\partial \widetilde{x}^k}{\partial x^i} \frac{\partial \widetilde{x}^l}{\partial x^j}\right)_{ij} = \begin{pmatrix} \frac{1}{(x^2)^2} & 0 \\ 0 & \frac{1}{(x^2)^2} \end{pmatrix}.$$

The result follows by Corollary 6.12.6. □

We are now ready to conclude that:

Theorem 6.15.13 *The Poincaré half plane is a model of non-Euclidean geometry.*

Proof We freely refer to Chap. 8 of [3], *Trilogy I*. We recall that our points are those of the Poincaré half plane, with the geodesics as lines. The incidence of a point and a line is just the membership relation. The "between" relation is that transposed from the real line via Lemma 6.15.8. The congruence relation for segments and angles is the congruence in terms of the Riemannian metric.

Proposition 6.15.6 attests the validity of the first incidence axiom; the other two incidence axioms are trivially satisfied. The first four axioms concerning the "between" relation follow at once from the corresponding properties of the real line, via Lemma 6.15.8. The last axiom on this "between" relation—the so-called *Pasch axiom*—is a routine exercise on lines and circles in the Euclidean plane, but is also an easy consequence of our Proposition 7.11.1. When a geodesic not containing a vertex of a triangle "enters" the triangle along one side, it must leave it (Proposition 7.11.1) along another side (Proposition 6.15.6).

The first five axioms concerning the congruence of segments or angles are immediate consequences of Lemma 6.15.8 and Proposition 6.15.4. Let us thus check the validity of the sixth congruence axiom: the so called *case of equality of two triangles*; writing $\equiv$ for the congruence relation:

If two triangles ABC and $A'B'C'$ are such that

$$AB \equiv A'B', \qquad \measuredangle ABC \equiv \measuredangle A'B'C', \qquad \measuredangle BAC \equiv \measuredangle B'A'C'$$

then

$$AC \equiv A'C', \qquad BC \equiv B'C', \qquad \measuredangle ACB \equiv \measuredangle A'C'B'.$$

To prove this, we shall show that the triangle ABC is congruent to a triangle $\tilde{A}\tilde{B}\tilde{C}$ "in canonical position", that is, a triangle such that:

- $\tilde{A} = (0, 1)$;
- $\tilde{B} = (0, b)$ with $b > 1$;
- $\tilde{C} = (a, c)$ with $a > 0$.

Analogously, the triangle $A'B'C'$ will be congruent to a triangle $\tilde{A}'\tilde{B}'\tilde{C}'$ in canonical position:

- $\tilde{A}' = (0, 1)$;
- $\tilde{B}' = (0, b')$ with $b' > 1$;
- $\tilde{C}' = (a', c')$ with $a' > 0$.

When this has been proved, since $AB \equiv A'B'$, necessarily $b = b'$. Thus

$$\tilde{A} = \tilde{A}', \qquad \tilde{B} = \tilde{B}'.$$

But since $\measuredangle ABC \equiv \measuredangle A'B'C'$, we also have $\measuredangle \tilde{A}\tilde{B}\tilde{C} \equiv \measuredangle \tilde{A}'\tilde{B}'\tilde{C}'$. Since moreover $\tilde{C}$ and $\tilde{C}'$ are both on the right hand side of the x^2-axis, the two geodesics through $\tilde{B}$, $\tilde{C}$ and $\tilde{B}'$, $\tilde{C}'$ coincide. Analogously, the two geodesics through $\tilde{A}$, $\tilde{C}$ and $\tilde{A}'$, $\tilde{C}'$ coincide. Eventually, the two triangles $\tilde{A}\tilde{B}\tilde{C}$ and $\tilde{A}'\tilde{B}'\tilde{C}'$ coincide. Since they are respectively congruent to the original triangles ABC and $A'B'C'$, the proof will be complete.

So we must now prove that every triangle ABC is congruent to a triangle in canonical position.

1. First, if it is not already the case, we force the geodesic AB to become a parallel to the x_2 axis. For this it suffices to perform an inversion of arbitrary power whose center P is one the two "intersection" points of the x^1-axis with the half circle through A and B. By Proposition 5.7.5 in [3], *Trilogy I*, the Euclidean (half)-circle through A and B becomes a Euclidean (half)-line perpendicular to the line joining the center P of inversion and the center of the circle, that is, perpendicular to the x^1-axis. By Proposition 6.15.12, the triangle ABC then becomes congruent to a triangle $A_1B_1C_1$ such that the geodesic A_1B_1 is a Euclidean parallel to the x^2-axis.

2. Second, if necessary, translate the triangle $A_1B_1C_1$ horizontally to get, by Proposition 6.15.10, a congruent triangle $A_2B_2C_2$ now with A_2 and B_2 on the x^2-axis.
3. Third, if necessary, perform an orthogonal symmetry around the x^2-axis to transform the triangle $A_2B_2C_2$ into a congruent triangle $A_3B_3C_3$ (see Proposition 6.15.11) now with C_3 on the right hand side of the x^2-axis.
4. Fourth, if this is not already the case, we shall force A_3 to become the point $(0, 1)$. If $A_3 = (\alpha, 0)$, it suffices to apply an inversion with center $(0, 0)$ and power α. By Proposition 6.15.12, we thus obtain a new triangle $A_4B_4C_4$, still congruent to ABC, now with $A_4 = (0, 1)$, B_4 still on the x^2-axis and C_4 still on the right hand side of this axis.
5. Finally, if it turns out that B_4 is below A_4, use Proposition 6.15.12 again and perform an inversion with center $(0, 0)$ and power 1 to obtain a triangle $\tilde{A}\tilde{B}\tilde{C}$ still congruent to ABC but now in canonical position.

This concludes the proof of the last congruence axiom.

The continuity axiom is an immediate consequence of Lemma 6.15.8. And the non-Euclidean axiom of parallels is Proposition 6.15.7. □

6.16 Embeddable Riemann Patches

We have seen that a plane curve can be "intrinsically" described—up to an isometry—by an arbitrary continuous function $\kappa(s)$: the curvature in terms of the arc length s (see Sect. 2.12). An analogous result holds for skew curves, this time using two sufficiently differentiable arbitrary functions $\kappa(s)$ and $\tau(s)$: the curvature and the torsion (see Sect. 4.6). Is there an analogous result for surfaces?

The *fundamental theorem of the theory of surfaces* tells us in a first approach that

> *A surface of $\mathbb{R}^3$ is entirely determined—up to an isometry—by the six coefficients E, F, G, L, M, N of its two fundamental quadratic forms.*

(See Definitions 5.4.5 and 5.8.6). However, in contrast to the case of curves, we can no longer expect six such arbitrary functions to always define a surface. Indeed in the case of a surface, these six functions are not "independent": we already know some specific properties relating them. Among other things:

- the coefficients E, F, G are those of a symmetric definite positive quadratic form (see Proposition 5.4.6);
- $LN - M^2 = R_{1212}$ (see Lemma 6.11.2) where R_{1212} can be written as a function of E, F, G (see Theorem 6.11.3).

And so on. So in fact, the *fundamental theorem of the theory of surfaces* must also answer the following question:

> *Give necessary and sufficient conditions on six functions E, F, G, L, M, N for being the coefficients of the two fundamental quadratic forms of a surface of $\mathbb{R}^3$.*

As we have just recalled, one necessary condition is the fact that E, F, G are the coefficients of a symmetric definite positive quadratic form: in other words, they must define a *Riemann patch* (see Definition 6.2.1). But this is certainly not sufficient, as our second observation $LN - M^2 = R_{1212}$ shows. Thus our question can be rephrased as

> *What are the additional conditions on a Riemann patch which will ensure that it is the patch associated with a surface of* $\mathbb{R}^3$?

This section is devoted to an answer to this question.

It is well-known that given a function $\psi(u, v, w)$ of class $\mathcal{C}^3$, the continuity of the partial derivatives forces the equality

$$\frac{\partial^3 \psi}{\partial u \partial v \partial w} = \frac{\partial^3 \psi}{\partial v \partial u \partial w}.$$

This simple fact is the key to solving our problem. As is often the case in the context of Riemannian geometry, such an easy formula can take an unexpectedly involved form. It will be convenient for us to switch back to the notation g_{ij} and h_{ij} of Definition 6.2.1 and Proposition 6.6.3.

Proposition 6.16.1 *Consider a regular parametric representation of class* $\mathcal{C}^3$ *of a surface*

$$f : U \longrightarrow \mathbb{R}^3.$$

The following equalities hold:

The Gauss Equations

$$\frac{\partial \Gamma_{jk}^l}{\partial x^i} - \frac{\partial \Gamma_{ik}^l}{\partial x^j} + \sum_m \left(\Gamma_{jk}^m \Gamma_{im}^l - \Gamma_{ik}^m \Gamma_{jm}^l\right) = \sum_m (h_{jk} h_{im} - h_{ik} h_{jm}) g^{lm}.$$

The Codazzi–Mainardi Equations

$$\sum_m \Gamma_{jk}^m h_{im} - \sum_m \Gamma_{ik}^m h_{jm} + \frac{\partial h_{jk}}{\partial x^i} - \frac{\partial h_{ik}}{\partial x^j} = 0.$$

Proof Write n for the normal vector to the surface (see Definition 5.5.7). Differentiating the equality $(n|n) = 0$ we get $(\frac{\partial n}{\partial x^i}|n) = 0$ proving that $\frac{\partial n}{\partial x^i}$ is in the tangent plane. Let us write

$$\frac{\partial n}{\partial x^i} = \alpha_i^1 \frac{\partial f}{\partial x^1} + \alpha_i^2 \frac{\partial f}{\partial x^2}.$$

Since n is orthogonal to $\frac{\partial f}{\partial x^j}$, differentiating the equality $(\frac{\partial f}{\partial x^j}|n) = 0$ with respect to x^i with respect to x^i we obtain

$$\left(\frac{\partial^2 f}{\partial x^i \partial x^j}\middle| n\right) + \left(\frac{\partial f}{\partial x^j}\middle|\frac{\partial n}{\partial x^i}\right) = 0$$

that is

$$\left(\frac{\partial^2 f}{\partial x^i \partial x^j}\middle| n\right) = -\left(\frac{\partial f}{\partial x^j}\middle|\frac{\partial n}{\partial x^i}\right).$$

From Definition 6.6.2, we then deduce

$$\begin{aligned} h_{ij} &= \left(\frac{\partial^2 f}{\partial x^i \partial x^j}\middle| n\right) \\ &= -\left(\frac{\partial f}{\partial x^j}\middle|\frac{\partial n}{\partial x^i}\right) \\ &= -\left(\frac{\partial f}{\partial x^j}\middle|\alpha_i^1 \frac{\partial f}{\partial x^1} + \alpha_i^2 \frac{\partial f}{\partial x^2}\right) \\ &= -\alpha_i^1 g_{j1} - \alpha_i^2 g_{j2}. \end{aligned}$$

This equality can be re-written as

$$\begin{pmatrix} h_{i1} \\ h_{i2} \end{pmatrix} = -\begin{pmatrix} g_{11} & g_{12} \\ g_{21} & g_{22} \end{pmatrix} \begin{pmatrix} \alpha_i^1 \\ \alpha_i^2 \end{pmatrix}$$

from which we deduce

$$\begin{pmatrix} \alpha_i^1 \\ \alpha_i^2 \end{pmatrix} = -\begin{pmatrix} g^{11} & g^{12} \\ g^{21} & g^{22} \end{pmatrix} \begin{pmatrix} h_{i1} \\ h_{i2} \end{pmatrix}$$

(see Definition 6.2.3), that is

$$\alpha_i^j = -\sum_k h_{ik} g^{jk}.$$

As a consequence

$$\frac{\partial n}{\partial x^i} = -\sum_{jk} h_{ik} g^{jk} \frac{\partial f}{\partial x^j}.$$

Let us now consider the third partial derivatives of f and introduce, for the needs of this proof, an explicit notation for its components:

$$\frac{\partial^3 f}{\partial x^i \partial x^j \partial x^k} = \Upsilon_{ijk}^1 \frac{\partial f}{\partial x^1} + \Upsilon_{ijk}^2 \frac{\partial f}{\partial x^2} + \Omega_{ijk} n.$$

Considering the definition of the Christoffel symbols of the second kind (Definition 6.6.2)

$$\frac{\partial f^2}{\partial x^j \partial x^k} = \Gamma_{jk}^1 \frac{\partial f}{\partial x^1} + \Gamma_{jk}^2 \frac{\partial f}{\partial x^2} + h_{jk} n$$

and differentiating this equality with respect to x^i we obtain

$$\begin{aligned}
\frac{\partial^3 f}{\partial x^i \partial x^j \partial x^k} &= \frac{\partial \Gamma_{jk}^1}{\partial x^i} \frac{\partial f}{\partial x^1} + \Gamma_{jk}^1 \frac{\partial^2 f}{\partial x^i \partial x^1} \\
&\quad + \frac{\partial \Gamma_{jk}^2}{\partial x^i} \frac{\partial f}{\partial x^2} + \Gamma_{jk}^2 \frac{\partial^2 f}{\partial x^i \partial x^2} \\
&\quad + \frac{\partial h_{jk}}{\partial x^i} n + h_{jk} \frac{\partial n}{\partial x^i} \\
&= \frac{\partial \Gamma_{jk}^1}{\partial x^i} \frac{\partial f}{\partial x^1} + \Gamma_{jk}^1 \left(\Gamma_{i1}^1 \frac{\partial f}{\partial x^1} + \Gamma_{i1}^2 \frac{\partial f}{\partial x^2} + h_{i1} n \right) \\
&\quad + \frac{\partial \Gamma_{jk}^2}{\partial x^i} \frac{\partial f}{\partial x^2} + \Gamma_{jk}^2 \left(\Gamma_{i2}^1 \frac{\partial f}{\partial x^1} + \Gamma_{i2}^2 \frac{\partial f}{\partial x^2} + h_{i2} n \right) \\
&\quad + \frac{\partial h_{jk}}{\partial x^i} n - h_{jk} \left(\sum_{lm} h_{im} g^{lm} \frac{\partial f}{\partial x^l} \right).
\end{aligned}$$

The three components of the third partial derivatives of f are then

$$\Upsilon_{ijk}^l = \frac{\partial \Gamma_{jk}^l}{\partial x^i} + \Gamma_{jk}^1 \Gamma_{i1}^l + \Gamma_{jk}^2 \Gamma_{i2}^l - h_{jk} \big(h_{i1} g^{l1} + h_{i2} g^{l2} \big)$$

and

$$\Omega_{ijk} = \Gamma_{jk}^1 h_{i1} + \Gamma_{jk}^2 h_{i2} + \frac{\partial h_{jk}}{\partial x^i}.$$

The Gauss equations translate simply as the equality $\Upsilon_{ijk}^l = \Upsilon_{jik}^l$, while the Codazzi–Mainardi equations translate as the equality $\Omega_{ijk} = \Omega_{jik}$. □

We are now ready to state the expected result. Its proof is highly involved and uses deep results from the theory of partial differential equations. This clearly runs outside the normal context for this introductory textbook. Therefore—even if not formally needed for the proof—we often rely on the intuition hidden behind the arguments involving the solutions of systems of partial differential equations.

Theorem 6.16.2 *A Riemann patch of class $\mathcal{C}^2$*

$$g_{ij} : U \longrightarrow \mathbb{R}, \quad 1 \le i, j \le 2$$

is, in a neighborhood V *of each point, induced by a regular parametric representation*

$$f: V \longrightarrow \mathbb{R}^3, \qquad (x^1, x^2) \mapsto f(x^1, x^2)$$

of a surface if and only if there exist three functions of class $\mathcal{C}^1$

$$h_{12}, \qquad h_{12} = h_{21}, \qquad h_{22}: V \longrightarrow \mathbb{R}$$

satisfying the Gauss–Codazzi–Mainardi *equations of Proposition* 6.16.1.

Proof Proposition 6.16.1 proves the necessity of the condition. Conversely, we are thus looking for a parametric representation $f(x^1, x^2)$, which—if it exists—will have two partial derivatives

$$\varphi_1(x^1, x^2) = \frac{\partial f}{\partial x^1}(x^1, x^2), \qquad \varphi_2(x^1, x^2) = \frac{\partial f}{\partial x^2}(x^1, x^2)$$

satisfying the requirements (see Definition 6.6.2)

$$\frac{\partial \varphi_j}{\partial x^i} = \Gamma_{ij}^1 \varphi_1 + \Gamma_{ij}^2 \varphi_2 + h_{ij}\mu$$

where μ will be the normal vector to the surface. From what we have observed in the proof of Proposition 6.16.1, we shall further have, if f exists,

$$\frac{\partial \mu}{\partial x^i} = -\sum_{lm} h_{lm} g^{lm} \frac{\partial f}{\partial x^l}.$$

We therefore consider the following system of partial differential equations, with three unknown functions

$$\varphi_1, \varphi_2, \mu: U \longrightarrow \mathbb{R}^3$$

that is, in terms of the components of these functions, nine functions $U \longrightarrow \mathbb{R}$:

$$\begin{cases} \frac{\partial \varphi_j}{\partial x^i} = \Gamma_{ij}^1 \varphi_1 + \Gamma_{ij}^2 \varphi_2 + h_{ij}\mu \\ \frac{\partial \mu}{\partial x^i} = -\sum_{lm} h_{lm} g^{lm} \varphi_l. \end{cases}$$

In these equations, g^{ij} is of course the inverse metric tensor of the Riemann patch and the Γ_{ij}^k are its Christoffel symbols. We are first interested in finding a solution $\varphi_1, \varphi_2, \mu$ of this system.

A general theorem on systems of partial differential equations (see Proposition B.4.1) asserts the existence of a solution of class $\mathcal{C}^2$ to the system above, provided some integrability conditions are satisfied. These conditions require that the given equations force the relations

$$\frac{\partial^2 \varphi_k}{\partial x^i \partial x^j} = \frac{\partial^2 \varphi_k}{\partial x^j \partial x^i}, \qquad \frac{\partial^2 \mu}{\partial x^i \partial x^j} = \frac{\partial^2 \mu}{\partial x^j \partial x^i}.$$

Let us prove that this is the case.

Since eventually, we want φ_1 and φ_2 to be the partial derivatives of the parametric representation f we are looking for, the first integrability condition should translate as the classical formula

$$\frac{\partial^3 f}{\partial x^i \partial x^j \partial x^k} = \frac{\partial^3 f}{\partial x^j \partial x^i \partial x^k}.$$

But this equality expresses exactly the *Gauss–Codazzi–Mainardi* equations, as we have seen in the proof of Proposition 6.16.1. Therefore the first integrability condition should follow from these equations. Let us observe that this is indeed the case.

From our system of partial differential equations, we obtain

$$\begin{aligned}
\frac{\partial^2 \varphi_k}{\partial x^i \partial x^j} &= \frac{\partial}{\partial x^i}\left(\Gamma_{jk}^1 \varphi_1 + \Gamma_{jk}^2 \varphi_2 + h_{jk}\mu\right) \\
&= \frac{\partial \Gamma_{jk}^1}{\partial x^i}\varphi_1 + \Gamma_{jk}^1 \frac{\partial \varphi_1}{\partial x^i} \\
&\quad + \frac{\partial \Gamma_{jk}^2}{\partial x^i}\varphi_2 + \Gamma_{jk}^2 \frac{\partial \varphi_2}{\partial x^i} \\
&\quad + \frac{\partial h_{jk}}{\partial x^i}\mu + h_{jk}\frac{\partial \mu}{\partial x^i} \\
&= \frac{\partial \Gamma_{jk}^1}{\partial x^i}\varphi_1 + \Gamma_{jk}^1\left(\Gamma_{i1}^1 \varphi_1 + \Gamma_{i1}^2 \varphi_2 + h_{i1}\mu\right) \\
&\quad + \frac{\partial \Gamma_{jk}^2}{\partial x^i}\varphi_2 + \Gamma_{jk}^2\left(\Gamma_{i2}^1 \varphi_1 + \Gamma_{i2}^2 \varphi_2 + h_{i2}\mu\right) \\
&\quad + \frac{\partial h_{kj}}{\partial x^i}\mu - h_{kj}\left(\sum_{l,m} h_{km} g^{lm} \varphi_l\right) \\
&= \left(\frac{\partial \Gamma_{jk}^1}{\partial x^i} + \sum_m \Gamma_{jk}^m \Gamma_{im}^1 - \sum_m h_{jk} h_{im} g^{1m}\right)\varphi_1 \\
&\quad + \left(\frac{\partial \Gamma_{jk}^2}{\partial x^i} + \sum_m \Gamma_{jk}^m \Gamma_{im}^2 - \sum_m h_{jk} h_{im} g^{2m}\right)\varphi_2 \\
&\quad + \left(\frac{\partial h_{kj}}{\partial x^i} + \sum_m \Gamma_{jk}^m h_{im}\right)\mu.
\end{aligned}$$

The corresponding formula for $\frac{\partial^2 \varphi_k}{\partial x^j \partial x^i}$ is obtained simply by permuting the indices i and j in the formula above. To prove the necessary equality of the two expressions (while we do not yet know the existence of φ_1, φ_2 and μ), it suffices to prove the equality of the respective coefficients of these unknown functions. But this is precisely what the *Gauss–Codazzi–Mainardi* equations say.

We still have to take care of the integrability condition concerning the function μ. Since we eventually want μ to become the normal vector to a surface whose parametric representation admits φ_1 and φ_2 as partial derivatives, we expect to have

$$\mu = \frac{\varphi_1 \times \varphi_2}{\|\varphi_1 \times \varphi_2\|}$$

where $\times$ indicates the cross product (see Sect. 1.7 in [4], *Trilogy II*). Therefore the permutability of the partial derivatives of μ should be a consequence of the permutability of the partial derivatives of φ_1, φ_2, that is, of the *Gauss–Codazzi–Mainardi* equations. It is indeed so.

We have at once

$$\begin{aligned}
\frac{\partial^2 \mu}{\partial x^j \partial x^i} &= -\sum_{lm}\left(\frac{\partial h_{im}}{\partial x^j} g^{lm}\varphi_l + h_{im}\frac{\partial g^{lm}}{\partial x^j}\varphi_l + h_{im}g^{lm}\frac{\partial \varphi_l}{\partial x^j}\right) \\
&= -\sum_{lm}\left(\frac{\partial h_{im}}{\partial x^j} g^{lm}\varphi_l + h_{im}\frac{\partial g^{lm}}{\partial x^j}\varphi_l + h_{im}g^{lm}\left(\Gamma_{jl}^1\varphi_1 + \Gamma_{jl}^2\varphi_2 + h_{jl}\mu\right)\right) \\
&= -\left(\sum_m \frac{\partial h_{im}}{\partial x^j} g^{1m} + \sum_m h_{im}\frac{\partial g^{1m}}{\partial x^j} + \sum_{lm} h_{im}g^{lm}\Gamma_{jl}^1\right)\varphi_1 \\
&\quad -\left(\sum_m \frac{\partial h_{im}}{\partial x^j} g^{2m} + \sum_m h_{im}\frac{\partial g^{2m}}{\partial x^j} + \sum_{lm} h_{im}g^{lm}\Gamma_{jl}^1\right)\varphi_2 \\
&\quad -\left(\sum_{lm} h_{im}g^{lm}h_{jl}\right)\mu.
\end{aligned}$$

We must therefore prove that the three coefficients of φ_1, φ_2 and μ are equal to those obtained when permuting the indices i and j. This is of course trivial for the coefficient of μ. We shall now prove the same for the coefficient of φ_1, the proof being analogous in the case of φ_2.

To achieve this, we first replace the coefficients g^{ij} by their values calculated in the proof of Proposition 6.2.4. We also replace the partial derivatives of the coefficients g^{ij} by their values given in Problem 6.18.6 (the proof is an easy routine calculation). Introducing all these values into the coefficient of φ_1, we obtain

$$\begin{aligned}
&\frac{1}{g_{11}g_{22} - g_{21}g_{12}}\left(\frac{\partial h_{j1}}{\partial x^i} g_{22} - \frac{\partial h_{j2}}{\partial x^i} g_{12} + 2h_{j1}\left(\Gamma_{2i}^1 g_{12} - \Gamma_{1k}^1 g_{22}\right)\right. \\
&\quad - h_{j2}\left(\Gamma_{1i}^2 g_{22} + \Gamma_{2i}^1 g_{11} - \Gamma_{1i}^1 g_{12} - \Gamma_{2i}^2 g_{12}\right) \\
&\quad \left. + h_{j1}g_{22}\Gamma_{i1}^1 - h_{j2}g_{12}\Gamma_{i1}^1 - h_{j1}g_{12}\Gamma_{i2}^1 + h_{j2}g_{11}\Gamma_{i2}^1\right)
\end{aligned}$$

$$= \frac{1}{g_{11}g_{22} - g_{21}g_{12}} \left(\frac{\partial h_{j1}}{\partial x^i} g_22 - \frac{\partial h_{j2}}{\partial x^i} g_{12} \frac{\partial h_{j1}}{\partial x^i} g_{22} - \frac{\partial h_{j2}}{\partial x^i} g_{12} \right.$$

$$\left. + h_{j1} g_{12} \Gamma_{2i}^1 - h_{j1} g_{22} \Gamma_{1i}^1 - h_{j2} g_{22} \Gamma_{1i}^2 + h_{j2} g_{12} \Gamma_{2i}^2 \right)$$

$$= \frac{1}{g_{11}g_{22} - g_{21}g_{12}} \left(g_{22} \left(\frac{\partial h_{j1}}{\partial x^i} - \sum_m h_{jm} \Gamma_{1i}^m \right) - g_{12} \left(\frac{\partial h_{j2}}{\partial x^i} - \sum_m h_{jm} \Gamma_{2i}^m \right) \right).$$

The last but one equality is obtained just by simplifying equal terms appearing with opposite signs. By the *Codazzi–Mainardi* equations, the coefficients of g_{22} and g_{12} in the last line are equal to those obtained when permuting the roles of the indices i and j. This concludes the proof of the integrability conditions.

Since the integrability conditions are satisfied, our system of partial differential equations admits solutions of class $\mathcal{C}^2$ in a neighborhood of each point. As usual, many solutions exist *a priori*, but we can force the uniqueness of the solution by imposing initial conditions at some fixed point $(x_0^1, x_0^2) \in U$. The idea is to choose initial conditions which force, at the given point (x_0^1, x_0^2), the properties that we eventually want to be satisfied, at all points, by the three functions φ_1, φ_2 and μ. More precisely, we want to have

$$\begin{cases} (\varphi_i(x_0^1, x_0^2) | \varphi_j(x_0^1, x_0^2)) = g_{ij}(x_0^1, x_0^2) \\ (\varphi_i(x_0^1, x_0^2) | \mu(x_0^1, x_0^2)) = 0 \\ (\mu(x_0^1, x_0^2) | \mu(x_0^1, x_0^2)) = 1 \end{cases}$$

since we want φ_1 and φ_2 to become the partial derivatives of a parametric representation, while μ should become the corresponding normal vector of length 1.

To force these requirement, let us first arbitrarily choose three vectors e_1, e_2, e_3 of $\mathbb{R}^3$ such that:

- e_3 is an arbitrary vector of length 1;
- e_1 and e_2 are perpendicular to e_3 and of lengths

$$\|e_1\| = \sqrt{g_{11}(x_0^1, x_0^2)}, \qquad \|e_2\| = \sqrt{g_{22}(x_0^1, x_0^2)};$$

- the angle θ between e_1 and e_2 is given by

$$\cos\theta = \frac{g_{12}(x_0^1, x_0^2)}{\sqrt{g_{11}(x_0^1, x_0^2)}\sqrt{g_{22}(x_0^1, x_0^2)}};$$

- the basis (e_1, e_2, e_3) of $\mathbb{R}^3$ has direct orientation

(see Definition 3.2.3 in [4], *Trilogy II*). Notice that all this makes sense by Proposition 6.2.2. Indeed $g_{11} > 0$ and $g_{22} > 0$; moreover $\cos^2\theta = 1$ would imply that the determinant of the metric tensor is zero, which is not the case; thus $\theta \neq k\pi$ and therefore e_1 and e_2 are not proportional.

The initial conditions that we impose on our system of partial differential equations are then simply

$$\varphi_1(x_0^1, x_0^2) = e_1, \qquad \varphi_2(x_0^1, x_0^2) = e_2, \qquad \mu(x_0^1, x_0^2) = e_3.$$

By Proposition B.4.1, we conclude the unique existence of three functions φ_1, φ_2 and μ of class $\mathcal{C}^2$, solutions of the system of partial differential equations above, and satisfying these initial conditions.

We are next interested in finding the expected function $f(x^1, x^2)$ such that

$$\frac{\partial f}{\partial x^1} = \varphi_1, \qquad \frac{\partial f}{\partial x^2} = \varphi_2.$$

This is another system of partial differential equations, which admits solutions in a neighborhood of our fixed point (x_0^1, x_0^2) as soon as the integrability conditions

$$\frac{\partial^2 f}{\partial x^i \partial x^j} = \frac{\partial^2 f}{\partial x^j \partial x^i}$$

are forced by the system of partial differential equations (see Proposition B.4.1 again). These integrability conditions thus mean

$$\frac{\partial \varphi_j}{\partial x^i} = \frac{\partial \varphi_i}{\partial x^j}$$

that is, considering the system of partial differential equations defining φ_1, φ_2 and μ

$$\Gamma_{ij}^1 \varphi_1 + \Gamma_{ij}^2 \varphi_2 + h_{ij}\mu = \Gamma_{ji}^1 \varphi_1 + \Gamma_{ji}^2 \varphi_2 + h_{ji}\mu.$$

These equalities hold by the assumption $h_{ij} = h_{ji}$ and because $\Gamma_{ij}^k = \Gamma_{ji}^k$, by Proposition 6.6.8.

But we know at once the general form of the solutions of the very simple system of partial differential equations defining f:

$$f(x^1, x^2) = \int_{x_0^1}^{x^1} \varphi_1(t, x^2)\,dt + v_0 = \int_{x_0^2}^{x^2} \varphi_2(x^1, t)\,dt + w_0$$

where v_0, w_0 are arbitrary constant vectors. (Of course, by taking the integral of a function with values in $\mathbb{R}^3$ we mean taking the integrals of its three components.) Fixing v_0 (or equivalently, w_0) as initial condition thus forces the uniqueness of the solution. Notice that since φ_1 and φ_2 are of class $\mathcal{C}^2$, f is of class $\mathcal{C}^3$.

Our next job is to prove that f is the parametric representation of a surface admitting precisely, as coefficients of its two fundamental quadratic forms, the coefficients g_{ij} and h_{ij}. For that, we need once more to rely on systems of partial differential equations.

We observe first that

$$\begin{aligned}
\frac{\partial(\varphi_i|\varphi_j)}{\partial x^k} &= \left(\frac{\partial\varphi_i}{\partial x^k}\middle|\varphi_j\right) + \left(\varphi_i\middle|\frac{\partial\varphi_j}{\partial x^k}\right)\\
&= \sum_l \Gamma_{ik}^l(\varphi_l|\varphi_j) + h_{ik}(\mu|\varphi_j) + \sum_l \Gamma_{jk}^l(\varphi_i|\varphi_l) + h_{jk}(\varphi_i|\mu)\\
\frac{\partial(\varphi_i|n)}{\partial x^k} &= \left(\frac{\partial\varphi_i}{\partial x^k}\middle|\mu\right) + \left(\varphi_i\middle|\frac{\partial\mu}{\partial x^k}\right)\\
&= \sum_l \Gamma_{ki}^l(\varphi_l|\mu) + h_{ki}(\mu|\mu) - \sum_{lm} h_{km}g^{lm}(\varphi_i|\varphi_l)\\
\frac{\partial(\mu|\mu)}{\partial x^k} &= 2\left(\frac{\partial\mu}{\partial x^k}\middle|\mu\right)\\
&= -2\sum_{lm} h_{km}g^{lm}(\varphi_l|\mu).
\end{aligned}$$

This proves that the functions

$$G_{ij} = (\varphi_i|\varphi_j), \qquad N_i = (\varphi_i|\mu), \qquad N = (\mu|\mu)$$

satisfy the system of partial differential equations

$$\begin{cases}
\frac{G_{ij}}{\partial x^k} = \sum_l \Gamma_{ik}^l G_{lj} + h_{ik}N_j + \sum_l \Gamma_{jk}^l G_{il} + h_{jk}N_i\\
\frac{\partial N_i}{\partial x^k} = \sum_l \Gamma_{ki}^l N_l + h_{ki}N - \sum_{lm} h_{km}g^{lm}G_{il}\\
\frac{\partial N}{\partial x^k} = -2\sum_{lm} h_{km}g^{lm}N_l.
\end{cases}$$

Observe that the initial conditions put on the system in φ_1, φ_2 and μ force precisely the satisfaction of the initial conditions:

$$G_{ij}\left(x_0^1, x_0^2\right) = g_{ij}\left(x_0^1, x_0^2\right), \qquad N_i\left(x_0^1, x_0^2\right) = 0, \qquad N\left(x_0^1, x_0^2\right) = 1.$$

Again Proposition B.4.1 on systems of partial differential equations asserts the uniqueness of a solution satisfying these initial conditions (this time there is no need to check the integrability conditions). Therefore to conclude that

$$\left(\frac{\partial f}{\partial x^i}\middle|\frac{\partial f}{\partial x^j}\right) = (\varphi_i|\varphi_j) = g_{ij}$$

it suffices to prove that

$$G_{ij} = g_{ij}, \qquad N_i = 0, \qquad N = 1$$

is also solution of the above system, satisfying the same initial conditions. By Proposition 6.6.8 and Definition 6.6.7

$$\sum_l \Gamma_{ik}^l g_{lj} + \sum_l \Gamma_{jk}^l g_{li} = \Gamma_{ikj} + \Gamma_{jki} = \frac{\partial g_{ij}}{\partial x^k}$$

and this takes care of the first equation. The second equation reduces to

$$0 = h_{ki} - \sum_{lm} h_{km} g^{lm} g_{il}$$

which reduces further to

$$0 = h_{ki} - h_{ki}$$

since the g^{lm} are the coefficients of the matrix inverse to that of the g_{im}. The third equation is trivially satisfied: it reduces to $\frac{\partial 1}{\partial x^k} = 0$. Moreover, the initial conditions indicated are trivially satisfied.

By uniqueness of the solution, we thus have at each point

$$g_{ij} = \left(\frac{\partial f}{\partial x^i}\middle|\frac{\partial f}{\partial x^j}\right), \qquad \left(\frac{\partial f}{\partial x^i}\middle|\mu\right) = 0, \qquad (\mu|\mu) = 1.$$

In particular, μ is at each point a vector of length 1 orthogonal to the partial derivatives of f.

When two vectors are linearly dependent, the matrix of their scalar products has zero determinant:

$$\begin{vmatrix} (\overrightarrow{u}\,|\,\overrightarrow{u}) & (\overrightarrow{u}\,|k\overrightarrow{u}) \\ (k\overrightarrow{u}\,|u) & (k\overrightarrow{u}\,|k\overrightarrow{u}) \end{vmatrix} = (u|u)^2 \begin{vmatrix} 1 & k \\ k & k^2 \end{vmatrix} = 0.$$

By Proposition 6.2.2, the condition

$$g_{ij} = \left(\frac{\partial f}{\partial x^i}\middle|\frac{\partial f}{\partial x^j}\right)$$

implies that the matrix of the scalar products of the partial derivatives of f is regular; these partial derivatives are thus linearly independent. Then by Proposition 5.2.4, f is a regular parametric representation of a surface. Since μ is of length 1 and orthogonal to the partial derivatives of f, it is the normal vector $\overrightarrow{n}$ to the surface.

We must still prove that the coefficients h_{ij} are those of the second fundamental quadratic form of f. We have already

$$\frac{\partial^2 f}{\partial x^i \partial x^j} = \frac{\partial \varphi_j}{\partial x^i} = \Gamma_{ij}\varphi_1 + \Gamma_{ij}\varphi_2 + h_{ij}\mu = \Gamma_{ij}\frac{\partial f}{\partial x^1} + \Gamma_{ij}\frac{\partial f}{\partial x^2} + h_{ij}\overrightarrow{n}$$

where the Γ_{ij}^k are the Christoffel symbols of the original Riemann patch and the h_{ij} are the functions given in the statement. Comparing with Definition 6.6.2, these equalities prove that the Γ_{ij}^k are also the Christoffel symbols of the surface represented by f, and the symbols h_{ij} are the coefficients of the second quadratic fundamental form. This concludes the proof of the existence of a surface f admitting the g_{ij} and h_{ij} of the statement as coefficients of its two fundamental quadratic forms. □

The proof of Theorem 6.16.2 indicates that the parametric representation f is unique for the given choices of initial conditions. Changing these initial conditions results simply in applying an isometry to the surface (see Problem 6.18.8). Thus the surface in Theorem 6.16.2 is in fact *unique up to an isometry*.

6.17 What Is a Riemann Surface?

The time has come to give an elegant solution to the problem raised in Sect. 5.1:

> *The sphere does not admit a parametric representation in the sense of Definition 5.1.1.*

In other words, once more thinking of the sphere as being the Earth, we cannot draw a "geographical map" of the whole Earth while respecting the requirements of Definition 5.1.1. Even if we were to forget the requirements in Definition 5.1.1, there is no particular practical interest in having a single geographical map of the whole Earth. Such a map would necessarily feature extreme distortions, so the sensible thing to do is to map the Earth using a full *atlas* of geographical maps. This is also how we define a surface in full generality.

Now let us be aware that every geographical map of a portion of the Earth—no matter how small—will necessarily have some distortions, because the Earth is not flat! If we have a full atlas of maps to describe the Earth, the same portion of the Earth may appear on several maps, with different distortions. As a consequence the "elastic rulers" called *metric tensors* will then be different for the same portion of the Earth on different maps. Nevertheless, these metric tensors will be equivalent, in order to calculate from the various maps the same actual result at the surface of the Earth.

Thus a surface should be a "universe" which can be mapped by an atlas of Riemann patches, in such a way that when two Riemann patches of the atlas describe a same portion of the "universe", they are equivalent as Riemann patches (see Definition 6.12.5). It remains to say what "universe" means: this is simply the very general notion of *topological space* (see Definition A.5.1). However, if you prefer not to enter into this level of generality, simply think of a "universe" as being a subset of $\mathbb{R}^3$ as in Chap. 5, provided with the usual notions of openness, continuity, and so on.

Definition 6.17.1 A *Riemann surface* of class $\mathcal{C}^k$ consists of:

1. a topological space $(X, \mathcal{T})$;
2. a covering $X = \bigcup_{i\in I} U_i$ of X by open subsets $U_i \in \mathcal{T}$;

3. for each index $i \in I$, a Riemann patch of class $\mathcal{C}^k$

$$g_{jl}\colon V_i \longrightarrow \mathbb{R}, \quad 1 \le j, l \le 2;$$

4. for each index $i \in I$, a homeomorphism

$$\varphi_i : V_i \longrightarrow U_i$$

which is called a *local map*.

These data must satisfy the following compatibility axiom. For every pair i, j of indices and every connected open subset $U \subseteq U_i \cap U_j$,

$$(\varphi_i)^{-1}(U) \xrightarrow{\varphi_i} U \xrightarrow{\varphi_j^{-1}} (\varphi_j)^{-1}(U)$$

is an equivalence of Riemann patches of class $\mathcal{C}^k$.

Extending the geographical terminology, the set of local maps is often called the *atlas* of local maps. In this book we shall mainly be interested in the following class of surfaces:

Definition 6.17.2 By a *Riemann surface in* $\mathbb{R}^3$ is meant a subset $X \subseteq \mathbb{R}^3$ provided with the induced topology (see Proposition A.5.4) and the structure of a Riemann surface (see Definition 6.17.1), in such a way that for each local map

$$\varphi_i : V_i \longrightarrow U_i$$

the corresponding metric tensor is

$$g_{jl} = \left(\frac{\partial \varphi_i}{\partial x_j} \middle| \frac{\partial \varphi_i}{\partial x_l} \right).$$

Notice that in Definition 6.17.2, the local map φ_i is a regular parametric representation of U_i viewed as an ordinary surface in $\mathbb{R}^3$ (see Definitions 5.1.1 and 5.2.1); the last requirement indicates that the Riemann structure on U_i is precisely that induced by the parametric representation (see Definition 6.1.1).

As expected:

Example 6.17.3 The sphere is a Riemann surface of $\mathbb{R}^3$.

Proof We know a parametric representation of the sphere of radius 1

$$x^2 + y^2 + z^2 = 1$$

punctured at its two poles $(0, 0, \pm 1)$ (see Example 5.1.6):

$$f(\theta, \tau) = (\cos\tau \cos\theta, \cos\tau \sin\theta, \sin\tau).$$

As we have seen, to be locally injective, this function must be considered on the open subset

$$V = \mathbb{R} \times \left]-\frac{\pi}{2}, +\frac{\pi}{2}\right[.$$

On this open subset of $\mathbb{R}^3$ we thus have four functions of class $\mathcal{C}^\infty$ defining at each point the metric tensor:

$$g_{ij}\colon V \longrightarrow \mathbb{R}, \quad 1 \leq i, j \leq 2$$

namely

$$\begin{pmatrix} g_{11} & g_{12} \\ g_{21} & g_{22} \end{pmatrix} = \begin{pmatrix} \cos^2 \tau & 0 \\ 0 & 1 \end{pmatrix}.$$

By Proposition A.9.4, each point $P_i \in V$ has a neighborhood V_i on which f is injective and moreover

$$f\colon V_i \longrightarrow f(V_i) = U_i$$

is a homeomorphism. Let us choose arbitrarily a family $(V_i)_{i\in I}$ of these V_i's such that the corresponding U_i cover the whole punctured sphere.

Let us now make some remarks which will help to support our intuition.

When producing a geographical atlas of the Earth, each map of the atlas generally corresponds to some fixed ranges of longitudes and latitudes, thus to some open rectangle in $\mathbb{R}^2$. Although it is not needed for the proof, to make the language more intuitive, we will freely choose each open subset V_i to be an open rectangle in $\mathbb{R}^2$.

Now we might be concerned that near the poles, the distortion of the maps "tends to infinity". First, we should be aware that such a distortion is mathematically not a problem at all, even if "geographically" it is certainly not recommended. Nevertheless if our intuition insists on avoiding excessive distortions, we are free to replace in what follows the "punctured sphere" by a "widely punctured sphere": for example,

$$f\colon \mathbb{R} \times]-\pi/3, \pi/3[\longrightarrow \mathbb{R}^3.$$

In such a case, we consider only those points whose latitude is "less than 60 degrees North or South".

Finally we might wonder how many maps we will have in our atlas. This depends on our choices, in particular on the size that we fix for each map. For example if we insist on covering the whole punctured sphere with individual maps whose distortion remains below some "geographically acceptable" bound, then near the poles, we will have to consider smaller and smaller maps, eventually ending up with infinitely many maps! Mathematically this is not a problem at all, since we do not have to physically print all these maps!

So long for this digression. Whatever our choice is: the punctured sphere or a "widely punctured" sphere, a finite or an infinite atlas, let us now observe that each U_i is an open subset of the sphere. By choice of the open subsets V_i (open rectangles in $\mathbb{R}^2$), U_i is thus the portion of the sphere situated between two meridians and two parallels. Join all points of the four edges of U_i (i.e. the portions of meridians and parallels limiting U_i) to the center of the sphere. The interior of the "generalized pyramid" obtained in this way is an open subset of $\mathbb{R}^3$, whose intersection with the sphere is precisely U_i. Therefore U_i is indeed an open subset of the sphere with respect to the induced topology, by Proposition A.5.4. (Of course V_i being a rectangle is unessential in this argument.)

Now the same point of the sphere can be written as

$$f(\theta_1, \tau_1) = f(\theta_2, \tau_2)$$

if and only if $\tau_1 = \tau_2$ while $\theta_1 = \theta_2 + 2k\pi$. Therefore if $U_i \cap U_j \neq \emptyset$, then the corresponding mapping

$$\varphi\colon f^{-1}(U_i \cap U_j) \xrightarrow{f} U_i \cap U_j \xrightarrow{f^{-1}} f^{-1}(U_i \cap U_j)$$

is simply given by

$$\varphi(\theta, \tau) = (\theta + 2k\pi, \tau).$$

This is a bijection of class $\mathcal{C}^\infty$ with inverse

$$\varphi^{-1}(\theta, \tau) = (\theta - 2k\pi, \tau)$$

still of class $\mathcal{C}^\infty$. This is thus a change of parameters of class $\mathcal{C}^\infty$ and therefore, by Proposition 6.12.2, it induces on each connected open subset an equivalence between the corresponding Riemann patches, as required by Definition 6.12.5.

This already presents the sphere, punctured (or widely punctured) at its two poles $(0, 0, \pm 1)$, as a Riemann surface in the sense of Definition 6.17.1.

In a perfectly analogous way, interchanging the roles of the second and the third components, the function

$$\widetilde{f}(\widetilde{\theta}, \widetilde{\tau}) = (\cos\widetilde{\tau}\cos\widetilde{\theta}, \sin\widetilde{\tau}, \cos\widetilde{\tau}\sin\widetilde{\theta})$$

is now a parametric representation of the sphere punctured (or widely punctured) at the two points $(0, \pm 1, 0)$. Just as above, this allows a presentation of this alternative punctured (or widely punctured) sphere as a Riemann surface. Let us write $\widetilde{V}_j$, $\widetilde{U}_j$ for the corresponding open subsets of $\mathbb{R}^2$ and of the sphere.

Considered together, these two punctured (or widely punctured) spheres cover the whole sphere. Therefore considered together, all the open subsets U_i and $\widetilde{U}_j$ cover the whole sphere. To conclude that we have so obtained a presentation of the sphere as a Riemann surface in the sense of Definition 6.17.1, it remains to prove the required compatibility condition when $U_i \cap \widetilde{U}_j \neq \emptyset$. Again by Proposition 6.12.2, this reduces to proving that the bijection

$$f^{-1}(U_i \cap \widetilde{U}_j) \xrightarrow{f} U_i \cap \widetilde{U}_j \xrightarrow{\widetilde{f}^{-1}} \widetilde{f}^{-1}(U_i \cap \widetilde{U}_j)$$

is a change of parameters of class $\mathcal{C}^\infty$, that is, is of class $\mathcal{C}^\infty$ as is its inverse.

So we must investigate the form of the change of parameters $\varphi(\theta,\tau)=(\widetilde{\theta},\widetilde{\tau})$ such that $f(\theta,\tau)=\widetilde{f}(\widetilde{\theta},\widetilde{\tau})$. This means

$$(\cos\tau\cos\theta,\cos\tau\sin\theta,\sin\tau)=(\cos\widetilde{\tau}\cos\widetilde{\theta},\sin\widetilde{\tau},\cos\widetilde{\tau}\sin\widetilde{\theta}).$$

But since in any case

$$-\frac{\pi}{2}<\tau<+\frac{\pi}{2},\qquad -\frac{\pi}{2}<\widetilde{\tau}<+\frac{\pi}{2}$$

we always necessarily have

$$\cos\tau\neq 0,\qquad \cos\widetilde{\tau}\neq 0.$$

Dividing the second component by the third one, on both sides of the equality, we then obtain

$$\tan\theta=\frac{\tan\widetilde{\tau}}{\cos\widetilde{\theta}}$$

while the third components give at once

$$\sin\tau=\cos\widetilde{\tau}\sin\widetilde{\tau}.$$

This proves that, at those points where both parametric representations are defined

$$(\theta,\tau)=\left(\arctan\frac{\tan\widetilde{\tau}}{\cos\widetilde{\theta}}\right),\qquad \arcsin(\cos\widetilde{\tau}\sin\widetilde{\theta}).$$

This is indeed a formula of class C^∞. An analogous proof holds for the inverse change of parameters, interchanging θ and $\widetilde{\theta}$, and analogously τ and $\widetilde{\tau}$. □

The example of the sphere should nevertheless not mislead the reader:

Warning 6.17.4 *Being a Riemann surface of $\mathbb{R}^3$ is a property which is neither stronger nor weaker than being a surface in the sense of Chap.* 5.

Proof The sphere is an example of a Riemann surface of $\mathbb{R}^3$ which does not admit a parametric representation in the sense of Chap. 5 (see Example 6.17.3).

On the other hand the surface of $\mathbb{R}^3$ (see Fig. 6.9) represented by

$$f\colon\mathbb{R}^2\longrightarrow\mathbb{R}^3,\qquad (u,v)\mapsto\left(\frac{u^2-1}{u^2+1},u\frac{u^2-1}{u^2+1},s\right)$$

is not a Riemann surface: at each "multiple" point $f(-1,v)=f(1,v)$ (a point where the surface "crosses itself"), every neighborhood of the point on the surface is constituted of two sheets, thus is not homeomorphic to an open subset of $\mathbb{R}^2$. □

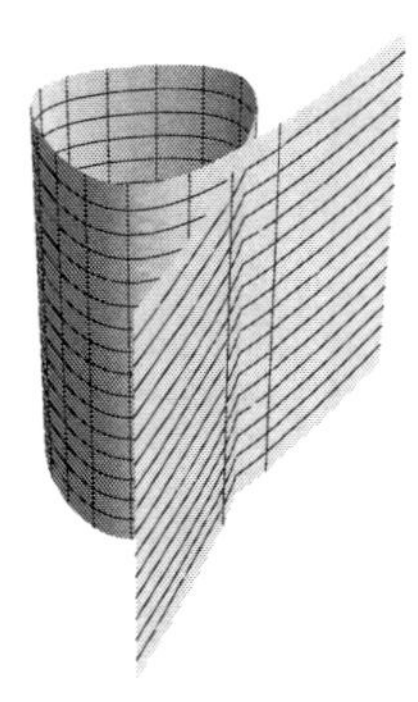

Fig. 6.9

Warning 6.17.5 *Even the support of an injective regular parametric representation*

$$f: U \longrightarrow \mathbb{R}^3$$

need not be a Riemann surface in $\mathbb{R}^3$.

Proof Observe that in the proof of Warning 6.17.4, restricting the parametric representation to the open subset

$$]-1, \infty[\times]-\infty, +\infty[$$

avoids having multiple points, but the problem remains the same at all points $f(1, v)$. For the topology induced by $\mathbb{R}^3$ on the support of the surface, each neighborhood of the point $f(1, v)$ is still comprised of two sheets. See also the comment at the end of Sect. 6.4. □

Let us conclude with a definition which gives evidence of the power of the notions and techniques developed in this chapter.

Definition 6.17.6 An n-dimensional Riemann patch of class $\mathcal{C}^k$ consists of a connected open subset $U \subseteq \mathbb{R}^n$, together with functions of class $\mathcal{C}^k$

$$g_{ij}: U \longrightarrow \mathbb{R}, \quad 1 \le i, j \le n$$

which, at each point $(x^1, \ldots, x^n)$, constitute a symmetric definite positive matrix.

The only differences with Definition 6.2.1 are:

- the replacement of $\mathbb{R}^2$ by $\mathbb{R}^n$;
- the fact that the indices vary from 1 to n.

If you want to develop *n-dimensional Riemannian geometry*, simply repeat all the definitions in dimension 2 by letting the indices vary from 1 to n. For example, if you are interested in Riemannian geometry of dimension 5, you will now have $5^3 = 125$ Christoffel symbols Γ_{ij}^k of the second kind, and $5^4 = 625$ components for the Riemann tensor. Nevertheless, the formulas remain "identical" to those in dimension 2: for example

$$R_{ijkl} = \frac{\partial \Gamma_{jli}}{\partial x^k} - \frac{\partial \Gamma_{jki}}{\partial x^l} + \sum_{\alpha=1}^{5} \left(\Gamma_{jk}^{\alpha} \Gamma_{li\alpha} - \Gamma_{jl}^{\alpha} \Gamma_{ki\alpha} \right), \quad 1 \le i, j, k, l \le 5.$$

Of course you can also transpose Definition 6.17.1 to dimension n, obtaining what is called a *Riemann manifold of dimension n*.

6.18 Problems

6.18.1 In a Riemann patch of class $\mathcal{C}^2$, prove that

$$R_{1212} = R_{2121} = -R_{1221} = -R_{2112}$$

while all other components of the Riemann tensor are equal to zero. Explain why your argument no longer works for Riemann patches of higher dimensions (see Definition 6.17.6).

6.18.2 Prove that the *Riemann tensor* is indeed a four times covariant tensor.

6.18.3 The *Riemann tensor* of Definition 6.11.4 is also called the *Riemann tensor of the second kind*. As you easily imagine, there is also a so-called *Riemann tensor of the first kind*:

$$R_{ijk}^{l} = \sum_{\alpha} g^{\alpha l} R_{\alpha ijk}.$$

Prove that this is indeed a tensor three times covariant and one time contravariant. Prove further that

$$R_{ijk}^{l} = \frac{\partial \Gamma_{ik}^{l}}{\partial x^j} - \frac{\partial \Gamma_{ij}^{l}}{\partial x^k} + \sum_{\alpha} \left(\Gamma_{ik}^{\alpha} \Gamma_{\alpha j}^{l} - \Gamma_{ij}^{\alpha} \Gamma_{\alpha k}^{l} \right)$$

while

$$R_{mijk} = \sum_{\alpha} g_{\alpha m} R_{ijk}^{\alpha}.$$

6.18.4 Prove that the Christoffel symbols, in a change of parameters, transform according to the formulas

$$\tilde{\Gamma}_{ijk} = \sum_{\gamma} \left(\sum_{\alpha,\beta} \Gamma_{\alpha\beta\gamma} \frac{\partial x^{\alpha}}{\partial \tilde{x}^i} \frac{\partial x^{\beta}}{\partial \tilde{x}^j} + \sum_{\alpha} g_{\alpha\gamma} \frac{\partial^2 x^{\alpha}}{\partial \tilde{x}^i \partial \tilde{x}^j} \right) \frac{\partial x^{\gamma}}{\partial \tilde{x}^k},$$

$$\tilde{\Gamma}_{ij}^{k} = \sum_{\gamma} \left(\sum_{\alpha,\beta} \Gamma_{\alpha\beta}^{\gamma} \frac{\partial x^{\alpha}}{\partial \tilde{x}^{i}} \frac{\partial x^{\beta}}{\partial \tilde{x}^{j}} + \frac{\partial^2 x^{\gamma}}{\partial \tilde{x}^{i} \partial \tilde{x}^{j}} \right) \frac{\partial \tilde{x}^{k}}{\partial x^{\gamma}}.$$

Therefore, they do not constitute a tensor.

6.18.5 With the comment at the end of Sect. 6.11 in mind, generalize Proposition 6.14.5 to express, in a system of geodesic coordinates, the Gaussian curvature of a Riemann patch.

6.18.6 Consider a Riemann patch of class $\mathcal{C}^2$

$$g_{ij}\colon U \longrightarrow \mathbb{R}, \quad 1 \le i, j \le 2.$$

Prove that

1. $\dfrac{\partial g_{ij}}{\partial x^k} = \Gamma_{ikj} + \Gamma_{jki} = \sum_l \Gamma_{ik}^l g_{lj} + \sum_l \Gamma_{jk}^l g_{li}$;
2. $\dfrac{\partial (g_{11}g_{22} - g_{21}g_{12})}{\partial x^k} = 2(\Gamma_{1k}^1 + \Gamma_{2k}^2)(g_{11}g_{22} - g_{21}g_{12})$;
3. $\dfrac{\partial g^{11}}{\partial x^k} = 2\dfrac{\Gamma_{2k}^1 g_{12} - \Gamma_{1k}^1 g_{22}}{g_{11}g_{22} - g_{21}g_{12}}, \quad \dfrac{\partial g^{22}}{\partial x^k} = 2\dfrac{\Gamma_{1k}^2 g_{21} - \Gamma_{2k}^2 g_{11}}{g_{11}g_{22} - g_{21}g_{12}},$

   $\dfrac{\partial g^{12}}{\partial x^k} = \dfrac{\partial g^{21}}{\partial x^k} = -\dfrac{\Gamma_{1k}^2 g_{22} + \Gamma_{2k}^1 g_{11} - \Gamma_{1k}^1 g_{21} - \Gamma_{2k}^2 g_{12}}{g_{11}g_{22} - g_{21}g_{12}}.$

(See Proposition 6.2.4 for the explicit values of the symbols g^{ij}.)

6.18.7 Prove that given a regular surface $f\colon U \longrightarrow \mathbb{R}^3$ of class $\mathcal{C}^2$ and an isometry $\varphi\colon \mathbb{R}^3 \longrightarrow \mathbb{R}^3$, the composite $\varphi \circ f$ is still a regular parametric representation of a surface. Prove that both fundamental quadratic forms of these surfaces have the same coefficients g_{ij} and h_{ij}.

6.18.8 Show that in the proof of Theorem 6.16.2:

- another choice of the vector v_0 (or equivalently, w_0) results in a translation of the surface;
- another choice of the vector e_3 keeps the point $f(x_0^1, x_0^2)$ fixed and results (via Theorem 4.12.4 in [4], *Trilogy II*) in a rotation of the surface around an axis passing through $f(x_0^1, x_0^2)$;
- another choice of the vectors e_1, e_2 in the plane perpendicular to e_3 results in a rotation of the surface around the axis of direction e_3 passing through $f(x_0^1, x_0^2)$;
- the choice of the inverse orientation for the basis (e_1, e_2, e_3) results in an orthogonal symmetry of the surface with respect to the plane passing through $f(x_0^1, x_0^2)$ and whose direction is that of the plane (e_1, e_2).

We conclude that the surface in Theorem 6.16.2 is defined *uniquely up to an isometry* (see Sect. 4.11 in [4], *Trilogy II*).

6.19 Exercises

6.19.1 Determine if the following pairs (U, g) are Riemann patches; if so, determine their class of differentiability.

1. $U = \mathbb{R}^2$; $g\colon U \longrightarrow \mathbb{R}^{2\times 2}$, $(x^1, x^2) \mapsto \begin{pmatrix} 1 & x^1 \\ -x^1 & 2 \end{pmatrix}$.
2. $U = \mathbb{R}^2 \setminus \{(0,0)\}$; $g\colon U \longrightarrow \mathbb{R}^{2\times 2}$, $(x^1, x^2) \mapsto \begin{pmatrix} \frac{1}{(x^1)^2+(x^2)^2} & 0 \\ 0 & 1 \end{pmatrix}$.
3. $U = \mathbb{R}^2$; $g\colon U \longrightarrow \mathbb{R}^{2\times 2}$, $(x^1, x^2) \mapsto \begin{pmatrix} \frac{1}{((x^1)^2+(x^2)^2+4)^2} & 0 \\ 0 & \frac{1}{((x^1)^2+(x^2)^2+4)^2} \end{pmatrix}$.
4. $U = \mathbb{R}^2$; $g\colon U \longrightarrow \mathbb{R}^{2\times 2}$, $(x^1, x^2) \mapsto \begin{pmatrix} |x^1| & \frac{1}{2}x^1 \\ \frac{1}{2}x^1 & |x^1| \end{pmatrix}$.
5. $U = \mathbb{R}^2$; $g\colon U \longrightarrow \mathbb{R}^{2\times 2}$, $(x^1, x^2) \mapsto \begin{pmatrix} x^1|x^1| & \frac{\sqrt{2}}{2}|x^1x^2| \\ \frac{\sqrt{2}}{2}|x^1x^2| & x^2|x^2| \end{pmatrix}$.

6.19.2 Construct a Riemann patch induced by:

1. the so-called *inverse plane*

$$\begin{cases} x = \frac{u}{u^2+v^2} \\ y = \frac{v}{u^2+v^2} \\ z = 0; \end{cases}$$

2. the hyperbolic paraboloid.

6.19.3 Consider the Riemann patch

$$U = \{(x^1, x^2) \,|\, x^1 < 0 \text{ or } x^2 \neq 0\};$$

$$g\colon U \to \mathbb{R}^{2\times 2}, \qquad (x^1, x^2) \mapsto \begin{pmatrix} 1+4(x^1)^2 & 4x^1x^2 \\ 4x^1x^2 & 1+4(x^2)^2 \end{pmatrix}.$$

Consider further

$$\widetilde{U} = \mathbb{R}^*_+ \times \,]0, 2\pi[\,;$$

$$\varphi\colon \widetilde{U} \to U, \qquad \varphi(\widetilde{x}^1, \widetilde{x}^2) = (\widetilde{x}^1 \cos \widetilde{x}^2, \widetilde{x}^1 \sin \widetilde{x}^2).$$

Determine $\widetilde{g}$ so that $\varphi\colon \widetilde{U} \longrightarrow U$ becomes an equivalence of Riemann patches between $(\widetilde{U}, \widetilde{g})$ and (U, g). What are the corresponding classes of differentiability?

6.19.4 Consider the cone $\mathcal{S}$ with parametric representation

$$f\colon U =]0, \infty[\times\mathbb{R} \longrightarrow \mathbb{R}^3, \qquad f(u, v) = (u \cos v, u \sin v, u).$$

1. Give a Riemann patch induced by $\mathcal{S}$.
2. In this Riemann patch, determine the tangent and the normal vector fields to the curves $x^1 = k$ and $x^2 = k$, with k a constant.

6.19.5 Compute the Riemann tensor of the Riemann patch defined in Exercise 6.19.1.3.

6.19.6 Compute the Christoffel symbols of the first and the second kind:

1. of the sphere of radius R centered at the origin;
2. of the cone.

6.19.7 On the circular cylinder

$$f : \mathbb{R}^2 \longrightarrow \mathbb{R}^3, \qquad f(u, v) = (\cos u, \sin u, v)$$

consider the curve $\mathcal{E}$ represented by

$$h : \mathbb{R} \longrightarrow \mathbb{R}^2, \qquad h(t) = (t, \sin t).$$

1. Show that this curve is an ellipse in $\mathbb{R}^3$.
2. Calculate the covariant derivative along $\mathcal{E}$ of the following vector fields, defined by their components with respect to the canonical basis of $\mathbb{R}^3$

$$\xi(t) = \begin{pmatrix} 0 \\ 0 \\ 1 \end{pmatrix}, \qquad \xi(t) = \begin{pmatrix} -\sin t \\ \cos t \\ 0 \end{pmatrix}.$$

6.19.8 On the sphere represented by

$$f(u, v) = (R \cos u \cos v, R \cos u \sin v, R \sin u), \qquad R > 0$$

consider the "parallel" $\mathcal{P}$ determined by $u = k$, $k \in \mathbb{R}$. Consider along $\mathcal{P}$ the vector field ξ admitting the components $\xi = \binom{1}{0}$ with repect to the basis of partial derivatives of f. Compute the covariant derivative of ξ along $\mathcal{P}$. Is ξ a parallel vector field?

6.19.9 Consider the helicoid with parametric representation

$$f : \mathbb{R}^2 \longrightarrow \mathbb{R}^3, \qquad f(u, v) = (u \cos v, u \sin v, v).$$

1. Construct a Riemann patch induced by the helicoid.
2. Compute the Christoffel symbols of the first and the second kind of the helicoid.
3. Consider the skew curve

$$f : \mathbb{R} \longrightarrow \mathbb{R}^3, \qquad f(t) = \left(t^2 \sin t, t^2 \cos t, \frac{\pi}{2} - t\right).$$

Is this curve a geodesic of the helicoid?

6.19.10 Let

$$g:]a, b[\longrightarrow \mathbb{R}^3, \qquad s \mapsto g(s)$$

be a normal parametric representation of class $\mathcal{C}^3$ of a 2-regular skew curve $\mathcal{C}$ (see Definition 4.1.3). Write $\mathbf{n}$ and $\mathbf{b}$ for the normal and the binormal vectors (see Definition 4.4.1). Given $a \in \mathbb{R}$, consider the surface $\mathcal{S}$ represented by

$$f:]a, b[\times \mathbb{R} \longrightarrow \mathbb{R}^3, \qquad f(s, u) = g(s) + u\big(\cos a \cdot \mathbf{n}(s) + \sin a \cdot \mathbf{b}(s)\big).$$

1. Determine a parametric representation of the surface $\mathcal{S}$ defined by choosing

$$a = 0, \qquad g(s) = \big(e^s \cos s, e^s \sin s, e^s\big).$$

2. Determine a constant a so that $\mathcal{C}$ becomes a geodesic of $\mathcal{S}$.
3. Construct a Riemann patch induced by $\mathcal{S}$ in the case $a = \frac{\pi}{2}$.
4. Choose again $a = \frac{\pi}{2}$. Compute the covariant derivative along $\mathcal{C}$ of the tangent vector field ξ having components $\xi = \left(\begin{smallmatrix} 1 \\ -1 \end{smallmatrix}\right)$ with respect to the basis of partial derivatives of f.

Chapter 7
Elements of the Global Theory of Surfaces

Up to now, most of the results concerning surfaces appearing in this book have referred only to what happens on some convenient neighborhood of a given point of the surface. This last chapter introduces some considerations which make sense only when one considers the surface globally.

First, we study some special families of surfaces, obtained by letting a curve revolve around an axis in its plane (the *surfaces of revolution*), by letting a line move continuously along a curve (the *ruled surfaces*), or by rolling up a piece of the plane (the *developable surfaces*). We also pay special attention to the surfaces with constant Gaussian curvature and in particular, to the sphere.

Of course, arriving at the end of this trilogy, we also "open some doors" to further fascinating developments of geometry. We achieve this by drawing the reader's attention to some striking results whose proofs often rely on some deep topological results which are beyond the scope of this book (such as the *Jordan curve theorem*).

We switch to the study of *curve polygons drawn on a surface*. Making clear which topological results we rely upon, we prove the famous *Gauss–Bonnet theorem* and we conclude with some first considerations on the *Euler–Poincaré* characteristics of a surface: an integer which gives information concerning the global shape of the surface.

7.1 Surfaces of Revolution

A first class of surfaces of interest is given by the *surfaces of revolution*:

Definition 7.1.1 A *surface of revolution* in three dimensional Euclidean space is one which can be obtained by revolving a plane curve $\mathcal{C}$ around a line contained in the plane of the curve, but not intersecting the curve (see Fig. 7.1).

Of course, there is no loss of generality in choosing the axis of revolution to be the z-axis. Then

F. Borceux, *A Differential Approach to Geometry*, DOI 10.1007/978-3-319-01736-5_7,

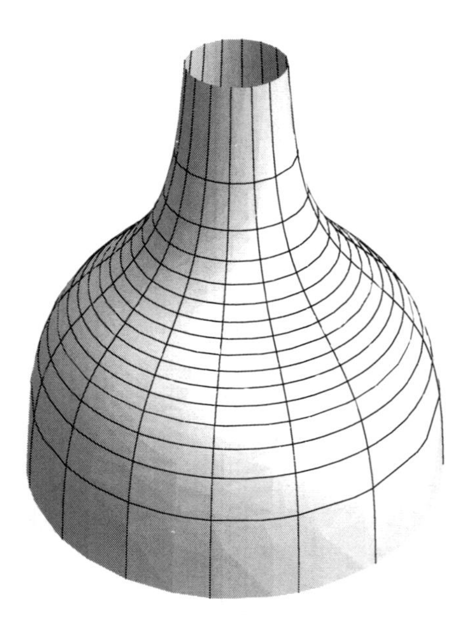

Fig. 7.1 A surface of revollution

Proposition 7.1.2 *In the xz-plane of three dimensional Euclidean space, consider a plane curve $\mathcal{C}$ of class $\mathcal{C}^k$ represented by*

$$f\colon]a,b[\longrightarrow \mathbb{R}^2, \qquad t \mapsto (x,z) = \big(f_1(t), f_2(t)\big)$$

and not intersecting the z-axis, i.e.

$$\forall t \in]a,b[\quad f_1(t) \neq 0.$$

Revolving the curve $\mathcal{C}$ about the z-axis yields a surface of revolution admitting the parametric representation of class $\mathcal{C}^k$

$$g\colon \mathbb{R} \times]a,b[\longrightarrow \mathbb{R}^3, \qquad (t,\theta) \mapsto (x,y,z) = \big(f_1(t)\cos\theta, f_1(t)\sin\theta, f_2(t)\big).$$

This representation g is regular as soon as f is regular.

Proof Of course, revolving $\mathcal{C}$ about the z-axis yields the "representation" g of the statement. Trivially when f is of class $\mathcal{C}^k$, so is g.

To have a parametric representation g of a surface, we must still prove that g is locally injective. But $g(t,\theta)$ lies in the plane π_θ containing the z-axis and making an angle θ with the xz-plane; moreover, $g(t,\theta)$ is never on the z-axis, because $f_1(t) \neq 0$ for all t. Thus two points $g(\theta_1, t_1)$ and $g(\theta_2, t_2)$, for two "close distinct values" θ_1, θ_2, lie in two different vertical planes $\pi_{\theta_1}, \pi_{\theta_2}$ but not in their intersection (the z-axis); these two points are thus distinct, whatever the values of t_1, t_2. On the other hand for θ_0 fixed, since f is locally injective, so is

$$t \mapsto \big(f_1(t)\cos\theta_0, f_1(t)\sin\theta_0, f_2(t)\big), \qquad \theta_0 \in \mathbb{R}$$

while at least one of the two quantities $\cos\theta_0$, $\sin\theta_0$ is non-zero. This proves that g is locally injective.

When f is regular, the two partial derivatives of g are

$$\frac{\partial g}{\partial t} = \big(f_1'(t)\cos\theta, f_1'(t)\sin\theta, f_2'(t)\big)$$
$$\frac{\partial g}{\partial \theta} = \big(-f_1(t)\sin\theta, f_1(t)\cos\theta, 0\big).$$

The cross product of these partial derivatives (see Sect. 1.7 in [4], *Trilogy II*), is then

$$\begin{aligned}&\big(-f_1(t)f_2'(t)\cos\theta, -f_1(t)f_2'(t)\sin\theta, f_1(t)f_1'(t)\big)\\ &\quad = f_1(t)\big(-f_2'(t)\cos\theta, -f_2'(t)\sin\theta, f_1'(t)\big).\end{aligned}$$

By Proposition 1.7.2 in [4], *Trilogy II*, we must prove that this vector is non-zero. By assumption, $f_1(t) \neq 0$, thus if $f_1'(t) \neq 0$ we are done. Otherwise, when $f_1'(t) = 0$, by regularity of f, we have $f_2'(t) \neq 0$; since $\cos\theta$, $\sin\theta$ never vanish together, the cross product is again non-zero. Let us recall that this regularity condition is already sufficient to force the local injectivity (see Proposition 5.2.4). □

The assumption that the curve $\mathcal{C}$ does not intersect the z-axis is essential in Proposition 7.1.2: if we had allowed $f(t_0) = 0$, then all pairs (t_0, θ) would have been mapped to the same point $g(t_0, \theta) = (0, 0, f_2(t_0))$, contradicting the local injectivity of g.

Of course the first example that we have in mind is the following.

Example 7.1.3 The *sphere*, punctured at its two poles, is a surface of revolution (see Fig. 7.2).

Proof The sphere is obtained by letting a half circle revolve about the diameter joining its two extremities. Omitting these two extremities gives the expected result. The parametric representation

$$f(\theta, \tau) = (r\cos\tau\cos\theta, r\cos\tau\sin\theta, r\sin\tau)$$

of Example 5.1.6 is in the expected form given by Proposition 7.1.2. □

In Example 7.1.3, it does not surprise us that we have to consider a *punctured sphere*, since the whole sphere does not admit a parametric representation in the sense of Definition 5.1.1 (see also Example 6.17.3). The following example is perhaps more enlightening in this respect:

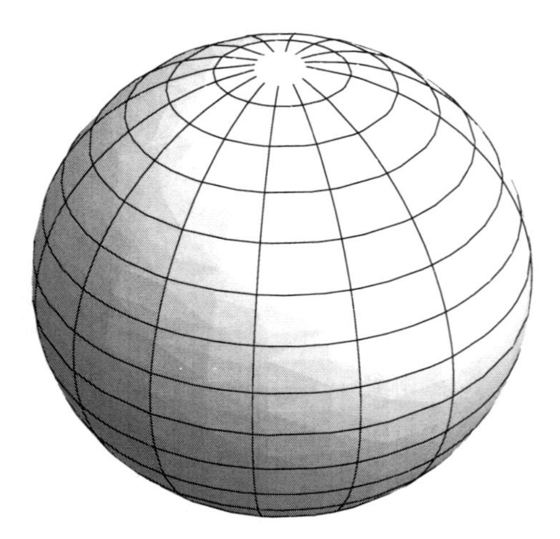

Fig. 7.2 The sphere

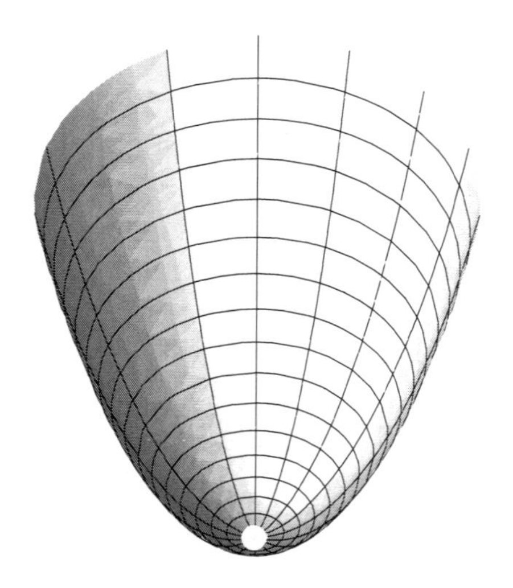

Fig. 7.3 A punctured circular elliptic paraboloid

Counterexample 7.1.4 The *circular elliptic paraboloid* (see Fig. 7.3) with equation

$$z = x^2 + y^2$$

is not a surface of revolution in the sense of Definition 7.1.1, but when it is punctured at the origin it becomes such a surface of revolution.

Proof Of course this is a regular surface of class $\mathcal{C}^\infty$ with parametric representation

$$f(x, y) = \big(x, y, x^2 + y^2\big)$$

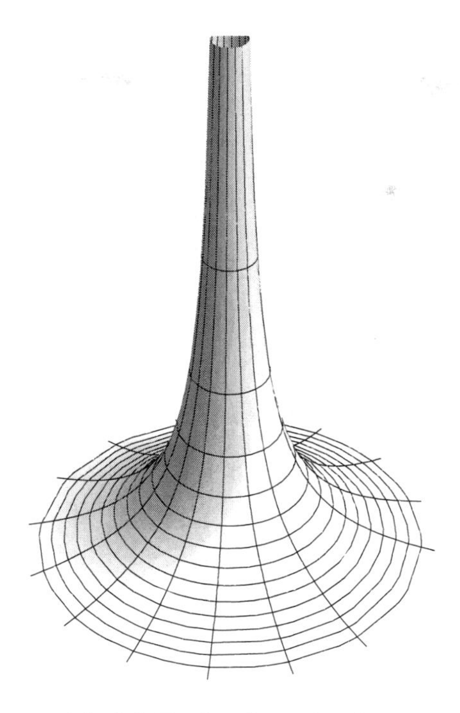

Fig. 7.4 The (hemi)-pseudo-sphere

(see Example 5.1.2). This surface is also obtained by revolving a parabola about its axis of symmetry, which yields the other "representation"

$$g(t,\theta) = \big(t\cos\theta, t\sin\theta, t^2\big).$$

But g is not a parametric representation in the sense of Definition 5.1.1, because it is not locally injective: all the points $(\theta, 0)$ are mapped onto the origin. Of course imposing $t > 0$ avoids the problem and presents the paraboloid punctured at the origin as a surface of revolution: the surface obtained by letting a half parabola revolve about the z-axis. □

One could of course decide to generalize our Definition 7.1.1 in order to recapture more examples of surfaces of revolution, such as the full paraboloid. We shall not do this. Let us instead review some other examples of interest:

Example 7.1.5 The *(hemi)-pseudo-sphere* with parametric representation

$$f(r,\theta) = \left(r\cos\theta, r\sin\theta, \int_r^1 \sqrt{\frac{1}{t^2} - 1}\, dt\right), \quad 0 < r < 1$$

(see Fig. 7.4 and Example 5.16.7) is a surface of revolution.

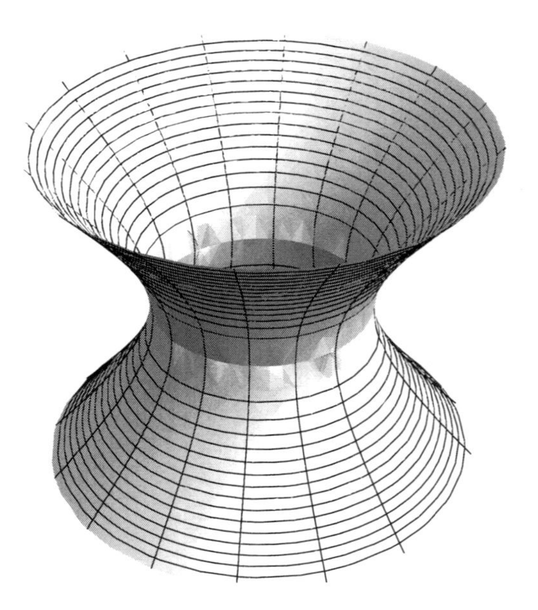

Fig. 7.5 The circular hyperboloid of one sheet

Proof The parametric representation f is already in the form given in Proposition 7.1.2, since $r > 0$. The example of the whole pseudo-sphere follows analogously (see Problem 5.17.15). □

Example 7.1.6 The *circular hyperboloid* of one sheet (see Fig. 7.5)

$$x^2 + y^2 - z^2 = 1$$

is a surface of revolution (see Sect. 1.14 in [4], *Trilogy II*).

Proof By Proposition 7.1.2, it suffices to consider the function f in the proof of Example 5.1.5. This surface is thus obtained by letting a branch of a hyperbola revolve about one of the axes of symmetry of the hyperbola. □

Example 7.1.7 The *torus* is a surface of revolution.

Proof By Example 5.1.7 (see Fig. 7.6), the torus admits the parametric representation

$$f(\tau, \theta) = \big((R + r\cos\tau)\cos\theta, (R + r\cos\tau)\sin\theta, r\sin\tau\big), \qquad R > r$$

and is thus a surface of revolution by Proposition 7.1.2. Let us recall that it is obtained by revolving a circle about an axis not intersecting the circle. □

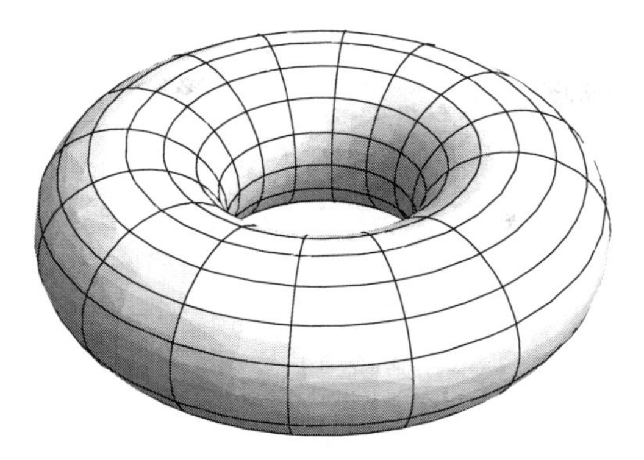

Fig. 7.6 The torus

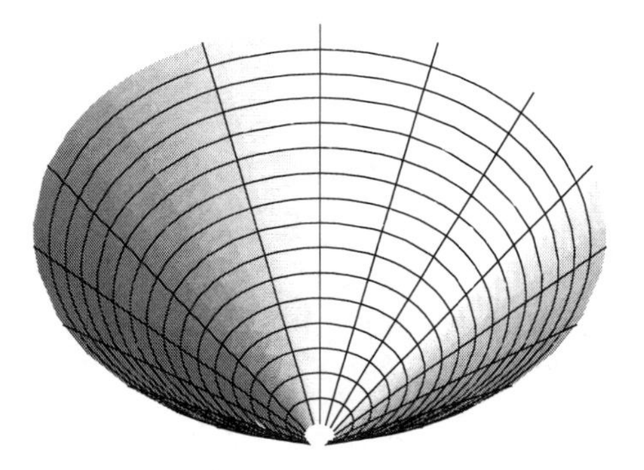

Fig. 7.7 A punctured circular semi-cone

Example 7.1.8 A *half circular cone* punctured at its vertex (Fig. 7.7) is a surface of revolution.

Proof Simply let the first half diagonal of the xz-plane revolve about the z-axis:

$$f(t,\theta) = (t\cos\theta, t\sin\theta, t), \quad t > 0$$

to obtain a parametric representation as in Proposition 7.1.2. This yields the upper half of the cone with equation

$$x^2 + y^2 - z^2 = 0$$

punctured at its vertex (see Sect. 1.14) in [4], *Trilogy II*. □

Example 7.1.9 A circular cylinder (see Fig. 7.8) is a surface of revolution.

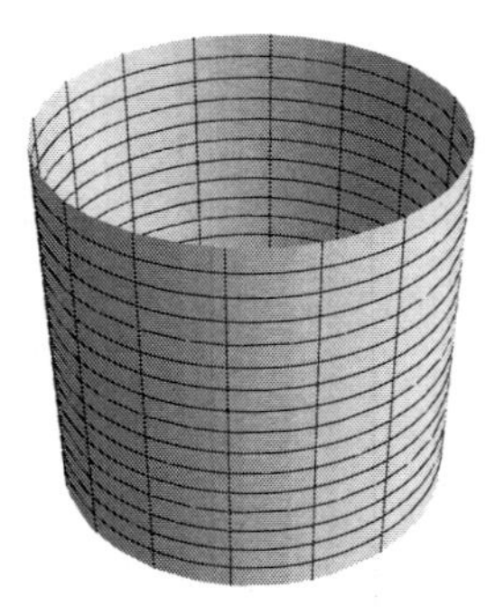

Fig. 7.8 The circular cylinder

Proof Simply let the vertical line $x = k$ ($k > 0$) of the xz-plane revolve around the z-axis:

$$f(t, \theta) = (k \cos\theta, k \sin\theta, t)$$

to get a parametric representation as in Proposition 7.1.2. □

It is common practice to make the further definition:

Definition 7.1.10 Given a surface of revolution as in Proposition 7.1.2, the *meridians* are the various positions of the curve $\mathcal{C}$ in the space, that is, the curves

$$t \mapsto \big(f_1(t)\cos\theta_0,\ f_1(t)\sin\theta_0,\ f_2(t)\big), \quad \theta_0 \in \mathbb{R}.$$

The *parallels* are the various circles, trajectories of a fixed point of the curve $\mathcal{C}$ during the revolution, that is, the curves

$$\theta \mapsto \big(f_1(t_0)\cos\theta,\ f_1(t_0)\sin\theta,\ f_2(t_0)\big), \quad t_0 \in\,]a, b[.$$

This clearly extends the classical terminology in the case of the sphere.

Proposition 7.1.11 *We consider a surface of revolution and use the notation and terminology of Proposition* 7.1.2 *and Definition* 7.1.11. *We assume that f is a normal parametric representation of class $\mathcal{C}^3$.*

1. *The meridians are geodesics and lines of curvature.*
2. *The parallels are lines of curvature.*
3. *The meridians are perpendicular to the parallels.*
4. *The normal curvature of each parallel is constant and equal to* $\frac{f_2'(t_0)}{f_1(t_0)}$.
5. *The Gaussian curvature of the surface is equal to* $-\frac{f_1''}{f_1}$.
6. *The coordinates (t, θ) as in Proposition* 7.1.2 *constitute a system of geodesic coordinates on each possible local map of the surface* (*see Definitions* 6.17.2 *and* 6.17.1).

Proof Up to a possible change of parameters, there is no loss of generality in assuming that $0 \in]a, b[$. As already observed in the proof of Proposition 7.1.2

$$\frac{\partial g}{\partial t} = \big(f_1'(t)\cos\theta,\, f_1'(t)\sin\theta,\, f_2'(t)\big)$$

$$\frac{\partial g}{\partial \theta} = \big(-f_1(t)\sin\theta,\, f_1(t)\cos\theta, 0\big),$$

from which we deduce at once

$$E = f_1'(t)^2 + f_2'(t)^2 = \big\|f'(t)\big\|^2 = 1, \qquad F = 0, \qquad G = f_1(t)^2.$$

The meridians are the curves $\theta = constant$ while the parallels are the curves $t = constant$. Since $E = 1$, the meridians are regular curves given in normal representation. Since $f_1(t) \neq 0$, we have $G \neq 0$ and the parallels are regular curves as well. Since $F = 0$, the meridians are perpendicular to the parallels.

The normal vector to the surface is the cross product of the two partial derivatives, divided by its norm. As observed in the proof of Proposition 7.1.2, this vector is proportional to

$$\overrightarrow{n} = \big(-f_2'(t)\cos\theta, -f_2'(t)\sin\theta,\, f_1'(t)\big)$$

and in fact is equal to that quantity, since the normality of f forces this vector $\overrightarrow{n}$ to be of norm 1.

The second partial derivatives of g are

$$\frac{\partial^2 g}{\partial t^2} = \big(f_1''(t)\cos\theta,\, f_1''(t)\sin\theta,\, f_2''(t)\big)$$

$$\frac{\partial^2 g}{\partial t \partial \theta} = \big(-f_1'(t)\sin\theta,\, f_1'(t)\cos\theta, 0\big)$$

$$\frac{\partial^2 f}{\partial \theta^2} = \big(-f_1(t)\cos\theta, -f_1(t)\sin\theta, 0\big).$$

It follows that

$$L = -f_2'(t)f_1''(t) + f_1'(t)f_2''(t), \qquad M = 0, \qquad N = f_1(t)f_2'(t).$$

But $(-f_2'(t), f_1'(t))$ is a vector perpendicular to $f'(t)$, with length 1 since f is in normal representation. This is thus a vector of length 1 parallel to $f''(t)$ (see Proposition 2.8.3). This forces

$$|L| = \big\|f''(t)\big\| = \kappa(t)$$

where κ indicates the curvature of the plane curve represented by f (see Definition 2.9.1).

With respect to the basis of partial derivatives in the tangent plane, the meridians are at each point the curves in the direction $(1, 0)$. Their normal curvature is thus simply (see Theorem 5.8.1)

$$\kappa_n(1,0) = L = \pm\kappa(t).$$

But since for a curve on a surface

$$\kappa^2 = \kappa_n^2 + \kappa_g^2$$

where κ_g is the geodesic curvature (see Proposition 6.9.3), we conclude that the geodesic curvature of the meridians is zero at each point. Thus the meridians are geodesics (see Definition 6.10.1).

Analogously the parallels are at each point in the direction (0, 1). Thus the normal curvature of the parallel $t = t_0$ is given by

$$\kappa_n(0,1) = \frac{N}{G} = \frac{f_1(t_0)f_2'(t_0)}{f_1(t_0)^2} = \frac{f_2'(t_0)}{f_1(t_0)}.$$

Let us now consider a non-umbilical point. Considering Proposition 5.10.5 and adopting its notation, choosing $\kappa = L$ shows that $(1,0)$ is a principal direction while choosing $\kappa = \frac{N}{G}$ shows that (0, 1) is a principal direction. This proves that the meridians and the parallels are lines of curvature (see Definition 5.14.1).

By Corollary 6.13.2, we are in a system of geodesic coordinates, on each possible local map.

The Gaussian curvature of the surface is given by (see Proposition 5.16.3)

$$\kappa_\tau = \frac{LN - M^2}{EG - F^2} = \frac{(-f_2'f_1'' + f_1'f_2'')(f_1 f_2')}{f_1^2} = \frac{-(f_2')^2 f_1'' + f_1' f_2'' f_2'}{f_1}.$$

By normality of f (see Definition 2.8.1) we have $(f_1')^2 + (f_2')^2 = 1$; differentiating this equality yields $f_2' f_2'' = -f_1' f_1''$. We thus obtain

$$\kappa_\tau = -\frac{f_1''}{f_1}\left((f_1')^2 + (f_2')^2\right) = -\frac{f_1''}{f_1}.$$

□

7.2 Ruled Surfaces

Roughly speaking, a *ruled surface* is one obtained by letting a straight line move continuously in space. Fixing arbitrarily a point on the line, its various positions in space constitute a trajectory which, most often, will satisfy the conditions for being a skew curve (see Definition 4.1.1). This suggests the following definition.

Definition 7.2.1 A *ruled surface* is one admitting a parametric representation of the form

$$f\colon U \longrightarrow \mathbb{R}^3, \qquad (s,t) \mapsto c(t) + s\,\xi(t)$$

where

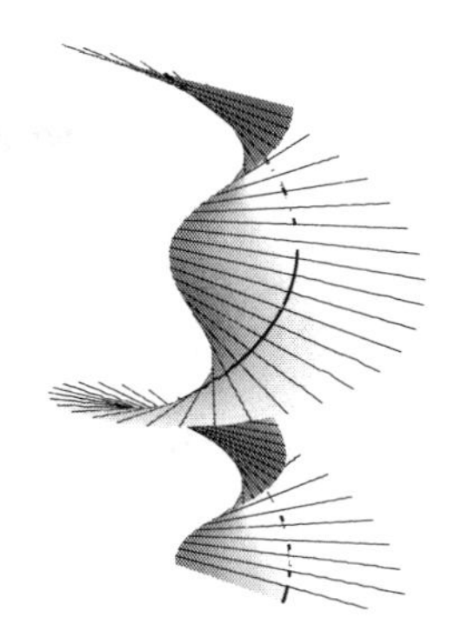

Fig. 7.9 A ruled surface

1. $c\colon]a,b[\longrightarrow \mathbb{R}^3$ is a parametric representation of a skew curve;
2. $\xi :]a,b[\longrightarrow \mathbb{R}^3$ is a continuous function such that $\xi(t) \neq 0$ for all t;
3. $U \subseteq \mathbb{R} \times]a,b[$ is a connected open subset containing points (s,t) for each $t \in]a,b[$.

The various lines (or portions of lines) on the surface, represented by

$$s \mapsto c(t) + s\,\xi(t), \qquad t \in]a,b[,$$

are called the *rulings* of the surface (see Fig. 7.9) and the curve c is called a *directrix* of the surface.

Let us stress that:

Counterexample 7.2.2 Given a skew curve c and a function ξ as in Definition 7.2.1, the function

$$(s,t) \mapsto f(s,t) = c(t) + s\xi(t)$$

is generally not a parametric representation of a surface, whatever the choice of the open subset U.

Proof Simply choose the curve to be the x-axis and ξ to be the constant function on $(1,0,0)$. Then each point $f(s,t)$ lies on the x-axis. □

So indeed, requiring that f is a parametric representation of a surface (or conditions implying this fact) must remain an explicit condition in Definition 7.2.1.

The first observation that we want to make concerning ruled surfaces is:

Proposition 7.2.3 *The Gaussian curvature of a regular ruled surface*

$$f(s,t) = c(t) + s\xi(t)$$

where c and s are of class $\mathcal{C}^2$ is always negative and given by

$$\kappa_\tau = \frac{-M^2}{EG - F^2}.$$

Proof Trivially, $\frac{\partial^2 f}{\partial s^2} = 0$, thus $N = 0$. The result follows by Proposition 5.16.3. □

The following property is also worth mentioning:

Proposition 7.2.4 *Consider a regular ruled surface represented by*

$$f \colon \mathbb{R} \times]a, b[\longrightarrow \mathbb{R}^3, \qquad (s, t) \mapsto c(t) + s\xi(t).$$

1. *The curve $c(t)$ is regular.*
2. *The tangent plane at every point with parameters (s, t) contains the ruling through that point.*
3. *The tangent plane is constant along the ruling*

$$s \mapsto c(t_0) + s\xi(t_0)$$

   if and only if the vectors $\xi(t_0)$, $\xi'(t_0)$ and $c'(t_0)$ are linearly dependent.
4. *When $c'(t_0)$, $\xi(t_0)$ and $\xi'(t_0)$ are linearly independent, the tangent plane to the surface at each point of this ruling rotates by a half turn around the ruling while s runs from $-\infty$ to $+\infty$.*

Proof As observed in Sect. 5.5, the tangent plane at a given point is the plane containing all the tangents to all the curves on the surface through this point. Since the tangent to a line is the line itself, the tangent plane at a point of a ruling contains this ruling.

The tangent plane at the point with parameters (s, t_0) has a direction generated by the two partial derivatives (see Definition 5.5.2) of f, that is

$$\frac{\partial f}{\partial t}(s, t_0) = c'(t_0) + s\,\xi'(t_0), \qquad \frac{\partial f}{\partial s}(s, t_0) = \xi(t_0).$$

By regularity of the surface, these two vectors are linearly independent for all values of s, thus in particular for $s = 0$. This proves already that $c'(t_0)$ and $\xi(t_0)$ are linearly independent, thus in particular that $c(t)$ is a regular curve (see Definition 2.2.1). But then the two partial derivatives generate the same direction plane, for all values of s, if and only if $\xi'(t_0)$ is a linear combination of the two linearly independent vectors $c'(t_0)$ and $\xi(t_0)$.

When the three vectors $c'(t_0)$, $\xi(t_0)$ and $\xi'(t_0)$ are linearly independent, distinct points of the ruling yield distinct directions of the tangent plane. When s runs from $-\infty$ to $+\infty$, the direction of the first partial derivative rotates through a half turn: from the direction of $-\xi'(t_0)$ at $-\infty$, to that of $+\xi'(t_0)$ at $+\infty$, passing through $c'(t_0)$ for $s = 0$. □

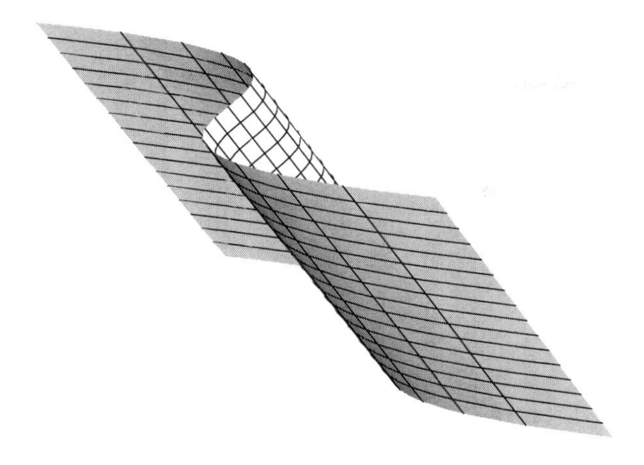

Fig. 7.10

Definition 7.2.5 Under the conditions of Proposition 7.2.4:

1. the ruling is called *singular* when the tangent plane is constant along that ruling;
2. when the ruling is not singular, the point of the ruling where the tangent plane is orthogonal to its limit position is called the *striction point* of the ruling.

In some cases, all rulings have a striction point and these striction points constitute a skew curve which can be taken as a *directrix* to describe the ruled surface. This is somehow a "canonical choice" of the directrix. But singular rulings will also turn out to play a crucial role later (see Theorem 7.6.1).

Our first example is that of a cylinder:

Example 7.2.6 Consider a regular plane curve C and a line ℓ not parallel to the plane of the curve. The figure comprising all the lines parallel to ℓ and passing through a point of C is called a (general) *cylinder* (see Fig. 7.10). Such a cylinder is a regular ruled surface.

Proof We consider again a regular plane curve represented by

$$c\colon\ ]a,b[\ \to \mathbb{R}^3, \qquad t \mapsto \bigl(c_1(t), c_2(t), 0\bigr)$$

and a vector (k,l,m) with $m \neq 0$. We define $\xi(t)$ to be the constant function on (k,l,m). The mapping

$$f: \mathbb{R} \times\ ]a,b[\ \longrightarrow \mathbb{R}^3, \qquad (s,t) \mapsto c(t) + s\xi(t) = \bigl(c_1(t)+sk, c_2(t)+sl, sm\bigr)$$

admits the partial derivatives

$$\frac{\partial f}{\partial t} = \bigl(c_1'(t), c_2'(t), 0\bigr), \qquad \frac{\partial f}{\partial s} = \bigl(k, l, m\bigr).$$

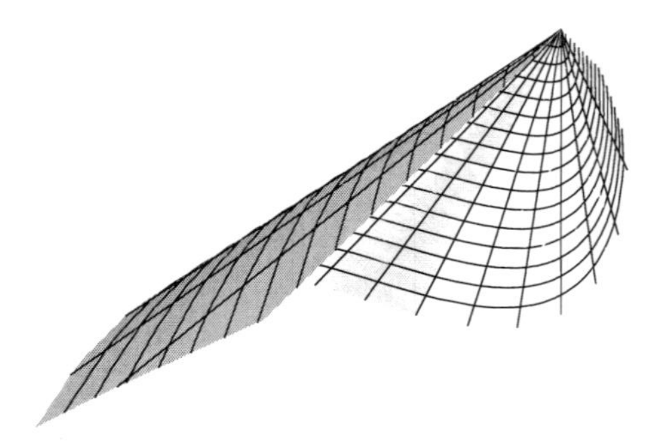

Fig. 7.11

Since c is regular and $m \neq 0$, this is a regular parametric representation (see Proposition 5.2.4) of a ruled surface (see Definition 7.2.1), namely, the cylinder indicated in the statement. □

Our next example is that of a cone:

Example 7.2.7 Consider a regular plane curve $\mathcal{C}$ and a point P not in the plane of $\mathcal{C}$. The figure comprising all the lines through P and a point of $\mathcal{C}$ is called a (general) *cone* (see Fig. 7.11). The point P is called the *vertex* of this cone. A half cone, punctured at its vertex, is a regular ruled surface.

Proof Let us consider the regular plane curve represented by

$$c\colon \left]a,b\right[\longrightarrow \mathbb{R}^3, \qquad t \mapsto \big(c_1(t), c_2(t), 0\big)$$

and the vertex $P = (k,l,m)$ with $m \neq 0$. For each value of t, we define $\xi(t)$ to be the vector joining P to $c(t)$:

$$\xi(t) = c(t) - (k,l,m).$$

We consider further the mapping

$$f(s,t) = c(t) + s\xi(t) = (1+s)c(t) - s(k,l,m).$$

The two partial derivatives of f are

$$\frac{\partial f}{\partial t} = (1+s)c'(t), \qquad \frac{\partial f}{\partial s} = c(t) - (k,l,m)$$

that is

$$\frac{\partial f}{\partial t} = \big((1+s)c_1'(t), (1+s)c_2'(t), 0\big), \qquad \frac{\partial f}{\partial s} = \big(c_1(t) - k, c_2(t) - l, m\big).$$

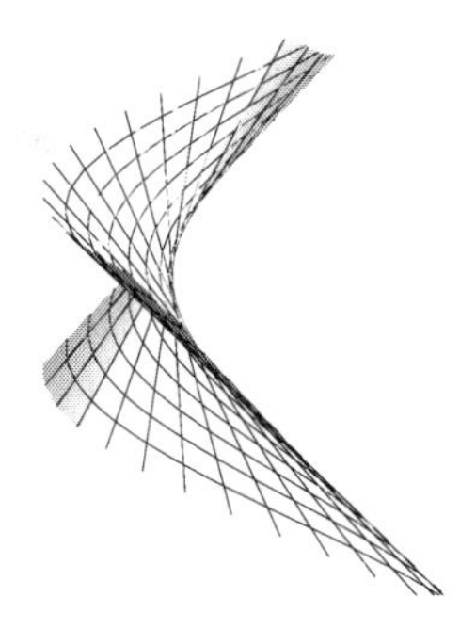

Fig. 7.12

Since $m \neq 0$ and c is regular, these two vectors are linearly independent as soon as $s \neq -1$. Thus by Proposition 5.2.4 and Definition 7.2.1,

$$f\colon]-1, \infty[\times]a, b[, \qquad (s, t) \mapsto c(t) + s\,\xi(t)$$

is a regular parametric representation of a ruled surface: namely, the half cone punctured at its vertex. □

Notice that this example of a cone also clearly shows why a ruled surface should not be defined by

$$f\colon \mathbb{R} \times]a, b[\longrightarrow \mathbb{R}^3, \qquad (s, t) \mapsto c(t) + s\xi(t).$$

The following example has already been considered in Exercise 5.18.19.

Example 7.2.8 Consider a 2-regular skew curve of class $\mathcal{C}^3$. All the half-tangents to this curve constitute a regular ruled surface called the *tangent surface* to the curve (see Fig. 7.12).

Proof Write

$$c\colon]a, b[\longrightarrow \mathbb{R}^3, \qquad t \mapsto c(t)$$

for a parametric representation of the curve. The figure composed of all the tangents to that curve is described by the function of class $\mathcal{C}^2$

$$f\colon \mathbb{R} \times]a, b[\longrightarrow \mathbb{R}^3, \qquad (s, t) \mapsto c(t) + sc'(t).$$

The partial derivatives of f are

$$\frac{\partial f}{\partial t} = c'(t) + s\,c''(t), \qquad \frac{\partial f}{\partial s} = c'(t).$$

For each $s \neq 0$, these two quantities are linearly independent, because c is 2-regular (see Definition 4.1.3). By Lemma 5.2.3, f is thus injective in a neighborhood of each

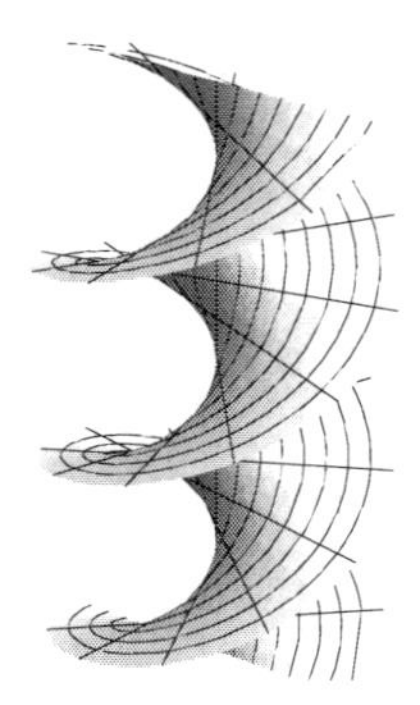

Fig. 7.13 The helicoid

point (s,t) with $s \neq 0$. We obtain the expected regular ruled surface represented by

$$f\colon]0,\infty[\times]a,b[\longrightarrow \mathbb{R}^3, \qquad (s,t)\mapsto c(t)+s\,c'(t).$$

Of course one could also work with $s \in]-\infty, 0[$.

Notice that for $s=0$, the partial derivatives of f both reduce to

$$\frac{\partial f}{\partial t}=c'(t), \qquad \frac{\partial f}{\partial s}=c'(t)$$

thus are never linearly independent. □

Let us now recall some examples that we have already met in Chap. 6:

Example 7.2.9 The *helicoid* (see Example 5.1.8 and Fig. 7.13) represented by

$$f\colon \mathbb{R}^2 \longrightarrow \mathbb{R}^3, \qquad (u,v)\mapsto (u\cos v, u\sin v, v)$$

is a ruled surface.

Proof Simply observe that

$$f(u,v)=(0,0,v)+u(\cos v,\sin v,0).$$

The curve c of Definition 7.2.1 is thus simply the z-axis. □

Example 7.2.10 The *Möbius strip* (see Example 5.1.9 and Fig. 7.14) represented by

$$f(t,\theta)=\big(R\cos\theta, R\sin\theta, 0\big)+t\left(\cos\frac{\theta}{2},\cos\frac{\theta}{2},\sin\frac{\theta}{2}\right)$$

is a ruled surface.

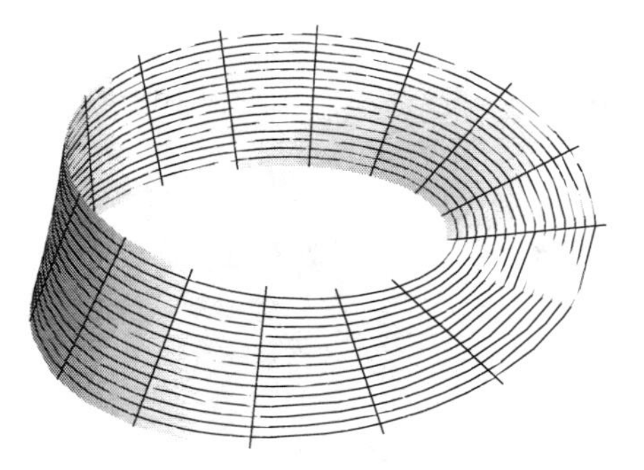

Fig. 7.14 The Möbius string

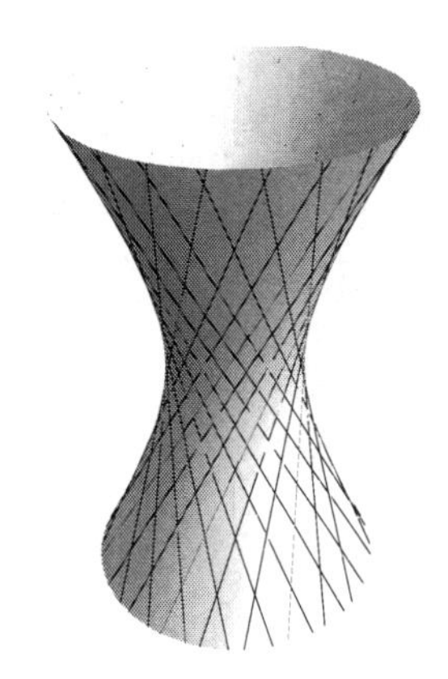

Fig. 7.15 The hyperboloid of one sheet

Proof This follows by Definition 7.2.1; here the curve c is a circle of radius R. □

Let us also recall the existence of *double ruled quadrics* (see Sect. 1.15 in [4], *Trilogy II*):

Example 7.2.11 The hyperboloid of one sheet (see Fig. 7.15)

$$\frac{x^2}{a^2} + \frac{y^2}{b^2} - \frac{z^2}{c^2} = 1$$

is a double ruled surface.

Proof As observed in the proof of Proposition 1.15.1 in [4], *Trilogy II*, through every point of the ellipse obtained by cutting the hyperboloid by the plane $z = 0$ pass two lines entirely contained in the hyperboloid. The parametric representation of this ellipse is given in Example 2.1.4. Thus the hyperboloid of one sheet should

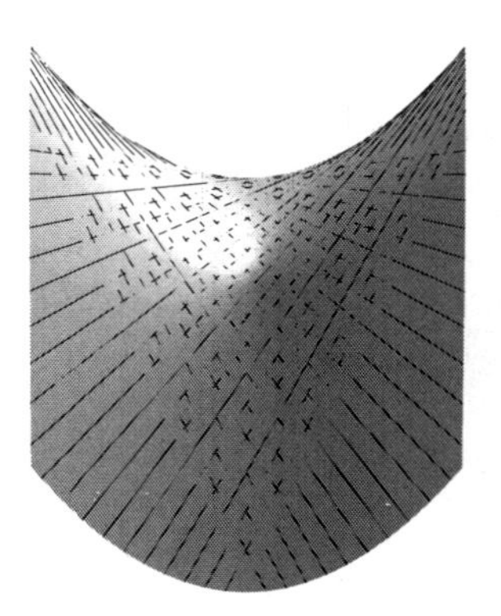

Fig. 7.16 The hyperbolic paraboloid

have two parametric representations of the form given in Definition 7.2.1:

$$f(t,\theta) = (a\cos\theta, b\sin\theta, 0) + t\big(\alpha(\theta), \beta(\theta), \gamma(\theta)\big).$$

It is quite easy to determine the three functions α, β, γ. Introducing the three components of f into the Cartesian equation of the statement we obtain

$$t^2\left(\frac{\alpha^2(\theta)}{a^2} + \frac{\beta^2(\theta)}{b^2} - \frac{\gamma^2(\theta)}{c^2}\right) + 2t\left(\frac{\alpha(\theta)\cos\theta}{a} + \frac{\beta(\theta)\sin\theta}{b}\right) = 0.$$

Since this must be the case for all values of t, both coefficients of this polynomial must be zero. Considering the coefficient of t yields at once

$$\alpha(\theta) = \frac{\sin\theta}{b}, \qquad \beta(\theta) = -\frac{\cos\theta}{a}$$

as solution (up to a scalar multiple); introducing these values into the coefficient of t^2 yields further

$$\gamma(\theta) = \pm\frac{c}{ab}.$$

This gives the two expected possible functions $\xi = (\alpha, \beta, \gamma)$ of Definition 7.2.1. □

Example 7.2.12 The hyperbolic paraboloid with equation

$$\frac{x^2}{a^2} - \frac{y^2}{b^2} = z$$

is a double ruled surface (see Fig. 7.16).

Proof Again as already observed in the proof of Proposition 1.15.2 in [4], *Trilogy II*, through every point of the hyperbolic paraboloid pass two lines contained in the surface.

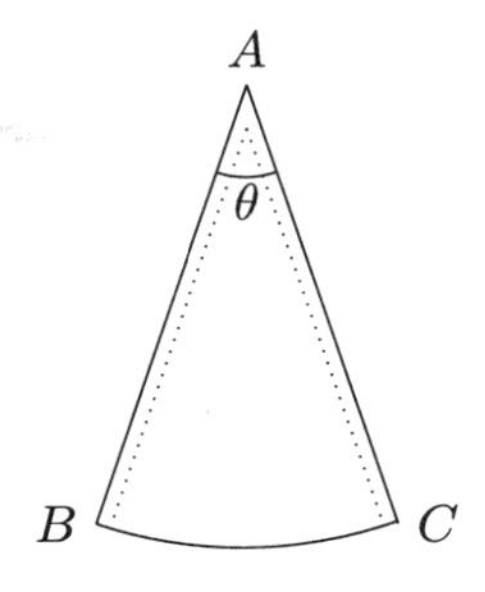

Fig. 7.17

Cutting the surface by the plane $y=0$ yields the parabola with equation $z=\frac{x^2}{a^2}$ which we take as curve c in Definition 7.2.1. We are thus looking for a parametric representation

$$f(t,x)=\left(x,0,\frac{x^2}{a^2}\right)+t\big(\alpha(x),\beta(x),\gamma(x)\big)$$

of the hyperbolic paraboloid. Introducing the three components of f into the Cartesian equation of the statement we obtain

$$t^2\left(\frac{\alpha^2(x)}{a^2}-\frac{\beta^2(x)}{b^2}\right)+t\left(\frac{2x\alpha(x)}{a^2}-\gamma(x)\right)=0.$$

Since this must hold for all values of t, both coefficients of this polynomial are zero. Considering the coefficient of t^2 yields at once $\alpha(x)=\pm a$ and $\beta(x)=\pm b$; introducing these values into the coefficient of t gives $\gamma(x)=\pm\frac{2x}{a}$. □

7.3 Applicability of Surfaces

In this section we formalize an important question, or more precisely, we observe that we have already formalized it. Let us first express our problem using a well-known example. Take a sheet of paper: everybody knows how to roll it up to get a cylinder, or to get a cone. As a consequence, putting these two operations together, you know how to transform a piece of a cylinder into a piece of a cone. This is of course done "without any stretching", since a piece of paper cannot be stretched!

Mathematically "without any stretching" means that if you consider the various curves that you can draw on the piece of paper, the lengths of all these curves are preserved when you roll up the sheet of paper to get a piece of a cylinder or a piece of a cone. One says that the piece of cylinder is *isometric* to the piece of cone.

You might convince yourself that although lengths are preserved by this process, angles are not. Take for example a piece of paper having the shape described in Fig. 7.17. Glue the segment AB onto the segment AC to get a half-cone with vertex A. On this cone, the two segments AB and AC coincide, thus their angle is zero, which was not the case in the plane! Thus angles are not preserved.

Although the arguments are correct, the conclusion is definitely false. Angles are preserved!

First, let us make clear that—for example—a whole cylinder cannot be unrolled and then rolled up again to obtain a cone: if you have glued a piece of paper to produce a cylinder, you first have to unglue it before being able to get a cone out of the same piece of paper. The problem that we have in mind thus concerns (possibly big) "pieces of surfaces obtained without gluing". Thus the problem should be stated for pieces of surfaces on which the parametric representation is at least injective. The most efficient way to achieve this is to consider local maps of Riemann surfaces in $\mathbb{R}^3$ (see Definition 6.17.1). Thus consider two Riemann surfaces of $\mathbb{R}^3$ admitting local maps

$$f\colon U \longrightarrow \mathbb{R}^3, \qquad \widetilde{f}\colon \widetilde{U} \longrightarrow \mathbb{R}^3.$$

Imagine that you have a homeomorphism (see Definition A.9.1)

$$\psi\colon f(U) \longrightarrow \widetilde{f}(\widetilde{U})$$

which preserves the length of every curve on the surfaces. The composite

$$\varphi\colon U \xrightarrow{f} f(U) \xrightarrow{\psi} \widetilde{f}(\widetilde{U}) \xrightarrow{\widetilde{f}^{-1}} \widetilde{U}$$

is then a homeomorphism, sufficiently differentiable as soon as ψ is. Saying that ψ preserves angles or distances on the surfaces in $\mathbb{R}^3$ is equivalent to saying that φ preserves angles or distances, computed in the corresponding Riemann patches in terms of the metric tensors (see Sect. 6.3). We then have:

Proposition 7.3.1 *Consider two Riemann patches of class $\mathcal{C}^0$*

$$g_{ij}\colon U \longrightarrow \mathbb{R}, \qquad \widetilde{g}_{kl}\colon \widetilde{U} \longrightarrow \mathbb{R}$$

and a change of parameters

$$\varphi\colon U \longrightarrow \widetilde{U}, \qquad (x^1, x^2) \mapsto (\widetilde{x}^1, \widetilde{x}^2)$$

of class $\mathcal{C}^1$, with inverse of class $\mathcal{C}^1$. The following conditions are equivalent:

1. *φ exhibits the equivalence of the two Riemann patches in the sense of Definition* 6.12.5;
2. *φ preserves the length of every curve, in the sense of Definition* 6.3.2;
3. *φ preserves the length of every curve and the angle between any two curves, in the sense of Definitions* 6.3.2 *and* 6.3.3.

Proof (1 ⇒ 3). Consider two regular curves

$$c\colon]a,b[\longrightarrow U, \qquad t \mapsto c(t) \qquad d\colon]p,q[\longrightarrow U, \qquad s \mapsto d(s)$$

such that $c(t_0) = d(s_0)$. Then by assumption

$$\begin{aligned}
&\big((\varphi \circ c)'(t_0) | (\varphi \circ d)'(s_0)\big)_{(\varphi \circ c)(t_0) = \varphi \circ d)(s_0)} \\
&= \sum_{kl} \widetilde{g}_{kl} \big((\varphi \circ c)^k\big)' \big((\varphi \circ d)^l\big)' \\
&= \sum_{kl} \widetilde{g}_{kl} \Big(\sum_i \frac{\partial \widetilde{x}^k}{\partial x^i} (c^i)'(t_0)\Big)\Big(\sum_j \frac{\partial \widetilde{x}^l}{\partial x^j} (d^j)'(s_0)\Big) \\
&= \sum_{ij} \Big(\sum_{kl} g_{kl} \frac{\partial \widetilde{x}^k}{\partial x^i} \frac{\partial \widetilde{x}^l}{\partial x^j}\Big) (c^i)'(t_0) (d^j)'(s_0) \\
&= \sum_{ij} g_{ij} (c^i)'(t_0) (d^j)'(s_0) \\
&= \big((c^i)'(t_0) \big| (d^j)'(s_0)\big)_{c(t_0) = d(s_0)}.
\end{aligned}$$

This proves the preservation of the scalar product of the tangent vector fields, thus also the preservation of the norm of each of them, just by choosing $c = d$. This forces condition 3 of the statement, which trivially implies condition 2.

($2 \Rightarrow 1$). Consider a point (x_0^1, x_0^2) in U and the corresponding point $\varphi(x_0^1, x_0^2) = (\widetilde{x}_0^1, \widetilde{x}_0^2)$ in $\widetilde{U}$. Consider in U the curve $c(x^1) = (x^1, x_0^2)$; this forces $c'(x^1) = (1, 0)$. Consider the corresponding curve $\varphi \circ c$ in $\widetilde{U}$, yielding

$$(\varphi \circ c)'(x^1) = \left(\frac{\partial \widetilde{x}^1}{\partial x^1}, \frac{\partial \widetilde{x}^2}{\partial x^1}\right)$$

since $(c^2)' = 0$. The preservation of the length of an arc of this curve reduces to the equality

$$\int_{x_0^1}^{x^1} g_{11}(x^1, x_0^2)\, dx^1 = \int_{x_0^1}^{x^1} \sum_{kl} \widetilde{g}_{kl}(\widetilde{x}^1, \widetilde{x}_0^2) \frac{\partial \widetilde{x}^k}{\partial x^1}(x^1, x_0^2) \frac{\partial \widetilde{x}^l}{\partial x^1}(x^1, x_0^2)\, dx^1.$$

Differentiating this equality with respect to x^1 yields

$$g_{11}(x^1, x_0^2) = \sum_{kl} \widetilde{g}_{kl}(\widetilde{x}^1, \widetilde{x}_0^2) \frac{\partial \widetilde{x}^k}{\partial x^1}(x^1, x_0^2) \frac{\partial \widetilde{x}^l}{\partial x^1}(x^1, x_0^2)$$

which is indeed the expected formula (see Definition 6.12.5).

An analogous argument holds of course for g_{22}, considering this time the curve $c(x^2) = (x_0^1, x^2)$.

Consider finally the curve $c(t) = (x_0^1 + t, x_0^2 + t)$ in U, which yields $c'(t) = (1, 1)$. The corresponding curve $\varphi \circ c$ in $\widetilde{U}$ now yields, with abbreviated notation,

$$(\varphi \circ c)' = \left(\sum_i \frac{\partial \widetilde{x}^1}{\partial x^i}, \sum_j \frac{\partial \widetilde{x}^2}{\partial x^j} \right).$$

The preservation of the length of an arc of this curve then yields, after differentiation with respect to t

$$\sum_{ij} g_{ij} = \sum_{klij} \widetilde{g}_{kl} \frac{\partial \widetilde{x}^k}{\partial x^i} \frac{\partial \widetilde{x}^l}{\partial x^j}.$$

Using the results already obtained for g_{11} and g_{22} and keeping in mind that the metric tensor is symmetric, this expression reduces to

$$g_{12} = \sum_{kl} \widetilde{g}_{kl} \frac{\partial \widetilde{x}^k}{\partial x^1} \frac{\partial \widetilde{x}^l}{\partial x^2}$$

which is again the expected formula, as in Definition 6.12.5. □

But then what about our "counterexample" of the cone? First of all, notice that the injectivity requirement to have a local map is certainly not satisfied, since the segment AB is identified with the segment AC. More importantly, note that this "couterexample" exhibits a problem at the *vertex* of the cone, and as already observed several times, a regular parametric representation of a piece of the cone can never reach the vertex.

A natural reaction might be to replace the segments AB and AC by two segments parallel to them but very close to them, inside the domain of definition of what now becomes a local map of the cone. These two segments intersect at a point close to A, but still inside the domain of definition of the local map. Since the two segments (in dotted lines in Fig. 7.17) can be chosen "infinitely close to the segments AB and BC", the corresponding angles on the cone should converge to zero when the new segments converge to AB and AC. Thus these angles cannot possibly remain constant and equal to the angle θ in Fig. 7.17.

You are assuming some form of continuity of this "angle function" at a point where it is not defined! Let us abandon this for the moment and consider the following. Imagine that your two new segments are parallel to AB and AC, at a distance ε. They therefore meet on the bisecting line of the angle BAC. Thus the two curves on the cone meet on the ruling exactly opposite to the ruling obtained from AB, AC. So when ε converges to zero, the two curves tend to proceed by a half turn, to the other side of the cone. A striking difference with the rulings obtained from AB and AC! It is no longer clear what happens to the angles. Certainly, no longer that clear that your continuity argument on angles works. Perhaps it would be better to rely on the proof of Proposition 7.3.1!

If you are not yet convinced and want to investigate your counterexample "concretely" on a parametric representation of the cone, consider the following hint.

What is the parametric representation of the cone that you are considering? Imagine that you have constructed a circular cone having the upper z-axis as axis, by rolling up the first quarter of $\mathbb{R}^2$, that is those points (x, y) with $x > 0$, $y > 0$. What happens to the point

$$(x, y) = (R\cos\theta, R\sin\theta) \in]0, \infty[\times]0, \infty[\ ?$$

The quarter of a circle of radius R in $\mathbb{R}^2$ becomes a "parallel" of the cone, thus a full circle four times smaller, that is, of radius $\frac{1}{4}R$. But this circle remains at a distance R from the vertex A, thus by Pythagoras' Theorem (see 4.3.5 in [4], *Trilogy II*) it lies at a height z such that

$$z^2 + \left(\frac{1}{4}R\right)^2 = R^2$$

that is $z = \frac{\sqrt{15}}{4}R$. The parametric representation thus maps the point (x, y) with polar coordinates (R, θ) to

$$\left(\frac{1}{4}R\cos 4\theta, \frac{1}{4}R\sin 4\theta, \frac{\sqrt{15}}{4}R\right).$$

Now keep in mind that

$$R^2 = x^2 + y^2, \qquad \cos\theta = \frac{x}{R}, \qquad \sin\theta = \frac{y}{R}$$

while

$$\cos 4\theta = 8\cos^4\theta - 8\cos^2\theta + 1, \qquad \sin 4\theta = 4\sin\theta\cos^3\theta - 4\sin^3\theta\cos\theta.$$

Putting together all these quantities, you get the parametric representation of the cone in terms of x and y. You are ready to compute angles using Proposition 5.4.4: a quantity converging to zero in the denominator will convince you that your "continuity argument" cannot possibly hold!

So long for this discussion: we have a proof of Proposition 7.3.1 and we shall rely on it. This proposition suggests the following definition:

Definition 7.3.2 Consider two local maps of two Riemann surfaces of $\mathbb{R}^3$

$$f: U \longrightarrow \mathbb{R}^3, \qquad h: V \longrightarrow \mathbb{R}^3.$$

These two local maps are *Riemann isometric* when the corresponding Riemann patches are equivalent (see Definitions 6.1.1 and 6.12.5).

More generally

Definition 7.3.3 Two Riemann surfaces of $\mathbb{R}^3$ are *locally Riemann isometric* when each point of each surface lies in a local map which is Riemann-isometric to a local map of the other surface.

Our example of the cylinder and the cone makes clear that in general, the two parametric representations f and h in Definition 7.3.2 are *not* equivalent in the sense of Definition 5.1.1: they represent in general *distinct* surfaces, with *distinct* supports and even, with supports having possibly very different shapes. On the other hand

Proposition 7.3.4 *Two Riemann isometric local maps of Riemann surfaces of class $\mathcal{C}^3$ in $\mathbb{R}^3$ yield the same Gaussian curvature at corresponding points.*

Proof With the notation of Definition 7.3.2, we have two local maps

$$f: U \longrightarrow \mathbb{R}^3, \qquad h: V \longrightarrow \mathbb{R}^3$$

and a change of parameters

$$\varphi: U \longrightarrow V$$

which is an equivalence of Riemann patches. The surface originally represented by h can equivalently be represented by $h \circ \varphi$. The metric tensor of this surface, when computed using the parametric representation $h \circ \varphi$, is thus the metric tensor computed via h and further transformed along φ, as in Proposition 6.12.1. By assumption, this is precisely the metric tensor of the other surface, computed using f. So the two surfaces represented by f and $h \circ \varphi$ have the same metric tensor and therefore the same total curvature at corresponding points (see Theorem 6.11.3). □

7.4 Surfaces with Zero Curvature

This section begins the systematic study of Riemann-isometric surfaces (see Definition 7.3.2) and, more generally, *locally Riemann-isometric surfaces* (see Definition 7.3.3). In Sect. 5.2.7 we introduced this problem via the case of the cylinder and the cone which—locally—are Riemann-isometric to a piece of the plane. Let us first investigate further such a situation in the very general context of Riemann surfaces.

Let us observe that a piece of the plane

$$f: U \longrightarrow \mathbb{R}^3, \qquad (x^1, x^2) \mapsto (x^1, x^2, 0)$$

viewed as a Riemann patch, is the connected open subset $U \subseteq \mathbb{R}^2$ provided with the corresponding metric tensor (see Definition 6.1.1), which simply takes the form of the identity matrix:

$$\begin{pmatrix} g_{11} & g_{12} \\ g_{21} & g_{22} \end{pmatrix} = \begin{pmatrix} 1 & 0 \\ 0 & 1 \end{pmatrix}.$$

The fundamental result is

Theorem 7.4.1 *Consider a Riemann surface of class $\mathcal{C}^3$. The following conditions are equivalent:*

1. *The surface is locally Riemann-isometric to the plane.*
2. *In a neighborhood of each point, there exists a system of coordinates such that the metric tensor takes the form*

$$\begin{pmatrix} \widetilde{g}_{11} & \widetilde{g}_{12} \\ \widetilde{g}_{21} & \widetilde{g}_{22} \end{pmatrix} = \begin{pmatrix} 1 & 0 \\ 0 & 1 \end{pmatrix}.$$

3. *The Gaussian curvature is zero at each point.*

Proof Conditions 1 and 2 of the statement are equivalent, just by definition.

($2 \Rightarrow 3$). If all the coefficients g_{ij} of the metric tensor are constant, all the Christoffel symbols of the first kind are zero (see Definition 6.6.7), hence so are the Christoffel symbols of the second kind (see Proposition 6.6.8) and eventually, all the coefficients of the Riemann tensor (see Definition 6.11.4). The result follows by Definition 6.11.5.

To prove ($3 \Rightarrow 2$), let us apply Theorem 6.13.1 and work locally in a Fermi system of geodesic coordinates, so that by Proposition 6.14.5, the Gaussian curvature is given by

$$\kappa_\tau = -\frac{1}{g_{22}} \frac{\partial^2 \sqrt{g_{22}}}{\partial (x^1)^2}.$$

The nullity of this quantity reduces to that of the second partial derivative, thus means that $\frac{\partial \sqrt{g_{22}}}{\partial x^1}$ is a constant function of x^1, for every fixed value x_0^2. Computing this partial derivative we obtain the constant function of x^1

$$\frac{\partial \sqrt{g_{22}}}{\partial x^1}(x^1, x_0^2) = \frac{1}{2\sqrt{g_{22}(x^1, x_0^2)}} \frac{\partial g_{22}}{\partial x^1}(x^1, x_0^2).$$

But in a Fermi system of geodesic coordinates

$$\frac{\partial g_{22}}{\partial x^1}(0, x_0^2) = 0$$

(see Theorem 6.13.1) so that, by constancy of the function above

$$\frac{\partial g_{22}}{\partial x^1}(x^1, x_0^2) = 0.$$

Once more this means that for every fixed value x_0^2, the function $g_{22}(x^1, x_0^2)$ is constant. But in a Fermi system of geodesic coordinates (see Theorem 6.13.1 again) $g_{22}(0, x^2) = 1$. By constancy of the function, $g_{22}(x^1, x_0^2) = 1$ for all values x_0^2 and x^1. Moreover, once more by Theorem 6.13.1, in a system of geodesic coordinates one always has $g_{11} = 1$ and $g_{12} = 0 = g_{21}$. □

Here is another interesting characterization of the surfaces which are locally Riemann-isometric to a plane:

Proposition 7.4.2 *Consider a Riemann surface of class $\mathcal{C}^3$. The following conditions are equivalent*:

1. *The surface is locally Riemann-isometric to a plane.*
2. *In a neighborhood of each point, the parallel transport of a vector along a piecewise regular curve is independent of the chosen curve.*

Of course by *parallel transport along a piecewise regular curve* we mean the parallel transport along the first side (see Theorem 6.8.4), followed by the parallel transport along the second side, and so on up to the last side.

Proof ($1 \Rightarrow 2$). By Theorem 7.4.1, we can choose the metric tensor to be the identity matrix. Since the coefficients g_{ij} of the metric tensor are constant, all the Christoffel symbols of the first kind are zero (see Definition 6.6.7), hence so are the Christoffel symbols of the second kind (see Proposition 6.6.8). This shows that the covariant derivative of a vector field (ξ^1, ξ^2) along a regular curve c is simply given by

$$\frac{\nabla \xi}{dt} = \left(\frac{d\xi^1}{dt}, \frac{d\xi^2}{dt}\right)$$

(see Definition 6.7.3). Thus the vector field ξ is parallel (see Theorem 6.8.4) precisely when

$$\frac{d\xi^1}{dt} = 0, \qquad \frac{d\xi^2}{dt} = 0$$

that is, when $\xi = (\xi^1, \xi^2)$ is constant. So the parallel transport of a vector along a regular curve—thus also along a piecewise regular curve—leaves that vector constant. It is thus independent of the chosen piecewise regular curve.

($2 \Rightarrow 1$). We apply Theorem 6.13.1 and work locally in a Fermi system of geodesic coordinates. Then $g_{11} = 1$ while $g_{12} = 0 = g_{21}$ and by Theorem 7.4.1, it remains to prove that $g_{22} = 1$. But Theorem 6.13.1 also tells us that $g_{22}(0, x^2) = 1$: thus it suffices to prove that g_{22} is a constant function of x^1, that is $\frac{\partial g_{22}}{\partial x^1} = 0$. Let us fix a point (x_0^1, x_0^2) in the Riemann patch.

At each point of the Riemann patch, let us consider the vector

$$\xi = \left(0, \frac{1}{\sqrt{g_{22}}}\right).$$

The restriction of ξ along the curve $x^1 = x_0^1$ is thus the tangent vector field of that curve (see Example 6.4.3).

Let us now consider the parallel transport of the vector $\xi(x_0^1, x_0^2)$ from the point (x_0^1, x_0^2) to the point $(x_0^1, 0)$

1. first, along the curve $x^1 = x_0^1$;
2. second, along the piecewise regular curve comprised of the succession of the three geodesics

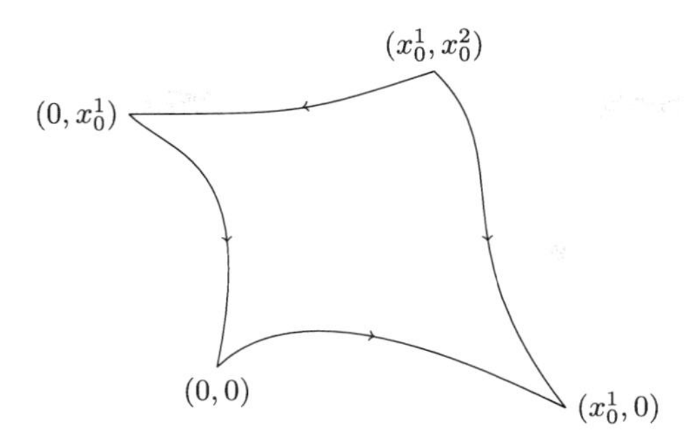

Fig. 7.18

- $x^2 = x_0^2$, to reach the point $(0, x_0^2)$;
- $x^1 = 0$, to reach the point $(0, 0)$;
- $x^2 = 0$, to reach the point $(x_0^1, 0)$,

(see Fig. 7.18). By assumption, both parallel transports must yield the same result.

Since the curve $x^2 = x_0^2$ is orthogonal to the curve $x^1 = x_0^1$, the vector $\xi(x_0^1, x_0^2)$, tangent to the second curve, is a vector of length 1 orthogonal to the curve $x^2 = x_0^2$. Since $x^2 = x_0^2$ is a geodesic in normal representation, its tangent vector field is a parallel vector field (see Proposition 6.10.2). By Proposition 6.8.3, the parallel transport of $\xi(x_0^1, x_0^2)$ along the curve $x^2 = x_0^2$ yields at $(0, x_0^2)$ a vector still of length 1 and still orthogonal to the curve $x^2 = x_0^2$. Since the curve $x^1 = 0$ is orthogonal to the curve $x^2 = x_0^2$, this is thus a vector of length 1 tangent to the curve $x^1 = 0$, that is, the tangent vector of that curve.

Since the curve $x^1 = 0$ is a geodesic in normal representation, again by Proposition 6.10.2, a further parallel transport along the curve $x^1 = 0$ up to the point $(0, 0)$ will simply produce the tangent vector to the curve $x^1 = 0$.

The argument, developed above for the curve $x^2 = x_0^2$, can be applied to the curve $x^2 = 0$ and proves that a further parallel transport of the vector to the point $(x_0^1, 0)$ yields a vector of length 1, orthogonal to the curve $x^2 = 0$ thus tangent to the curve $x^1 = x_0^1$. This is thus the vector $\xi(x_0^1, 0)$.

Applying the above argument twice

- from (x_0^1, x_0^2) to $(x_0^1, 0)$;
- from $(x_0^1, 0)$ to (x_0^1, x_1^2),

we conclude that the parallel transport of the vector $\xi(x_0^1, x_0^2)$ to an arbitrary point (x_0^1, x_1^2) yields the vector $\xi(x_0^1, x_1^2)$. The tangent vector field ξ to the curve $x^1 = x_0^1$ is thus a parallel one. Notice that by Proposition 6.10.2, the curve $x^1 = x_0^1$ is thus a geodesic.

The differential equations for the vector field

$$\xi = \left(0, \frac{1}{\sqrt{g_{22}}}\right)$$

being a parallel vector field along the curve

$$c(x^2) = (x_0^1, x^2)$$

then reduce to

$$\begin{cases} \dfrac{d\frac{1}{\sqrt{g_{22}}}}{dx^2} + \frac{1}{\sqrt{g_{22}}} \Gamma_{22}^2 = 0 \\ \frac{1}{\sqrt{g_{22}}} \Gamma_{22}^1 = 0. \end{cases}$$

The second equation shows that $\Gamma_{22}^1 = 0$ at each point (x_0^1, x^2), that is, everywhere since x_0^1 has been arbitrarily fixed. But by Proposition 6.6.8 and keeping in mind that $g_{12} = 0 = g_{21}$, thus $g^{12} = 0 = g^{21}$

$$\Gamma_{22}^1 = g^{11} \Gamma_{221} = -\frac{1}{2} g^{11} \frac{\partial g_{22}}{\partial x^1}.$$

Since $g_{11} = 1$, $g^{11} = \frac{1}{g_{11}} = 1$ so that we obtain $\frac{\partial g_{22}}{\partial x^1} = 0$ as expected. □

7.5 Developable Surfaces

Let us now formalize the problem used to introduce the problem of applicability of surfaces (see Sect. 7.3): *Which surfaces of $\mathbb{R}^3$ can be obtained by rolling up a piece of paper?*

Definition 7.5.1 A Riemann surface of class C^3 is *developable* when it admits a parametric representation

$$f : U \longrightarrow \mathbb{R}^3$$

which preserves the length of every curve.

Let us first observe that developable surfaces are precisely those studied in Sect. 7.4.

Proposition 7.5.2 *For a Riemann surface of class C^3, the following conditions are equivalent:*

1. *the surface is developable;*
2. *the surface is locally Riemann-isometric to the plane;*
3. *the total curvature is zero at each point.*

Proof By Theorem 7.4.1, conditions 2 and 3 are equivalent. Proposition 7.3.1 shows that condition 2 is equivalent to the local preservation of lengths. Of course global preservation of lengths (i.e. condition 1) trivially implies the local preservation. It remains to prove that the local preservation of lengths implies the global preservation.

Let us thus write $f\colon U \longrightarrow \mathbb{R}^3$ for a parametric representation of the surface such that each point of U admits a neighborhood on which f preserves the length of every curve. Let us further consider a regular curve $h\colon]a,b[\longrightarrow U$ and two points $a < c < d < b$. We must prove that the length of the arc of the curve h on $[c,d]$ is preserved by f.

For each $c \leq t \leq d$, there exists an open neighborhood V_t of $h(t)$ on which f preserves lengths. Then $h^{-1}(V_t)$ is an open subset of $\mathbb{R}$, thus a union of open intervals (see Sect. A.1); one of these intervals contains t: let us call it I_t. By definition, $h(I_t) \subseteq V_t$. The various open intervals I_t thus cover the interval $[c,d]$, which is compact; therefore finitely many of them—let us say, I_{t_0} to I_{t_n}—already suffice to cover $[c,d]$ (see Sect. A.3). But then we can find finitely many points

$$c < s_1 < s_2 < \cdots < s_{n-1} < s_n < d$$

such that each interval $[s_i, s_{i+1}]$ is contained in one of the intervals I_{t_j}. Since the length of the curve represented by h is preserved by f on each piece $[s_i, s_{i+1}]$, by adding the finitely many pieces, the length on the full interval $[c,d]$ is preserved. □

Example 7.5.3 In $\mathbb{R}^3$, every cylinder of class $\mathcal{C}^3$ is developable.

Proof A cylinder has a parametric representation of the form

$$f(s,t) = c(t) + s\xi$$

where ξ is a constant vector (see Example 7.2.6). It follows at once that $\frac{\partial f}{\partial s} = \xi$, thus $\frac{\partial^2 f}{\partial t \partial s} = 0$. Therefore $M = 0$ (see Theorem 5.8.2) and the total curvature of the cylinder is zero by Proposition 7.2.3. The result follows by Proposition 7.5.2. □

Example 7.5.4 In $\mathbb{R}^3$, every cone of class $\mathcal{C}^3$ is developable.

Proof A cone has a parametric representation of the form

$$f(s,t) = c(t) + s\big(c(t) - \xi\big)$$

where ξ is a constant vector (see Example 7.2.7). It follows at once that

$$\frac{\partial f}{\partial t} = c'(t) + sc'(t), \qquad \frac{\partial f}{\partial s} = c(t) - \xi, \qquad \frac{\partial^2 f}{\partial t \partial s} = c'(t).$$

In particular $\frac{\partial^2 f}{\partial t \partial s}$ is parallel to $\frac{\partial f}{\partial t}$ and therefore, is in the direction of the tangent plane. It is thus orthogonal to the normal vector $\overrightarrow{n}$ to the surface, proving that

$M = 0$ (see Theorem 5.8.2). The total curvature of the cone is then zero by Proposition 7.2.3. The result follows by Proposition 7.5.2. □

Example 7.5.5 In $\mathbb{R}^3$, every tangent surface of class $\mathcal{C}^3$ of a curve is developable.

Proof A tangent surface has a parametric representation of the form

$$f(s,t) = c(t) + sc'(t)$$

(see Example 7.2.8). It follows at once that

$$\frac{\partial f}{\partial t} = c'(t) + sc''(t), \qquad \frac{\partial f}{\partial s} = c'(t), \qquad \frac{\partial^2 f}{\partial t \partial s} = c''(t).$$

Subtracting the two partial derivatives shows that $c''(t)$ is in the direction of the tangent plane, thus is orthogonal to the normal vector $\overrightarrow{n}$ to the surface. This proves that $M = 0$ (see Theorem 5.8.2). The total curvature of the tangent surface is then zero by Proposition 7.2.3. The result follows by Proposition 7.5.2. □

7.6 Classification of Developable Surfaces

The key to a classification of the developable surfaces of $\mathbb{R}^3$ is the following result (see Exercise 7.17.5 concerning the assumption on the absence of planar points).

Proposition 7.6.1 *Consider a Riemann surface in $\mathbb{R}^3$, without any planar point, and given by a regular parametric representation of class $\mathcal{C}^3$. The following conditions are equivalent*:

1. *The surface is developable.*
2. *The surface is locally Riemann-isometric to a ruled surface all of whose rulings are singular.*

Proof Assume first that the surface is developable. By Proposition 7.5.2, we know that the Gaussian curvature is zero at each point; thus at least one principal curvature is zero at each point (see Definition 5.16.1). But since by assumption there are no planar points (see Definition 5.16.9), there are no umbilical points and thus both principal curvatures cannot vanish together. Let us simply write κ for the non-zero principal curvature at each point.

Let us now apply Theorem 5.6.2 and, in a neighborhood of P, choose a parametric representation $f(s,t)$ such that:

- $f(0,0) = P$;
- at each point, $\frac{\partial f}{\partial t}$ is oriented along the principal direction of principal curvature κ;
- at each point, $\frac{\partial f}{\partial s}$ is oriented along the principal direction of principal curvature 0.

Performing an additional change of coordinates separately on the two axes

$$\varphi(s,t) = \big(\varphi_1(t), \varphi_2(s)\big)$$

does not affect the orientation of the partial derivatives, thus there is no loss of generality in assuming that

- the curve $s \mapsto f(s,0)$ is in normal representation;
- the curve $t \mapsto f(0,t)$ is in normal representation.

Since the partial derivatives of f are oriented along the principal directions, they are orthogonal at each point (see Theorem 5.10.1), thus $F(s,t) = 0$ at each point. Furthermore, since the two curves $t = 0$, $s = 0$ are in normal representation, we have in fact

$$E(0,t) = 1, \qquad F(s,t) = 0, \qquad G(s,0) = 1.$$

Moreover, one has

$$L(s,t) = 0, \qquad M(s,t) = 0, \qquad N(s,t) = \kappa(s,t)G(s,t).$$

Indeed the principal directions are at each point given by $(1,0)$ and $(0,1)$ with respect to the basis of partial derivatives. Introducing these directions into the formula of Proposition 5.8.4 gives the first and the third equalities. The second equality follows from the first equation in Proposition 5.10.5, when choosing the direction $(0,1)$ corresponding to the principal curvature 0.

Proving that f has the form

$$f(s,t) = c(t) + s\,\xi(t)$$

(that is, we have a ruled surface) is equivalent to proving that f is linear in the variable s, which by derivation and integration, is equivalent to proving that $\frac{\partial^2 f}{\partial s^2} = 0$ at each point. We shall achieve this part of the proof by showing that $\frac{\partial^2 f}{\partial s^2} = 0$ is orthogonal to all three vectors of the orthonormal basis

$$\vec{\varepsilon_1} = \frac{\frac{\partial f}{\partial s}}{\|\frac{\partial f}{\partial s}\|} = \frac{\frac{\partial f}{\partial s}}{\sqrt{E}}, \qquad \vec{\varepsilon_2} = \frac{\frac{\partial f}{\partial t}}{\|\frac{\partial f}{\partial t}\|} = \frac{\frac{\partial f}{\partial t}}{\sqrt{G}}, \qquad \vec{n} = \varepsilon_1 \times \varepsilon_2.$$

Of course, $\vec{n}$ is at each point the normal vector to the surface.

Since $(\vec{n} \,|\, \frac{\partial f}{\partial t}) = 0$, we obtain

$$\left(\frac{\partial \vec{n}}{\partial t}\,\middle|\,\frac{\partial f}{\partial t}\right) + \left(\vec{n}\,\middle|\,\frac{\partial^2 f}{\partial t^2}\right) = 0$$

that is

$$\left(\frac{\partial \vec{n}}{\partial t}\,\middle|\,\frac{\partial f}{\partial t}\right) = -N = \kappa G.$$

Analogously we have

$$\left(\frac{\partial \overrightarrow{n}}{\partial s}\middle|\frac{\partial f}{\partial s}\right) = -L = 0$$

and this proves already that $\frac{\partial^2 f}{\partial s^2} = 0$ is orthogonal to $\overrightarrow{n}$. Finally

$$\left(\frac{\partial \overrightarrow{n}}{\partial t}\middle|\frac{\partial f}{\partial s}\right) = \left(\frac{\partial \overrightarrow{n}}{\partial s}\middle|\frac{\partial f}{\partial t}\right) = -M = 0.$$

Differentiating the equality $(\overrightarrow{n}|\overrightarrow{n}) = 1$ indicates that the partial derivatives of $\overrightarrow{n}$ are orthogonal to $\overrightarrow{n}$. Expressed with respect to the orthonormal basis $(\varepsilon_1, \varepsilon_2)$ they are thus given by (see Proposition 4.6.2)

$$\begin{aligned}
\frac{\partial \overrightarrow{n}}{\partial t} &= \left(\frac{\partial \overrightarrow{n}}{\partial t}\middle|\varepsilon_1\right)\varepsilon_1 + \left(\frac{\partial \overrightarrow{n}}{\partial t}\middle|\varepsilon_2\right)\varepsilon_2 \\
&= -\frac{M}{\sqrt{EG}}\frac{\partial f}{\partial s} - \frac{N}{G}\frac{\partial f}{\partial t} \\
&= -\kappa\frac{\partial f}{\partial t} \\
\frac{\partial \overrightarrow{n}}{\partial s} &= \left(\frac{\partial \overrightarrow{n}}{\partial s}\middle|\varepsilon_1\right)\varepsilon_1 + \left(\frac{\partial \overrightarrow{n}}{\partial s}\middle|\varepsilon_2\right)\varepsilon_2 \\
&= -\frac{L}{E}\frac{\partial f}{\partial s} - \frac{M}{\sqrt{EG}}\frac{\partial f}{\partial t} \\
&= 0.
\end{aligned}$$

In particular this forces the first equality below:

$$\left(\frac{\partial^2 f}{\partial s^2}\middle|\frac{\partial f}{\partial t}\right) = \left(\frac{\partial^2 f}{\partial s^2}\middle| -\frac{1}{\kappa}\frac{\partial \overrightarrow{n}}{\partial t}\right) = \frac{1}{\kappa}\left(\frac{\partial f}{\partial s}\middle|\frac{\partial^2 \overrightarrow{n}}{\partial t \partial s}\right) = 0.$$

The second equality is obtained by differentiating the equality

$$\left(\frac{\partial f}{\partial s}\middle|\frac{\partial \overrightarrow{n}}{\partial t}\right) = \left(\frac{\partial f}{\partial s}\middle| -\kappa\frac{\partial f}{\partial t}\right) = 0$$

and the third one holds since $\frac{\partial \overrightarrow{n}}{\partial v} = 0$. All this proves that $\frac{\partial^2 f}{\partial s^2}$ is orthogonal to ε_2.

To prove the orthogonality to ε_1, we observe that

$$\frac{\partial}{\partial t}\left(\frac{\partial f}{\partial s}\middle|\frac{\partial f}{\partial s}\right) = 2\left(\frac{\partial f}{\partial s}\middle|\frac{\partial^2 f}{\partial t \partial s}\right) = -2\left(\frac{\partial^2 f}{\partial s^2}\middle|\frac{\partial f}{\partial t}\right) = 0;$$

the last but one equality is obtained by differentiating the equality $(\frac{\partial f}{\partial s}|\frac{\partial f}{\partial t}) = 0$. These equalities prove that $(\frac{\partial f}{\partial s}|\frac{\partial f}{\partial s}) = E^2$ is a constant function of s. Since

$E(0, s) = 1$, we conclude that $E(s, t) = 1$ at all points. Differentiating the equality $(\frac{\partial f}{\partial s}|\frac{\partial f}{\partial s}) = 1$ we obtain $2(\frac{\partial^2 f}{\partial s^2}|\frac{\partial f}{\partial s}) = 0$ which proves that $\frac{\partial^2 f}{\partial s^2}$ is orthogonal to ε_1.

We have thus proved that $\frac{\partial^2 f}{\partial s^2} = 0$ and we know therefore that f has the expected form

$$f(s, t) = c(t) + s\xi(t)$$

for being a ruled surface. The rulings are the curves $t = k$, with k a constant. But since $\frac{\partial \vec{n}}{\partial s} = 0$, the normal vector to the surface is constant along each ruling. So the tangent plane to the surface is constant along each ruling and thus each ruling is singular (see Definition 7.2.5).

Conversely, consider a ruled surface represented by

$$f(s, t) = c(t) + s\xi(t)$$

and assume that all rulings are singular. This means that the normal vector $\vec{n}$ is constant along each ruling $t = k$, k a constant, that is, $\frac{\partial \vec{n}}{\partial s} = 0$. Differentiating the equality $(\frac{\partial f}{\partial t}|\vec{n}) = 0$ we then get $(\frac{\partial^2 f}{\partial t \partial s}|\vec{n}) = (\frac{\partial f}{\partial t}|\frac{\partial \vec{n}}{\partial s}) = 0$. In other words, $M = 0$. By Proposition 7.2.3, the Gaussian curvature is zero. The result follows by Theorem 7.6.1. □

Let us now prove that every developable surface is obtained by "gluing together" pieces of cylinders, cones and tangent surfaces (see Examples 7.5.3, 7.5.4, 7.5.5).

Theorem 7.6.2 *Consider a developable Riemann surface in $\mathbb{R}^3$, without any planar points, and given by a regular parametric representation of class $\mathcal{C}^3$. In a neighborhood U of each point, there exists a dense sub-neighborhood $V \subseteq U$ such that, in a neighborhood of each point of V the surface has one of the following three forms*:

- *a cylinder* (*see Example* 7.2.6);
- *a cone* (*see Example* 7.2.7);
- *a tangent surface to a curve* (*see Example* 7.2.8).

Proof In a neighborhood of P, we freely use the parametric representation

$$f: U \longrightarrow \mathbb{R}^3, \qquad (s, t) \mapsto f(s, tr)$$

already considered in the proof of Proposition 7.6.1, together with the various results established in that proof. Of course we can at once restrict our attention to a neighborhood of the form $I \times J$, where I and J are two open intervals of the real line. We also freely use the basic topological notions and results of Appendix A.

First we notice that

$$(\xi'|\vec{n}) = \left(\frac{\partial^2 f}{\partial t \partial s}\middle|\vec{n}\right) = M = 0.$$

Thus ξ' lies in the tangent plane. On the other hand

$$(\xi'|\xi) = \left(\frac{\partial^2 f}{\partial t \partial s}\middle|\frac{\partial f}{\partial s}\right) = -\left(\frac{\partial f}{\partial t}\middle|\frac{\partial^2 f}{\partial s^2}\right) = 0;$$

the last but one equality is obtained by differentiating the equality $(\frac{\partial f}{\partial t}|\frac{\partial f}{\partial s}) = 0$ and the last one holds because $\frac{\partial^2 f}{\partial s^2} = 0$. All this proves that ξ' is orthogonal to $\xi = \frac{\partial f}{\partial s}$ in the tangent plane, thus ξ' is parallel to $\frac{\partial f}{\partial t}$. Observing further that

$$\frac{\partial f}{\partial t} = c'(t) + s\,\xi'(t)$$

we conclude that c' is parallel to $\frac{\partial f}{\partial t}$, thus to ξ'.

The curve c is the curve $s = 0$: by choice of f, it is in normal representation. Thus $\|c'\| = 1$ and since ξ' is parallel to c',

$$\xi' = (\xi'|c')c'$$

(see Proposition 4.6.2 in [3], *Trilogy I*). The rest of our argument is based on the consideration of the real valued function

$$\mu(t) = (\xi'(t)|c'(t))$$

which remains of class $\mathcal{C}^2$.

By continuity, if μ or μ' takes a non-zero value at some point, it takes a non-zero value on a whole neighborhood of that point. Therefore the following three subsets of U are open and, trivially, pairwise disjoint:

$$V_1 = \left\{ t_0 \in I \;\middle|\; \begin{array}{l}\text{there exists a neighborhood } W \text{ of } t_0 \\ \text{such that for all } t \in W\ \mu(t) = 0\end{array} \right\},$$

$$V_2 = \left\{ t_0 \in I \;\middle|\; \begin{array}{l}\mu(t_0) \neq 0 \\ \text{there exists a neighborhood } W \text{ of } t_0 \\ \text{such that for all} t \in W\ \mu'(t) = 0\end{array} \right\},$$

$$V_3 = \left\{ t_0 \in I \;\middle|\; \begin{array}{l}\text{there exists a neighborhood } W \text{ of } t_0 \\ \text{such that for all } t \in W\ \mu(t) \neq 0 \text{ and } \mu'(t) \neq 0\end{array} \right\}.$$

It is immediate that the complement of $V_1 \cup V_2 \cup V_3$ in I is given by $W_1 \cup W_2$, where

$$W_1 = \left\{ t_0 \in I \;\middle|\; \begin{array}{l}\text{every neighborhood } W \text{ of } t_0 \\ \text{contains a point } t \text{ such that } \mu(t) \neq 0\end{array} \right\},$$

$$W_2 = \left\{ t_0 \in I \;\middle|\; \begin{array}{l}\text{every neighborhood } W \text{ of } t_0 \\ \text{contains a point } t \text{ such that } \mu'(t) \neq 0\end{array} \right\}.$$

We now want to prove that the open subset $V_1 \cup V_2 \cup V_3$ is dense in I. For that we must prove that every neighborhood of every point of I intersects $V_1 \cup V_2 \cup V_3$

(see Proposition A.6.3). The result is obvious for the points of $V_1 \cup V_2 \cup V_3$, thus it remains to prove the assertion for the points of $W_1 \cup W_2$.

Let W be a neighborhood in I of a point $t_0 \in W_1$. By definition of W_1, there exists a $t \in W$ such that $\mu(t) \neq 0$.

- If $\mu'(t) \neq 0$, by continuity this is the case on a neighborhood of t and then $t \in V_3$.
- If $\mu'(t_0) = 0$, two possibilities occur:

 1. $\mu' = 0$ on a neighborhood of t, in which case $t \in V_2$;
 2. every neighborhood of t contains a point t' such that $\mu'(t') \neq 0$; restricting further our attention to a smaller neighborhood W' of t on which μ remains everywhere non-zero, we conclude that $t' \in V_3$.

Let W be a neighborhood in I of a point $t_0 \in W_2$. By definition of W_1, there exists a $t \in W$ such that $\mu'(t) \neq 0$.

- If $\mu(t) \neq 0$, by continuity this is the case on a neighborhood of t and then $t \in V_3$.
- If $\mu(t_0) = 0$, two possibilities occur *a priori*, but in fact only one is acceptable:

 1. $\mu = 0$ on a whole neighborhood of t, implying $\mu'(t) = 0$, which is not the case;
 2. thus every neighborhood of t contains a point t' such that $\mu(t') \neq 0$; restricting further our attention to a smaller neighborhood W' of t on which μ' remains everywhere non-zero, we conclude that $t' \in V_3$.

This concludes the proof of the density of $V_1 \cup V_2 \cup V_3$ in I and as a consequence, it follows at once that

$$(V_1 \times J) \cup (V_2 \times J) \cup (V_3 \times J)$$

is dense in the open subset $I \times J$ on which we consider f. We shall prove that the three pieces of this union correspond to the three cases of developable surfaces mentioned in the statement.

If $t_0 \in V_1$, since $\xi'(t) = \mu(t) \cdot c'(t)$, we obtain $\xi'(t) = 0$ on a neighborhood of t_0. Thus $\xi(t)$ is constant on a neighborhood I' of t_0 and the surface is a cylinder on $I' \times J$ (see Example 7.2.6).

If $t_0 \in V_2$, then $\mu(t_0) \neq 0$ and μ is constant on a neighborhood of t_0, because its derivative is zero. Therefore on that neighborhood, $\xi'(t) = \mu(t_0) \cdot c'(t)$. Integrating this equality, we get

$$\xi(t) - \xi(t_0) = \mu(t_0)\big(c(t) - c(t_0)\big).$$

We can then re-write the parametric representation as

$$\begin{aligned} f(s,t) &= c(t) + s\big(\xi(t_0) + \mu(t_0)\big(c(t) - c(t_0)\big)\big) \\ &= c(t) + s\mu(t_0)\big(c(t) - \big(c(t_0) - \xi(t_0)\big)\big). \end{aligned}$$

Performing further the change of parameter $s' = s\mu(t_0)$, we recapture the parametric representation of a cone (see Example 7.2.7).

Finally we consider the case $t_0 \in V_3$. By definition of V_3, $\mu(t)$ and $\mu'(t)$ are non-zero on some neighborhood of t_0. It thus makes sense to consider, on such a neighborhood, the function

$$\lambda(t) = c(t) - \frac{\xi(t)}{\mu(t)}.$$

One computes at once that

$$\lambda' = c' - \frac{\xi'\mu - \xi\mu'}{\mu^2} = \frac{\xi'}{\mu} - \frac{\xi'}{\mu} + \xi\frac{\xi'}{\mu^2} = \xi\frac{\xi'}{\mu^2}$$

$$\lambda'' = \xi'\frac{\mu'}{\mu^2} + \xi\frac{\mu''\mu^2 - 2\mu'^2\mu}{\mu^4}.$$

Since ξ' is orthogonal to ξ and ξ' is non-zero, it follows that λ' and λ'' are linearly independent, thus λ is a 2-regular parametric representation of a curve (see Proposition 4.1.4 and Definition 4.1.3).

Let us now re-write the parametric representation of the surface in terms of this curve λ:

$$\begin{aligned} c(t) + s\xi(t) &= \lambda(t) + \frac{\xi(t)}{\mu(t)} + s\,\xi(t) \\ &= \lambda(t) + \xi(t)\left(\frac{1}{\mu(t)} + s\right) \\ &= \lambda(t) + \lambda'(t)\frac{\mu(t)^2}{\mu'(t)}\left(\frac{1}{\mu(t)} + s\right) \\ &= \lambda(t) + \lambda'(t)\left(\frac{\mu(t)}{\mu'(t)} + s\frac{\mu(t)^2}{\mu'(t)}\right). \end{aligned}$$

It remains to perform the change of parameters

$$\widetilde{t} = t, \qquad \widetilde{s} = \frac{\mu(t)}{\mu'(t)} + s\frac{\mu(t)^2}{\mu'(t)}$$

to obtain as expected (see Example 7.2.8) a representation of the form

$$\lambda(\widetilde{t}) + \widetilde{s}\lambda'(\widetilde{t}).$$

That this is a good change of parameters is attested by the *Local Inverse Theorem* (see Theorem 1.3.1), since the matrix of partial derivatives has the form

$$\begin{pmatrix} \frac{\partial\widetilde{t}}{\partial t} & \frac{\partial\widetilde{t}}{\partial s} \\ \frac{\partial\widetilde{s}}{\partial t} & \frac{\partial\widetilde{s}}{\partial s} \end{pmatrix} = \begin{pmatrix} 1 & 0 \\ \frac{\partial\widetilde{s}}{\partial t} & \frac{\mu(t)^2}{\mu'(t)} \end{pmatrix}.$$

The determinant of this matrix is simply $\frac{\mu(t)^2}{\mu'(t)}$, which is always non-zero since $t_0 \in V_3$. □

7.7 Surfaces with Constant Curvature

Developable surfaces have zero Gaussian curvature; since all of them are locally Riemann-isometric to a plane, any two of them are locally Riemann-isometric to each other (see Sects. 7.4, 7.5 and 7.6). This section generalizes these considerations to the case of surfaces with arbitrary constant Gaussian curvature.

Theorem 7.7.1 *Consider two Riemann surfaces of class $\mathcal{C}^3$ in $\mathbb{R}^3$, both with constant Gaussian curvature. The following conditions are equivalent:*

1. *The two surfaces have the same Gaussian curvature.*
2. *The two surfaces are locally Riemann-isometric* (*see Definition* 7.3.3).

Proof One implication is attested by Proposition 7.3.4. So let us assume that both surfaces have constant Gaussian curvature κ. We consider the *Fermi* system of coordinates (see Theorem 6.13.1) in a neighborhood of a point P of the first surface, based on some arbitrarily chosen geodesic passing through P. The constant Gaussian curvature is given by

$$\kappa = -\frac{1}{\sqrt{g_{22}}}\frac{\partial^2\sqrt{g_{22}}}{\partial(x^1)^2}$$

(see Proposition 6.14.5). For each fixed value x_0^2, this provides the equality

$$\frac{\partial^2\sqrt{g_{22}(x^1,x_0^2)}}{\partial(x^1)^2} = -\kappa\sqrt{g_{22}(x^1,x_0^2)}.$$

Moreover, still by Theorem 6.13.1, we have the two initial conditions

$$\sqrt{g_{22}(0,x^2)} = 1, \qquad \frac{\partial\sqrt{g_{22}(0,x_0^2)}}{\partial x^1} = 0.$$

This can be viewed as a second order differential equation involving the function $g_{22}(x^1,x_0^2)$ and the variable x^1 (see Sect. B.2). Integrating this system is immediate and provides the solutions

$$\begin{aligned}
\sqrt{g_{22}\big(x^1,x_0^2\big)} &= \cos\big(\sqrt{\kappa}x^1\big) && \text{if } \kappa > 0\\
\sqrt{g_{22}\big(x^1,x_0^2\big)} &= 1 && \text{if } \kappa = 0\\
\sqrt{g_{22}\big(x^1,x_0^2\big)} &= \cosh\big(\sqrt{-\kappa}x^1\big) && \text{if } \kappa < 0.
\end{aligned}$$

In Fermi coordinates, whatever the surface with constant Gaussian curvature κ, whatever the point P chosen as origin and whatever the base geodesic through P, the metric tensor, in a neighborhood of P, takes the form

$$\begin{pmatrix} 1 & 0 \\ 0 & \cos^2(\sqrt{\kappa}\,x^1) \end{pmatrix} \quad \text{if } \kappa > 0$$

$$\begin{pmatrix} 1 & 0 \\ 0 & 1 \end{pmatrix} \quad \text{if } \kappa = 0$$

$$\begin{pmatrix} 1 & 0 \\ 0 & \cosh^2(\sqrt{-\kappa}\,x^1) \end{pmatrix} \quad \text{if } \kappa < 0.$$

Thus in Fermi coordinates, the two surfaces of the statement admit exactly the same metric tensor in a neighborhood of two arbitrary points and therefore are locally Riemann-isometric. □

Notice that the proof of Theorem 7.7.1 establishes a more precise result:

Corollary 7.7.2 *Consider two Riemann surfaces of class $\mathcal{C}^3$ in $\mathbb{R}^3$, both with the same constant Gaussian curvature κ. Fix a point P of the first surface and choose a geodesic through P; analogously fix a point Q of the second surface and choose a geodesic through Q. Then P and Q belong to respective local maps which are equivalent via an equivalence which maps P to Q and transforms into each other the two chosen geodesics.*

Corollary 7.7.3 *Locally, a Riemann surface of class $\mathcal{C}^3$ in $\mathbb{R}^3$ with constant Gaussian curvature is Riemann-isometric to a sphere, a plane or a pseudo-sphere.*

Proof This follows by Theorem 7.7.1 and Examples 5.16.5, 5.16.4 and 5.16.7. □

Observe that Corollary 7.7.2 tells us in particular that two arbitrary points of the same pseudo-sphere always admit Riemann-isometric neighborhoods. It also tells us that fixing a point P of the pseudo-sphere and two geodesics through that point, we can "rotate" on the pseudo-sphere some neighborhood of P so that the first geodesic is moved onto the second one. These results are far from being obvious when just observing the shape of a pseudo-sphere, as in Fig. 5.13. Nevertheless, it follows from our Theorem 7.7.1: if you had a (of course non-flat) piece of paper having exactly the shape of a piece of the pseudo-sphere, you would be able to run that piece of paper everywhere along the pseudo-sphere, all the time maintaining full contact between the piece of paper and the pseudo-sphere. Thus for two Riemann surfaces of $\mathbb{R}^3$, *having the same constant Gaussian curvature* does not imply *having the same shape* as subsets of $\mathbb{R}^3$.

Now what about the surfaces with constant positive curvature? Clearly enough, two pieces (of the same size) of a sphere always have the *same shape* as pieces of $\mathbb{R}^3$. But are there other surfaces than the sphere which admit a constant strictly positive Gaussian curvature?

Example 7.7.4 There exist surfaces of revolution, other than the sphere, with constant positive Gaussian curvature.

Proof With the notation of Proposition 7.1.11, to obtain a surface of revolution with constant Gaussian curvature $\frac{1}{R^2}$, it suffices to rotate a plane curve given in normal parametric representation by

$$f(t) = \bigl(f_1(t), f_2(t)\bigr)$$

and such that

$$\frac{1}{R^2} = -\frac{f_1''}{f_1}.$$

This last equality can be regarded as a differential equation

$$f_1'' = -\frac{1}{R^2} f_1.$$

The general solution of this equation is

$$f_1(t) = a\cos\left(\frac{t}{R} + b\right)$$

where a and b are constants. Observing that

$$f_1'(t) = -\frac{a}{R}\sin\left(\frac{t}{R} + b\right)$$

one obtains f_2' from the normality requirement $(f_1')^2 + (f_2')^2 = 1$

$$f_2' = \sqrt{1 - \frac{a^2}{R^2}\sin^2\left(\frac{t}{R} + b\right)}$$

and thus

$$f_2(t) = \int_0^t \sqrt{1 - \frac{a^2}{R^2}\sin^2\left(\frac{t}{R} + b\right)}\,dt.$$

Observe that when $a = R$, we get simply

$$f_2(t) = \int_0^t \cos t = \sin t + k$$

where k is a constant. The parametric representation

$$f(t,\theta) = \left(a\cos\left(\frac{t}{R} + b\right)\cos\theta, a\cos\left(\frac{t}{R} + b\right)\sin\theta, \sin t + k\right)$$

is that of a sphere.

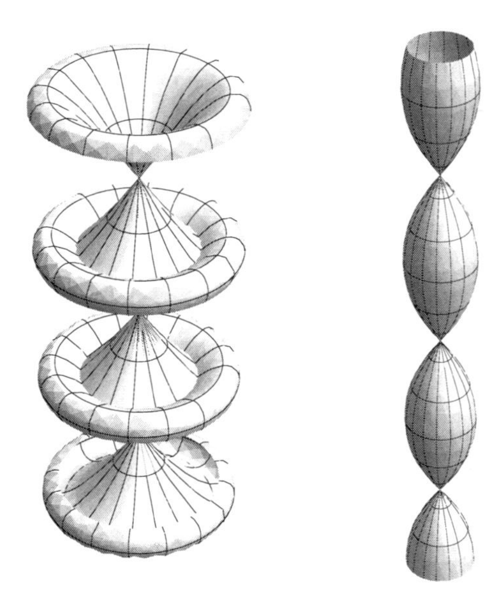

Fig. 7.19 Surfaces of revolution with constant Gaussian curvature

When $a \neq R$, the integral defining f_2 becomes an elliptic integral and must be computed numerically in order to exhibit the shape of the surface. Figure 7.19 corresponds respectively to the cases $a = 2R$ and $a = \frac{R}{2}$. Clearly, the local injectivity fails at various places so that to get actual parametric representations, only convenient portions of these surfaces should be considered. □

The case of surfaces of revolution with constant negative curvature is analogous.

7.8 The Sphere

Example 7.7.4 shows that the sphere is by no means the only surface of $\mathbb{R}^3$ with constant positive curvature. Nevertheless it is the only *compact connected* surface of $\mathbb{R}^3$ with constant positive Gaussian curvature, as attested by *Liebmann's Theorem.* We freely refer to Appendix A for the topological arguments.

Let us begin with a first characterization of the sphere:

Proposition 7.8.1 *A connected Riemann surface of class $\mathcal{C}^3$ in $\mathbb{R}^3$, all of whose points are umbilical, necessarily has constant Gaussian curvature and is contained in a plane or a sphere.*

Proof On each local map, we use the notation of Theorem 5.13.1, establishing the Rodrigues formula. We write κ_n for the normal curvature which, at each point, is by assumption the same in all directions. The Rodrigues formula tells us that

$$\frac{\partial \overrightarrow{n}}{\partial u} = -\kappa_n \frac{\partial f}{\partial u}, \qquad \frac{\partial \overrightarrow{n}}{\partial v} = -\kappa_n \frac{\partial f}{\partial v}.$$

Differentiating these equalities we obtain

$$\frac{\partial^2 \overrightarrow{n}}{\partial u \partial v} = -\frac{\partial \kappa_n}{\partial v}\frac{\partial f}{\partial u} - \kappa_n \frac{\partial^2 f}{\partial u \partial v}$$

$$\frac{\partial^2 \overrightarrow{n}}{\partial v \partial u} = -\frac{\partial \kappa_n}{\partial u}\frac{\partial f}{\partial v} - \kappa_n \frac{\partial^2 f}{\partial v \partial u}.$$

Comparing these two equalities yields at once

$$\frac{\partial \kappa_n}{\partial v}\frac{\partial f}{\partial u} = \frac{\partial \kappa_n}{\partial u}\frac{\partial f}{\partial v}.$$

Since the partial derivatives are linearly independent (see Definition 5.2.1), this forces

$$\frac{\partial \kappa_n}{\partial v} = 0, \qquad \frac{\partial \kappa_n}{\partial u} = 0.$$

This means precisely that on each local map, κ_n is constant, hence so is the Gaussian curvature which is equal to κ_n^2 (see Definition 5.16.1).

If $\kappa_n = 0$, the Rodrigues formula becomes simply

$$\frac{\partial \overrightarrow{n}}{\partial u} = 0, \qquad \frac{\partial \overrightarrow{n}}{\partial v} = 0$$

which shows that $\overrightarrow{n}$ is constant. Choosing an orthonormal basis $(P; e_1, e_2, e_3)$ in $\mathbb{R}^3$, with P a point of the local map and $e_3 = \overrightarrow{n}$, the parametric representation

$$f(u,v) = \big(f_1(u,v), f_2(u,v), f_3(u,v)\big)$$

of the surface has at each point its partial derivatives in the plane (e_1, e_2), thus

$$\frac{\partial f_3}{\partial u} = 0, \qquad \frac{\partial f_3}{\partial v} = 0.$$

This proves that $f_3(u,v) = k$, with k a constant. The local map of the surface is then contained in the plane with equation $x_3 = k$.

If $\kappa_n \neq 0$, by path-connectedness of the domain of the local map (see Theorem A.10.7), let us choose a path $c(t)$ in U joining a fixed point P of the local map to an arbitrary point Q of the local map. The Rodrigues formula yields

$$\frac{d\overrightarrow{n}(c(t))}{dt} = -\kappa_n \frac{df(c(t))}{dt}$$

where κ_n is constant. Integrating this equality yields

$$\vec{n}\left(c(t)\right) = -\kappa_n f\left(c(t)\right) + (a_1, a_2, a_3)$$

where (a_1, a_2, a_3) is a constant vector. This equality can be re-written as

$$\left(\frac{a_1}{\kappa_n}, \frac{a_2}{\kappa_n}, \frac{a_3}{\kappa_n}\right) - f\left(c(t)\right) = \frac{\vec{n}\,(c(t))}{\kappa_n}.$$

Since $\vec{n}$ has length 1, this proves that all points $f(c(t))$ are at a distance $\frac{1}{|\kappa_n|}$ from the fixed point $(\frac{a_1}{\kappa_n}, \frac{a_2}{\kappa_n}, \frac{a_3}{\kappa_n})$. In particular, every point Q of the local map is on the sphere with center $(\frac{a_1}{\kappa_n}, \frac{a_2}{\kappa_n}, \frac{a_3}{\kappa_n})$ and radius $\frac{1}{|\kappa_n|}$.

Now given two points Q, Q' in two distinct local maps, by path-connectedness of the Riemann surface (see Corollary A.10.8), there is a continuous path joining them. To pass from Q to Q', the curve possibly crosses various local maps. By continuity, to pass from one local map to another, this curve passes each time through a point P in the intersection of these two consecutive local maps (see Corollary A.10.4). Fixing this point P and working separately in the two local maps, the considerations above show that both local maps are contained in the same plane or in the same sphere, according to the case. □

The key to *Liebmann's Theorem* is the following lemma, due to *Hilbert*.

Lemma 7.8.2 (Hilbert's Lemma) *Consider a regular surface of class $\mathcal{C}^3$ in $\mathbb{R}^3$ and its two principal curvature functions, written κ_1 and κ_2. Suppose that at some point P:*

- $\kappa_1(P)$ *reaches a local maximum;*
- $\kappa_2(P)$ *reaches a local minimum;*
- $\kappa_1(P) > \kappa_2(P)$.

Then $\kappa_\tau(P) \leq 0$, where κ_τ indicates the Gaussian curvature.

Proof Let us apply Theorem 5.14.2 and work with a parametric representation such that all lines of coordinates $x^1 = k$, $x^2 = l$, with k, l constant, are lines of curvature; there is no loss of generality in assuming further that P is the point with parameters $(0, 0)$. The principal curvature κ_1 is thus at each point the normal curvature in the direction $(1, 0)$, while κ_2 is the normal curvature in the direction $(0, 1)$. By Theorem 5.8.2, at all points,

$$\kappa_1 = \frac{L}{E}, \qquad \kappa_2 = \frac{N}{G}.$$

Still by Theorem 5.14.2, we also know that $F = 0$ and $M = 0$ at all points.

We switch now to the notation of Riemannian geometry. Keeping in mind that $g_{12} = 0 = g_{21}$, Proposition 6.6.5 yields

$$\frac{1}{2}\frac{\partial g_{11}}{\partial x^1} = \Gamma_{111}$$

$$\frac{1}{2}\frac{\partial g_{11}}{\partial x^2} = -\Gamma_{112} = \Gamma_{121} = \Gamma_{211}$$

$$\frac{1}{2}\frac{\partial g_{22}}{\partial x^2} = \Gamma_{222}$$

$$\frac{1}{2}\frac{\partial g_{22}}{\partial x^1} = -\Gamma_{221} = \Gamma_{122} = \Gamma_{212}.$$

Since

$$\begin{pmatrix} g^{11} & g^{12} \\ g^{21} & g^{22} \end{pmatrix} = \begin{pmatrix} \frac{1}{g_{11}} & 0 \\ 0 & \frac{1}{g_{22}} \end{pmatrix}$$

Proposition 6.6.4 yields further

$$\Gamma_{11}^1 = \frac{1}{2}\frac{1}{g_{11}}\frac{\partial g_{11}}{\partial x^1}$$

$$\Gamma_{11}^2 = -\frac{1}{2}\frac{1}{g_{22}}\frac{\partial g_{11}}{\partial x^2}$$

$$\Gamma_{12}^1 = \Gamma_{21}^1 = \frac{1}{2}\frac{1}{g_{11}}\frac{\partial g_{11}}{\partial x^2}$$

$$\Gamma_{12}^2 = \Gamma_{21}^2 = \frac{1}{2}\frac{1}{g_{22}}\frac{\partial g_{22}}{\partial x^1}$$

$$\Gamma_{22}^1 = -\frac{1}{2}\frac{1}{g_{11}}\frac{\partial g_{22}}{\partial x^1}$$

$$\Gamma_{22}^2 = \frac{1}{2}\frac{1}{g_{22}}\frac{\partial g_{22}}{\partial x^2}.$$

We shall now use the *Codazzi–Mainardi* equations of Proposition 6.16.1, keeping in mind that $h_{12} = h_{21} = M = 0$. Choosing $i = 1 = k$ and $j = 2$, the equation reduces to

$$\Gamma_{21}^1 h_{11} - \Gamma_{11}^2 h_{22} - \frac{\partial h_{11}}{\partial x^2} = 0.$$

Using the values of the Christoffel symbols calculated above, as well as the values of the two principal curvatures, this gives

$$\frac{\partial h_{11}}{\partial x^2} = \frac{1}{2}\frac{\partial g_{11}}{\partial x^2}\left(\frac{h_{11}}{g_{11}} + \frac{h_{22}}{g_{22}}\right) = \frac{1}{2}\frac{\partial g_{11}}{\partial x^2}(\kappa_1 + \kappa_2).$$

In a perfectly analogous way, one computes that

$$\frac{\partial h_{22}}{\partial x^1} = \frac{1}{2}\frac{\partial g_{22}}{\partial x^1}(\kappa_1 + \kappa_2).$$

Using these equalities, one computes at once that

$$\frac{\partial \kappa_1}{\partial x^2} = \frac{\partial \frac{h_{11}}{g_{11}}}{\partial x^2} = \frac{1}{2g_{11}} \frac{\partial g_{11}}{\partial x^2}(\kappa_2 - \kappa_1)$$

and analogously

$$\frac{\partial \kappa_2}{\partial x^1} = \frac{\partial \frac{h_{22}}{g_{22}}}{\partial x^1} = \frac{1}{2g_{22}} \frac{\partial g_{22}}{\partial x^1}(\kappa_2 - \kappa_1).$$

By assumption, at the point $P = (0, 0)$, the two principal curvatures reach an extremum, thus their partial derivatives are zero at that point:

$$\frac{\partial \kappa_i}{\partial x^j}(0, 0) = 0.$$

Since moreover, still by assumption, $\kappa_2(0, 0) \neq \kappa_1(0, 0)$, we conclude that

$$\frac{\partial g_{11}}{\partial x^2}(0, 0) = 0, \qquad \frac{\partial g_{22}}{\partial x^1}(0, 0) = 0.$$

Let us differentiate a second time the principal curvature functions. First

$$\frac{\partial^2 \kappa_1}{\partial (x^2)^2} = \frac{1}{2} \frac{\frac{\partial^2 g_{11}}{\partial (x^2)^2} g_{11} - (\frac{\partial g_{11}}{\partial x^2})^2}{g_{11}^2}(\kappa_2 - \kappa_1) + \frac{1}{2g_{11}} \frac{\partial g_{11}}{\partial x^2}\left(\frac{\partial \kappa_2}{\partial x^2} - \frac{\partial \kappa_1}{\partial x^2}\right).$$

At the point $P = (0, 0)$, where $\frac{\partial \kappa_1}{\partial x^2}$ vanishes, we obtain

$$\frac{\partial^2 \kappa_1}{\partial (x^2)^2} = \frac{1}{2g_{11}(0, 0)} \frac{\partial^2 g_{11}}{\partial (x^2)^2}(0, 0)\big(\kappa_2(0, 0) - \kappa_1(0, 0)\big).$$

Since $\kappa_1(0, x^2)$ has a local *maximum* at $x^2 = 0$, we also have

$$\frac{\partial^2 \kappa_1}{\partial (x^2)^2}(0, 0) \leq 0.$$

Since moreover $\kappa_1(0, 0) > \kappa_2(0, 0)$ while $g_{11}(0, 0) > 0$, as norm of the first partial derivative, we conclude that

$$\frac{\partial^2 g_{11}}{\partial (x^2)^2}(0, 0) \geq 0.$$

Analogously

$$\frac{\partial^2 \kappa_2}{\partial (x^1)^2} = \frac{1}{2g_{22}(0, 0)} \frac{\partial^2 g_{22}}{\partial (x^1)^2}(0, 0)\big(\kappa_1(0, 0) - \kappa_2(0, 0)\big)$$

and

$$\frac{\partial^2\kappa_2}{\partial(x^1)^2}(0,0) \geq 0$$

since $\kappa_2(x^1, 0)$ has a local *minimum* at $x^1 = 0$. This forces as above

$$\frac{\partial^2 g_{22}}{\partial(x^1)^2}(0,0) \geq 0.$$

The Gaussian curvature is given by the general formula

$$\kappa_\tau = \frac{R_{1212}}{g_{11}g_{22} - g_{12}g_{21}}$$

(see Definition 6.11.5). Since the denominator is always strictly positive (see Definition 6.2.1), the sign of the Gaussian curvature is that of R_{1212}. At the point $P = (0,0)$, the formula of Theorem 6.11.3, in view of the various quantities already calculated above, reduces simply to

$$\begin{aligned} R_{1212}(0,0) &= \frac{\partial \Gamma_{221}}{\partial x^1}(0,0) - \frac{\partial \Gamma_{211}}{\partial x^2}(0,0) \\ &= -\frac{1}{2}\frac{\partial^2 g_{22}}{\partial(x^1)^2}(0,0) - \frac{1}{2}\frac{\partial^2 g_{11}}{\partial(x^2)^2}(0,0) \leq 0 \end{aligned}$$

which concludes the proof of the lemma. □

We are now ready to characterize the spheres:

Theorem 7.8.3 (Liebmann) *The only compact connected Riemann surfaces of* $\mathbb{R}^3$ *with constant positive Gaussian curvature are the spheres.*

Proof We retain the notation of the *Hilbert lemma* (see Lemma 7.8.2) and consider the function

$$k(Q) = \kappa_1(Q) - \kappa_2(Q)$$

where Q runs through all the points of the surface. Since the surface is compact by assumption, the function k^2, which is of course positive, reaches a maximum (see Corollary A.8.4). Let us prove first that this maximum is zero.

If this is not the case, there exists a point P on the surface where $\kappa_1(P) \neq \kappa_2(P)$; let us say, $\kappa_1(P) > \kappa_2(P)$. Then $k(P) > 0$ and thus by continuity, $k(Q) > 0$ in a neighborhood of P. Therefore, since $k^2(Q)$ reaches its maximum at P, so does $k(Q)$. But $\kappa_1\kappa_2$ is the total curvature (see Definition 5.16.1), thus is a constant strictly positive function by assumption. Therefore if κ_1 increases in a neighborhood of P, necessarily κ_2 decreases and the difference $k = \kappa_1 - \kappa_2$ becomes greater. This cannot be the case, since k admits a local maximum at P. In other words, κ_1 itself passes through a local maximum at P; and since $\kappa_1\kappa_2$ is constant, κ_2 passes through a local minimum at P. Thus Hilbert's lemma applies (see Lemma 7.8.2) and the total curvature at P is negative. This contradicts our assumptions.

We have thus proved that the maximum of the function k^2 is necessarily zero. In other words, $\kappa_1 = \kappa_2$ at each point and thus each point is umbilical. By Proposition 7.8.1, the Riemann surface is contained in a sphere of radius $\frac{1}{\sqrt{\kappa}}$.

It remains to prove that the Riemann surface $\mathcal{R}$ is the full sphere $\mathcal{S}$. Choose an arbitrary point $P \in \mathcal{R}$, belonging to some local map

$$f : U_P \longrightarrow \mathcal{R} \subseteq \mathcal{S} \subseteq \mathbb{R}^3.$$

There is no loss of generality in choosing U_P sufficiently small so that $f(U_P)$ is a small "open circle" around P in $\mathcal{R}$; writing B_P for the corresponding small open ball centered at P, we have

$$f(U_P) = B_P \cap \mathcal{R} = B_P \cap \mathcal{S}.$$

Putting $V = \bigcup_{P \in \mathcal{R}} B_P$, we obtain an open subset $V \subseteq \mathbb{R}^3$ such that $V \cap \mathcal{S} = \mathcal{R}$.

But $\mathcal{R}$ is compact by assumption, thus it is closed in $\mathbb{R}^3$ and thus also in $\mathcal{S}$, by Proposition A.7.4. Therefore $\mathcal{S} \setminus \mathcal{R}$ is open in $\mathcal{S}$ and thus $\mathcal{S} \setminus \mathcal{R} = W \cap \mathcal{S}$ with W an open subset of $\mathbb{R}^3$. But then

$$\mathcal{S} = (V \cap \mathcal{S}) \cup (W \cap \mathcal{S}) = \mathcal{R} \cup (\mathcal{S} \setminus \mathcal{R})$$

is a covering of the sphere by two disjoint open subsets of $\mathcal{S}$. The sphere is trivially path-connected, thus connected (see Corollary A.10.8). By connectedness, the sphere is thus already contained in one of these two open subsets, while the other one is empty. Since $\mathcal{R}$ is not empty, it is $\mathcal{S} \setminus \mathcal{R}$ which is empty and $\mathcal{R} = \mathcal{S}$. □

7.9 A Counterexample

The results of Sects. 7.4 and 7.7 could give the false impression that two surfaces with the same Gaussian curvature are necessarily locally Riemann isometric. This is by no means the case, when the Gaussian curvature is not constant. Here is a counterexample.

Counterexample 7.9.1 There exist surfaces with the same Gaussian curvature, which are not locally Riemann isometric.

Proof Consider the two parametric representations (see Fig. 7.20)

$$f(x^1, x^2) = (x^1 \cos x^2, x^1 \sin x^2, \log x^1)$$
$$f(\widetilde{x}^1, \widetilde{x}^2) = (\widetilde{x}^1 \cos \widetilde{x}^2, \widetilde{x}^1 \sin \widetilde{x}^2, \widetilde{x}^2)$$

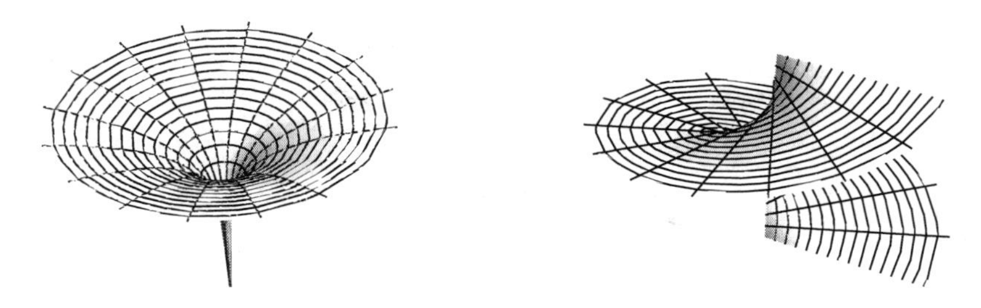

Fig. 7.20 Two surfaces with the same Gaussian curvature

defined for all values $x^1 > 0$, $\widetilde{x}^1 > 0$. The corresponding metric tensors are

$$(g_{ij})_{ij} = \begin{pmatrix} 1+\frac{1}{(x^1)^2} & 0 \\ 0 & (x^1)^2 \end{pmatrix}, \qquad (\widetilde{g}_{ij})_{ij} = \begin{pmatrix} 1 & 0 \\ 0 & 1+(\widetilde{x}^1)^2 \end{pmatrix}$$

and the corresponding Gaussian curvatures are

$$\kappa_\tau = \frac{-1}{1+(x^1)^2}, \qquad \widetilde{\kappa}_\tau = \frac{-1}{1+(\widetilde{x}^1)^2}.$$

These two surfaces thus have the same Gaussian curvature function.

Let us now prove that no change of parameter

$$\widetilde{x}^1 = \widetilde{x}^1(x^1, x^2), \qquad \widetilde{x}^2 = \widetilde{x}^2(x^1, x^2)$$

can exhibit a local Riemann isometry between these two surfaces. If this were the case, then by Proposition 7.3.4 this change of parameters would preserve the Gaussian curvature and thus would yield

$$\frac{-1}{1+(x^1)^2} = \frac{-1}{1+(\widetilde{x}^1)^2}.$$

Since $x^1 > 0$ and $\widetilde{x}^1 > 0$, this is equivalent to $x^1 = \widetilde{x}^1$. This would imply

$$\begin{pmatrix} \frac{\partial \widetilde{x}^1}{\partial x^1} & \frac{\partial \widetilde{x}^2}{\partial x^1} \\ \frac{\partial \widetilde{x}^1}{\partial x^2} & \frac{\partial \widetilde{x}^2}{\partial x^2} \end{pmatrix} = \begin{pmatrix} 1 & \frac{\partial \widetilde{x}^2}{\partial x^1} \\ 0 & \frac{\partial \widetilde{x}^2}{\partial x^2} \end{pmatrix}.$$

Since such a matrix is regular (see Proposition 5.2.2), we have $\frac{\partial \widetilde{x}^2}{\partial x^2} \neq 0$.

Moreover, the existence of a Riemann isometry (see Definition 6.12.5) would yield further

$$g_{ij} = \sum_{kl} \widetilde{g}_{kl} \frac{\partial \widetilde{x}^k}{\partial x^i} \frac{\partial \widetilde{x}^l}{\partial x^j}$$

that is, introducing the equality $x^1 = \widetilde{x}^1$ into the values of the $\widetilde{g}_{kl}$

$$1 + \frac{1}{(x^1)^2} = 1 + \left(1 + (x^1)^2\right)\left(\frac{\partial \widetilde{x}^2}{\partial x^1}\right)^2$$

$$0 = \frac{\partial \widetilde{x}^2}{\partial x^1} \cdot \frac{\partial \widetilde{x}^2}{\partial x^2}$$

$$(x^1)^2 = \left(1 + (x^1)^2\right)\left(\frac{\partial \widetilde{x}^2}{\partial x^2}\right)^2.$$

Since $\frac{\partial \widetilde{x}^2}{\partial x^2} \neq 0$, the second of these equalities forces $\frac{\partial \widetilde{x}^2}{\partial x^1} = 0$. Introducing this value into the first equality then yields

$$1 + \frac{1}{(x^1)^2} = 1$$

which is impossible. □

7.10 Rotation Numbers

Our next purpose in this chapter is to prove a very deep and fundamental theorem of surface theory: the generalization of the *Hopf Umlaufsatz* (see Theorem 2.15.2) to the case of a piecewise regular simple closed curve on a surface. To achieve this program, we need some preliminaries presented in this and the following sections.

Let us first investigate, for closed curves in a Riemann patch, the notion of *rotation number* as studied in Theorems 2.13.5, 2.14.8 and 2.15.2. First, we generalize Corollary 2.12.2.

Proposition 7.10.1 *Consider a regular curve of class C^2 in a Riemann patch of class C^2*

$$c\colon]a,b[\to U, \qquad g_{ij}\colon U \longrightarrow \mathbb{R}, \quad i,j = 1,2.$$

Assume that the Riemann patch is given in a geodesic system of coordinates (see Theorem 6.13.1) and that the curve is given in normal representation. Write

$$e^1(x^1,x^2) = (1,0), \qquad e^2(x^1,x^2) = \frac{1}{\sqrt{g_{22}(x^1,x^2)}}(0,1)$$

for the two vector fields constituting at each point an orthonormal basis with respect to the metric tensor, oriented along the lines of coordinates $x^1 = k$, $x^2 = l$, with k, l constant.

1. *There exists a function of class C^1*

$$\theta\colon]a,b[\longrightarrow \mathbb{R}$$

such that $\theta(s)$ is at each point, up to a multiple of 2π, the angle (in the sense of the metric tensor) between $e^1(c(s))$ and the tangent vector $c'(s)$, that is,

$$c'(s) = \cos\theta(s)e_1\big(c(s)\big) + \sin\theta(s)e_2\big(c(s)\big).$$

2. *Two possible such functions θ differ only by a constant multiple of 2π.*

Such a function is called an angular function *of the curve.*

Proof Saying that $\|c'(s)\| = 1$ means

$$\big(c_1'\big)^2 + g_{22}\big(c_2'\big)^2 = 1$$

that is

$$\big(c_1'\big)^2 + \big(\sqrt{g_{22}}c_2'\big)^2 = 1.$$

Applying Lemma 2.12.1, we get the expected function θ satisfying

$$\cos\theta = c_1', \qquad \sin\theta = \sqrt{g_{22}}c_2'$$

that is

$$c' = c_1'(1,0) + c_2'(0,1) = \cos\theta(1,0) + \frac{\sin\theta}{\sqrt{g_{22}}}(0,1) = \cos\theta e_1 + \sin\theta e_2. \qquad \square$$

The notion of angular function is easily generalized to piecewise regular curves in Riemann patches:

Definition 7.10.2 Consider a piecewise regular curve $c(s)$ of class $\mathcal{C}^2$ in a Riemann patch of class $\mathcal{C}^2$. Suppose that the curve is given in normal representation and the Riemann patch in a system of geodesic coordinates. An *angular function* for this curve consists of angular function θ_i on each side of the curve, defined in such a way that at each corner $c(s_i)$, the corresponding external angle α_i is given by

$$\alpha_i = \theta_i(s_i) - \theta_{i-1}(s_i).$$

Lemma 7.10.3 *Consider a piecewise regular curve $c(s)$ of class $\mathcal{C}^2$ in a Riemann patch of class $\mathcal{C}^2$. Suppose that the curve is given in normal representation and the Riemann patch in a system of geodesic coordinates. This curve admits an angular function. Two angular functions differ by a constant multiple of 2π.*

Proof By Proposition 7.10.1, on each side we have infinitely many possible angular functions, one for each multiple of 2π. Fixing the value of θ_i at one point thus fixes its value on the whole side. Choose θ_{i_0} arbitrarily on an arbitrarily chosen side of the curve. Then proceed side by side, using the requirement

$$\alpha_i = \theta_i(s_i) - \theta_{i-1}(s_i)$$

to fix the initial value of the function θ_i when passing from one side to the next. $\square$

Proposition 7.10.4 *Consider a closed and piecewise regular plane curve of class $\mathcal{C}^2$ in a Riemann patch of class $\mathcal{C}^2$:*

$$c\colon \mathbb{R} \longrightarrow U, \qquad g_{ij}\colon U \longrightarrow \mathbb{R}, \quad i, j = 1, 2.$$

Suppose that the curve is given in normal representation and the Riemann patch, in a geodesic system of coordinates. Write:

- ω *for the minimal period of* c;
- $s_0 < s_1 < \cdots < s_{n-1} < s_n = s_0 + \omega$ *for the values of the parameter corresponding to the corners of the curve*;
- α_i *for the external angle of the curve at the point* $c(s_i)$, *computed of course in terms of the metric tensor* (*see Sect.* 6.2);
- θ *for an angular function along the curve* (*see Definition* 7.10.2), *so that* θ_i *is the angular function of the* i*-th side* $c([s_i, s_{i+1}])$ *of the curve.*

Under these conditions,

$$\sum_{i=0}^{n-1} \bigl(\theta_i(s_{i+1}) - \theta_i(s_i)\bigr) + \sum_{i=0}^{n-1} \alpha_i = 2k\pi, \quad k \in \mathbb{Z}.$$

The integer k *does not depend on the choice of the angular function* θ *and is called the* rotation number *of the curve.*

Proof With the usual convention $\theta_{-1} = \theta_{n-1}$

$$\begin{aligned}
&\sum_{i=0}^{n-1} \bigl(\theta_i(s_{i+1}) - \theta_i(s_i)\bigr) + \sum_{i=0}^{n-1} \alpha_i \\
&\quad = \sum_{i=0}^{n-1} \bigl(\theta_i(s_{i+1}) - \theta_i(s_i)\bigr) + \sum_{i=0}^{n-1} \bigl(\theta_i(s_i) - \theta_{i-1}(s_i)\bigr) \\
&\quad = \sum_{i=0}^{n-1} \bigl(\theta_i(s_{i+1}) - \theta_{i-1}(s_i)\bigr) \\
&\quad = \theta_{n-1}(s_n) - \theta_{-1}(s_0).
\end{aligned}$$

These last two values represent the same angle between $e_1(c(s_0))$ and $c'(s_0)$, computed for two values of the parameter which differ simply by a period. The difference is thus necessarily a multiple of 2π. Notice further that another choice of the angular function θ_i would not affect the value of k, since all functions θ_i would simply be modified by adding a constant multiple of 2π: a quantity which disappears when taking the differences. □

It should be observed that Proposition 7.10.4, in contrast to the situation in Theorem 2.14.8, does not tell us anything about a possible relation with the "curvature"

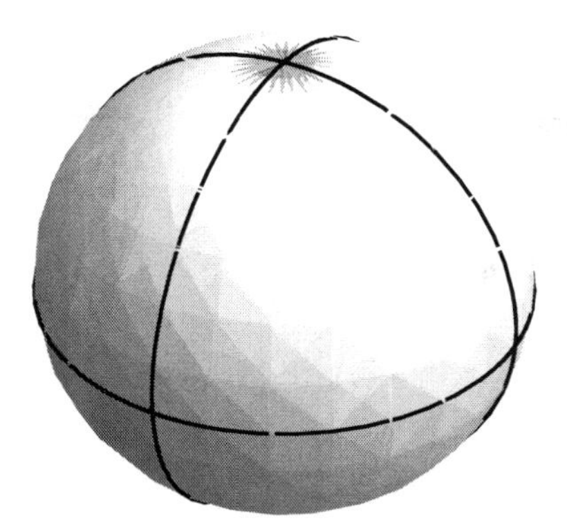

Fig. 7.21

of the curve. In fact, it is easy to convince oneself that just replacing the curvature, in Theorem 2.14.8, by the geodesic curvature, in Proposition 7.10.4, cannot possibly work. Simply consider a sphere and cut it by the equator and by two meridians perpendicular to each other (see Fig. 7.21). One obtains eight triangles. The three sides of each triangle are geodesics (see Example 6.10.4) and the three exterior angles are each equal to $\frac{\pi}{2}$ (or to $-\frac{\pi}{2}$, depending on the orientation). Thus in this case

$$\int_{\text{triangle}} \kappa_g + \sum \alpha_i = \pm \frac{3\pi}{2}.$$

This is definitely not a multiple of 2π.

The more precise relation of the geodesic curvature with the rotation number k of Proposition 7.10.4 will be precisely the content of the *Gauss–Bonnet Theorem* 7.13.1: in fact, both the geodesic curvature *and* the Gaussian curvature will enter into the story.

We next transpose Theorem 2.15.2.

Proposition 7.10.5 *The rotation number of a piecewise, regular and simple closed curve of class $\mathcal{C}^2$, in a Riemann patch of class $\mathcal{C}^2$ given in geodesic coordinates, is equal to ± 1.*

Proof We use the notation of Proposition 7.10.4. On U, let us consider, for every real number $t \in [0, 1]$, the metric tensor

$$\begin{pmatrix} g_{11}^{(t)} & g_{12}^{(t)} \\ g_{21}^{(t)} & g_{22}^{(t)} \end{pmatrix} = \begin{pmatrix} 1 & 0 \\ 0 & t + (1-t)g_{22} \end{pmatrix}$$

where $(g_{ij})_{ij}$ indicates the original metric tensor. Since $g_{22} > 0$ (see Theorem 6.13.1) we have at once $g_{22}^{(t)} > 0$ and thus the $(g_{ij}^{(t)})_{ij}$ constitute a new metric

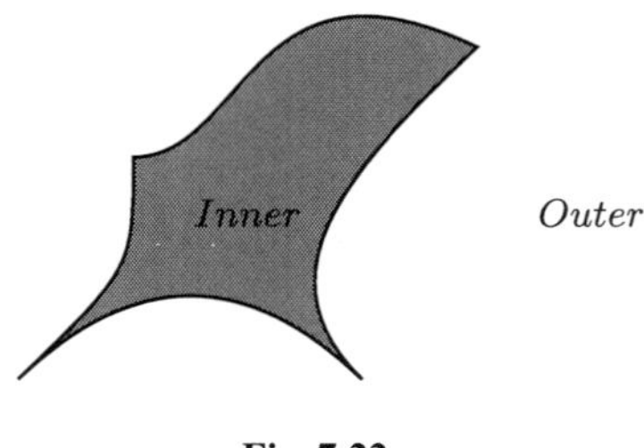

Fig. 7.22

tensor. Notice that when $t = 0$ we recapture the original metric tensor, while for $t = 1$ we obtain the identity matrix, that is, the ordinary metric tensor of the plane.

Let us write further $\theta_i^{(t)}$ and $\alpha_i^{(t)}$, respectively, for the angular functions and the exterior angles computed using the metric tensor $g_{ij}^{(t)}$. It is immediate that the functions $\theta_i^{(t)}$ and $\alpha_i^{(t)}$ are continuous in t. By Proposition 7.10.4, we know that for each value of t

$$\sum_{i=0}^{n-1}\left[\theta_i^{(t)}(s_{i+1}) - \theta_i^{(t)}(s_i)\right] + \sum_{i=0}^{n-1}\alpha_i^{(t)}$$

is a multiple of 2π. This proves that the function

$$t \mapsto \frac{1}{2\pi}\left(\sum_{i=0}^{n-1}\left[\theta_i^{(t)}(s_{i+1}) - \theta_i^{(t)}(s_i)\right] + \sum_{i=0}^{n-1}\alpha_i^{(t)}\right)$$

takes only integer values. Since this function is continuous, it must therefore be constant. Furthermore, since for $t = 1$ the corresponding Riemann patch is that of a plane, by Theorem 2.15.2, this constant integer is equal to ± 1. □

7.11 Polygonal Domains

It is easy to see that a triangle, a circle, a quadrilateral, or indeed any simple closed curve that you draw concretely on a sheet of paper (see Fig. 7.22), delimits an *inner* and an *outer* domain in the plane, although in some tricky situations, such as that of Fig. 7.23, it can take some time to decide if a given point P is in the inner or the outer domain of the polygon.

The reader may wonder why we have insisted so much on considering not just a simple closed curve as before, but also its inner domain. This is because, when passing to the case of surfaces, the existence of an inner domain delimited by a closed curve is no longer valid in general. To understand this point, consider Fig. 7.24, which presents various curves "on the torus". The depicted situations are clearly of very different natures. In the upper case, it clearly makes sense to speak of the "inner

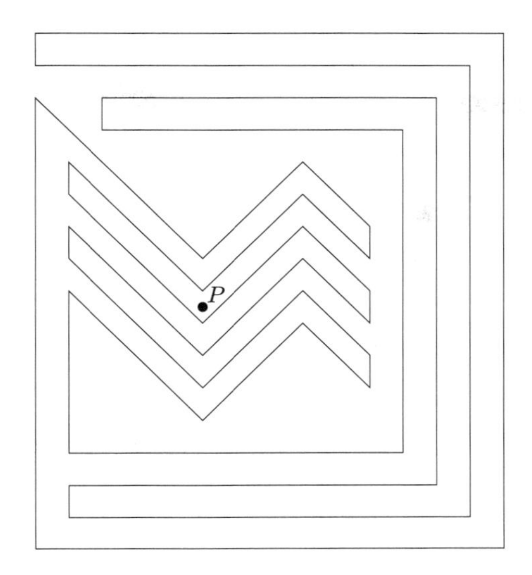

Fig. 7.23

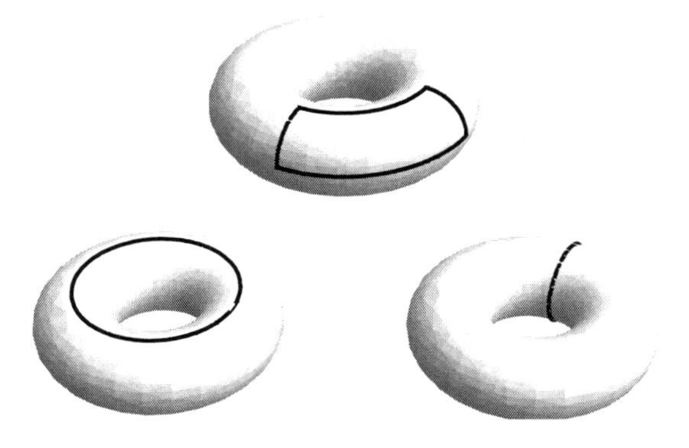

Fig. 7.24

domain" delimited on the torus by the quadrilateral, but in the two lower cases, this is definitely not the case: the "holes" of the torus somehow "lie inside" the circle.

Contrary to what is sometimes thought, proving in full generality that a simple closed curve divides the plane into an inner and an outer domain is a rather easy task: but proving the properties of the inner domain is a very difficult task.

Proposition 7.11.1 *Every compact subset $\mathcal{C} \subseteq \mathbb{R}^2$ in the plane determines a unique partition of the plane*

$$\mathbb{R}^2 = \mathsf{Inner}(\mathcal{C}) \cup \mathcal{C} \cup \mathsf{Outer}(\mathcal{C})$$

having the following properties:

- $\mathsf{Inner}(\mathcal{P})$ *is a bounded open subset called the* inner domain *of* $\mathcal{C}$.
- $\mathsf{Outer}(\mathcal{P})$ *is a connected unbounded open subset called the* outer domain *of* $\mathcal{C}$.
- *Every continuous path (see Definition A.8.2) joining a point of the inner domain to a point of the outer domain necessarily intersects* $\mathcal{C}$.

Proof By compactness, $\mathcal{C}$ is contained in some closed ball $B(O,R)$ (see Definition A.3.1). Fix a point P outside this closed ball and define:

- $\mathsf{Inner}(\mathcal{C})$ to be the set of those points $A \in \mathbb{R}^2 \setminus \mathcal{C}$ such that every continuous path joining A to P intersects $\mathcal{C}$;
- $\mathsf{Outer}(\mathcal{C})$ to be the set of those points $A \in \mathbb{R}^2 \setminus \mathcal{C}$ such that there exists a continuous path joining A to P and not intersecting $\mathcal{C}$.

Trivially, this definition forces at once the partition into three disjoint subsets

$$\mathbb{R}^2 = \mathsf{Inner}(\mathcal{C}) \cup \mathcal{C} \cup \mathsf{Outer}(\mathcal{C}).$$

Fix $A \in \mathbb{R}^2 \setminus \mathcal{C}$. Then the function

$$\mathcal{C} \longrightarrow \mathbb{R}, \qquad B \mapsto d(A,B)$$

expressing the distance between A and a point $B \in \mathcal{C}$ is continuous. By Corollary A.8.4, this function reaches a minimum ε; this minimum is not zero, because A is not in $\mathcal{C}$. Thus the open ball $B(A,\varepsilon)$ does not intersect the compact set $\mathcal{C}$. Of course every point of this open ball can be joined by a segment to the center A of the ball. This implies at once that if $A \in \mathsf{Outer}(\mathcal{C})$, then all points of the open ball are still in $\mathsf{Outer}(\mathcal{C})$. In other words, $\mathsf{Outer}(\mathcal{C})$ is open (see Definition A.1.2). A perfectly analogous argument shows that $\mathsf{Inner}(\mathcal{C})$ is open.

Trivially, every two points outside the original closed ball containing $\mathcal{C}$ can be joined by a continuous path not intersecting $\mathcal{C}$. In particular, all these points belong to $\mathsf{Outer}(\mathcal{C})$. Thus $\mathsf{Outer}(\mathcal{C})$ is unbounded. Since $\mathsf{Inner}(\mathcal{C})$ is disjoint from $\mathsf{Outer}(\mathcal{C})$, it follows further that $\mathsf{Inner}(\mathcal{C})$ is contained in the closed ball $\overline{B}(O,r)$, thus is bounded.

Trivially also, $P \in \mathsf{Outer}(\mathcal{C})$. Since two points of $\mathsf{Outer}(\mathcal{C})$ can be joined to P by continuous paths entirely contained in $\mathsf{Outer}(\mathcal{C})$, it follows at once that $\mathsf{Outer}(\mathcal{C})$ is path-connected, thus connected (see Sect. A.10). Of course if a point A in $\mathsf{Inner}(\mathcal{C})$ is joined by a continuous path to a point $B \in \mathsf{Outer}(\mathcal{C})$, further join B to P by a continuous path not intersecting $\mathcal{C}$. By definition of $\mathsf{Inner}(\mathcal{P})$, the path from A to P intersects the curve and this is thus at a point on the path between A and B.

Let us now prove the uniqueness of a partition as in the statement. Let

$$\mathsf{Inner}'(\mathcal{C}) \cup \mathcal{C} \cup \mathsf{Outer}'(\mathcal{C})$$

be another partition with the same properties. Since we have two partitions with a common component $\mathcal{C}$, it suffices to prove that

$$\mathsf{Outer}'(\mathcal{C}) \subseteq \mathsf{Outer}(\mathcal{C}), \qquad \mathsf{Inner}'(\mathcal{C}) \subseteq \mathsf{Inner}(\mathcal{C}).$$

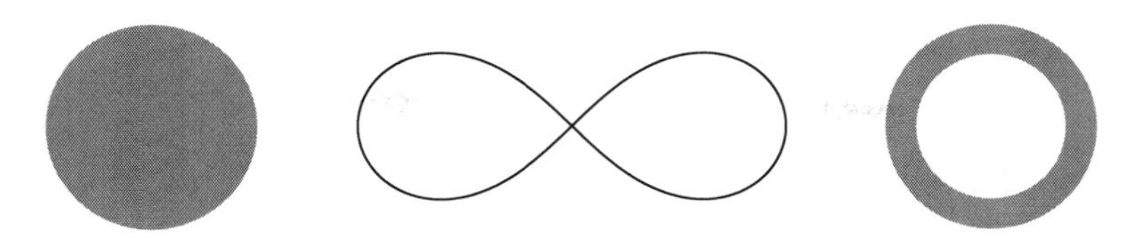

Fig. 7.25

Let us recall that by Theorem A.10.7, $\mathsf{Outer}(\mathcal{C})$ and $\mathsf{Outer}'(\mathcal{C})$ are path connected.

Since $\mathsf{Outer}'(\mathcal{C})$ is unbounded, it contains a point $Q \notin B(O, R)$, the closed ball containing $\mathcal{C}$. By path connectedness every point $A \in \mathsf{Outer}'(\mathcal{C})$ can be joined to Q by a continuous path contained in $\mathsf{Outer}'(\mathcal{C})$, thus not intersecting $\mathcal{C}$; and trivially Q can be joined to P by a continuous path not intersecting the closed ball $B(O, R)$, thus not intersecting $\mathcal{C}$. This yields a continuous path joining A to P and not intersecting $\mathcal{C}$; therefore by definition of $\mathsf{Outer}(\mathcal{C})$, $A \in \mathsf{Outer}(\mathcal{C})$.

Conversely choose a point $B \in \mathsf{Inner}'(\mathcal{C})$. Consider an arbitrary continuous path joining B to P and follow it by a continuous path from P to Q, not intersecting $\mathcal{C}$. Since $B \in \mathsf{Inner}'(\mathcal{C})$ and $Q \in \mathsf{Outer}'(\mathcal{C})$, by assumption the composite path intersects $\mathcal{C}$. Since the path from P to Q does not intersect $\mathcal{C}$, we conclude that the path from B to P intersects $\mathcal{C}$. Since this continuous path is arbitrary, this proves that $B \in \mathsf{Inner}(\mathcal{C})$, by definition of $\mathsf{Inner}(\mathcal{C})$. □

Proposition 7.11.2 *In $\mathbb{R}^2$, the union of a compact subset $\mathcal{C}$ and its inner domain is still a compact subset.*

Proof By Proposition 7.11.1, the union in the statement is the complement of $\mathsf{Outer}(\mathcal{C})$, thus it is a closed subset (see Proposition A.2.2). Since $\mathcal{C}$ is bounded by assumption and its inner domain is bounded by Proposition 7.11.1, their union is bounded as well. Therefore this union is compact (see Definition A.3.1). □

Of course, Proposition 7.11.1 applies in particular when the compact subset $\mathcal{C}$ is the support of a closed curve. Indeed if a closed curve is represented by a periodic continuous function $f : \mathbb{R} \longrightarrow \mathbb{R}^2$, its support is the continuous image under f of the closed interval $[0, \omega]$, with ω the minimal period of the curve. By Proposition A.8.3, this is indeed a compact subset of $\mathbb{R}^2$.

But let us make clear that Proposition 7.11.1 does not tell us much—and *cannot* tell us much—about the inner domain of a closed curve, simple or not. Not even the fact that it is non-empty! Furthermore it can reveal nothing about the possible connectedness of this inner domain, nor anything about the *boundary* of the inner or the outer domain (see Definition A.6.4). Indeed in Proposition 7.11.1 choose as compact subset a closed ball, a lemniscate (see Example 2.13.4) or a "closed ring", as in Fig. 7.25. Respectively, the inner domain is empty, is constituted of two pieces or does not admit the closed ring as boundary.

The additional properties just mentioned turn out to be valid when the compact subset in Proposition 7.11.1 is the support of a *simple closed curve*. But this becomes a very deep, difficult and famous theorem in topology: the so-called *Jordan curve theorem*.

Theorem 7.11.3 (Jordan Curve Theorem) *Every simple closed curve $\mathcal{C}$ in $\mathbb{R}^2$ determines a unique partition of the plane*

$$\mathbb{R}^2 = \mathsf{Inner}(\mathcal{C}) \cup \mathcal{C} \cup \mathsf{Outer}(\mathcal{C})$$

such that:

1. $\mathsf{Inner}(\mathcal{C})$ *is a connected bounded open subset called the* inner domain *of the curve.*
2. $\mathsf{Outer}(\mathcal{C})$ *is a connected unbounded open subset called the* outer domain *of the curve.*
3. *Every continuous path joining a point of the inner domain and a point of the outer domain necessarily intersects the curve.*
4. *The curve $\mathcal{C}$ is the* boundary *of both its inner domain and its outer domain.*

In view of Proposition 7.11.2, it is sensible to make the following definition:

Definition 7.11.4 A *polygonal domain* of class $\mathcal{C}^k$ in $\mathbb{R}^2$ consists of the union of a piecewise regular simple closed curve of class $\mathcal{C}^k$ and its inner domain.

More generally:

Definition 7.11.5 A *polygonal domain* $\mathcal{P}$ of class $\mathcal{C}^k$ on a Riemann surface $\mathcal{S}$ of class $\mathcal{C}^k$ in $\mathbb{R}^3$ is the image $\mathcal{P} = f(\mathcal{P}')$ of a polygonal domain $\mathcal{P}'$ of class $\mathcal{C}^k$ in $\mathbb{R}^2$, contained in the domain of a local map of the surface

$$f_i : U_i \longrightarrow \mathcal{S} \subseteq \mathbb{R}^3, \qquad \mathcal{P}' \subseteq U_i \subseteq \mathbb{R}^2.$$

By the *Jordan curve Theorem* 7.11.3 we then have:

Proposition 7.11.6 *In $\mathbb{R}^2$, the polygonal domain determined by a piecewise regular simple closed curve $\mathcal{C}$ is compact and admits the curve $\mathcal{C}$ as border.*

Proof This follows by Proposition 7.11.2 and Theorem 7.11.3. □

Proposition 7.11.7 *Under the conditions of Definition* 7.11.5, *the polygonal domain $\mathcal{P}$ is compact and the boundary $\partial\mathcal{P}$ of $\mathcal{P}$ on the surface is the image under f of the boundary $\partial\mathcal{P}'$ of $\mathcal{P}'$ in $\mathbb{R}^2$.*

Proof $\mathcal{P}$ is compact as a continuous image of the compact subset $\mathcal{P}'$ (see Propositions 7.11.6 and A.8.3).

For every $A \in \mathcal{P}'$ there exists an open ball $B(A, \varepsilon_A) \subseteq U_i$. Then the various open balls $B(A, \frac{\varepsilon_A}{2})$ cover $\mathcal{P}'$. Since $\mathcal{P}'$ is compact by Definition 7.11.4, finitely many of these ball already cover $\mathcal{P}'$ (see Theorem A.3.3):

$$\mathcal{P}' \subseteq B\left(A_1, \frac{\varepsilon_{A_1}}{2}\right) \cup \cdots \cup B\left(A_1, \frac{\varepsilon_{A_1}}{2}\right).$$

The corresponding closed balls satisfy

$$\overline{B}\left(A_1, \frac{\varepsilon_{A_1}}{2}\right) \cup \cdots \cup \overline{B}\left(A_1, \frac{\varepsilon_{A_1}}{2}\right) \subseteq B(A_1, \varepsilon_1) \cup \cdots \cup B(A_n, \varepsilon_n) \subseteq U_i.$$

The finite union of these closed balls trivially remains compact, therefore the restriction of f to this union becomes a homeomorphism (see Proposition A.9.3). Therefore the restriction of f to the corresponding union of open balls is a homeomorphism as well. Then of course, this homeomorphism preserves the notions of interior, closure and thus boundary (see Appendix A). □

Of course it would be possible, in the considerations above, to avoid any reference to the Jordan curve theorem: it would suffice to cheat a little bit and somehow include the Jordan curve theorem in the definition of a polygonal domain. In other words, one could define a polygonal domain in $\mathbb{R}^2$ as a compact subset, whose boundary is a piecewise regular simple closed curve and whose interior (see Definition A.5.7) is the inner part of that curve. However, as the next sections will show, this does not allow us to get very far before explicitly needing the Jordan curve theorem for other purposes.

7.12 Polygonal Decompositions

The results of Sect. 7.10 assume that we are working in a system of geodesic coordinates; we know that such systems exist locally (see Theorem 6.13.1). We must ask what these results become in the case of polygonal domains not contained in a system of geodesic coordinates. In order to answer this question we will "split" an arbitrary polygonal domain into finitely many smaller pieces, each of these now being a polygonal domain contained in a system of geodesic coordinates.

Our arguments will use some very intuitive results, closely related to the—also very intuitive—*Jordan Curve Theorem* 7.11.3. Nevertheless, a rigorous proof of these intuitive results can only be achieved via an involved application of the Jordan curve theorem: this is clearly beyond the scope of the present book. We shall thus take these intuitive results as granted, making clear at which point of the proof we do so.

Definition 7.12.1 Let K be a compact subset of $\mathbb{R}^2$ or more generally, of a Riemann surface of $\mathbb{R}^3$. A *polygonal decomposition* of K consists of a covering of K

$$K = K_1 \cup \cdots \cup K_n$$

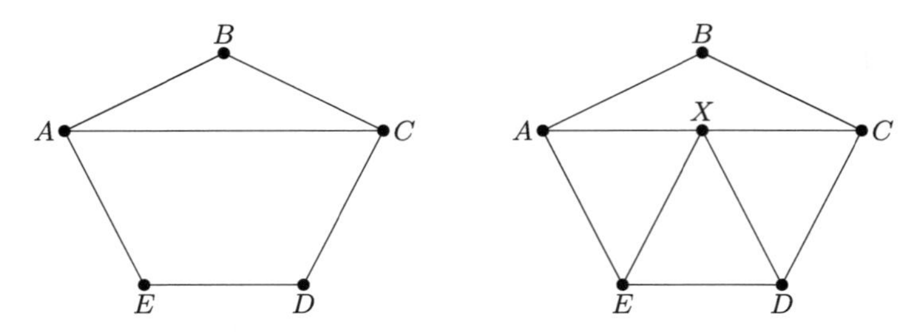

Fig. 7.26

by finitely many polygonal domains $K_i \subseteq K$, each having at least three sides, in such a way that, for all indices $i \neq j$, one of the following three situations hold:

- the intersection of K_i and K_j is empty;
- the intersection of K_i and K_j is reduced to a common corner of both of them;
- the intersection of K_i and K_j is reduced to a common side of both of them.

When all polygonal domains K_i have exactly three sides, the decomposition is called a *triangulation*.

The following terminology is classical.

Definition 7.12.2 Consider a polygonal decomposition in the sense of Definition 7.12.1.

- The various polygonal domains K_i are called the *faces* of the decomposition.
- The various sides of the various polygonal domains K_i are called the *edges* of the decomposition.
- The various corners of the various polygonal domains K_i are called the *vertices* of the decomposition.

Figure 7.26 gives two examples of a polygonal decomposition of a pentagon $ABCDE$. Observe that the triangle ABC is a face of the first polygonal decomposition, but in the second polygonal decomposition, it must now be considered as a quadrilateral $ABCX$. Indeed X is a vertex of the triangle DEX. Since the intersection of the two faces ABC and DEX is a single point X, this point must be a corner of both faces, by Definition 7.12.1. Thus X must now be considered as a vertex of $ABCX$.

Proposition 7.12.3 *Given a polygonal decomposition of a polygonal domain in* $\mathbb{R}^2$ *or on a Riemann surface, write*

- f *for the number of faces*;
- e *for the number of edges*;
- v *for the number of vertices*.

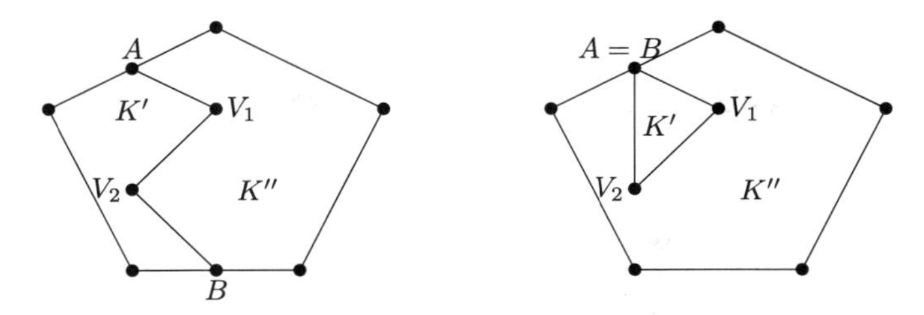

Fig. 7.27

One always has

$$\mathsf{f} - \mathsf{e} + \mathsf{v} = 1.$$

Proof Let $K = K_1 \cup \cdots \cup K_n$ be a polygonal decomposition of a polygonal domain K. We shall reconstruct this decomposition step by step, dividing each time one of the faces into two smaller faces, and observing that at each step the equality in the statement remains valid. Let us stress that we do not claim that all steps correspond to intermediate polygonal decompositions in the sense of Definition 7.12.1.

Let us begin with the given polygonal domain K, viewed as a polygonal decomposition of itself. It thus has 1 face and—let us say—n vertices; it therefore has n edges. In this trivial case we have

$$\mathsf{f} - \mathsf{e} + \mathsf{v} = 1 - n + n = 1.$$

If the decomposition has more than one face, consider a face K_i containing a point P of the border of K. Of course the border of K_i is not entirely contained in the border of K, otherwise one would have $K_i = K$. Following the border of K_i (let us say, in the positive orientation), we will eventually leave the border of K. By Definition 7.12.1 of a polygonal decomposition, this must be when leaving a vertex A of K_i. But since we have to come back to P, we must come back (one or several times) to the border of K, again at some vertex of K_i. Write B for the vertex of K_i at which we return for the first time to the border of K (see Fig. 7.27). We have described a "polygonal path", joining two points A and B of the border of K and not having any other point in common with this border. Notice that possibly, $A = B$.

If $A = B$, the polygonal path that we have just described is a simple closed curve contained in K, with $A = B$ as the only point on ∂K. If $A \neq B$, there are two ways to join A and B following the border of K; together with the polygonal path joining A and B, this provides two simple closed curves, thus by the Jordan curve Theorem 7.11.3, two corresponding polygonal domains K' and K''. In both cases, a refined application of the Jordan curve Theorem allows us to prove the rather intuitive fact that we have obtained a covering $K = K' \cup K''$ of K, where K' and K'' are each obtained as unions of some of the original K_i's while $K \cap K'$ is the polygonal line joining A and B.

Let us write

$$A = V_0, V_1, \ldots, V_{m-1}, V_m = B$$

for the various vertices of the original decomposition which lie on the "polygonal path" from A to B. Consider now the "pseudo-decomposition" comprising the two "pseudo-faces" K', K'' and all the vertices and sides of the original decomposition which lie on the border of K or on the "polygonal path" from A to B. (This is generally not a polygonal decomposition in the sense of Definition 7.12.1, since K' and K'' have a whole polygonal path as intersection.) With respect to the first step, one "pseudo-face" has been added (two pseudo-faces instead of one face). Let us keep writing f for the number of pseudo-faces.

- If A and B were already vertices of K, $m-1$ new vertices $V_1, \ldots, V_{m-1}$ have been added to the decomposition, as well as the m edges of the polygonal path. The quantity in the statement thus becomes

$$(\mathsf{f}+1)-(\mathsf{e}+m)+(\mathsf{v}+m-1)=\mathsf{f}-\mathsf{e}+\mathsf{v}$$

  that is, remains unchanged.
- If A was already a vertex of K but B was not, then m new vertices $V_1, \ldots, V_m$ have been added to the decomposition, as well as the m edges of the polygonal path; but introducing the new vertex B on one edge of K also splits this edge into two edges, thus adding one additional edge to the decomposition. The quantity in the statement then becomes

$$(\mathsf{f}+1)-(\mathsf{e}+m+1)+(\mathsf{v}+m)=\mathsf{f}-\mathsf{e}+\mathsf{v}$$

  and again, remains unchanged.
- If A and B are both new vertices, an analogous observation indicates that the quantity in the statement becomes

$$(\mathsf{f}+1)-(\mathsf{e}+m+2)+(\mathsf{v}+m+1)=\mathsf{f}-\mathsf{e}+\mathsf{v}$$

  when $A \neq B$ and

$$(\mathsf{f}+1)-(\mathsf{e}+m+1)+(\mathsf{v}+m)=\mathsf{f}-\mathsf{e}+\mathsf{v}$$

  when $A=B$. Again this quantity remains unchanged.

So in this pseudo-decomposition $K=K' \cup K''$, we still have $\mathsf{f}-\mathsf{e}+\mathsf{v}=1$.

If $K=K' \cup K''$ is the decomposition in the statement, we are done. Otherwise apply separately to K' and K the process just described for K: we obtain a polygonal pseudo-decomposition in more pieces and such that the quantity $\mathsf{f}-\mathsf{e}+\mathsf{v}$ remains unchanged, thus is still equal to 1. After at most $n-1$ steps, all the pseudo-faces will have been reduced to faces of the original decomposition and the proof will be complete. □

An important use of polygonal decompositions is the following result:

Proposition 7.12.4 *Every polygonal domain on a Riemann surface admits a polygonal decomposition such that all faces are contained in a system of geodesic coordinates.*

Proof Let $\mathcal{P} \subseteq \mathcal{S}$ be a polygonal domain on the Riemann surface $\mathcal{S}$. As in Definition 7.11.5, we obtain $\mathcal{P}$ as the image of a polygonal domain $\mathcal{P}'$ in $\mathbb{R}^2$:

$$f: U \longrightarrow \mathcal{S}, \qquad \mathcal{P}' \subseteq U \subseteq \mathbb{R}^2.$$

By Theorem 6.13.1, in a neighborhood of each point, there exists a system of geodesic coordinates. Thus for each point $P \in U$ there exists an open ball $B(P, r_P) \subseteq U$ on which the corresponding piece of the surface admits a system of geodesic coordinates. The various open balls $B(P, \frac{r_P}{2})$ cover the compact subset $\mathcal{P}'$, thus finitely many of them already do (see Theorem A.3.3):

$$\mathcal{P}' \subseteq B\left(P_1, \frac{r_{P_1}}{2}\right) \cup \cdots \cup B\left(P_1, \frac{r_{P_n}}{2}\right).$$

Write

$$r = \min\left\{\frac{r_{P_1}}{2}, \ldots, \frac{r_{P_n}}{2}\right\}.$$

Then for each point $P \in \mathcal{P}'$, the open ball $B(P, r)$ is contained in one of the open balls $B(P_i, \frac{r_{P_i}}{2})$ and therefore the corresponding portion $f(B(P, r))$ of the surface admits a system of geodesic coordinates.

It then remains to choose a polygonal decomposition of $\mathcal{P}'$ into pieces sufficiently small so that each is contained in an open ball of radius r. Once more the existence of such a decomposition is rather intuitive, but a precise proof requires a refined application of the Jordan curve Theorem 7.11.3. □

7.13 The Gauss–Bonnet Theorem

The *Gauss–Bonnet* theorem is a consequence of the *Green–Riemann* formula (see Theorem 1.12.3): it relates the integral of the Gaussian curvature on a polygonal domain K of a surface with the integral of the geodesic curvature on the border ∂K of K.

Theorem 7.13.1 (Gauss–Bonnet) *Consider a Riemann surface of class $\mathcal{C}^3$ in $\mathbb{R}^3$ and a polygonal domain $\mathcal{P}$ on it. Orient the border positively. In that case*

$$\iint_{\mathcal{P}} \kappa_\tau + \int_{\partial\mathcal{P}} \kappa_g + \sum_{i=1}^{n} \alpha_i = 2\pi$$

where:

- κ_τ *is the Gaussian curvature of the surface*;
- κ_g *is the geodesic curvature of the border $\partial\mathcal{P}$ of $\mathcal{P}$*;
- α_i *are the external angles of the border $\partial\mathcal{P}$.*

Proof In a first approach, let us assume that the polygonal domain is contained in a portion of the surface admitting a system of geodesic coordinates.

We consider a parametric representation of class $\mathcal{C}^3$ of this portion of the surface

$$f\colon U \longrightarrow \mathbb{R}^3, \qquad (x^1, x^2) \mapsto f(x^1, x^2)$$

of the surface and we assume at once that it has been chosen so that (x^1, x^2) is a system of geodesic coordinates. We write $\mathcal{P}'$ for the polygonal domain $\mathcal{P}' \subseteq U$ whose image under f is the polygonal domain $\mathcal{P}$ on the surface. We also choose a parametric representation of the border of $\mathcal{P}$ which restricts as a normal parametric representation of each "side" of $\mathcal{P}$. More precisely, if $S_0, \ldots, S_n = S_0$ are the successive corners of $\mathcal{P}'$, we choose real numbers $s_0 < \cdots < s_n$ and a function

$$c\colon [s_0, s_n] \longrightarrow U$$

so that on each interval $[s_i, s_{i+1}]$, $f \circ c$ is a normal representation of the side $f(s_i)\, f(s_{i+1})$ of $\mathcal{P}$.

First, we compute the integral of the Gaussian curvature:

$$\begin{aligned}
\iint_{\mathcal{P}} \kappa_\tau &= \iint_{\mathcal{P}'} \kappa_\tau \left\| \frac{\partial f}{\partial x^1} \times \frac{\partial f}{\partial x^2} \right\| dx^1\, dx^2 \\
&\qquad \text{(by the formula for computing a curve integral)} \\
&= \iint_{\mathcal{P}'} \kappa_\tau \sqrt{g_{11} g_{22} - g_{12} g_{21}}\, dx^1\, dx^2 \\
&\qquad \text{(see Proposition 5.4.7)} \\
&= \iint_{\mathcal{P}'} \kappa_\tau \sqrt{g_{22}}\, dx^1\, dx^2 \\
&\qquad \text{(by Theorem 6.13.1)} \\
&= \iint_{\mathcal{P}'} -\frac{\partial^2 \sqrt{g_{22}}}{\partial (x^1)^2}\, dx^1\, dx^2 \\
&\qquad \text{(by Proposition 6.14.5)} \\
&= \iint_{\mathcal{P}'} -\frac{\partial}{\partial x^1} \left(\frac{1}{2\sqrt{g_{22}}} \frac{\partial g_{22}}{\partial x^1} \right) dx^1\, dx^2 \\
&\qquad \text{(differentiation of a composite)} \\
&= -\int_{\partial \mathcal{P}'} \frac{1}{2\sqrt{g_{22}}} \frac{\partial g_{22}}{\partial x^1}\, dx^2 \\
&\qquad \text{(by the Green–Riemann formula 1.12.3)} \\
&= -\int_{s_0}^{s_n} \frac{1}{2\sqrt{g_{22}}} \frac{\partial g_{22}}{\partial x^1} \frac{dc^2}{ds}\, ds \\
&\qquad \text{(formula for curve integrals).}
\end{aligned}$$

Let us now compute the integral of the geodesic curvature along the border of $\mathcal{P}$. For this we apply Corollary 6.14.4. Of course as usual for piecewise regular curves, to avoid any ambiguity, we use the notation θ_i to indicate, along the i-side of the curve, the angle between the curve $x^2 = c^2(s)$ and the curve $c(s)$.

$$\begin{aligned}
\int_{\partial\mathcal{P}} \kappa_g &= \int_{s_0}^{s_n} \kappa_g \left\|(f \circ c)'\right\| ds \\
&\quad \text{(formula for curve integrals)} \\
&= \int_{s_0}^{s_n} \kappa_g \, ds \\
&\quad \text{(because } f \circ c \text{ is normal on each side)} \\
&= \sum_{i=0} n-1 \int_{s_i}^{s_{i+1}} \frac{d\theta_i}{ds} + \frac{1}{2\sqrt{g_{22}}} \frac{\partial g_{22}}{\partial x^1} \partial dc^2 ds \\
&\quad \text{(as computed above)} \\
&= \sum_{i=0}^{n-1} \big[\theta_i(s_{i+1}) - \theta_i(s_i)\big] + \int_{s_0}^{s_n} \frac{1}{2\sqrt{g_{22}}} \frac{\partial g_{22}}{\partial x^1} \partial dc^2 ds.
\end{aligned}$$

Comparing the two formulas above for the integral of κ_τ and that of κ_g, we conclude that it remains to prove that

$$\sum_{i=0}^{n-1} \big[\theta_i(s_{i+1}) - \theta_i(s_i)\big] + \sum_{i=0}^{n-1} \alpha_i = 2\pi.$$

This is precisely the content of Proposition 7.10.5.

We now have to remove the assumption of being in a system of geodesic coordinates. For that we apply Proposition 7.12.4, use its notation, and consider a decomposition of the polygonal domain into **f** faces

$$\mathcal{P} = \mathcal{P}_1 \cup \cdots \cup \mathcal{P}_{\mathsf{f}}$$

with each $\mathcal{P}_t$ a polygonal domain contained in a system of geodesic coordinates. By the first part of the proof we have, for each index t

$$\iint_{\mathcal{P}_t} \kappa_\tau + \int_{\partial\mathcal{P}_t} \kappa_g + \sum_{i_t=1}^{n_t} \alpha_{i_t}^{(t)} = 2\pi$$

where the $\alpha_i^{(t)}$ are the external angles of $\mathcal{P}_t$ and n_t is the number of vertices. To obtain the expected result for the polygonal domain $\mathcal{P}$, it "suffices" to add all these equalities.

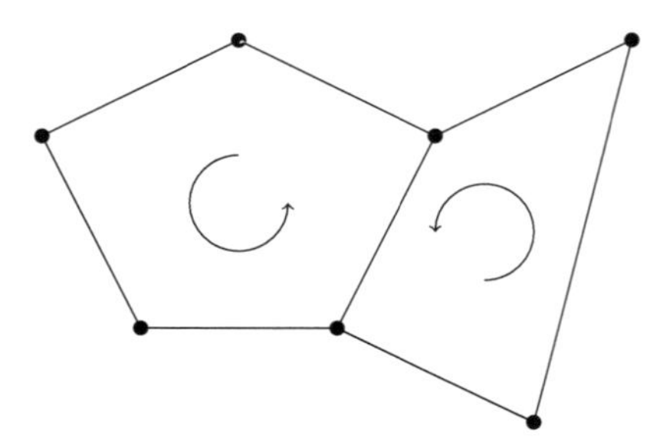

Fig. 7.28

First of all,

$$\iint_{\mathcal{P}} \kappa_\tau = \sum_{t=1}^{\mathbf{f}} \iint_{\mathcal{P}_t} \kappa_\tau$$

because $\mathcal{P} = \mathcal{P}_1 \cup \cdots \cup \mathcal{P}_{\mathbf{f}}$ while the corresponding integrals on each $\mathcal{P}_t \cup \mathcal{P}_{t'}$ $(t \neq t')$ are zero, by definition of a polygonal decomposition (see Definition 7.12.1).

Next, we take care of the integral of the geodesic curvature. Each edge of the decomposition

- is an edge of a unique $\mathcal{P}_t$ when it is part of the border $\partial\mathcal{P}$;
- is a common edge of exactly two faces $\mathcal{P}_t$, $\mathcal{P}_{t'}$, $(t \neq t')$ otherwise.

But in the second case, since we travel around the border of all faces of the decomposition in the positive sense, we travel around the common edge positively for one of the two faces and negatively for the other one (see Fig. 7.28). The two integrals of the geodesic curvature along this edge thus have opposite signs and their sum is therefore zero. Eventually, only the terms corresponding to edges lying on $\partial\mathcal{P}$ remain and we obtain

$$\sum_{t=1}^{\mathbf{f}} \int_{\partial\mathcal{P}_t} \kappa_g = \int_{\partial\mathcal{P}} \kappa_g.$$

Thus adding all the Gauss–Bonnet equalities for the individual polygonal domains $\mathcal{P}_i$ we obtain

$$\iint_{\mathcal{P}} \kappa_\tau + \int_{\partial\mathcal{P}} \kappa_g + \sum_{t=1}^{\mathbf{f}} \sum_{i_t=1}^{n_t} \alpha_{i_t}^{(t)} = 2\pi\mathbf{f}.$$

To conclude the proof, it thus remains to show that

$$\sum_{t=1}^{\mathbf{f}} \sum_{i_t=1}^{n_t} \alpha_{i_t}^{(t)} = \sum_{i=1}^{n} \alpha_i + 2\pi(\mathbf{f} - 1).$$

Writing $\beta_{i_t}^{(t)}$ for the internal angles of $\mathcal{P}_t$, we have

$$\sum_{t=1}^{\mathbf{f}}\sum_{i_t=1}^{n_t}\alpha_{i_t}^{(t)} = \sum_{t=1}^{\mathbf{f}}\sum_{i_t=1}^{n_t}\left(\pi - \beta_{i_t}^{(t)}\right) = \sum_{t=1}^{\mathbf{f}} n_t\pi - \sum_{t=1}^{\mathbf{f}}\sum_{i_t=1}^{n_t}\beta_{i_t}^{(t)}.$$

Instead of grouping these angles face by face, let us group them instead vertex by vertex:

- at a vertex located in the interior of $\mathcal{P}$, the sum of all the internal angles of all faces containing this vertex is equal to 2π;
- at a vertex on the border $\partial\mathcal{P}$, but not a vertex of the original polygonal domain $\mathcal{P}$, the sum of all the internal angles of all faces containing this vertex is equal to π;
- at a vertex of the original polygonal domain $\mathcal{P}$, the sum of all the internal angles of all faces containing this vertex is equal to β_i, the internal angle of $\mathcal{P}$ at this vertex. Again $\beta_i = \pi - \alpha_i$, with α_i the corresponding external angle.

Let us write $\mathbf{v}_{\mathsf{int}}$ for the number of vertices of the decomposition which lie in the interior of $\mathcal{P}$. Since $\mathcal{P}$ has n vertices, there remain $\mathbf{v} - \mathbf{v}_{\mathsf{int}} - n$ vertices of the decomposition which lie on $\partial\mathcal{P}$ without being a vertex of $\mathcal{P}$. We thus obtain

$$\sum_{t=1}^{\mathbf{f}}\sum_{i_t=1}^{n_t}\beta_{i_t}^{(t)} = \mathbf{v}_{\mathsf{int}}\cdot 2\pi + (\mathbf{v} - \mathbf{v}_{\mathsf{int}} - n)\cdot\pi + n\pi - \sum_{i=1}^{n}\alpha_i = (\mathbf{v}_{\mathsf{int}} + \mathbf{v})\pi - \sum_{i=1}^{n}\alpha_i.$$

But counting all the vertices of all faces is the same as counting all the faces appearing at all vertices. As we have seen, according to the fact that the vertex is inside $\mathcal{P}$ or on its boundary, this number of faces at a vertex equals the number of edges at this vertex or this number of edges minus 1. This "minus 1" possibility occurs $\mathbf{v} - \mathbf{v}_{\mathsf{int}}$ times. Since moreover each edge appears twice, at two distinct vertices, we infer that

$$\sum_{t=1}^{\mathbf{f}} n_t = 2\mathbf{e} - (\mathbf{v} - \mathbf{v}_{\mathsf{int}}).$$

Putting all these results together and applying Proposition 7.12.3 we obtain,

$$\begin{aligned}\sum_{t=1}^{\mathbf{f}}\sum_{i_t=1}^{n_t}\alpha_{i_t}^{(t)} &= \left(\sum_{t=1}^{\mathbf{f}}\sum_{i_t=1}^{n_t}\right)\pi - \sum_{t=1}^{\mathbf{f}}\sum_{i_t=1}^{n_t}\beta_{i_t}^{(t)} \\ &= (2\mathbf{e} - \mathbf{v} + \mathbf{v}_{\mathsf{int}})\pi - (\mathbf{v}_{\mathsf{int}} + \mathbf{v})\pi + \sum_{i=1}^{n}\alpha_i \\ &= 2(\mathbf{e} - \mathbf{v})\pi + \sum_{i=1}^{n}\alpha_i\end{aligned}$$

$$= 2(\mathbf{f} - 1)\pi + \sum_{i=1}^{n} \alpha_i .$$

This is precisely the expected equality which concludes the proof. □

7.14 Geodesic Triangles

The study of *Euclid's Elements* in [3], *Trilogy I*, has provided evidence that the triangle is the very basic ingredient of the properties on which Greek geometry of the plane is based. *Hilbert's axiomatization* of the plane (see Chap. 8 in [3], *Trilogy I*)—Euclidean or non-Euclidean—again grants a crucial role to the triangle. The generalization of the notion of "triangle" to the context of a surface is easy:

Definition 7.14.1 By a *geodesic triangle* on a Riemann surface of class $\mathcal{C}^3$ in $\mathbb{R}^3$ is meant a polygonal domain with exactly three sides, these sides being geodesics of the surface.

The following result is then a special case of the Gauss–Bonnet Theorem:

Theorem 7.14.2 (Theorema Elegantissimum; Gauss) *On a Riemann surface of class $\mathcal{C}^3$ in $\mathbb{R}^3$, the sum of the interior angles of a geodesic triangle $\mathcal{T}$ is equal to*

$$\pi + \iint_{\mathcal{T}} \kappa_{\tau} .$$

Proof In the Gauss–Bonnet theorem 7.13.1, the geodesic curvature of the boundary is now zero (see Definition 6.10.1). Writing α_i for the exterior angles and $\beta_i = \pi - \alpha_i$ for the interior angles, the Gauss–Bonnet formula becomes

$$2\pi = \iint_{\mathcal{T}} \kappa_{\tau} + \sum_{i=1}^{3} \alpha_i = \iint_{\mathcal{T}} \kappa_{\tau} + 3\pi - \sum_{i=1}^{3} \beta_i .$$

This is the expected result. □

Corollary 7.14.3 *Consider a Riemann surface of class $\mathcal{C}^3$ in $\mathbb{R}^3$ and its Gaussian curvature κ_{τ}.*

- *If $\kappa_{\tau} < 0$ at each point, then the sum of the interior angles of every geodesic triangle is strictly less than π.*
- *If $\kappa_{\tau} = 0$ at each point, then the sum of the interior angles of every geodesic triangle is equal to π.*
- *If $\kappa_{\tau} > 0$ at each point, then the sum of the interior angles of every geodesic triangle is strictly greater than π.*

In view of Sect. 7.3 in [3], *Trilogy I*, the following definition is natural:

Definition 7.14.4 Consider a Riemann surface of class $\mathcal{C}^3$ in $\mathbb{R}^3$ and its Gaussian curvature κ_τ.

- The surface is called *hyperbolic* when $\kappa_\tau < 0$ at each point.
- The surface is called *elliptic* when $\kappa_\tau > 0$ at each point.

Observe that this terminology is perfectly consistent with that of Definition 5.16.9.

7.15 The Euler–Poincaré Characteristic

Let us conclude this book—and this *Trilogy*—by opening a door onto algebraic topology. The *Euler–Poincaré* characteristic of a surface is just an integer, but an integer which tells us something interesting about the shape of the surface. This is the spirit of algebraic topology: to a given topological space (here, a surface) one associates some useful algebraic object. The whole power of algebra can then be applied to infer algebraic properties of this algebraic object and translate them into topological properties of the original topological space.

A very important result which is often proved using the techniques of algebraic topology is the following:

Theorem 7.15.1 *Every compact Riemann surface of class $\mathcal{C}^2$ in $\mathbb{R}^3$ is orientable.*

Clearly, by "orientable" we mean that one can choose an orientation on each local map (see Definition 5.7.1) in such a way that these orientations coincide on every intersection of two local maps.

On the other hand, adapting first the argument developed in the "proof" of Proposition 7.12.4 and using a rather strong version of the *Jordan Curve Theorem* 7.11.3, one also proves that:

Proposition 7.15.2 *Every compact Riemann surface of class $\mathcal{C}^3$ in $\mathbb{R}^3$ admits a polygonal decomposition (see Definition* 7.12.1).

Applying the Gauss–Bonnet Theorem, we then infer:

Theorem 7.15.3 *Given a compact Riemann surface $\mathcal{S}$ of class $\mathcal{C}^3$ in $\mathbb{R}^3$, the quantity*

$$\chi = \frac{1}{2\pi} \int\int_{\mathcal{S}} \kappa_\tau$$

is an integer, called the Euler–Poincaré *characteristic of the surface. Given a polygonal decomposition of the surface, with* **f** *faces,* **e** *edges and* **v** *vertices, the Euler–Poincaré characteristic is also equal to*

$$\chi = \mathbf{f} - \mathbf{e} + \mathbf{v}.$$

Proof The proof is an easy adaptation of the last part of the proof of the Gauss–Bonnet Theorem 7.13.1. Applying Proposition 7.12.4 and using its notation, we consider a decomposition of the surface into **f** faces

$$\mathcal{S} = \mathcal{P}_1 \cup \cdots \cup \mathcal{P}_{\mathbf{f}}.$$

The Gauss–Bonnet Theorem 7.13.1 yields for each index t

$$\iint_{\mathcal{P}_t} \kappa_\tau + \int_{\partial \mathcal{P}_t} \kappa_g + \sum_{i_t=1}^{n_t} \alpha_{i_t}^{(t)} = 2\pi$$

where the $\alpha_i^{(t)}$ are the external angles of $\mathcal{P}_t$ and n_t is the number of vertices (or edges) of $\mathcal{P}_t$. To conclude, it "suffices" to add all these equalities.

First of all,

$$\iint_{\mathcal{S}} \kappa_\tau = \sum_{t=1}^{\mathbf{f}} \iint_{\mathcal{P}_t} \kappa_\tau$$

because $\mathcal{S} = \mathcal{P}_1 \cup \cdots \cup \mathcal{P}_{\mathbf{f}}$ while the corresponding integrals on each $\mathcal{P}_t \cap \mathcal{P}_{t'}$ $(t \neq t')$ are zero, by definition of a polygonal decomposition (see Definition 7.12.1).

Next, we take care of the integral of the geodesic curvature. Since we have a covering of the whole surface, every edge (which is contained in a local map, by Definition 7.11.5) necessarily appears as an edge of two different faces. Moreover, since the surface is orientable, if we travel positively along the borders of these two faces, we travel along the mutual edge in opposite directions. Thus the two corresponding terms annihilate each other and all the terms involving the geodesic curvature disappear when summing over t.

Thus adding all the Gauss–Bonnet equalities for the individual polygonal domains $\mathcal{P}_i$ we obtain

$$\iint_{\mathcal{S}} \kappa_\tau + \sum_{t=1}^{\mathbf{f}} \sum_{i_t=1}^{n_t} \alpha_{i_t}^{(t)} = 2\pi \mathbf{f}.$$

To conclude the proof, it remains to show that

$$\sum_{t=1}^{\mathbf{f}} \sum_{i_t=1}^{n_t} \alpha_{i_t}^{(t)} = 2\pi(\mathbf{e} - \mathbf{v}).$$

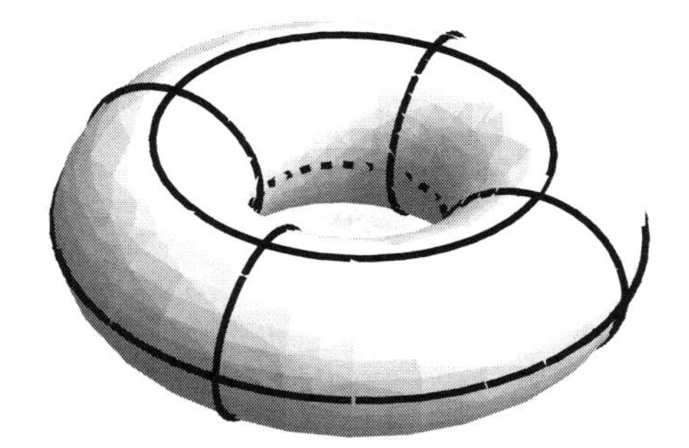

Fig. 7.29

As in the proof of the Gauss–Bonnet Theorem 7.13.1, we write $\beta_{i_t}^{(t)}$ for the interior angles of $\mathcal{P}_t$ and obtain

$$\sum_{t=1}^{\mathbf{f}}\sum_{i_t=1}^{n_t}\alpha_{i_t}^{(t)}=\sum_{t=1}^{\mathbf{f}}\sum_{i_t=1}^{n_t}\left(\pi-\beta_{i_t}^{(t)}\right)=\sum_{t=1}^{\mathbf{f}}n_t\pi-\sum_{t=1}^{\mathbf{f}}\sum_{i_t=1}^{n_t}\beta_{i_t}^{(t)}.$$

This time, since we have a decomposition of the whole surface, each vertex is an "interior" vertex of $\mathcal{P}$ and the sum of the interior angles at this vertex equals 2π. Therefore

$$\sum_{t=1}^{\mathbf{f}}\sum_{i_t=1}^{n_t}\beta_{i_t}^{(t)}=2\pi\mathbf{v}.$$

On the other hand $\sum_{t=1}^{\mathbf{f}}n_t$ is the sum of the number of edges of the various faces: since each edge appears in exactly two faces, this sum is simply $2\mathbf{e}$. This implies

$$\sum_{t=1}^{\mathbf{f}}\sum_{i_t=1}^{n_t}\alpha_{i_t}^{(t)}=2\pi\mathbf{e}-2\pi\mathbf{v}$$

as expected. □

Theorem 7.15.3 is a very powerful one: it allows us to compute the integral $\iint_{\mathcal{S}}\kappa_\tau$ just by counting faces, edges and vertices. Let us do this for some easy compact surfaces:

Example 7.15.4 The Euler–Poincaré characteristic χ of:

- the *sphere* (see Fig. 7.29) is 2;
- the *torus* (see Fig. 7.29) is 0;
- the *torus with two holes* (see Fig. 7.30) is -2;
- the *torus with n holes* is $2(1-n)$.

 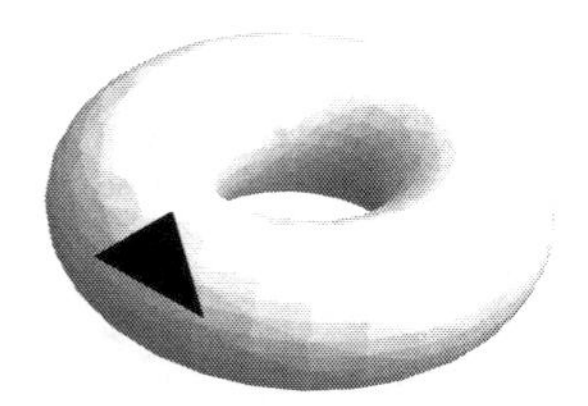

Fig. 7.30

Proof In the first three cases, just "draw" these surfaces together with a polygonal decomposition and apply Theorem 7.15.3. In Fig. 7.29, we have cut the sphere and the torus by the three coordinate planes. For the sphere we get:

- 8 triangular faces;
- 12 edges;
- 6 vertices,

thus

$$\chi = 8 - 12 + 6 = 2.$$

For the torus we obtain:

- 16 quadrilateral faces;
- 24 edges;
- 8 vertices,

thus

$$\chi = 16 - 24 + 8 = 0.$$

Now how can you construct a torus with n holes? First take a torus with $n-1$ holes, and a torus with one hole. Cut a small triangle out of each of these and glue the two pieces together along these triangles (see Fig. 7.30). Then choose polygonal decompositions of the two original tori:

- with f_{n-1} faces, e_{n-1} edges and v_{n-1} vertices in the case of the torus with $n-1$ holes;
- with f_1 faces, e_1 edges and v_1 vertices in the case of the torus with one hole.

Choose these decompositions so that the triangle involved in the cut-and-glue process is a face of each decomposition. Then consider the resulting decomposition of the torus with n holes; it has f_{n-1} faces, e_{n-1} edges and v_{n-1} vertices and

$$\mathsf{f}_n = (\mathsf{f}_{n-1} - 1) + (\mathsf{f}_1 - 1), \qquad \mathsf{e}_n = \mathsf{e}_{n-1} + \mathsf{e}_1 - 3, \qquad \mathsf{v}_n = \mathsf{v}_{n-1} + \mathsf{v}_1 - 3.$$

This implies

$$\begin{aligned}\mathsf{f}_n - \mathsf{e}_n + \mathsf{v}_n &= (\mathsf{f}_{n-1} - \mathsf{e}_{n-1} + \mathsf{v}_{n-1}) + (\mathsf{f}_1 - \mathsf{e}_1 + \mathsf{v}_1) - 2\\ &= (\mathsf{f}_{n-1} - \mathsf{e}_{n-1} + \mathsf{v}_{n-1}) - 2\end{aligned}$$

since the torus has Euler–Poincaré characteristic 0. Writing χ_n for the Euler–Poincaré characteristic of the torus with n holes, this proves that $\chi_n = \chi_{n-1} - 2$. Together with the initial value $\chi_1 = 0$, this forces the conclusion by induction. □

The conclusion is rather clear: the Euler–Poincaré characteristic χ recaptures the information on "the number of holes" in the surface. In the tori of Example 7.15.4, the number of holes is equal to $1 - \frac{\chi}{2}$. So indeed, just knowing the number χ already gives us some pertinent information on the shape of the surface.

Of course, we have not given a precise definition of what a "hole" in a surface is. The answer leads us into the fascinating world of algebraic topology.

7.16 Problems

7.16.1 Consider a regular curve c of class $\mathcal{C}^2$ on an orientable regular surface f of class $\mathcal{C}^2$. Write $\overrightarrow{n}$ for the normal vector to the surface. The curve c is a line of curvature on the surface if and only if the ruled surface

$$g(s,t) = c(t) + s\overrightarrow{n}\big(c(t)\big)$$

is developable.

7.16.2 The tangent plane along a ruling of the tangent surface to a regular curve of class $\mathcal{C}^2$ coincides with the osculating plane to the curve at the corresponding point.

7.16.3 A regular surface of class $\mathcal{C}^2$ has constant zero curvature if and only if, in the neighborhood of each point, there exist two families of geodesics which intersect at a constant angle.

7.16.4 On a regular surface of class $\mathcal{C}^2$ with constant Gaussian curvature, prove that the area of a geodesic polygon is entirely determined by the knowledge of its interior angles.

7.16.5 An *ovaloid* is a compact surface of class $\mathcal{C}^2$ in $\mathbb{R}^3$ whose Gaussian curvature is strictly positive at each point. Such a surface is *convex*, that is, at each point, lies entirely on one side of the tangent plane at that point. (Compare with Theorem 2.16.4.)

7.16.6 Two ovaloids which are locally Riemann isometric are necessarily isometric as subsets of $\mathbb{R}^3$.

7.16.7 On a Riemann surface of class $\mathcal{C}^2$ in $\mathbb{R}^3$, with strictly negative Gaussian curvature, a closed geodesic necessarily intersects itself.

7.16.8 On a connected Riemann surface of class $\mathcal{C}^2$ in $\mathbb{R}^3$, with strictly positive Gaussian curvature, two different closed geodesics necessarily intersect each other.

Fig. 7.31 A curve with constant compass direction

7.17 Exercises

7.17.1 Show that the curve

$$c\colon \,]0,\infty[\, \to \mathbb{R}^2, \qquad t \mapsto (2\arctan t, \log t)$$

intersects all the meridians of the sphere

$$f(\varphi,\theta) = (\cos\theta\sin\varphi, \sin\theta\sin\varphi, \cos\varphi)$$

at a constant angle $\frac{\pi}{4}$. (This curve of "constant compass-direction" *spirals* around the North and the South pole; see Fig. 7.31.)

7.17.2 For each of the two families of rulings, find the striction lines of the hyperboloid of one sheet and the hyperbolic paraboloid.

7.17.3 Find the striction lines of the helicoid (Example 5.1.8) and of the Möbius strip (Example 5.1.9).

7.17.4 The *catenoid* is the surface of revolution obtained by rotating a catenary (see Sect. 3.17); see Fig. 7.32:

$$f\left(x^1,x^2\right) = \left(k\cosh\frac{x^1}{k}\cos x^2, k\cosh\frac{x^1}{k}\sin x^2, x^1\right).$$

The *catenoid* is locally Riemann isometric to the *helicoid* of Example 5.1.8

$$g\left(\widetilde{x}^1,\widetilde{x}^2\right) = \left(\widetilde{x}^1\cos\widetilde{x}^2, \widetilde{x}^1\sin\widetilde{x}^2, k\widetilde{x}^2\right)$$

via the change of parameters

$$\widetilde{x}^1 = a\sinh\frac{x^1}{k}, \qquad \widetilde{x}^2 = x^2.$$

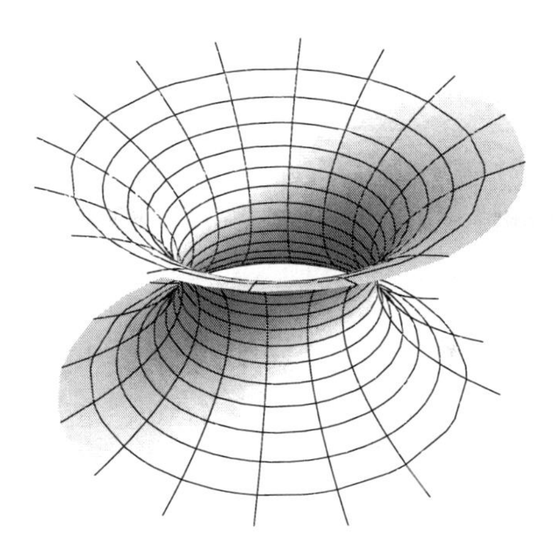

Fig. 7.32 The catenoid

7.17.5 Consider a ruled surface of class $\mathcal{C}^2$ represented by

$$f(s,t) = c(t) + s\xi(t).$$

Prove that the surface is developable if and only if

$$\left(c' \times \xi \,\middle|\, \xi'\right) = 0.$$

Prove that this is also equivalent to saying that $\frac{\partial^2 f}{\partial s \partial t}$ is a linear combination of $\frac{\partial f}{\partial s}$ and $\frac{\partial f}{\partial t}$.

7.17.6 Consider the following data

$$\begin{pmatrix} g_{11} & g_{12} \\ g_{21} & g_{22} \end{pmatrix} = \begin{pmatrix} 1 & 0 \\ 0 & 1 \end{pmatrix}$$

and

$$\begin{pmatrix} h_{11} & h_{12} \\ h_{21} & h_{22} \end{pmatrix} = \begin{pmatrix} \frac{e^{-(\frac{1\pm v}{u})^2}}{1\pm v} & \mp \frac{u\, e^{-(\frac{1\pm v}{u})^2}}{(1\pm v)^2} \\ \mp \frac{u\, e^{-(\frac{1\pm v}{u})^2}}{(1\pm v)^2} & \frac{u^2\, e^{-(\frac{1\pm v}{u})^2}}{(1\pm v)^3} \end{pmatrix}$$

where the upper sign must be chosen for $u \geq 0$ and the lower sign for $u \leq 0$. These data satisfy the Codazzi–Mainardi equations. The corresponding surface has constant zero Gaussian curvature, but in a neighborhood of $(0,0)$ it is not locally Riemann isometric to a ruled surface.

7.17.7 Prove that the surface of revolution

$$f(t,\theta) = \big(\alpha(t)\cos\theta, \alpha(t)\sin\theta, \beta(t)\big)$$

where

$$\alpha(t) = a\cos\frac{t}{k} + b\sin\frac{t}{k}, \qquad \beta(t) = \int \sqrt{1+\alpha'(t)^2}\,dt$$

has constant positive Gaussian curvature $\frac{1}{k^2}$. When is such a surface a sphere?

7.17.8 Prove that the surface of revolution

$$f(t,\theta) = \big(\alpha(t)\cos\theta, \alpha(t)\sin\theta, \beta(t)\big)$$

where

$$\alpha(t) = ae^{\frac{t}{k}} + be^{-\frac{t}{k}}, \qquad \beta(t) = \int \sqrt{1+\alpha'(t)^2}\,dt$$

has constant negative Gaussian curvature $-\frac{1}{k^2}$.

Appendix A
Topology

An elegant presentation of the general theory of surfaces requires mastering—at least—the topology of finite dimensional real spaces, but preferably also some basic notions of general topology. This appendix is devoted to a quick review of these notions.

A.1 Open Subsets in Real Spaces

We are thus interested here in some basic aspects of what is called the *topology* of the spaces $\mathbb{R}^n$ (and feel free to assume $n \neq 0$, if you want to avoid confusing trivialities). First, a well-known notion:

Definition A.1.1 Consider a point $P \in \mathbb{R}^n$ and a real number $\varepsilon > 0$. The *open ball* with center $P \in \mathbb{R}^n$ and radius ε is

$$B(P, \varepsilon) = \{A \in \mathbb{R}^n \,|\, d(P, A) < \varepsilon\} \subseteq \mathbb{R}^n$$

where $d(P, A)$ indicates the distance between A and P.

A subset $U \subseteq \mathbb{R}^n$ is *open* when, given a point $P \in U$, "all the points around P are still in U". More precisely:

Definition A.1.2 A subset $U \subseteq \mathbb{R}^n$ is *open* when, for every point $P \in U$, there exists a real number $\varepsilon > 0$ such that the open ball $B(P, \varepsilon)$ with center P and radius ε is entirely contained in U.

For example, in $\mathbb{R}^2$, the "open upper half plane"

$$U = \{(x, y) \in \mathbb{R}^2 \,|\, y > 0\}$$

F. Borceux, *A Differential Approach to Geometry*, DOI 10.1007/978-3-319-01736-5,

Fig. A.1

is open because given $P = (a, b)$ with $b > 0$, the open ball with center (a, b) and radius $\frac{b}{2}$ is still entirely contained in U (see Fig. A.1). But the "closed upper half plane"

$$\overline{U} = \{(x, y) \in \mathbb{R}^2 | y \geq 0\}$$

is not open, because given a point $Q = (c, 0)$ in it, every open ball with center $(c, 0)$ and radius ε contains the point $(c, -\frac{\varepsilon}{2})$ which is not in $\overline{U}$.

Here are the stability properties of open subsets:

Proposition A.1.3 *In* $\mathbb{R}^n$:

1. *an arbitrary union of open subsets is open*;
2. *a finite union of open subsets is open.*

In particular, $\mathbb{R}^n$ *itself and the empty subset* $\emptyset \subseteq \mathbb{R}^n$ *are open.*

Proof The reader familiar with logical arguments will have noticed at once that $\mathbb{R}^n$ can be viewed as the intersection of an empty family of open subsets, while the empty subset is the union of an empty family of open subsets. To avoid logical "contortions", let us treat these two particular cases separately.

Given a point $P \in \mathbb{R}^n$ and an arbitrary real number $\varepsilon > 0$, the open ball $B(P, \varepsilon)$ is of course contained in $\mathbb{R}^n$. Thus $\mathbb{R}^n$ is open.

The empty subset is open, because the condition for being open vanishes in this case (there is no point in the empty set, thus no condition to check).

Given a non-empty family $(U_i)_{i \in I}$ of open subsets together with a point $P \in \bigcup_{i \in I} U_i$, there exists an index i_0 such that $P \in U_{i_0}$. Therefore there exists an open ball $B(P, \varepsilon)$ contained in U_{i_0}, thus also in the union.

Given a non-empty finite family $(U_i)_{i \in I}$ of open subsets and a point $P \in \bigcap_{i \in I} U_i$, for each index i we get a real number $\varepsilon_i > 0$ such that the open ball $B(P, \varepsilon_i)$ is contained in U_i. Choosing for ε the smallest of these ε_i (which exists, since I is finite), the open ball $B(P, \varepsilon)$ is contained in each U_i, thus in their intersection. □

Let us also mention a classical point of terminology:

Definition A.1.4 Let P be a point of $\mathbb{R}^n$.

1. An open subset U containing P is called an *open neighborhood* of P.
2. An arbitrary neighborhood of P is an arbitrary subset $V \subseteq X$ containing an open neighborhood of P.

A.2 Closed Subsets in Real Spaces

We now investigate the case of closed subsets of $\mathbb{R}^n$. Let us recall that in $\mathbb{R}^n$, every Cauchy sequence of points converges.

Definition A.2.1 A subset $A \subseteq \mathbb{R}^n$ is *closed* when the limit of every Cauchy sequence of points in A still belongs to A.

Of course a closed ball with center P and radius $\varepsilon > 0$

$$\overline{B}(P, \varepsilon) = \{Q \in \mathbb{R}^n | d(P, Q) \leq \varepsilon\}$$

or a closed box

$$\prod_{i=1}^{n} [a_n, b_n], \quad a_i < b_i$$

are examples of closed subsets.

Closed subsets can easily be characterized in terms of open sets:

Proposition A.2.2 *A subset $A \subseteq \mathbb{R}^n$ is closed if and only if its complement $\complement A \subseteq \mathbb{R}^n$ is open.*

Proof Consider a closed subset $A \subseteq \mathbb{R}^n$ and a point $P \in \complement A$. We must prove the existence of an open ball $B(P, \varepsilon)$ contained in $\complement A$. If this is not the case, for every integer $n > 0$, the open ball $B(P, \frac{1}{n})$ is not contained in $\complement A$, thus we can find a point

$$P_n \in A \cap B\left(P, \frac{1}{n}\right).$$

But then $(P_n)_{n \in \mathbb{N}}$ is a Cauchy sequence of points $P_n \in A$ converging to $P \notin A$, which contradicts the assumption that A is closed.

Conversely, suppose that $\complement A$ is open. Let $(P_n)_{n \in \mathbb{N}}$ be a Cauchy sequence of points in A with limit P. We must prove that $P \in A$. If not, $P \in \complement A$ and since $\complement A$ is open, there exists an open ball $B(P, \varepsilon)$ contained in $\complement A$. But then this open ball does not contain any of the points P_n, which is a contradiction since the sequence $(P_n)_{n \in \mathbb{N}}$ converges to P. □

A.3 Compact Subsets in Real Spaces

The notion of a *compact* subset is more involved:

Definition A.3.1 Consider a subset $A \subseteq \mathbb{R}^n$.

1. A is *bounded* when it is contained in some closed ball $\overline{B}(P, \varepsilon)$.

2. A is *compact* when it is bounded and closed.

Of course, closed balls and closed boxes are again typical examples of compact subsets of $\mathbb{R}^n$. We also have:

Proposition A.3.2 *For a subset $A \subseteq \mathbb{R}^n$, the following conditions are equivalent*:

1. *A is bounded*;
2. *A is contained in some compact subset B.*

Proof When A is bounded, it is contained in some closed ball $B(P, \varepsilon)$ (Definition A.3.1) and such a closed ball is a compact subset. Conversely if A is contained is some compact subset B, then B is contained in a closed ball $B(P, \varepsilon)$, thus *a fortiori* A is also contained in this ball, and hence is bounded. □

The following characterization is crucial.

Theorem A.3.3 (Heine–Borel) *For an arbitrary subset $A \subseteq \mathbb{R}^n$, the following conditions are equivalent*:

1. *A is compact*;
2. *if A is contained in the union of a family of open subsets, it is already contained in the union of a finite subfamily of these.*

Proof Suppose that A is compact and $A \subseteq \bigcup_{i \in I} U_i$ with each U_i open. We develop the proof by *reductio ad absurdum* and assume that A is not contained in any finite union

$$A \subseteq U_{i_1} \cup \cdots \cup U_{i_m}.$$

Since A is compact, it is bounded, thus contained in a closed ball which is itself contained in a closed box Box_0:

$$A \subseteq \overline{B}(Q, r) \subseteq \prod_{k=1}^{n} [a_k, b_k] = \mathsf{Box}_0.$$

The complement $\complement A$ of A is open (see Proposition A.2.2), thus the various U_i together with $\complement A$ constitute an open covering of the closed box Box_0 (and even of the whole space $\mathbb{R}^n$). Of course no finite sub-covering already covers the closed box Box_0, otherwise the corresponding finitely many U_i would cover A.

Now divide each interval $[a_k, b_k]$ into two equal pieces, which results in a division of the original box Box_0 into 2^n smaller boxes, with sides half the length of the original. If each of these small boxes could be covered by a finite sub-covering of $\bigcup_{i \in I} U_i \cup \complement A$, the same conclusion would apply to their union Box_0, since there are only finitely many small boxes. Thus at least one small box cannot be covered by a finite sub-covering of $\bigcup_{i \in I} U_i \cup \complement A$. Let us choose one such small box and call it Box_1. Observe that this argument uses in an essential way the finiteness of the dimension n of the space: otherwise, one would have infinitely many smaller boxes and the argument would fail.

Repeat the same division process with Box_1, and so on, to get an infinite sequence of closed boxes

$$\mathsf{Box}_0 \supseteq \mathsf{Box}_1 \supseteq \mathsf{Box}_2 \cdots \supseteq \mathsf{Box}_l \supseteq \cdots$$

where at each stage the length of all sides is divided by 2 and none of these boxes is contained in a finite subcovering of $\bigcup_{i\in I} U_i \cup \complement A$.

Choose now a point $P_s \in \mathsf{Box}_s$ in each of these boxes. Since the length of the sides is each time divided by 2, the sequence $(P_s)_{s\in\mathbb{N}}$ is a Cauchy sequence. Therefore it converges to a point $P \in \mathbb{R}^n$. Certainly $P \in \mathsf{Box}_0$, because that box is closed and contains the various elements P_s of the sequence (see Definition A.2.1). Notice that the same argument, applied to the indices $s \geq s_0$, proves that $P \in \mathsf{Box}_{s_0}$ for each s_0. Now since P belongs to Box_0, it belongs to one of the elements of the covering $\bigcup_{i\in I} U_i \cup \complement A$; let us write W for this element (thus W is one of the U_i's or $W = \complement A$). Since W is open and $P \in W$, there exists some open ball $B(P, \varepsilon)$ still contained in W. But for s sufficiently big, each "tiny" box Box_s (which contains P) is contained in $B(P, \varepsilon)$, thus in W. Therefore each such tiny Box_s is contained in a single element of the covering $\bigcup_{i\in I} U_i \cup \complement A$, which is a contradiction. This proves $1 \Rightarrow 2$.

Conversely, let us cover A with open balls of radius 1 centered at each point of A:

$$A \subseteq \bigcup_{P\in A} B(P, 1).$$

By assumption, A is already contained in a finite sub-covering. Trivially, a finite union of balls of radius 1 is bounded, thus A itself is bounded.

To prove that A is closed, it suffices to prove that $\complement A$ is open (see Proposition A.2.2). Fix $P \in \complement A$. For every $Q \in A$ consider the two open balls

$$B(Q, \varepsilon_Q), \quad B(P, \varepsilon_Q), \quad \varepsilon_Q = \frac{1}{3}d(P, Q)$$

where $d(P, Q)$ indicates the distance between P and Q. The various $B(Q, \varepsilon_Q)$, for all $Q \in A$, cover A, thus already finitely many of them do:

$$A \subseteq B(Q_1, \varepsilon_{Q_1}) \cup \cdots \cup B(Q_t, \varepsilon_{Q_t}), \quad t \in \mathbb{N}.$$

Consider then

$$\varepsilon = \min\{\varepsilon_{Q_1}, \ldots, \varepsilon_{Q_t}\}.$$

The open ball $B(P, \varepsilon)$ does not meet any of the balls $B(Q_j, \varepsilon_{Q_j})$ which cover A, thus $B(P, \varepsilon) \subseteq \complement A$. This proves that $\complement A$ is open (see Definition A.1.2). □

Let us also observe that:

Proposition A.3.4 *In $\mathbb{R}^n$, every neighborhood of a point contains a compact neighborhood of this point. (One says that $\mathbb{R}^n$ is* locally compact.*)*

Proof A neighborhood V of a point $P \in \mathbb{R}^n$ (see Definition A.1.4) contains an open subset U containing P. But U itself contains an open ball centered at P (Definition A.1.2). It suffices then to consider

$$P \in B\left(P, \frac{\varepsilon}{3}\right) \subseteq \overline{B}\left(P, \frac{\varepsilon}{2}\right) \subseteq B(P, \varepsilon) \subseteq U \subseteq W.$$

This concludes the proof because every closed ball is compact. □

A.4 Continuous Mappings of Real Spaces

Let us recall that given a subset $X \subseteq \mathbb{R}^n$, a mapping $f: X \longrightarrow \mathbb{R}^m$ is *continuous* when

$$\forall P \in X\ \forall \varepsilon > 0\ \exists \delta > 0\ \forall Q \in X \quad d(P, Q) < \delta \quad \Longrightarrow \quad d\big(f(P), f(Q)\big) < \varepsilon.$$

The mapping f is *uniformly continuous* on X when the following stronger property holds

$$\forall \varepsilon > 0\ \exists \delta > 0\ \forall P \in X\ \forall Q \in X \quad d(P, Q) < \delta \quad \Longrightarrow \quad d\big(f(P), f(Q)\big) < \varepsilon$$

that is, when δ can be chosen the same whatever the point P. This definition can equivalently be rephrased in terms of open subsets, instead of distances:

Proposition A.4.1 *Let $X \subseteq \mathbb{R}^n$ be a subset. For a mapping $f: X \longrightarrow \mathbb{R}^m$, the following conditions are equivalent, when X is viewed as a topological subspace of $\mathbb{R}^n$:*

1. *f is continuous;*
2. *the inverse image under f of an open subset of $\mathbb{R}^m$ is an open subset of X.*

Proof If f is continuous and $U \subseteq \mathbb{R}^m$ is open, we must prove that for each $P \in f^{-1}(U)$, there exists some $\delta > 0$ such that $B(P, \delta) \cap X \subseteq f^{-1}(U)$ (see Definition A.1.2). But since $f(P) \in U$ and U is open, there exists an $\varepsilon > 0$ such that $B(f(P), \varepsilon) \subseteq U$. Choosing δ as in the definition of continuity forces the conclusion.

Conversely assume that $f^{-1}(U)$ is open in X for each open subset $U \subseteq \mathbb{R}^m$. Given P and ε as in the definition of continuity, put $U = B(f(P), \varepsilon)$ and simply choose δ such that

$$B(P, \delta) \cap X \subseteq f^{-1}\big(B\big(f(P), \varepsilon\big)\big);$$

such a δ exists since $f^{-1}(B(f(P), \varepsilon))$ is open by assumption. □

Moreover, in the compact case:

Theorem A.4.2 (Heine–Cantor theorem) *When $X \subseteq \mathbb{R}^n$ is a compact subset, every continuous mapping $f\colon X \longrightarrow \mathbb{R}^m$ is uniformly continuous.*

Proof Fix $\varepsilon > 0$. By continuity, for every $P \in X$ choose $\delta_P > 0$ such that

$$f\big(B(P, \delta_P)\big) \subseteq B\left(f(P), \frac{\varepsilon}{2}\right).$$

The various open balls $B(P, \frac{\delta_P}{2})$ cover the compact subset X, thus a finite number

$$B\left(P_1, \frac{\delta_{P_1}}{2}\right), \dots, B\left(P_n, \frac{\delta_{P_n}}{2}\right),$$

of them already cover X. Write

$$\delta = \frac{1}{2} \min\{\delta_1, \dots, \delta_n\}.$$

If $A, B \in X$ and $d(A, B) \leq \delta$, by the *Minkowski identity* (see Proposition 4.3.2 in [4], *Trilogy II*) the two points A, B are necessarily in some open ball $B(P_i, \delta_i)$. Then

$$d\big(f(A), f(P_i)\big) \leq \frac{\varepsilon}{2}, \qquad d\big(f(B), f(P_i)\big) \leq \frac{\varepsilon}{2}$$

and again by the Minkowski identity

$$d\big(f(A), f(B)\big) \leq \varepsilon. \qquad \square$$

A.5 Topological Spaces

With Propositions A.1.3 and A.2.2 in mind, we make the following definition.

Definition A.5.1 A *topological space* is a pair $(X, \mathcal{T})$ where:

- X is a set;
- $\mathcal{T}$ is a family of subsets of X, called *open subsets*.

These data must satisfy the following axioms:

1. an arbitrary union of open subsets is open;
2. a finite intersection of open subsets is open.

In particular, X itself and the empty subset are always open.

The subset $X \subseteq X$ is indeed the intersection of the empty (thus finite) family of open subsets, while $\emptyset \subseteq X$ is its union.

Definition A.5.2 In a topological space, a neighborhood of a point P is a subset V containing an open subset U containing P: $P \in U \subseteq V$.

Proposition A.5.3 *For a subset $U \subseteq X$ of a topological space $(X, \mathcal{T})$, the following conditions are equivalent*:

1. *U is open*;
2. *U is a neighborhood of each of its points.*

Proof ($1 \Rightarrow 2$) is obvious by Definition A.5.2. Conversely, for each point $P \in U$ there exists an open subset $P \in U_P \subseteq U$ and U is open, as a union of the open subsets U_P (see Definition A.5.1). □

The examples of topological spaces of interest for this book are essentially obtained via the following proposition:

Proposition A.5.4 *Let $(X, \mathcal{T})$ be a topological space and $Y \subseteq X$ an arbitrary subset. Define*

$$\mathcal{S} = \{Y \cap U \,|\, U \in \mathcal{T}\}.$$

Then $(Y, \mathcal{S})$ is a topological space, called a topological sub-space *of $(X, \mathcal{T})$, and $\mathcal{S}$ is the induced topology on Y.*

Proof $(Y, \mathcal{S})$ is a topological space simply because

$$Y \cap \left(\bigcup_{i \in I} U_i\right) = \bigcup_{i \in I}(Y \cap U_i), \qquad Y \cap \left(\bigcap_{i \in I} U_i\right) = \bigcap_{i \in I}(Y \cap U_i).$$ □

Since each space $\mathbb{R}^n$ is a topological space by Proposition A.1.3, Proposition A.5.4 yields several examples closely related to the theory of surfaces:

Example A.5.5 Given a parametric representation

$$f: U \mapsto \mathbb{R}^3$$

of a surface, the support $f(U)$ of this surface is a topological sub-space of $\mathbb{R}^3$. □

We also have:

Example A.5.6 Every quadric (see Sect. 1.14 in [4], *Trilogy II*) is a topological sub-space of $\mathbb{R}^3$. □

Let us conclude with another classical notion:

Proposition A.5.7 *For every subset $A \subseteq X$ of a topological space $(X, \mathcal{T})$, there exists a greatest open subset $\overset{\circ}{A}$ contained in A, called the* interior *of A.*

Proof By Definition A.5.1, it suffices to define the interior $\overset{\circ}{A}$ as the union of all the open subsets contained in A. □

A.6 Closure and Density

With Proposition A.2.2 in mind, we make the following definition:

Definition A.6.1 Let $(X, \mathcal{T})$ be a topological space. A subset $A \subseteq X$ is *closed* when its complement $\complement A$ is open.

Proposition A.6.2 *For every subset $A \subseteq X$ of a topological space $(X, \mathcal{T})$, there exists a smallest closed subset $\overline{A}$ containing A. It is called the* closure *of A.*

Proof Since a union of open subsets is open (see Definition A.5.1), an intersection of closed subsets is closed. Therefore $\overline{A}$ is simply the intersection of all the closed subsets containing A. □

Proposition A.6.3 *The closure of a subset $A \subseteq X$ of a topological space $(X, \mathcal{T})$ is the set of those points of X all of whose neighborhoods intersect A.*

Proof Write $\widetilde{A}$ for the set of those points all of whose neighborhoods intersect A. Trivially, $A \subseteq \widetilde{A}$. Let us first prove that $\widetilde{A}$ is closed, that is, its complement $\complement\widetilde{A}$ is open.

If $P \in \complement\widetilde{A}$, there exists a neighborhood V of P which does not intersect A and by Definition A.5.2, there is no loss of generality in assuming that V is open. But then V is a neighborhood of each point $Q \in V$ which does not interect A. So $Q \in \complement\widetilde{A}$ as well and finally $V \subseteq \complement\widetilde{A}$. Thus $\complement\widetilde{A}$ is a neighborhood of each of its points and therefore, it is open (see Proposition A.5.3).

Consider now $A \subseteq C$ with C an arbitrary closed subset; we must prove that $\widetilde{A} \subseteq C$ that is, $\complement C \subseteq \complement\widetilde{A}$. If $P \in \complement C$, then $\complement P$ is an open subset containing P and not intersecting A. Thus $P \in \complement\widetilde{A}$.

All this proves that $\widetilde{A} = \overline{A}$ is the closure of A (see Proposition A.6.2). □

We conclude this section with two other classical notions:

Definition A.6.4 The boundary of a subset $A \subseteq X$ of a topological space $(X, \mathcal{T})$ is the set theoretical difference $\overline{A} \setminus \overset{\circ}{A}$ between its closure (see Proposition A.6.2) and its interior (see Proposition A.5.7).

Definition A.6.5 Consider subsets $A \subseteq B$ in a topological space $(X, \mathcal{T})$. The subset A is *dense* in B when $B \subseteq \overline{A}$, that is, B is contained in the closure of A (see Proposition A.6.2).

A.7 Compactness

Following the *Heine–Borel Theorem* (see Theorem A.3.3), we define further:

Definition A.7.1 Let $(X, \mathcal{T})$ be a topological space. A subset $A \subseteq X$ is *compact* when, from every covering $A \subseteq \bigcup_{i \in I} U_i$ of A by open subsets U_i, one can extract a finite sub-covering

$$A \subseteq U_{i_1} \cup \cdots \cup U_{i_m}.$$

Following Proposition A.3.2, we also define:

Definition A.7.2 In a topological space $(X, \mathcal{T})$, a subset $A \subseteq X$ is *bounded* when it is contained in a compact subset.

Let us stress that the equivalence (see Theorem A.3.3)

$$\textit{compact} \quad \Leftrightarrow \quad \textit{bounded closed}$$

is not valid in arbitrary topological spaces. Nevertheless, it true in the so-called *Hausdorff* topological spaces: a class of spaces of great interest.

Definition A.7.3 A *Hausdorff* space $(X, \mathcal{T})$ is a topological space in which two distinct points can always be included in two disjoint open subsets.

$$P \neq Q \in X \quad \Longrightarrow \quad \exists U, V \in \mathcal{T} \quad U \cap V = \emptyset \qquad P \in U \qquad Q \in V.$$

Of course, the spaces $\mathbb{R}^n$ are Hausdorff spaces. The observant reader will have noticed that we used this *Hausdorff* property in the last part of the proof of the *Heine–Borel Theorem* (see Theorem A.3.3). More generally we have:

Proposition A.7.4 *For a subset $A \subseteq X$ of a given Hausdorff topological space $(X, \mathcal{T})$, the following conditions are equivalent*:

1. *A is compact*;
2. *A is bounded and closed.*

Proof ($1 \Rightarrow 2$). By Definition A.7.2, every compact subset is bounded. To prove that A is closed, we prove that its complement $\complement A$ is open (see Definition A.6.1). We must therefore prove that $\complement A$ is a neighborhood of each of its points (see Proposition A.5.3). Fix $P \in \complement A$. By Hausdorffness, for every point $Q \in A$, there exist disjoint open subsets $V_Q \ni Q$, $W_Q \ni P$. The various V_Q cover A thus by compactness, finitely many of them already cover A (see Definition A.7.1):

$$A \subseteq V_{Q_1} \cup \cdots \cup V_{Q_n}.$$

Then

$$W_{Q_1} \cap \cdots \cap W_{Q_n}$$

is an open subset (see Definition A.5.1) containing P and not intersecting A.

($2 \Rightarrow 1$). Let A be closed and $A \subseteq C$ with C compact. If $(U_i)_{i \in I}$ is an open covering of A, adding the open subset $\complement A$ to the U_i's, we obtain an open covering of the compact subset C. One can thus extract a finite open sub-covering and the corresponding U_i's then constitute a finite open covering of A. □

A.8 Continuous Mappings

Proposition A.4.1 immediately suggests the following definition:

Definition A.8.1 A *continuous mapping*

$$f : (X, \mathcal{T}) \longrightarrow (Y, \mathcal{S})$$

between topological spaces is a mapping $f : X \longrightarrow Y$ such that

$$U \in \mathcal{S} \quad \Longrightarrow \quad f^{-1}(U) \in \mathcal{T}$$

that is, the inverse image of an open subset is open.

Let us emphasize the following point of terminology:

Definition A.8.2 By a *continuous path* joining two points A and B in a topological space $(X, \mathcal{T})$ is meant a continuous function

$$f : [a, b] \longrightarrow (X, \mathcal{T}), \qquad f(a) = A, \qquad f(b) = B, \quad a, b \in \mathbb{R}.$$

Given a continuous mapping, the inverse image of an open subset is open (Definition A.8.1), thus the inverse image of a closed subset is again closed (see Definition A.5.1). What about compact subsets? Perhaps unexpectedly we have the following:

Proposition A.8.3 *Let $f : (X, \mathcal{T}) \longrightarrow (Y, \mathcal{S})$ be a continuous function between topological spaces. Then the* direct *image of a compact subset is a compact subset.*

Proof Let $A \subseteq X$ be compact. Consider an open covering $f(A) \subseteq \bigcup_{i \in I} U_i$ in $(Y, \mathcal{S})$. Then $A \subseteq \bigcup_{i \in I} f^{-1}(U_i)$ is an open covering in $(X, \mathcal{T})$. By compactness of A, we can extract a finite sub-covering

$$A \subseteq f^{-1}(U_{i_1}) \cup \cdots \cup f^{-1}(U_{i_m})$$

from which

$$f(A) \subseteq \left(f \circ f^{-1}\right)(U_{i_1}) \cup \cdots \cup \left(f \circ f^{-1}\right)(U_{i_m}) \subseteq U_{i_1} \cup \cdots \cup U_{i_m}. \qquad \square$$

Corollary A.8.4 *Let $(X,\mathcal{T})$ be a topological space and $f\colon (X,\mathcal{T}) \longrightarrow \mathbb{R}$ be a continuous function. When X is compact, the function f is bounded and attains its bounds.*

Proof By Proposition A.8.3, $f(X)$ is compact. By Theorem A.3.3 and Definition A.3.1, $f(X)$ is bounded and closed in $\mathbb{R}$. Let us thus write

$$s = \sup\{f(x)|x \in X\}, \qquad i = \min\{f(x)|x \in X\}$$

for the supremum and the infimum of $f(X)$. For each $0 \neq n \in \mathbb{N}$, there is thus an element $x_n \in X$ such that $s - f(x_n) < \frac{1}{n}$. The sequence $f(x_n) \in f(X)$ is then a Cauchy sequence converging to s; but then $s \in f(X)$ because $f(X)$ is closed (see Definition A.2.1); thus $s = f(x)$ for some $x \in X$. An analogous argument holds for i. □

A.9 Homeomorphisms

Among the continuous mappings, the "isomorphisms" of topological spaces play a special role.

Definition A.9.1 A *homeomorphism* $f\colon (X,\mathcal{T}) \longrightarrow (Y,\mathcal{S})$ between topological spaces is a continuous mapping admitting a continuous inverse

$$f^{-1}\colon (Y,\mathcal{S}) \longrightarrow (X,\mathcal{T}).$$

Proposition A.9.2 *A homeomorphism $f\colon (X,\mathcal{T}) \longrightarrow (Y,\mathcal{S})$ induces a bijection*

$$f\colon \mathcal{T} \longrightarrow \mathcal{S}, \quad U \mapsto f(U)$$

between open subsets.

Proof Writing g for the inverse of f, f^{-1} and g^{-1} are inverse bijections between $\mathcal{T}$ and $\mathcal{S}$, and $f(U)$ as in the statement is thus exactly $g^{-1}(U)$. □

Proposition A.9.3 *A continuous bijection between two compact Hausdorff spaces is necessarily a homeomorphism.*

Proof Let $f\colon (X,\mathcal{T}) \longrightarrow (Y,\mathcal{S})$ be a continuous bijection between compact Hausdorff spaces (see Sect. A.7). We must prove that $(f^{-1})^{-1}$—that is, f—transforms an open subset of X into an open subset of Y. Of course it is equivalent to prove that f maps a closed subset onto a closed subset (see Definition A.6.1). But in a compact space, every subset is bounded (see Definition A.7.2). Thus in Hausdorff spaces, being closed is the same as being compact (see Proposition A.7.4), and indeed f transforms compact subsets into compact subsets, by Corollary 5.1.2. □

The homeomorphisms of interest in this book are the following:

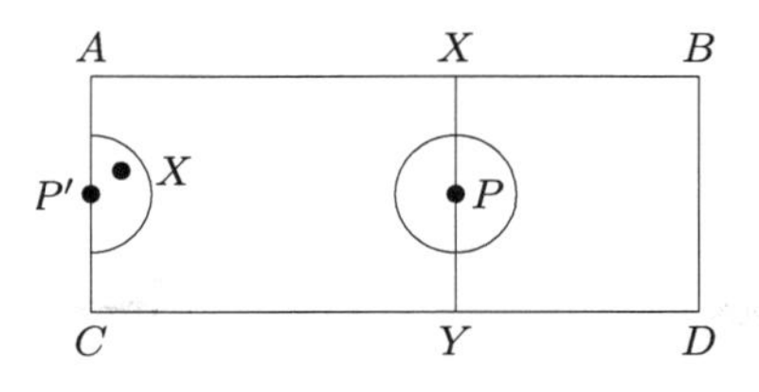

Fig. A.2

Proposition A.9.4 *Let $f : V \longrightarrow \mathbb{R}^3$ be a regular parametric representation of a surface. Each point $P \in V$ has an open neighborhood $U \subseteq V$ on which f is injective and*

$$f : U \longrightarrow f(U)$$

is a homeomorphism.

Proof Consider a point $P = (u_0, v_0) \in V$. Since f is locally injective, it is injective on some neighborhood of P and there is no loss of generality in choosing this neighborhood to be an open ball $B(P, \varepsilon)$ (see Sect. A.1). Consider then

$$U = B\left(P, \frac{\varepsilon}{2}\right) \subseteq \overline{B}\left(P, \frac{\varepsilon}{2}\right) \subseteq B(P, \varepsilon)$$

where $\overline{B}$ indicates the closed ball. Since a closed ball is compact (see Definition A.3.1), Proposition A.9.3 implies that

$$f : \overline{B}\left(P, \frac{\varepsilon}{2}\right) \longrightarrow f\left(\overline{B}\left(P, \frac{\varepsilon}{2}\right)\right)$$

is a homeomorphism, since it is bijective and continuous. Therefore the further restriction

$$f : U \longrightarrow f(U)$$

is a homeomorphism as well. □

The reader should be aware that in general, under the conditions of Proposition A.9.4—however U is an open subset of V—$f(U)$ generally has no reason to be an open subset of $f(V)$. Not even when f is injective on V.

For a counterexample, consider in the plane $\mathbb{R}^2$ a *closed* rectangle as in Fig. A.2. Then "roll" the left hand part in order to "glue" in $\mathbb{R}^3$ the side AC onto the segment XY (see Fig. A.3). The corresponding *open* rectangle has now been mapped injectively and continuously in $\mathbb{R}^3$; let us write Rect for this open rectangle. This yields a surface in the sense of Definition 5.1.1, with an injective parametric representation. But if you consider an arbitrary small open circle $U \subseteq \mathsf{Rect}$ around a point P of the segment XY, $f(U)$ will never be open on the surface. If this were the case, there would be an open subset W of $\mathbb{R}^3$ such that $f(U) = W \cap f(\mathsf{Rect})$ (see

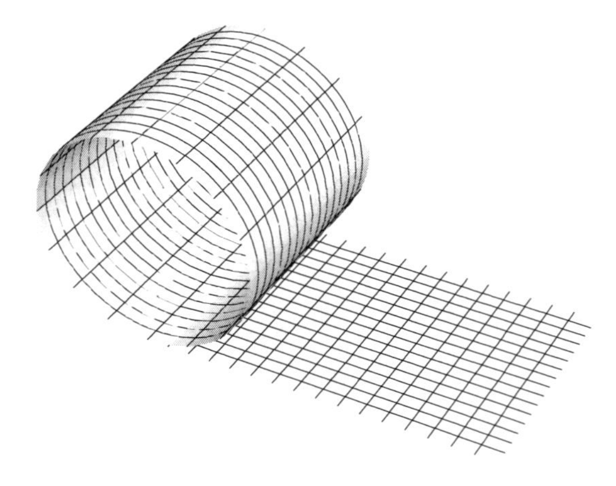

Fig. A.3

Proposition A.5.4). But by Definition A.1.2, W would contain an open ball of some radius ε centered at $f(P)$, and such an open ball would necessarily contain points of the form $f(X)$, with X "near" the corresponding point P' on the segment AC. This contradicts the fact that

$$B(P,\varepsilon) \cap f(\mathsf{Rect}) \subseteq W \cap f(\mathsf{Rect}) = f(U).$$

A.10 Connectedness

In this book, we are mainly interested in *open subsets of one piece* and *surfaces of one piece*. The mathematical term for "being of one piece" is *connected*. There are many possible definitions, non-equivalent in the most general context, but nevertheless generally equivalent in the more precise contexts in which we use them.

The straightforward way of saying that an open subset is "of one piece" is to say ... that it is not constituted of two pieces! Let us put this definition in its most general context, not just for open subsets.

Definition A.10.1

- A topological space $(X,\mathcal{T})$ is *connected* when it is non-empty and cannot be written as the union of two disjoint non-empty open subsets.
- An arbitrary subset $A \subseteq X$ of a topological space $(X,\mathcal{T})$ is *connected* when, provided with the induced topology (see Proposition A.5.4), it is a connected topological space.

A second way of saying that an open subset is "of one piece" is to say that, remaining always inside of it, you can move continuously from one point to any other one. Again we state this definition in its most general setting:

Definition A.10.2

- A topological space $(X, \mathcal{T})$ is *path-connected* when it is non-empty and any two points P, Q in X can be joined by a continuous path.
- A subset $A \subseteq X$ of a topological space $(X, \mathcal{T})$ is *path-connected* when, provided with the induced topology (see Proposition A.5.4), it is a path-connected topological space.

Let us first exhibit the relation between these two notions.

Lemma A.10.3 *Every closed interval* $[a, b]$ *of the real line is connected.*

Proof Suppose that $[a, b] \subseteq U_1 \cup U_2$ with U_1 and U_2 disjoint non-empty open subsets of $[a, b]$. There is no loss of generality in assuming that $a \in U_1$; thus by disjointness, $a \notin U_2$. Consider

$$c = \inf\{t \in [a, b] \mid t \in U_2\}.$$

If $c \in U_2$, then by disjointness $c \neq a$. By openness of U_2, a whole neighborhood of c in $[a, b]$ is still in U_2. This contradicts the infimum property of c.

The case $c \in U_1$ and $c = b$ would imply $[a, b] = U_1$, by definition of c. This would force U_2 to be empty, which is not the case.

Finally if $c \in U_1$ and $c \neq b$, by openness of U_1, a whole neighborhood of c in $[a, b]$ is still in U_1. Since by disjointness, no point of U_2 can be in this neighborhood, this again contradicts the infimum property of c. □

Corollary A.10.4 *Let* U_1 *and* U_2 *be two open subsets of an arbitrary topological space* $(X, \mathcal{T})$. *Given* $a, b \in \mathbb{R}$ *and a continuous function*

$$f : [a, b] \longrightarrow U_1 \cup U_2 \subseteq X, \qquad f(a) \in U_1, \qquad f(b) \in U_2,$$

there exists a point $c \in [a, b]$ *such that* $f(c) \in U_1 \cap U_2$.

Proof The open subsets $f^{-1}(U_1)$ and $f^{-1}(U_2)$ of $[a, b]$ (see Definition A.8.1) are non-empty and cover the closed interval $[a, b]$. Thus they cannot be disjoint, by Lemma A.10.3. □

Proposition A.10.5 *Every path-connected topological space is connected.*

Proof Let $(X, \mathcal{T})$ be a path-connected topological space. Suppose that $X = U_1 \cup U_2$, where U_1 and U_2 are disjoint non-empty open subsets. Choose

$$P \in U_1, \qquad P \notin U_2; \qquad Q \in U_2, \qquad Q \notin U_1.$$

By path-connectedness, choose further a continuous function

$$f : [a, b] \longrightarrow (X, \mathcal{T}), \qquad f(a) = P, \qquad f(b) = Q.$$

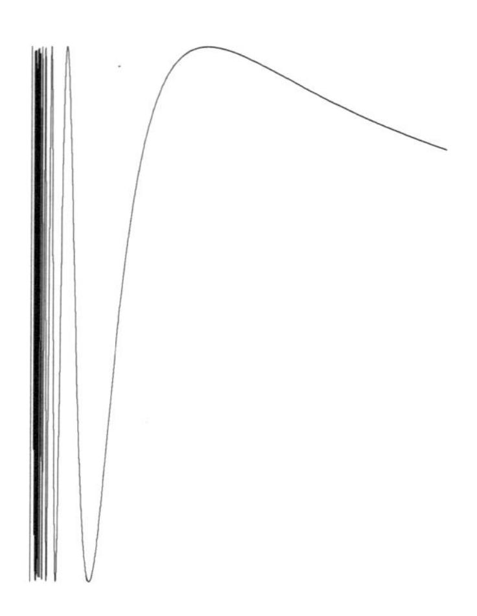

Fig. A.4

It follows that

$$[a,b] = f^{-1}(U_1) \cup f^{-1}(U_2)$$

where $f^{-1}(U_1)$ and $f^{-1}(U_2)$ are two disjoint non-empty open subsets of the interval $[a,b]$ (see Definition A.8.1). This is a contradiction, because the interval $[a,b]$ is connected (see Lemma A.10.3). □

The converse implication is false:

Counterexample A.10.6 Consider the topological subspace of $\mathbb{R}^2$ defined by

$$X = X_0 \cup X_+$$

$$X_0 = \{(0,y) | -1 \leq y \leq +1\}, \qquad X_+ = \left\{\left(x, \sin\frac{1}{x}\right) \middle| x > 0\right\}.$$

This space is connected, but not path-connected (see Fig. A.4).

Proof The space X is not path-connected, because there is no continuous function joining a point $(x, \sin\frac{1}{x})$ to a point $(0,y)$. Indeed, suppose that such a continuous function

$$f\colon [a,b] \longrightarrow X, \qquad t \mapsto \bigl(f_1(t), f_2(t)\bigr), \qquad f(a) = (1, \sin 1), \qquad f(b) = (0,0)$$

exists. Put

$$c = \inf\bigl\{t \in [a,b] \big| f(t) \in X_0\bigr\}.$$

Then

- $f([a,c]) \cap X_0 = \emptyset$ if $f(c) \in X_+$;

- $f([a,c]) \cap X_0 = \{f(c)\}$ if $f(c) \in X_0$.

On the other hand every neighborhood of a point $(0, y) \in X_0$ contains infinitely many points of the form $(x, \sin\frac{1}{x})$ with $0 < x < 1$; thus each such point is in the closure of $f([a,c])$, i.e.

- $X_0 \subseteq \overline{f([a,c])}$

(see Proposition A.6.3). Thus $f([a,c])$ is not equal to its closure, that is, is not closed. By Proposition A.7.4, it is not compact. But this is a contradiction, because $[a,c]$ is compact (see Proposition A.8.3). This concludes the proof that X is not path-connected.

But X is connected. Indeed a partition $X = U_1 \cup U_2$ into two disjoint non-empty open subsets would restrict to a partition of both X_0 and X_+ into two disjoint open subsets. Since both X_0 and X_+ are trivially path-connected, they are connected by Proposition A.10.5 and the induced partition should be trivial: the whole of X_0 or X_+ and an empty subset. In other words, $X_0 \subseteq U_i$ and $X_+ \subseteq U_j$ $(i \neq j)$. But again, any open subset which contains points of X_0 contain infinitely many points of X_+. Thus U_i contains a point of X_+ thus a point of U_j, which is a contradiction. □

Nevertheless we have the following:

Theorem A.10.7 *For an open subset $U \subseteq \mathbb{R}^n$, the following conditions are equivalent:*

1. *U is connected;*
2. *U is path connected.*

Proof By Proposition A.10.5, it remains to prove $(1 \Rightarrow 2)$. If U is connected, fix a point $P \in U$ and write:

- U_1 for the set of those points in U which can be joined to P by a continuous path in U;
- U_2 for the set of those points in U which cannot be joined to P by a continuous path in U.

Given $Q \in U_1$, there is a whole open ball centered at Q and still contained in U (see Definition A.1.2). But Q can then further be joined by a segment to every point R of this open ball, a segment which is entirely contained in the ball, thus in U. Putting together this segment and a path in U from P to Q, we conclude that $R \in U_1$. So U_1 contains an open ball centered at each of its points Q and therefore, U_1 is open by Definition A.1.2.

The same argument, used *ad absurdum*, shows that U_2 is open as well. Since U_1 and U_2 are two disjoint open subsets covering U, by connectedness, one of them must be empty. Since trivially $P \in U_1$, we get $U_2 = \emptyset$ and thus $U = U_1$. □

Corollary A.10.8 *For a Riemann surface $\mathcal{S}$ of $\mathbb{R}^3$, the following conditions are equivalent:*

1. $\mathcal{S}$ *is connected*;
2. $\mathcal{S}$ *is path-connected.*

Proof Again by Proposition A.10.5, it suffices to prove $(1 \Rightarrow 2)$. It suffices to adapt the proof of Theorem A.10.7.

- U_1 is the set of those points in $\mathcal{S}$ which can be joined to P by a continuous path in $\mathcal{S}$;
- U_2 is the set of those points in $\mathcal{S}$ which cannot be joined to P by a continuous path in $\mathcal{S}$.

Every point $Q \in \mathcal{S}$ lies in some local map, which is thus homeomorphic to a connected and therefore path-connected open subset of $\mathbb{R}^2$ (see Definitions 6.17.1, 6.2.1 and Theorem A.10.7). As in Theorem A.10.7 this forces U_1 and U_2 to be open and thus finally $U_2 = \emptyset$ and $U = U_1$. □

Example A.10.9 For a subset A of the real line, the following conditions are equivalent:

1. A is a generalized interval;
2. A is connected;
3. A is path-connected.

Proof By "generalized interval" we mean an interval of one of the forms

$$]a, b[, \quad]a, b], \quad [a, b[, \quad [a, b], \quad a, b \in \mathbb{R}$$

but possibly also

$$]-\infty, b[, \quad]-\infty, b], \quad [a, +\infty[, \quad]a, +\infty[, \quad]-\infty, +\infty[, \quad a, b \in \mathbb{R}.$$

Trivially, the generalized intervals are path connected. This proves $(1 \Rightarrow 3)$.

For the converse, for two points $a < b$, a path-connected subset X of $\mathbb{R}$ necessarily contains the full interval $[a, b]$. Thus X is a generalized interval with $\inf X$ and $\sup X$ (possibly infinite) as extremities. This proves $(3 \Rightarrow 1)$.

$(3 \Rightarrow 2)$ is attested by Proposition A.10.5. Finally if $X \subseteq \mathbb{R}$ is connected and $a < b$ are in X, necessarily all points $a < c < b$ are in X as well, otherwise X would admit a covering by two disjoint open subsets of $\mathbb{R}$

$$X \subseteq]-\infty, c[\cup]c, +\infty[.$$

Considering the intersections with X would contradict the connectedness of X. But then again, with $a < b$, X contains the whole interval $[a, b]$ and X is a generalized interval. □

Example A.10.10 In the real plane $\mathbb{R}^2$

- an open ball $B(P, \varepsilon)$,

- an open rectangle $]a, b[\times]c, d[$,
- an open half plane $\mathbb{R} \times]a, \infty[$,
- an open strip $\mathbb{R} \times]a, b[$

are all examples of connected open subsets.

Proof Proving path-connectedness is in each case obvious, and the result follows by Theorem A.10.7. □

Appendix B
Differential Equations

This appendix states—without any proofs—the theorems on systems of differential equations and systems of partial differential equations which have been used in this volume. We have deliberately chosen to rely only on these few basic results and not on more sophisticated and diverse results which could have simplified some arguments. In this appendix, all functions are real valued functions with real variable(s).

B.1 First Order Differential Equations

The simplest case (the so-called *linear one*), which we have used several times in this book, is:

Proposition B.1.1 *Consider functions of class* $\mathcal{C}^k$ $(k \geq 1)$

$$\alpha_{ij}\colon \mathbb{R} \longrightarrow \mathbb{R}, \quad 1 \leq i, j \leq n$$

and the system of differential equations

$$f_i' = \sum_{j=1}^{n} \alpha_{ij} f_j$$

with unknown functions

$$f_1, \ldots, f_n\colon \mathbb{R} \longrightarrow \mathbb{R}.$$

For arbitrary constants t_0 *and* $a_1, \ldots, a_n$, *consider further the initial conditions*

$$f_1(t_0) = a_1, \ldots, f_n(t_0) = a_n.$$

There exists a neighborhood of t_0 *on which the system admits a unique solution of class* $\mathcal{C}^{k+1}$

$$f_1(t), \ldots, f_n(t)$$

F. Borceux, *A Differential Approach to Geometry*, DOI 10.1007/978-3-319-01736-5,

satisfying the prescribed initial conditions.

We also need a *non-linear* result. Its basic form is:

Proposition B.1.2 *Consider functions of class* $\mathcal{C}^k$ $(k \geq 1)$

$$\alpha_i, \ldots, \alpha_n : \mathbb{R}^n \longrightarrow \mathbb{R}$$

and the system of differential equations

$$f_i' = \alpha_i(f_1, \ldots, f_n)$$

with unknown functions

$$f_1, \ldots, f_n : \mathbb{R} \longrightarrow \mathbb{R}.$$

For arbitrary constants t_0 *and* $a_1, \ldots, a_n$, *consider further the initial conditions*

$$f_1(t_0) = a_1, \ldots, f_n(t_0) = a_n.$$

There exists a neighborhood of t_0 *on which the system admits a unique solution of class* $\mathcal{C}^{k+1}$

$$f_1(t), \ldots, f_n(t)$$

satisfying the prescribed initial conditions.

Of course, Proposition B.1.1 is a special case of Proposition B.1.2.

B.2 Second Order Differential Equations

A classical "trick" allows us to pass from differential equations of order 1 to differential equations of order 2:

Proposition B.2.1 *Consider functions of class* $\mathcal{C}^k$ $(k \geq 1)$

$$\alpha_i, \ldots, \alpha_n : \mathbb{R}^{2n} \longrightarrow \mathbb{R}$$

and the system of differential equations

$$f_i'' = \alpha_i\bigl(f_1, \ldots, f_n, f_1', \ldots, f_n'\bigr)$$

with unknown functions

$$f_1, \ldots, f_n : \mathbb{R} \longrightarrow \mathbb{R}.$$

For arbitrary constants

$$t_0, \qquad a_1, \ldots, a_n, \qquad b_1, \ldots, b_n$$

consider further the initial conditions

$$f_1(t_0)=a_1,\ldots,f_n(t_0)=a_n,\qquad f_1'(t_0)=b_1,\ldots,f_n'(t_0)=b_n.$$

There exists a neighborhood of t_0 on which the system admits a unique solution of class $\mathcal{C}^{k+1}$

$$f_1(t),\ldots,f_n(t)$$

satisfying the prescribed initial conditions.

Proof Putting

$$g_1=f_1',\ldots,g_n=f_n'$$

the system above can be re-written as a system of $2n$-differential equations

$$\begin{cases} f_i'=g_i \\ g_i'=\alpha_i(f_1,\ldots,f_n,g_1,\ldots,g_n) \end{cases}$$

with the initial conditions

$$f_1(t_0)=a_1,\ldots,f_n(t_0)=a_n,\qquad g_1(t_0)=b_1,\ldots,g_n(t_0)=b_n.$$

The result follows by Proposition B.1.2. □

B.3 Variable Initial Conditions

An important refinement that we need is the "continuity of the solution of a system of differential equations with respect to the initial conditions". That is, if we let the initial conditions vary continuously, the solution of the system varies continuously as well.

Proposition B.3.1 *Consider functions of class $\mathcal{C}^k$ $(k\geq 1)$*

$$\alpha_i,\ldots,\alpha_n\colon\mathbb{R}^n\longrightarrow\mathbb{R}$$

and the system of differential equations

$$f_i'=\alpha_i(f_1,\ldots,f_n)$$

with unknown functions

$$f_1,\ldots,f_n\colon\mathbb{R}\longrightarrow\mathbb{R}.$$

Consider now functions of class $\mathcal{C}^{k+1}$

$$a_i\colon\mathbb{R}\longrightarrow\mathbb{R},\qquad s\mapsto a_i(s),\quad 1\leq i\leq n.$$

For a fixed value t_0 and each fixed value of the parameter s, consider the initial conditions

$$f_1(t_0) = a_1(s), \ldots, f_n(t_0) = a_n(s)$$

and the solution, for each fixed value of s,

$$f_1^{(s)}, \ldots, f_n^{(s)}$$

given by Proposition B.1.2 in a neighborhood of t_0. For each fixed value s_0 of s, there exists a neighborhood of (t_0, s_0) in $\mathbb{R}^2$ on which the functions

$$f_1(t,s) = f_1^{(s)}(t), \ldots, f_n(t,s) = f_n^{(s)}(t)$$

are of class $\mathcal{C}^{k+1}$.

Of course Proposition B.3.1 applies in particular in the context of Proposition B.2.1.

Proposition B.3.2 *Consider functions of class $\mathcal{C}^k$ $(k \geq 1)$*

$$\alpha_i, \ldots, \alpha_n : \mathbb{R}^n \longrightarrow \mathbb{R}$$

and the system of differential equations

$$f_i'' = \alpha_i\big(f_1, \ldots, f_n, f_1', \ldots, f_n'\big)$$

with unknown functions

$$f_1, \ldots, f_n : \mathbb{R} \longrightarrow \mathbb{R}.$$

Consider now functions of class $\mathcal{C}^{k+1}$

$$a_i : \mathbb{R} \longrightarrow \mathbb{R}, \qquad s \mapsto a_i(s), \quad 1 \leq i \leq n;$$

$$b_i : \mathbb{R} \longrightarrow \mathbb{R}, \qquad s \mapsto b_i(s), \quad 1 \leq i \leq n.$$

For a fixed value t_0 and each fixed value of the parameter s, consider the initial conditions

$$f_1(t_0) = a_1(s), \ldots, f_n(t_0) = a_n(s), \qquad f_1'(t_0) = b_1(s), \ldots, f_n'(t_0) = b_n(s)$$

and the solution, for each fixed value of s,

$$f_1^{(s)}, \ldots, f_n^{(s)}$$

given by Proposition B.2.1 *in a neighborhood of t_0. For each fixed value s_0 of s, there exists a neighborhood of (t_0, s_0) in $\mathbb{R}^2$ on which the functions*

$$f_1(t,s) = f_1^{(s)}(t), \ldots, f_n(t,s) = f_n^{(s)}(t)$$

are of class $\mathcal{C}^{k+1}$.

B.4 Systems of Partial Differential Equations

For systems of partial differential equations, we need the following result, which is in fact simply a disguised result on systems of differential equations (see the comment following the proof).

Proposition B.4.1 *Consider functions of class* $\mathcal{C}^k$ $(k \geq 1)$

$$\alpha_{ij}, \beta_{ij} \colon \mathbb{R}^2 \longrightarrow \mathbb{R}, \quad 1 \leq i, j \leq n$$

and the system of partial differential equations

$$\begin{cases} \dfrac{\partial f_i}{\partial x} = \displaystyle\sum_{j=1}^{n} \alpha_{ij}(x, y) f_j(x, y) \\ \dfrac{\partial f_i}{\partial y} = \displaystyle\sum_{j=1}^{n} \beta_{ij}(x, y) f_j(x, y) \end{cases}$$

with unknown functions

$$f_1, \ldots, f_n \colon \mathbb{R}^2 \longrightarrow \mathbb{R}.$$

Consider further the initial conditions

$$f_1(x_0, y_0) = a_1, \ldots, f_n(x_0, y_0) = a_n$$

for fixed constants $x_0, y_0, a_1, \ldots, a_n$. *When the equations above force the compatibility conditions*

$$\frac{\partial^2 f_i}{\partial x \partial y} = \frac{\partial^2 f_i}{\partial y \partial x}$$

(see below for a precise meaning of this sentence), there exists a neighborhood of (x_0, y_0) *on which the system admits a unique solution of class* $\mathcal{C}^{k+1}$

$$f_1(x, y), \ldots, f_n(x, y)$$

satisfying the prescribed initial conditions. The uniqueness condition does not require the compatibility conditions.

Proof Fix $y = y_0$ in the first equations, which we thus view as a system of differential equations with unknown functions $f_i(x, y_0)$ and initial conditions $f_i(x_0, y_0) = a_i$. By Proposition B.1.1 we obtain a unique solution $\varphi_i(x)$ on a neighborhood of x_0. Choose x_1 in this neighborhood and fix $x = x_1$ in the second equations. We obtain another system of differential equations with unknown functions $f_i(x_1, y)$ and initial conditions $f_i(x_1, y_0) = \varphi_i(x_1)$. Again by Proposition B.1.1 we obtain a unique solution on a neighborhood of y_0. By Proposition B.3.1, we obtain on a neighborhood of (x_0, y_0), a unique solution $f(x, y)$ of class $\mathcal{C}^{k+1}$ satisfying the first equations for $y = y_0$ and all of the second equations.

Observe that in particular this forces the *uniqueness* of a solution $f(x, y)$ of class $\mathcal{C}^{k+1}$ satisfying *all* the equations and the initial conditions. (Of course, even if we had not assumed the compatibility conditions up to now, they are certainly satisfied by our function f since f is of class $\mathcal{C}^1$.)

Now of course, permuting the roles of x and y, we obtain, on a neighborhood of (x_0, y_0), a unique solution $g(x, y)$ of class $\mathcal{C}^{k+1}$ satisfying the second equations for $x = x_0$ and all the first equations. To conclude, it thus "suffices" to prove that $f(x, y) = g(x, y)$ on a neighborhood of (x_0, y_0). This is where the compatibility conditions enter the story. □

Let us now make clear what the compatibility conditions precisely mean. Formally differentiating the first equation(s) with respect to y we obtain

$$\begin{aligned}\frac{\partial^2 f_i}{\partial y \partial x} &= \sum_j \left(\frac{\partial \alpha_{ij}}{\partial x} f_j + \alpha_{ij} \frac{\partial f_j}{\partial x} \right) \\ &= \sum_j \frac{\partial \alpha_{ij}}{\partial x} f_j + \sum_{jl} \alpha_{ij} \beta_{jl} f_l \\ &= \sum_l \left(\frac{\partial \alpha_{il}}{\partial x} + \sum_j \alpha_{ij} \beta_{jl} \right) f_l.\end{aligned}$$

Analogously

$$\frac{\partial^2 f_i}{\partial y \partial x} = \sum_l \left(\frac{\partial \beta_{il}}{\partial x} + \sum_j \beta_{ij} \alpha_{jl} \right) f_l.$$

One says that the expected compatibility conditions are "forced by the equations" as soon as, for each l, we have the equality of the respective coefficients of f_l in these expressions. That is,

In Proposition B.4.1, the *compatibility conditions* mean that the coefficients of the partial differential equations satisfy:

$$\frac{\partial \alpha_{il}}{\partial x} + \sum_j \alpha_{ij} \beta_{jl} = \frac{\partial \beta_{il}}{\partial x} + \sum_j \beta_{ij} \alpha_{jl}$$

for all indices i, l.

References and Further Reading

1. T. Apostol, *Mathematical Analysis* (Addison-Wesley, Reading, 1974)
2. F. Borceux, *Invitation à la Géométrie* (CIACO, Louvain-la-Neuve, 1986)
3. F. Borceux, *An Axiomatic Approach to Geometry*. Geometric Trilogy, vol. I (Springer, Berlin, 2014)
4. F. Borceux, *An Algebraic Approach to Geometry*. Geometric Trilogy, vol. II (Springer, Berlin, 2014)
5. C.B. Boyer, *A History of Mathematics* (Wiley, New York, 1968)
6. S.S. Chern, *Differential Geometry*. Lect. Notes (Dpt. Math. Univ., Chicago, 1954)
7. E. Coddington, N. Levinson, *Theory of Ordinary Differential Equations* (McGraw-Hill, New York, 1955)
8. J.P. Collette, *Histoire des Mathématiques*, two vols. (Editions du Renouveau Pédagogique, 1973–1979)
9. H. Coxeter, *Non-Euclidean Geometry* (University of Toronto Press, Toronto, 1942)
10. P. Dupont, *Introduction à la Géométrie* (De Boeck, Reading, 2002)
11. L. Eisenhart, *Riemannian Geometry* (Princeton University Press, Princeton, 1964)
12. P.D. Goetz, *Introduction to Differential Geometry* (Addison-Wesley, Reading, 1970)
13. M.J. Greenberg, *Euclidean and Non-Euclidean Geometries* (Freeman, New York, 1974)
14. H. Hopf, Über die Drehung der Tangenten und Sehnen Ebener Kurven. Compos. Math. **2**, 50–62 (1935)
15. F. John, *Partial Differential Equations*. Applied Mathematical Sciences (Springer, Berlin, 1982)
16. Y. Kerbrat, J.M. Braemer, *Géométrie des Courbes et des Surfaces* (Hermann, Paris, 1976)
17. M. Kline, *Mathematical Thoughts from Ancient to Modern Times* (Oxford University Press, London, 1972)
18. W. Klingenberg, *A Course in Differential Geometry*. Graduate Texts in Math., vol. 51 (Springer, Berlin, 1978)
19. D. Laugwitz, *Differential and Riemannian Geometry* (Academic Press, San Diego, 1965)
20. D. Lehmann, R. Bkouche, *Initiation à la Géométrie* (Presses Universitaires de France, Paris, 1988)
21. J. Lelong-Ferrand, *Géométrie Différentielle* (Masson, Paris, 1963)
22. M. Lipschutz, *Differential Geometry*. Schaum's Outline Series (McGraw-Hill, New York, 1969)
23. E.H. Lockwood, *A Book of Curves* (Cambridge University Press, Cambridge, 1961)
24. G.E. Martin, *The Foundations of Geometry and the Non-Euclidean Plane*. Undergraduate Texts in Math. (Springer, Berlin, 1975)
25. J. Mawhin, *Analyse* (De Boeck, New York, 1997)
26. S.M. Nikol'skii, *Mathematical Analysis* (Springer, Berlin, 2002)

F. Borceux, *A Differential Approach to Geometry*, DOI 10.1007/978-3-319-01736-5,

27. B. O'Neill, *Elementary Differential Geometry* (Academic Press, San Diego, 1966)
28. N. Rouche, J. Mawhin, *Equations Différentielles Ordinaires*, two vols. (Masson, Paris, 1973)
29. L.C. Piccini, G. Stampacchia, G. Vidossich, *Ordinary Differential Equations in* $\mathbb{R}^n$. Applied Mathematical Sciences, vol. 39 (Springer, Berlin, 1984)
30. M. Spivak, *Differential Geometry*, Vols. 2 and 3 (Publish or Perish, New York, 1979)
31. J. Steiner, Einfache Beweise der Isoperimetrischen Hauptsätze. J. Reine Angew. Math. **18**, 289–296 (1838)
32. J. Thorpe, *Elementary Topics in Differential Geometry*. Graduate Texts in Math. (Springer, Berlin, 1979)

Index

F. Borceux, *A Differential Approach to Geometry*, DOI 10.1007/978-3-319-01736-5,